NONLINEAR PROGRAMMING

MOS-SIAM Series on Optimization

This series is published jointly by the Mathematical Optimization Society and the Society for Industrial and Applied Mathematics. It includes research monographs, books on applications, textbooks at all levels, and tutorials. Besides being of high scientific quality, books in the series must advance the understanding and practice of optimization. They must also be written clearly and at an appropriate level.

Series Volumes

Biegler, Lorenz T., *Nonlinear Programming: Concepts, Algorithms, and Applications to Chemical Processes*
Shapiro, Alexander, Dentcheva, Darinka, and Ruszczyński, Andrzej, *Lectures on Stochastic Programming: Modeling and Theory*
Conn, Andrew R., Scheinberg, Katya, and Vicente, Luis N., *Introduction to Derivative-Free Optimization*
Ferris, Michael C., Mangasarian, Olvi L., and Wright, Stephen J., *Linear Programming with MATLAB*
Attouch, Hedy, Buttazzo, Giuseppe, and Michaille, Gérard, *Variational Analysis in Sobolev and BV Spaces: Applications to PDEs and Optimization*
Wallace, Stein W. and Ziemba, William T., editors, *Applications of Stochastic Programming*
Grötschel, Martin, editor, *The Sharpest Cut: The Impact of Manfred Padberg and His Work*
Renegar, James, *A Mathematical View of Interior-Point Methods in Convex Optimization*
Ben-Tal, Aharon and Nemirovski, Arkadi, *Lectures on Modern Convex Optimization: Analysis, Algorithms, and Engineering Applications*
Conn, Andrew R., Gould, Nicholas I. M., and Toint, Phillippe L., *Trust-Region Methods*

NONLINEAR PROGRAMMING

Concepts, Algorithms, and Applications to Chemical Processes

Lorenz T. Biegler

Carnegie Mellon University
Pittsburgh, Pennsylvania

Society for Industrial and Applied Mathematics
Philadelphia

Mathematical Optimization Society
Philadelphia

10 9 8 7 6 5 4 3 2

Library of Congress Cataloging-in-Publication Data

Biegler, Lorenz T.
Nonlinear programming : concepts, algorithms, and applications to chemical processes / Lorenz T. Biegler.
p. cm.
Includes bibliographical references and index.
ISBN 978-0-898717-02-0
1. Chemical processes. 2. Nonlinear programming. I. Title.
TP155.75.B54 2010
519.7'6–dc22 2010013645

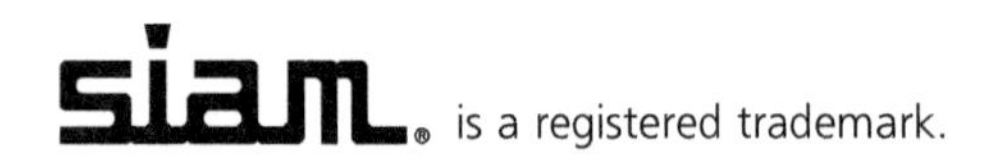

In memory of my father

To my mother

To Lynne and to Matthew

To all my students

Contents

Preface

Chemical engineering applications have been a source of challenging optimization problems for over 50 years. For many chemical process systems, detailed steady state and dynamic behavior can now be described by a rich set of detailed nonlinear models, and relatively small changes in process design and operation can lead to significant improvements in efficiency, product quality, environmental impact, and profitability. With these characteristics, it is not surprising that systematic optimization strategies have played an important role in chemical engineering practice. In particular, over the past 35 years, nonlinear programming (NLP) has become an indispensable tool for the optimization of chemical processes. These tools are now applied at research and process development stages, in the design stage, and in the online operation of these processes. More recently, the scope of these applications is being extended to cover more challenging, large-scale tasks including process control based on the optimization of nonlinear *dynamic* models, as well as the incorporation of nonlinear models into strategic planning functions.

Moreover, the ability to solve large-scale process optimization models cheaply, even online, is aided by recent breakthroughs in nonlinear programming, including the development of modern barrier methods, deeper understanding of line search and trust region strategies to aid global convergence, efficient exploitation of second derivatives in algorithmic development, and the availability of recently developed and widely used NLP codes, including those for barrier methods [81, 391, 404], sequential quadratic programming (SQP) [161, 159], and reduced gradient methods [119, 245, 285]. Finally, the availability of optimization modeling environments, such as AIMMS, AMPL, and GAMS, as well as the NEOS server, has made the formulation and solution of optimization accessible to a much wider user base. All of these advances have a huge impact in addressing and solving process engineering problems previously thought intractable. In addition to developments in mathematical programming, research in process systems engineering has led to optimization modeling formulations that leverage these algorithmic advances, with specific model structure and characteristics that lead to more efficient solutions.

This text attempts to make these recent optimization advances accessible to engineers and practitioners. Optimization texts for engineers usually fall into two categories. First, excellent mathematical programming texts (e.g., [134, 162, 294, 100, 227]) emphasize fundamental properties and numerical analysis, but have few specific examples with relevance to real-world applications, and are less accessible to practitioners. On the other hand, equally good engineering texts (e.g., [122, 305, 332, 53]) emphasize applications with well-known methods and codes, but often without their underlying fundamental properties. While their approach is accessible and quite useful for engineers, these texts do not aid in a deeper understanding of the methods or provide extensions to tackle large-scale problems efficiently.

To address the modeling and solution of large-scale process optimization problems, it is important for optimization practitioners to understand

- which NLP methods are best suited for specific applications,
- how large-scale problems should be formulated and what features should be emphasized, and
- how existing methods can be extended to exploit specific structures of large-scale optimization models.

This book attempts to fill the gap between the math programming and engineering texts. It provides a firm grounding in fundamental algorithmic properties but also with relevance to real-world problem classes through case studies. In addressing an engineering audience, it covers state-of-the-art gradient-based NLP methods, summarizes key characteristics and advantages of these methods, and emphasizes the link between problem structure and efficient methods. Finally, in addressing a broader audience of math programmers it also deals with steady state and dynamic models derived from chemical engineering.

The book is written for an audience consisting of

- engineers (specifically chemical engineers) interested in understanding and applying state-of-the-art NLP algorithms to specific applications,
- experts in mathematical optimization (in applied math and operations research) who are interested in understanding process engineering problems and developing better approaches to solving them, and
- researchers from both fields interested in developing better methods and problem formulations for challenging engineering problems.

The book is suitable for a class on continuous variable optimization, but with an emphasis on problem formulation and solution methods. It is intended as a text for graduate and advanced undergraduate classes in engineering and is also suitable as an elective class for students in applied mathematics and operations research. It is also intended as a reference for practitioners in process optimization in the area of design and operations, as well as researchers in process engineering and applied mathematics.

The text is organized into eleven chapters, and the structure follows that of a short course taught to graduate students and industrial practitioners which has evolved over the past 20 years. Included from the course are a number of problems and computer projects which form useful exercises at the end of each chapter.

The eleven chapters follow sequentially and build on each other. Chapter 1 provides an overview of nonlinear programming applications in process engineering and sets the motivation for the text. Chapter 2 defines basic concepts and properties for nonlinear programming and focuses on fundamentals of unconstrained optimization. Chapter 3 then develops Newton's method for unconstrained optimization and discusses basic concepts for globalization methods; this chapter also develops quasi-Newton methods and discusses their characteristics.

Chapter 4 then follows with fundamental aspects of constrained optimization, building on the concepts in Chapter 2. Algorithms for equality constrained optimization are derived in Chapter 5 from a Newton perspective that builds on Chapter 3. Chapter 6 extends this approach to inequality constraints and discusses algorithms and problem formulations that are essential for developing large-scale optimization models. Chapter 7 then discusses steady state process optimization and describes the application of NLP methods to modular and equation-oriented simulation environments.

Chapter 8 introduces the emerging field of dynamic modeling and optimization in process systems. A survey of optimization strategies is given and current applications are summarized. The next two chapters deal with two strategies for dynamic optimization. Chapter 9 develops optimization methods with embedded differential-algebraic equation (DAE) solvers, while Chapter 10 describes methods that embed discretized DAE models within the optimization formulation itself. Chapter 11 concludes this text by presenting complementarity models that can describe a class of discrete decisions. Embedded within nonlinear programs, these lead to mathematical programs with complementarity constraints (MPCC) that apply to both steady state and dynamic process optimization models.

Finally, it is important to mention what this book does not cover. As seen in the table of contents, the book is restricted to methods and applications centered around gradient-based nonlinear programming. Comprising the broad area of optimization, the following topics are not considered, although extensive citations are provided for additional reading:

- *Optimization in function spaces.* While the text provides a practical treatment of DAE optimization, this is developed from the perspective of finite-dimensional optimization problems. Similarly, PDE-constrained optimization problems [51, 50] are beyond the scope of this text.

- *Iterative linear solvers.* Unlike methods for PDE-based formulations, indirect linear solvers are almost never used with chemical process models. Hence, the NLP strategies in this book will rely on direct (and often sparse) linear solvers with little coverage of iterative linear solvers.

- *Optimization methods for nondifferentiable functions.* These methods are not covered, although some nondifferentiable features may be addressed through reformulation of the nonlinear program. Likewise, derivative-free optimization methods are not covered. A recent treatment of this area is given in [102].

- *Optimization problems with stochastic elements.* These problems are beyond the scope of this text and are covered in a number of texts including [57, 216].

- *Optimization methods that ensure global solutions.* The NLP methods covered in the text guarantee only local solutions unless the appropriate convexity conditions hold. Optimization methods that ensure global solutions for nonconvex problems are not covered here. Extensive treatment of these methods can be found in [144, 203, 379].

- *Optimization methods for problems with integer variables.* These methods are beyond the scope of this book. Resources for these mixed integer problems can be found in [53, 143, 295].

Nevertheless, the NLP concepts and algorithms developed in this text provide useful background and a set of tools to address many of these areas.

Acknowledgments
There are many people who deserve my thanks in the creation of this book. I have been privileged to have been able to work with outstanding graduate students and research colleagues. I am grateful to all of them for the inspiration that they brought to this book. In particular, many thanks to Nikhil Arora, Brian Baumrucker, Antonio Flores-Tlacuahuac, Shiva Kameswaran, Carl Laird, Yi-dong Lang, Nuno Oliveira, Maame Yaa Poku, Arvind Raghunathan, Lino Santos, Claudia Schmid, Andreas Wächter, Dominique Wolbert, and Victor Zavala for their research contributions that were incorporated into the chapters. In addition, special thanks go to Brian Baumrucker, Vijay Gupta, Shiva Kameswaran, Yi-dong Lang, Rodrigo Lopez Negrete, Arvind Raghunathan, Andreas Wächter, Kexin Wang, and Victor Zavala for their careful reading of the manuscript.

I am also very grateful to my research colleagues at Carnegie Mellon (Ignacio Grossmann, Nick Sahinidis, Art Westerberg, and Erik Ydstie) and at Wisconsin (Mike Ferris, Christos Maravelias, Harmon Ray, Jim Rawlings, Ross Swaney, Steve Wright) for their advice and encouragement during the writing of this book. Also, I very much appreciate the discussions and advice from Mihai Anitescu, John Betts, Georg Bock, Steve Campbell, Andy Conn, Tim Kelley, Katya Kostina, Sven Leyffer, Pu Li, Wolfgang Marquardt, Hans Mittlemann, Jorge Nocedal, Sachin Pawardhan, Danny Ralph, Zhijiang Shao, and Philippe Toint. My thanks also go to Lisa Briggeman, Sara Murphy and Linda Thiel at SIAM for their advice and many suggestions.

Finally, it is a privilege to acknowledge the support of the Fulbright Foundation and the Hougen Visiting Professorship for very fruitful research stays at the University of Heidelberg and the University of Wisconsin, respectively. Without these opportunities this book could not have been completed.

Lorenz T. Biegler

Chapter 1

Introduction to Process Optimization

Most things can be improved, so engineers and scientists optimize. While designing systems and products requires a deep understanding of influences that achieve desirable performance, the need for an *efficient and systematic decision-making approach* drives the need for optimization strategies. This introductory chapter provides the motivation for this topic as well as a description of applications in chemical engineering. Optimization applications can be found in almost all areas of engineering. Typical problems in chemical engineering arise in process design, process control, model development, process identification, and real-time optimization. The chapter provides an overall description of optimization problem classes with a focus on problems with continuous variables. It then describes where these problems arise in chemical engineering, along with illustrative examples. This introduction sets the stage for the development of optimization methods in the subsequent chapters.

1.1 Scope of Optimization Problems

From a practical standpoint, we define the *optimization* task as follows: given a system or process, find the best solution to this process within constraints. This task requires the following elements:

- An *objective function* is needed that provides a scalar quantitative performance measure that needs to be minimized or maximized. This can be the system's cost, yield, profit, etc.
- A predictive model is required that describes the behavior of the system. For the optimization problem this translates into a set of equations and inequalities that we term *constraints*. These constraints comprise a feasible region that defines limits of performance for the system.
- *Variables* that appear in the predictive model must be adjusted to satisfy the constraints. This can usually be accomplished with multiple instances of variable values, leading to a feasible region that is determined by a subspace of these variables. In many engineering problems, this subspace can be characterized by a set of *decision variables* that can be interpreted as *degrees of freedom* in the process.

Optimization is a fundamental and frequently applied task for most engineering activities. However, in many cases, this task is done by trial and error (through case study). To avoid such tedious activities, we take a systematic approach to this task, which is as efficient as possible and also provides some guarantee that a better solution cannot be found.

The systematic determination of optimal solutions leads to a large family of methods and algorithms. Moreover, the literature for optimization is dynamic, with hundreds of papers published every month in dozens of journals. Moreover, research in optimization can be observed at a number of different levels that necessarily need to overlap but are often considered by separate communities:

- At the *mathematical programming*[1] level, research focuses on understanding fundamental properties of optimization problems and algorithms. Key issues include existence of solutions, convergence of algorithms, and related issues such as stability and convergence rates.

- The *scientific computing* level is strongly influenced by mathematical properties as well as the implementation of the optimization method for efficient and "practical" use. Here research questions include numerical stability, ill-conditioning of algorithmic steps, and computational complexity and performance.

- At the level of *operations research*, attention is focused on formulation of the optimization problem and development of solution strategies, often by using well-established solution methods. Many of the problems encountered at this level consider well-structured models with linear and discrete elements.

- At the *engineering* level, optimization strategies are applied to challenging, and often poorly defined, real-world problems. Knowledge of optimization at this level is engaged with the efficiency and reliability of applicable methods, analysis of the solution, and diagnosis and recovery from failure of the solution method.

From the above description of optimization research, it is clear that successful development of an optimization strategy within a given level requires a working knowledge of the preceding levels. For instance, while it is important at the mathematical programming level to develop the "right" optimization *algorithm*, at the engineering level it is even more important to solve the "right" optimization *problem formulation*. On the other hand, as engineers need to consider optimization tasks on a regular basis, a systematic approach with a fundamental knowledge of optimization formulations and algorithms is essential. It should be noted that this requires not only knowledge of existing software, which may have limited application to particularly difficult problems, but also knowledge of the underlying algorithmic principles that allow challenging applications to be addressed. In the next section we begin with a classification of mathematical programming problems. This is followed by examples of optimization problems in chemical engineering that will be addressed in this text. Finally, a simple example is presented to motivate the development of optimization methods in subsequent chapters.

[1] The term *mathematical programming* was coined in the 1940s and is somewhat unrelated to computer programming; it originally referred to the more general concept of optimization in the sense of optimal planning.

1.2 Classification of Optimization Problems

Optimization is a key enabling tool for decision making in chemical engineering. It has evolved from a methodology of academic interest into a technology that continues to significant impact in engineering research and practice. Optimization algorithms form the core tools for (a) experimental design, parameter estimation, model development, and statistical analysis; (b) process synthesis, analysis, design, and retrofit; (c) model predictive control and real-time optimization; and (d) planning, scheduling, and the integration of process operations into the supply chain for manufacturing and distribution.

As shown in Figure 1.1, optimization problems that arise in chemical engineering can be classified in terms of continuous and discrete variables. When represented in algebraic form, the formulation of discrete/continuous optimization problems can be written as *mixed integer optimization problems*. The most general of these is the mixed integer nonlinear program (MINLP) of the form

$$\begin{aligned} \min_{x,y} \quad & f(x,y) \\ \text{s.t.} \quad & h(x,y) = 0, \\ & g(x,y) \leq 0, \\ & x \in \mathbb{R}^n, \quad y \in \{0,1\}^t, \end{aligned} \tag{1.1}$$

where $f(x,y)$ is the objective function (e.g., cost, energy consumption, etc.), $h(x,y) = 0$ are the equations that describe the performance of the system (e.g., material balances, production rates), and the inequality constraints $g(x,y) \leq 0$ can define process specifications or constraints for feasible plans and schedules. Note that the operator max $f(x,y)$ is equivalent to min $-f(x,y)$. We define the real n-vector x to represent the continuous variables while the t-vector y represents the discrete variables, which, without loss of generality, are often restricted to take $0/1$ values to define logical or discrete decisions, such as assignment of equipment and sequencing of tasks. (These variables can also be formulated to take on other integer values as well.) Problem (1.1) corresponds to an MINLP when any of the

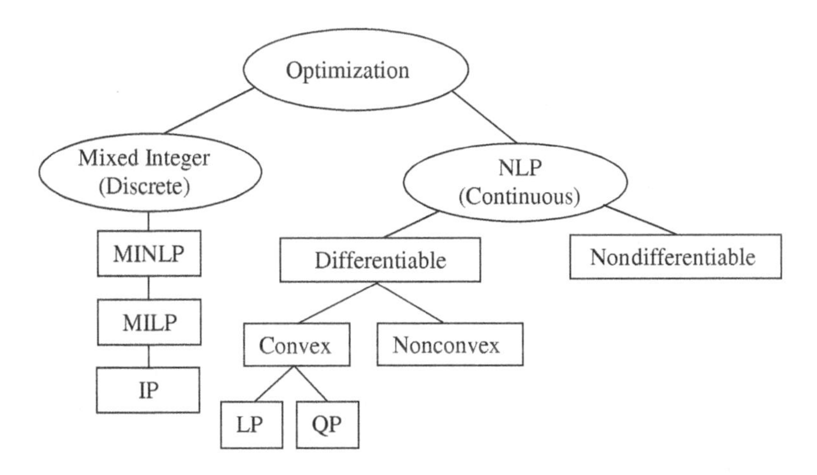

Figure 1.1. *Classes of optimization problems.*

functions involved are nonlinear. If the functions $f(x,y)$, $g(x,y)$, and $h(x,y)$ are linear (or vacuous), then (1.1) corresponds to a mixed integer linear program (MILP). Further, for MILPs, an important case occurs when all the variables are integer; this gives rise to an integer programming (IP) problem. IP problems can be further classified into a number of specific problems (e.g., assignment, traveling salesman, etc.), not shown in Figure 1.1.

If there are no $0/1$ variables, then problem (1.1) reduces to the nonlinear program (1.2) given by

$$\begin{aligned} \min_{x \in \mathbb{R}^n} \quad & f(x) \\ \text{s.t.} \quad & h(x) = 0, \\ & g(x) \le 0. \end{aligned} \tag{1.2}$$

This general problem can be further classified. First, an important distinction is whether the problem is assumed to be differentiable or not. In this text, we will assume that the functions $f(x)$, $h(x)$, and $g(x)$ have continuous first and second derivatives. (In many cases, nonsmooth problems can be reformulated into a smooth form of (1.2).)

Second, a key characteristic of (1.2) is whether it is convex or not, i.e., whether it has a convex objective function, $f(x)$, and a convex feasible region. This can be defined as follows.

- A set $S \in \mathbb{R}^n$ is convex if and only if all points on the straight line connecting any two points in this set are also within this set. This can be stated as

$$x(\alpha) = \alpha x_1 + (1-\alpha)x_2 \in S \quad \text{for all } \alpha \in (0,1) \text{ and } x_1, x_2 \in S. \tag{1.3}$$

- A function $\phi(x)$ is convex if its domain X is convex and

$$\alpha\phi(x_1) + (1-\alpha)\phi(x_2) \ge \phi(x(\alpha)) \tag{1.4}$$

  holds for all $\alpha \in (0,1)$ and all points $x_1, x_2 \in X$. (Strict convexity requires that the inequality (1.4) be strict.)

- Convex feasible regions require $g(x)$ to be a convex function and $h(x)$ to be *linear*.

- A function $\phi(x)$ is (strictly) concave if the function $-\phi(x)$ is (strictly) convex.

If (1.2) is a convex problem, then any local solution (for which a better, feasible solution cannot be found in a neighborhood around this solution) is guaranteed to be a global solution to (1.2); i.e., no better solution exists. On the other hand, nonconvex problems may have multiple local solutions, i.e., feasible solutions that minimize the objective function only within some neighborhood about the solution.

Further specializations of problem (1.2) can be made if the constraint and objective functions satisfy certain properties, and specialized algorithms can be constructed for these cases. In particular, if the objective and constraint functions in (1.2) are linear, then the resulting linear program (LP) can be solved in a finite number of steps. Methods to solve LPs are widespread and well implemented. Currently, state-of-the-art LP solvers can handle millions of variables and constraints, and the application of further decomposition methods leads to the solution of problems that are two or three orders of magnitude larger than this. Quadratic programs (QPs) represent a slight modification of LPs through the addition of a quadratic term in the objective function. If the objective function is convex, then the resulting convex QP can also be solved in a finite number of steps. While QP models are generally not as large or widely applied as LP models, a number of solution strategies have been created to solve large-scale QPs very efficiently. These problem classes are covered in Chapter 4.

Finally, we mention that while the nonlinear programming (NLP) problem (1.2) is given as a finite-dimensional representation, it may result from a possible large-scale discretization of differential equations and solution profiles that are distributed in time and space.

1.3 Optimization Applications in Chemical Engineering

Optimization has found widespread use in chemical engineering applications, especially in the engineering of process systems. Problems in this domain often have many alternative solutions with complex economic and performance interactions, so it is often not easy to identify the optimal solution through intuitive reasoning. Moreover, the economics of the system often indicate that finding the optimum solution translates into large savings, along with a large economic penalty for sticking to suboptimal solutions. Therefore, optimization has become a major technology that helps the chemical industry to remain competitive.

As summarized in Table 1.1, optimization problems are encountered in all facets of chemical engineering, from model and process development to process synthesis and design, and finally to process operations, control, scheduling, and planning.

Process development and modeling is concerned with transforming experimental data and fundamental chemical and physical relationships into a predictive process model. Incorporating the tasks of experimental design, parameter estimation, and discrimination of competing models, this step gives rise to NLP and MINLP problems.

At the next level, process synthesis and design incorporates these predictive models to devise a process that is both technically and economically feasible. The synthesis step is largely focused on establishing the structure of the process flowsheet (usually with simplified models) and systematic strategies have been developed for particular subsystems that involve reaction, separation, energy management, and waste reduction. On the other hand, the design step is concerned with establishing equipment parameters and nominal operating conditions for the flowsheet. As a result, synthesis problems can be addressed with a wide range of optimization formulations, while the design step requires more detailed models that give rise to NLP and MINLP problems.

In the operation of chemical processes, there is widespread interest in improving the scheduling and planning of these operations. These tasks are usually formulated as LP and MILP problems which are less detailed, as most operations are described through time requirements and activities. On the other hand, the development of large-scale NLP tools has led to the application of real-time optimization which interacts with the process and responds to changes in the production schedule and from other inputs. The results of the real-time optimization also feed operating conditions or *setpoints* to the control system. The distinction between these two levels is that real-time optimization describes steady state behavior, while the control system responds essentially to the process dynamics. The control task addresses maintenance of optimal production levels and rejection of disturbances. For multivariable, constrained systems, model predictive control (MPC) is now accepted as a widely used optimization-based strategy. MPC models allow various levels of detail to handle linear and nonlinear dynamics as well as discontinuities that occur in hybrid systems.

Table 1.1. *Mathematical programming in process systems engineering.*

	LP	MILP	QP	NLP	MINLP
Process Model Building				X	X
Process Design & Synthesis					
Heat Exchangers	X	X		X	X
Mass Exchangers	X	X		X	X
Separations		X		X	X
Reactors	X			X	X
Flowsheeting				X	X
Process Operations					
Scheduling	X	X			X
Supply Chain	X	X			X
Real-Time Optimization	X		X	X	
Process Control					
Model Predictive Control	X		X		
Nonlinear MPC			X	X	
Hybrid MPC		X			

Table 1.1 summarizes model types that have been formulated for process engineering applications. Design is dominated by NLP and MINLP models due to the need for the explicit handling of performance equations, although simpler targeting models for synthesis give rise to LP and MILP problems. Operations problems, in contrast, tend to be dominated by linear models, LPs and MILPs, for planning, scheduling, and supply chain problems. Nonlinear Programming, however, plays a crucial role at the level of real-time optimization. Finally, process control has traditionally relied on LP and NLP models, although MILPs are being increasingly used for hybrid systems. It is also worth noting that the applications listed in Table 1.1 have been facilitated not only by progress in optimization algorithms, but also by modeling environments such as GAMS [71] and AMPL [148].

This book focuses on the nonlinear programming problem (1.2) and explores methods that locate local solutions efficiently. While this approach might first appear as a restricted form of optimization, NLPs have broad applications, particularly for large-scale engineering models. Moreover, while the study of NLP algorithms is important on its own, these algorithms also form important components of strategies for MINLP problems and for finding the global optimum of nonconvex problems.

1.4 Nonlinear Programming Examples in Chemical Engineering

At the heart of any of the tasks in Table 1.1 lies a performance model of the process. Whether based on first principles or empirical relations, the model represents the behavior of reaction, separation, and mass, heat and momentum transfer in the process. The model equations for these phenomena typically consist of conservation laws (based on mass, heat, and momentum), physical and chemical equilibrium among species and phases, and additional constitutive equations that describe the rates of chemical transformation or transport of mass and energy.

Chemical process models are often represented by a collection of individual unit models (the so-called unit operations) that usually correspond to major pieces of process equipment. Unit models are assembled within a process flowsheet that describes the interaction of equipment either for steady state or dynamic behavior. As a result, models can be described by algebraic or differential equations. For example, steady state process flowsheets are usually described by lumped parameter models described by algebraic equations. Similarly, dynamic process flowsheets are described by lumped parameter models described by differential-algebraic equations (DAEs). Models that deal with spatially distributed models are frequently considered at the unit level, with partial differential equations (PDEs) that model fluid flow, heat and mass transfer, and reactions. On the other hand, distributed models are usually considered too expensive to incorporate within an overall process model (although recent research [421, 243] is starting to address these issues).

Process models may also contain stochastic elements with uncertain variables. While these features are beyond the scope of this text, Chapter 6 considers a limited treatment of uncertainty through the formulation and efficient solution of multiple scenario optimization problems. These formulations link multiple instances of process models together with common variables. Similarly, models can also be linked to include multiple processes over multiple time scales. As a result, interactions between different operating levels (see Table 1.1) or among spatially distributed processes can be exploited through an optimization formulation.

To illustrate the formulation of NLPs from process models we introduce three examples from design, real-time optimization, and control. While solution strategies and results are deferred to later chapters, some detail is provided here to demonstrate both the characteristics of the resulting NLPs and some challenges in their solution.

1.4.1 Design of a Small Heat Exchanger Network

Consider the optimization of the small process network shown in Figure 1.2 with two process streams and three heat exchangers. Using temperatures defined by $T_{in} > T_{out}$ and $t_{out} > t_{in}$, the "hot" stream with a fixed flow rate F and heat capacity C_p needs to be cooled from T_{in} to T_{out}, while the "cold" stream with fixed flow rate f and heat capacity c_p needs to be heated from t_{in} to t_{out}. This is accomplished by two heat exchangers; the heater uses steam at temperature T_s and has a heat duty Q_h, while the cooler uses cold water at temperature T_w and has a heat duty Q_c. However, considerable energy can be saved by exchanging heat between the hot and cold streams through the third heat exchanger with heat duty Q_m and hot and cold exit temperatures, T_m and t_m, respectively.

The model for this system is given as follows:

- The energy balance for this system is given by

$$Q_c = FC_p(T_m - T_{out}), \tag{1.5}$$

$$Q_h = fc_p(t_{out} - t_m), \tag{1.6}$$

$$Q_m = fc_p(t_m - t_{in}) = FC_p(T_{in} - T_m). \tag{1.7}$$

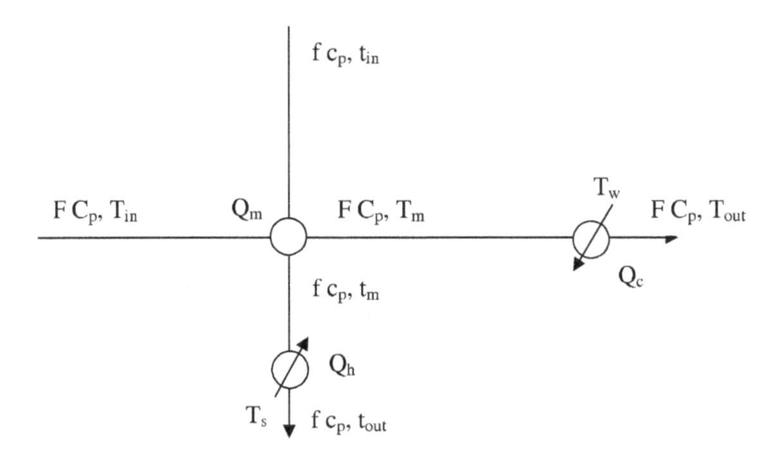

Figure 1.2. *Example of simple heat exchanger network.*

- Each heat exchanger also has a capital cost that is based on its area A_i, $i \in \{c,h,m\}$, for heat exchange. Here we consider a simple countercurrent, shell and tube heat exchanger with an overall heat transfer coefficient, U_i, $i \in \{c,h,m\}$. The resulting area equations are given by

$$Q_i = U_i A_i \Delta T_{lm}^i, \quad i \in \{c,h,m\}. \tag{1.8}$$

- The log-mean temperature difference ΔT_{lm}^i is given by

$$\Delta T_{lm}^i = \frac{\Delta T_a^i - \Delta T_b^i}{\ln(\Delta T_a^i / \Delta T_b^i)}, \quad i \in \{c,h,m\}, \tag{1.9}$$

and

$$\begin{aligned} \Delta T_a^c &= T_m - T_w, \quad \Delta T_b^c = T_{out} - T_w, \\ \Delta T_a^h &= T_s - t_m, \quad \Delta T_b^h = T_s - t_{out}, \\ \Delta T_a^m &= T_{in} - t_m, \quad \Delta T_b^m = T_m - t_{in}. \end{aligned}$$

Our objective is to minimize the total cost of the system, i.e., the energy cost as well as the capital cost of the heat exchangers. This leads to the following NLP:

$$\min \quad \sum_{i \in \{c,h,m\}} (\hat{c}_i Q_i + \bar{c}_i A_i^\beta) \tag{1.10}$$

$$\text{s.t.} \quad (1.5)\text{–}(1.9), \tag{1.11}$$

$$Q_i \geq 0, \; \Delta T_a^i \geq \epsilon, \; \Delta T_b^i \geq \epsilon, \quad i \in \{c,h,m\}, \tag{1.12}$$

where the cost coefficients $\hat{c}_i$ and $\bar{c}_i$ reflect the energy and amortized capital prices, the exponent $\beta \in (0,1]$ reflects the economy of scale of the equipment, and a small constant $\epsilon > 0$ is selected to prevent the log-mean temperature difference from becoming undefined.

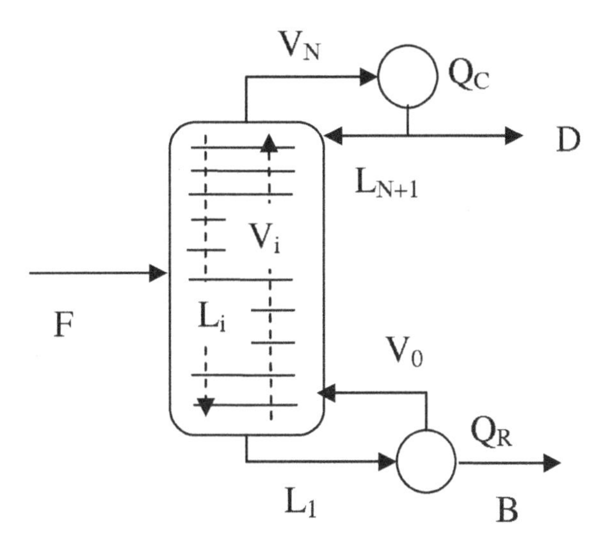

Figure 1.3. *Distillation column example.*

This example has one degree of freedom. For instance, if the heat duty Q_m is specified, then the hot and cold stream temperatures and all of the remaining quantities can be calculated.

1.4.2 Real-Time Optimization of a Distillation Column

Distillation is the most common means for separation of chemical components and lies at the heart of petroleum refining; it has no moving parts and scales easily and economically to all production levels. However, distillation is highly energy intensive and can consume 80%–90% of the total energy in a typical chemical or petrochemical process. As a result, optimization of distillation columns is essential for the profitability of these processes. Moreover, because distillation feeds, product demands and even ambient conditions change over time, the real-time optimization in response to these changes is also a key contributor to successful operation. Consider the distillation column shown in Figure 1.3 with N trays. As seen in the figure, liquid and vapor contact each other and approach equilibrium (i.e., boiling) on each tray. Moreover, the countercurrent flow of liquid and vapor provides an enrichment of the volatile (light) components in the top product and the remaining components in the bottom product. Two heat exchangers, the top condenser and the bottom reboiler, act as sources for the condensed liquid vapor and boiled-up vapor, respectively. The hydrocarbon feed contains chemical components given by the set

$$\mathcal{C} = \{propane,\ isobutane,\ n\text{-}butane,\ isopentane,\ n\text{-}pentane\}.$$

The column is specified to recover most of the *n-butane* (the light key) in the top product and most of the *isopentane* (the heavy key) in the bottom product. We assume a total condenser and partial reboiler, and that the liquid and vapor phases are in equilibrium. A tray-by-tray distillation column model is constructed as follows using the MESH (Mass-Equilibrium-Summation-Heat) equations:

Total Mass Balances

$$B + V_0 - L_1 = 0, \tag{1.13}$$

$$L_i + V_i - L_{i+1} - V_{i-1} = 0, \quad i \in [1, N], \ i \notin \mathcal{S}, \tag{1.14}$$

$$L_i + V_i - L_{i+1} - V_{i-1} - F = 0, \quad i \in \mathcal{S}, \tag{1.15}$$

$$L_{N+1} + D - V_N = 0. \tag{1.16}$$

Component Mass Balances

$$Bx_{0,j} + V_0 y_{0,j} - L_1 x_{1,j} = 0, \quad j \in \mathcal{C}, \tag{1.17}$$

$$L_i x_{i,j} + V_i y_{i,j} - L_{i+1} x_{i+1,j} - V_{i-1,j} y_{i-1,j} = 0, \quad j \in \mathcal{C}, \quad i \in [1, N], \ i \notin \mathcal{S}, \tag{1.18}$$

$$L_i x_{i,j} + V_i y_{i,j} - L_{i+1} x_{i+1,j} - V_{i-1} y_{i-1,j} - F x_{F,j} = 0, \quad j \in \mathcal{C}, \ i \in \mathcal{S}, \tag{1.19}$$

$$(L_{N+1} + D) x_{N+1,j} - V_N y_{N,j} = 0, \quad j \in \mathcal{C}, \tag{1.20}$$

$$x_{N+1,j} - y_{N,j} = 0, \quad j \in \mathcal{C}. \tag{1.21}$$

Enthalpy Balances

$$BH_B + V_0 H_{V,0} - L_1 H_{L,1} - Q_R = 0, \tag{1.22}$$

$$L_i H_{L,i} + V_i H_{V,i} - L_{i+1} H_{L,i+1} - V_{i-1} H_{V,i-1} = 0, \quad i \in [1, N], \ i \notin \mathcal{S}, \tag{1.23}$$

$$L_i H_{L,i} + V_i H_{V,i} - L_{i+1} H_{L,i+1} - V_{i-1} H_{V,i-1} - F H_F = 0, \quad i \in \mathcal{S}, \tag{1.24}$$

$$V_N H_{V,N} - (L_{N+1} + D) H_{L,D} - Q_C = 0. \tag{1.25}$$

Summation, Enthalpy, and Equilibrium Relations

$$\sum_{j=1}^{m} y_{i,j} - \sum_{j=1}^{m} x_{i,j} = 0, \quad i = 0, \ldots, N+1, \tag{1.26}$$

$$y_{i,j} - K_{i,j}(T_i, P, x_i) x_{i,j} = 0, \quad j \in \mathcal{C}, \ i = 0, \ldots, N+1, \tag{1.27}$$

$$H_{L,i} = \varphi_L(x_i, T_i), H_{V,i} = \varphi_V(y_i, T_i), \quad i = 1, \ldots, N, \tag{1.28}$$

$$H_B = \varphi_L(x_0, T_0), H_F = \varphi_L(x_F, T_F), H_{N+1} = \varphi_L(x_{N+1}, T_{N+1}), \tag{1.29}$$

where

i	tray index numbered starting from reboiler (= 1)
$j \in \mathcal{C}$	components in the feed. The most volatile (lightest) is *propane*
P	pressure in the column
$\mathcal{S} \in [1, N]$	set of feed tray locations in column, numbered from the bottom
F	feed flow rate
L_i / V_i	flow rate of liquid/vapor leaving tray i
T_i	temperature of tray i
H_F	feed enthalpy
$H_{L,i} / H_{V,i}$	enthalpy of liquid/vapor leaving tray i
x_F	feed composition
$x_{i,j}$	mole fraction j in liquid leaving tray i
$y_{i,j}$	mole fraction j in vapor leaving tray i
$K_{i,j}$	nonlinear vapor/liquid equilibrium constant
φ_V / φ_L	nonlinear vapor/liquid enthalpy function
D/B	distillate/bottoms flow rate
Q_R / Q_C	heat load on reboiler/condenser
$lk/hk \in \mathcal{C}$	light/heavy key components that determine the separation.

The feed is a saturated liquid with component mole fractions specified in the order given above. The column is operated at a constant pressure, and we neglect pressure drop across the column. This problem has 2 degrees of freedom. For instance, if the flow rates for V_0 and L_{N+1} are specified, all of the other quantities can be calculated from (1.13)–(1.27). The objective is to minimize the reboiler heat duty which accounts for a major portion of operating costs, and we specify that the mole fraction of the light key must be 100 times smaller in the bottom than in the top product. The optimization problem is therefore given by

$$\min Q_R \tag{1.30a}$$

$$\text{s.t.} \quad (1.13)\text{–}(1.27) \tag{1.30b}$$

$$x_{bottom,lk} \le 0.01 x_{top,lk}, \tag{1.30c}$$

$$L_i, V_i, T_i \ge 0, \quad i = 1, \ldots, N+1, \tag{1.30d}$$

$$D, Q_R, Q_C \ge 0, \tag{1.30e}$$

$$y_{i,j}, x_{i,j} \in [0,1], \quad j \in \mathcal{C}, \quad i = 1, \ldots, N+1. \tag{1.30f}$$

Distillation optimization is an important and challenging industrial problem and it also serves as a very useful testbed for nonlinear programming. This challenging application can be scaled up in three ways; through the number of trays to increase overall size, through the number of components to increase the size of the equations per tray, and through phase equilibrium relations, which also increase the nonlinearity and number of equations on each tray. An extended version of this problem will be considered in Chapter 11.

1.4.3 Model Predictive Control

Model predictive control (MPC) is a widely used optimization-based control strategy because it applies to general multivariable systems and allows constraints both on the input

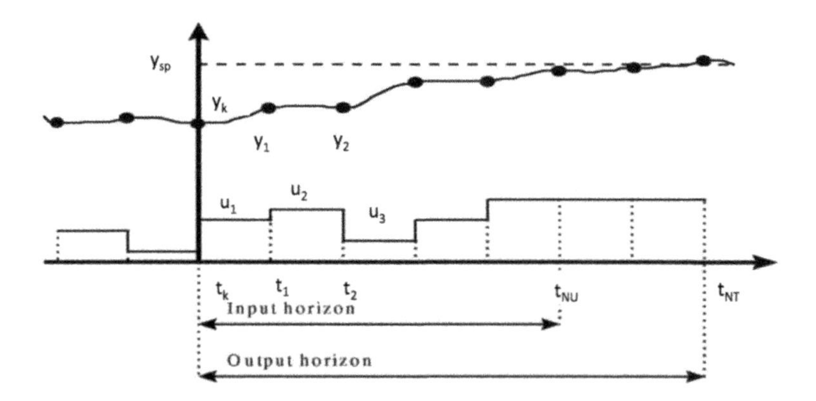

Figure 1.4. *Schematic for model predictive control.*

and output variables. In its most common form, a linear model is used to represent the dynamic process within separate time periods. A quadratic objective function is used to drive the output (or controlled) variables back to their desired values (or setpoints) and also to stabilize the control (or manipulated) variable profiles. As a result, a so-called quadratic programming problem is solved online. As seen in Figure 1.4 the control variables are allowed to vary in the first several time periods (H_u known as the input horizon) and the process is simulated for several more time periods (H_p known as the output horizon). The output horizon must be long enough so that the process will come back to steady state at the end of the simulation. After a horizon subproblem is solved, the control variables in the first time period are implemented on the process. At the next time period, the MPC model is updated with new measurements from the process and the next horizon subproblem is set up and solved again. The input horizon can be adjusted within the limits, $1 \le H_u \le H_p$. As long as the time sampling periods are short enough and the process does not stray too far from the setpoint (about which the system can be linearized), then a linear model usually leads to good controller performance [47]. On the other hand, for processes undergoing large changes including periods of start-up, shut-down, and change of steady states, the linear model for the process is insufficient, nonlinear models must be developed, and, as described in Chapter 9, a more challenging nonlinear program needs to be solved.

For most MPC applications, a linear time invariant (LTI) model is used to describe the process dynamics over time interval Δt; this is of the following form:

$$x^{k+1} = Ax^k + Bu^k, \tag{1.31}$$

where $x^k \in \mathbb{R}^{n_x}$ is the vector of state variables at $t = t_0 + k\Delta t$, $u^k \in \mathbb{R}^{n_u}$ is the vector of manipulated variables at $t = t_0 + k\Delta t$, $A \in \mathbb{R}^{n_x \times n_x}$ is the state space model matrix, $B \in \mathbb{R}^{n_x \times n_u}$ is the manipulated variable model matrix, $\Delta t \in \mathbb{R}$ is the sampling time, and $t_0 \in \mathbb{R}$ is the current time. A set of output (or controlled) variables is selected using the following linear transformation:

$$y^k = Cx^k + d^k, \tag{1.32}$$

where $y^k \in \mathbb{R}^{n_y}$ is the vector of output variables at $t = t_0 + k\Delta t$, $d^k \in \mathbb{R}^{n_y}$ is the vector of known disturbances at $t = t_0 + k\Delta t$, and $C \in \mathbb{R}^{n_y \times n_x}$ is the controlled variable mapping matrix.

The control strategy for this process is to find values for manipulated variables u^k, $k = 1,\dots,H_u$, over an input horizon H_u such that (1.31)–(1.32) hold over the output horizon H_p for $k = 1,\dots,H_p$ and $y^{H_p} \approx y^{\text{sp}}$ (i.e., endpoint constraint). To accomplish these goals, the following MPC QP is set up and solved:

$$\min_{u^k,x^k,y^k} \sum_{k=1}^{H_p} (y^k - y^{\text{sp}})^T Q_y (y^k - y^{\text{sp}}) + \sum_{k=1}^{H_u} (u^k - u^{k-1})^T Q_u (u^k - u^{k-1}) \quad (1.33a)$$

$$\text{s.t.} \quad x^k = Ax^{k-1} + Bu^{k-1} \quad \text{for } k = 1,\dots,H_p, \quad (1.33b)$$

$$y^k = Cx^k + d^k \quad \text{for } k = 1,\dots,H_p, \quad (1.33c)$$

$$u^k = u^{H_u} \quad \text{for } k = H_u + 1,\dots,H_p \quad (1.33d)$$

$$-\hat{b} \le \hat{A}^T u^k \le \hat{b} \quad \text{for } k = 1,\dots,H_p, \quad (1.33e)$$

$$u^L \le u_k \le u^U \quad \text{for } k = 1,\dots,H_p \quad (1.33f)$$

$$-\Delta u^{\max} \le u^k - u^{k-1} \le +\Delta u^{\max} \quad \text{for } k = 1,\dots,H_p, \quad (1.33g)$$

where $u^0 = u(t_0) \in \mathbb{R}^{n_u}$ is the initial value of manipulated variables, $y^{\text{sp}} \in \mathbb{R}^{n_y}$ is the vector of setpoints for output variables, $Q_y \in \mathbb{R}^{n_y \times n_y}$ is the diagonal penalty matrix for output variables, $Q_u \in \mathbb{R}^{n_u \times n_u}$ is the diagonal penalty matrix for manipulated variables, $\hat{b} \in \mathbb{R}^{n_u}$ are bounds for state constraints, $\hat{A} \in \mathbb{R}^{n_u \times n_u}$ is the transformation matrix for state constraints, and $\Delta u^{\max} \in \mathbb{R}^{n_u}$ is the maximum change allowed for the actuators.

This particular objective function (1.33a) allows the resulting MPC controller to drive the outputs to their setpoints and the terms for u^k help to regularize the QP and dampen wild changes in the manipulated variables. The constraints (1.33b) and (1.33c) are the process model and controlled variable selection equations, respectively. The constraints (1.33b)–(1.33d) form the dynamic model and are common to every MPC application. In particular, the constraints (1.33d) fix the control variables after the end of the input horizon H_u. The remaining constraints (1.33e)–(1.33g) are specific to the process. The ability of MPC to handle multivariable dynamic models along with additional bound constraints distinguishes this popular controller over its alternatives.

1.5 A Motivating Application

The above NLP problems can be solved with the algorithms presented in later chapters. However, to motivate the solution of these optimization problems we consider a smaller example shown in Figure 1.5. The cylindrical vessel has a fixed volume (V) given by $V = (\pi/4)D^2L$, and we would like to determine the length (L) and diameter (D) that minimize the cost of tank by solving the nonlinear program:

$$\min_{L,D} f \equiv c_s \pi DL + c_t(\pi/2)D^2 \quad (1.34a)$$

$$\text{s.t.} \quad V = (\pi/4)D^2L, \quad (1.34b)$$

$$D \ge 0, \quad L \ge 0. \quad (1.34c)$$

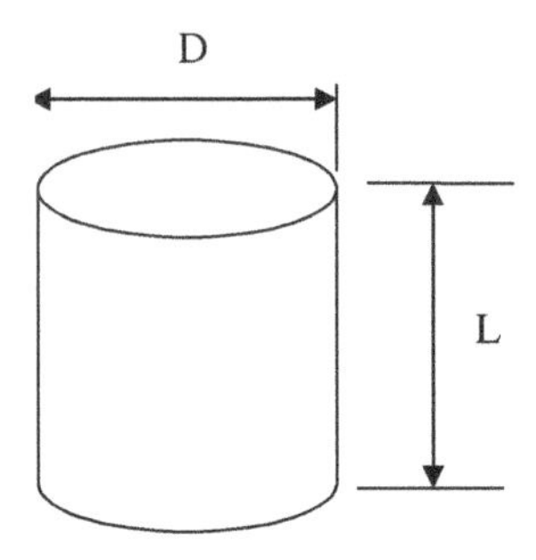

Figure 1.5. *Cylindrical tank optimization.*

Note that the cost of the tank is related to its surface area (i.e., the amount of material required) and c_s is the per unit area cost of the tank's side, while c_t is the per unit area cost of the tank's top and bottom.

This problem can be simplified by neglecting the bound constraints and eliminating L from the problem, leading to the unconstrained problem:

$$\min f \equiv 4c_s V/D + c_t(\pi/2)D^2. \tag{1.35}$$

Differentiating this objective with respect to D and setting the derivative to zero gives

$$\frac{df}{dD} = -4c_s V/D^2 + c_t \pi D = 0, \tag{1.36}$$

yielding $D = (\frac{4c_s V}{\pi c_t})^{1/3}$, $L = (\frac{4V}{\pi})^{1/3}(\frac{c_t}{c_s})^{2/3}$ and the aspect ratio $L/D = c_t/c_s$.

The optimal result helps to explain why soft drink cans are long and thin while storage tanks are short and fat. Moreover, we see that the solution of this problem is easily derived from simple differential calculus. Nevertheless, it is easy to see that the problem can be generalized so that an analytical solution may be much more difficult:

- The inequality constraints are neglected, and because L and D are both positive, they were satisfied (and inactive) at the solution. If this were not the case, more complicated optimality conditions would be applied, as derived in Chapter 4.
- The equality constraint that related L and D allowed an easy elimination of one of the variables. For nonlinear equations this is often not possible and an implicit elimination is needed, as described in Chapters 5 and 6.
- The variable D could be solved directly from the nonlinear equation (1.36) to yield a closed form solution. In many more applications, this is not possible and an indirect solution procedure must be required, as described in Chapter 3.

Dealing with these issues, especially as they apply to much larger problems, requires the numerical algorithms developed in this text. In particular, the last aspect in characterizing and finding solutions to unconstrained optimization problems is described in the next two chapters.

1.6 Summary and Notes for Further Reading

This chapter introduces the scope of optimization problems along with its conceptual components. Optimization covers a broad area, with a wide collection of problem classes. All of these can be found in chemical engineering applications. After a brief description of these areas, we focus on nonlinear programming problems and discuss specific application process examples of NLP. A small example is then covered in detail in order to motivate some of the issues in developing concepts and algorithms for NLP problems and to set the stage for the next chapters.

The overview in this chapter provides only a brief survey of process optimization applications. Modeling, analysis, design, and operations in chemical engineering are described in hundreds of textbooks and thousands of research sources. A reader unfamiliar with this literature is invited to sample *Perry's Chemical Engineers' Handbook*, known as the "chemical engineer's bible" [121], for a quick immersion into this area. Moreover, there is a growing list of textbooks that deal with optimization in chemical engineering, including the undergraduate texts [53, 122, 332]. All of these give a good overview of a broad set of optimization applications in chemical engineering.

In addition, optimization algorithms have been incorporated into a wide variety of optimization modeling platforms, including MATLAB®, GAMS [71], and AMPL [148], as well as widely used commercial chemical process simulators such as ASPEN, gPROMS, HySyS, Pro/II, and ROMeo. More information can be found on the websites of each vendor.

Finally, recent reviews [52, 175] provide a broader scope of the wealth of optimization problems in chemical engineering beyond NLP. These include MINLPs [53, 143, 295], global optimization algorithms and applications [144, 379], and stochastic programming [57, 216].

1.7 Exercises

1. Examine whether the following functions are convex or not:
   - $x^2 + ax + b$, $x \in \mathbb{R}$,
   - x^3, $x \in \mathbb{R}$,
   - x^4, $x \in \mathbb{R}$,
   - $\log(x)$, $x \in (0, 1]$.

2. Consider the minimization problem $\min\ |x| + |y|$ s.t. $x^2 + y^2 = 1$. By introducing binary decision variables to handle the absolute value terms, reformulate this problem as an MILP of the form (1.1).

3. Show that the nonsmooth optimization problem $\min\ \max(x, y)$ s.t. $x + y = 1, x, y \in \mathbb{R}$ has the same solution as

$$\begin{aligned} \min \quad & z \\ \text{s.t.} \quad & z \geq x, \\ & z \geq y, \\ & x + y = 1. \end{aligned}$$

4. Show that as $\Delta T_a^i \to \Delta T_b^i$, the log-mean temperature difference ΔT_{lm}^i given by (1.9) converges to the limit $\Delta T_{lm}^i = \Delta T_a^i$.

5. Modify the motivating example (1.34) and consider finding the minimum cost of a parallelepiped with a fixed volume. Show that the optimal solution must be a cube.

6. Consider the distillation column optimization (1.30) and introduce binary decision variables to formulate the optimal location of the feed tray as an MINLP of the form (1.1).

Chapter 2

Concepts of Unconstrained Optimization

Unconstrained optimization is a useful tool for many engineering applications, particularly in the estimation of parameters and states from data, to assess trade-offs for economic analysis and to analyze a variety of chemical engineering systems. Moreover, an understanding of unconstrained optimization concepts is essential for optimization of constrained problems. These underlying concepts form the basis for many of the developments in later chapters. This chapter describes the formulation of unconstrained optimization problems, discusses fundamentals including continuity, differentiability, and convexity, and develops first and second order conditions for local optimality. The chapter includes several examples and motivates the need for iterative algorithms in the following chapter.

2.1 Introduction

We begin by considering a scalar, real objective function in n variables, i.e., $f(x):\mathbb{R}^n \to \mathbb{R}$. The unconstrained optimization problem can be written as

$$\min_{x\in\mathbb{R}^n} \quad f(x). \tag{2.1}$$

Note that this problem is equivalent to

$$\max_{x\in\mathbb{R}^n} \quad -f(x) \tag{2.2}$$

and that the objective function values are related by

$$-\max_{x}\,(-f(x)) = \min_{x}\, f(x). \tag{2.3}$$

To motivate the formulation and solution of unconstrained optimization problems, we consider the following data fitting example problem.

Example 2.1 Consider the reaction of chemical species A, B, and C in an isothermal batch reactor according to the following first order reactions: $A \xrightarrow{k_1} B$ and $A \xrightarrow{k_2} C$. The equations

Table 2.1. *Concentration data for first order parallel reactions.*

i	time, t_i	$\hat{a}(t_i)$	$\hat{b}(t_i)$	$\hat{c}(t_i)$
1	0.1	0.913	0.0478	0.0382
2	0.2	0.835	0.0915	0.0732
3	0.3	0.763	0.1314	0.1051
4	0.4	0.697	0.1679	0.1343
5	0.5	0.637	0.2013	0.1610
6	0.6	0.582	0.2318	0.1854
7	0.7	0.532	0.2596	0.2077
8	0.8	0.486	0.2851	0.2281
9	0.9	0.444	0.3084	0.2467
10	1	0.406	0.3296	0.2637

that describe this reaction are given by

$$\frac{da}{dt} = -(k_1 + k_2)a(t), \quad a(0) = 1, \tag{2.4}$$

$$\frac{db}{dt} = k_1 a(t), \quad b(0) = 0, \tag{2.5}$$

$$a(0) = a(t) + b(t) + c(t), \tag{2.6}$$

with the dimensionless concentrations $a(t) = [A](t)/[A](0)$, $b(t) = [B](t)/[A](0)$, and $c(t) = [C](t)/[A](0)$, normalized by the initial concentration of A, i.e., $[A](0)$. These equations can be solved to get

$$\begin{aligned} a(t) &= \exp(-(k_1 + k_2)t), \\ b(t) &= \frac{k_1}{k_1 + k_2}(1 - \exp(-(k_1 + k_2)t)), \\ c(t) &= \frac{k_2}{k_1 + k_2}(1 - \exp(-(k_1 + k_2)t)). \end{aligned} \tag{2.7}$$

In Table 2.1 we consider $N = 10$ experimental data points for this system. We define the kinetic parameters as variables $x^T = [k_1, k_2]$ and a least squares function

$$f(x) = \sum_{i=1}^{N} \left[(\hat{a}(t_i) - a(t_i))^2 + (\hat{b}(t_i) - b(t_i))^2 + (\hat{c}(t_i) - c(t_i))^2 \right]$$

that is summed over a set of concentration measurements, $\hat{a}$, $\hat{b}$, and $\hat{c}$, obtained at normalized time, t_i. Figure 2.1 presents a contour plot in the space of the kinetic parameters, with level sets for different values of $f(x)$. The lowest value of $f(x) = 2.28 \times 10^{-6}$, and from the figure we see that this occurs when $k_1 = 0.50062$ and $k_2 = 0.40059$. This corresponds to the best fit of the kinetic model (2.7) to the experimental data. ∎

This example leads us to ask two questions.

- What criteria characterize the solution to the unconstrained optimization problem?
- How can one satisfy these criteria and therefore locate this solution quickly and reliably?

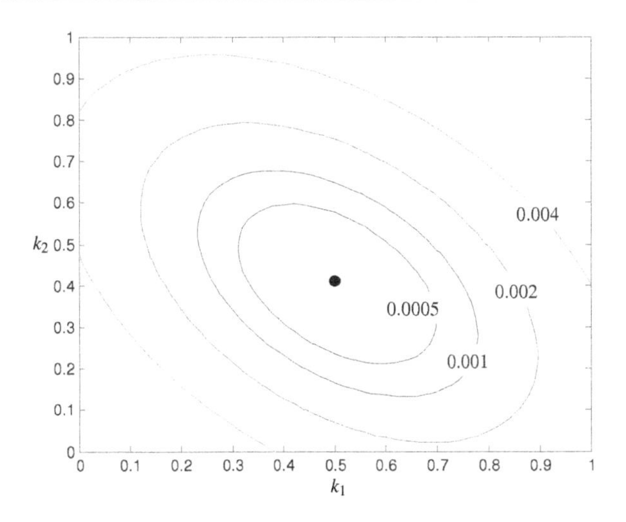

Figure 2.1. *Contour plot of parameter estimation example.*

The first question will be considered in this chapter while the second question will be explored in the next. But before considering these questions, we first introduce some background material.

2.2 Basic Concepts

Some background is presented briefly in this section to support the concepts presented later in this chapter. This material is covered in a number of standard texts, and more detail on these concepts can also be found in [110, 134, 166, 226, 227, 294]. We begin with a brief review of linear algebra, leading to quadratic forms and concluding with a classification of functions.

2.2.1 Vectors and Matrices

Because of the need to introduce matrices and vectors at this stage, a brief review of some basic definitions and properties is helpful here. In general, we adopt the convention that scalar constants and variables will generally be represented by Greek letters, e.g., α, β, γ, vectors will be represented by lowercase letters, x, y, z, and matrices by uppercase letters, X, Y, Z.

Definition 2.2 Operations with matrices can be given as follows:

- Multiplying two matrices $A \in \mathbb{R}^{n\times m}$ and $B \in \mathbb{R}^{m\times p}$ leads to the product $C \in \mathbb{R}^{n\times p}$ with elements $\{C\}_{ij} = \sum_{k=1}^{m}\{A\}_{ik}\{B\}_{kj}$. This operation is defined only if the number of columns in A and rows of B is the same.

- The *transpose* of $A \in \mathbb{R}^{n\times m}$ is $A^T \in \mathbb{R}^{m\times n}$ with the rows and columns of A interchanged, i.e., $\{A^T\}_{ij} = \{A\}_{ji}$.

- A *symmetric* matrix $A \in \mathbb{R}^{n\times n}$ satisfies $A = A^T$.

- A *diagonal* matrix $A \in \mathbb{R}^{n\times n}$ has nonzero elements only on the diagonal, i.e., $\{A\}_{ij} = 0,\ i \neq j$.
- The *identity* matrix $I \in \mathbb{R}^{n\times n}$ is defined as

$$\{I\}_{ij} = \begin{cases} 1 \text{ if } i = j, \\ 0 \text{ otherwise.} \end{cases}$$

Vectors and matrices can be characterized and described by a number of scalar measures. The following definitions introduce these.

Definition 2.3 Determinant of a Matrix.

- A *determinant* is a scalar associated to a square matrix, i.e., $\det(A)$ for $A \in \mathbb{R}^{n\times n}$, and can be defined recursively by $\det(A) = \sum_i (-1)^{i+j}\{A\}_{ij}\bar{A}_{ij}$ for any column j, or $\det(A) = \sum_j (-1)^{i+j}\{A\}_{ij}\bar{A}_{ij}$ for any row i, where $\bar{A}_{ij}$ is the determinant of an $(n-1)$-order matrix with row i and column j removed.
- Some properties for the determinant include

$$\begin{aligned} &\det(AB) = \det(A)\det(B) \\ &\det(A) = \det(A^T) \\ &\det(\alpha A) = \alpha \det(A) \\ &\det(I) = 1. \end{aligned}$$

- A *singular matrix* is a square matrix A with $Av = 0$ for some $v \neq 0$. In this case, $\det(A) = 0$.

Definition 2.4 Rank of a Matrix.

- The rank of a matrix $r(A)$ is the order of the largest submatrix in A that has a nonzero determinant.
- A matrix $A \in \mathbb{R}^{n\times m}$ has full row (column) rank if none of the row (column) vectors can be written as a linear combination of the other row (column) vectors. The rank of such a matrix is $r(A) = \min(n, m)$.
- A matrix that does not have full rank is *rank deficient*.

Definition 2.5 Inverse of a Matrix.

- If the matrix $A \in \mathbb{R}^{n\times n}$ is nonsingular, i.e., has full rank, then an *inverse* A^{-1} exists that satisfies $AA^{-1} = A^{-1}A = I$.
- The linear system $Ax = b$ with nonsingular $A \in \mathbb{R}^{n\times n}$, $x \in \mathbb{R}^n$ and $b \in \mathbb{R}^n$, has a unique solution $x = A^{-1}b$.
- From Definition 2.3, $\det(A^{-1}) = 1/\det(A)$.
- A matrix $Q \in \mathbb{R}^{n\times n}$ with the property $Q^{-1} = Q^T$ is called an *orthogonal matrix*.

Definition 2.6 Eigenvectors and Eigenvalues.

- For a square matrix $A \in \mathbb{R}^{n\times n}$, the *eigenvalue* λ and the corresponding *eigenvector* $v \neq 0$ satisfy $Av = \lambda v$.
- Because $(A-\lambda I)v = 0$, $(A-\lambda I)$ is singular, and we can define a characteristic nth degree polynomial given by $\det(A-\lambda I) = 0$ to find the n eigenvalues.
- $\det(A) = \prod_{i=1}^{n} \lambda_i$.
- Eigenvalues of symmetric matrices are always real. Moreover, by collecting eigenvectors and eigenvalues, a symmetric matrix $A \in \mathbb{R}^{n\times n}$ can be represented by $A = V\Lambda V^T$, where V is the orthogonal matrix of eigenvectors, $V = [v_1, v_2, \ldots, v_n]$, and Λ is a diagonal matrix of eigenvalues with $\{\Lambda\}_{ii} = \lambda_i$.
- Nonsymmetric matrices can have complex eigenvalues. Also, eigenvalues are not defined for nonsquare matrices. Instead, *singular values* $\sigma_i \geq 0$, $i = 1,\ldots,m$, are often useful to characterize $A \in \mathbb{R}^{n\times m}$ with $m \leq n$. These are derived from the eigenvalues of (A^TA) with $\sigma_i = [\lambda_i(A^TA)]^{\frac{1}{2}}$.
- Consider the matrix $A \in \mathbb{R}^{n\times n}$ with all real eigenvalues. Such matrices can be further classified as follows:
  - Matrices with $\lambda_i > 0$, $i = 1,n$, are said to be *positive definite*; i.e., $y^TAy > 0$ for all $y \neq 0$.
  - Matrices with $\lambda_i < 0$, $i = 1,n$, are said to be *negative definite*; i.e., $y^TAy < 0$ for all $y \neq 0$.
  - Matrices with $\lambda_i \geq 0$, $i = 1,n$, are said to be *positive semidefinite*; i.e., $y^TAy \geq 0$ for all $y \neq 0$.
  - Matrices with $\lambda_i \leq 0$, $i = 1,n$, are said to be *negative semidefinite*; i.e., $y^TAy \leq 0$ for all $y \neq 0$.
  - Matrices with both positive and negative eigenvalues are *indefinite*.
  - A matrix with $\lambda_i = 0$, for some i, is *singular*.

Definition 2.7 Vector and Matrix Norms.

- A *vector norm* $\|\cdot\|$ is a measure of the length of a vector and it maps from $\mathbb{R}^n$ to a nonnegative real that has the following properties:

  $$\|x\| = 0 \implies x = 0 \text{ for } x \in \mathbb{R}^n,$$
  $$\|\alpha x\| = |\alpha|\|x\| \text{ for all } x \in \mathbb{R}^n,\ \alpha \in \mathbb{R},$$
  $$\|x+y\| \leq \|x\| + \|y\| \text{ for all } x,y \in \mathbb{R}^n.$$

- The p-norm is defined by $\|x\|_p = (\sum_{i=1}^{n} |x_i|^p)^{1/p}$, where $p \geq 1$. Examples and properties include

  $p = 1$, $\|x\|_1 = \sum_{i=1}^{n} |x_i|$, the 1-norm,

  $p = 2$, $\|x\|_2 = (\sum_{i=1}^{n} x_i^2)^{\frac{1}{2}}$, the Euclidean norm,

$p = \infty$, $\|x\|_\infty = \max_i |x_i|$, the max-norm,

$|x^T y| \le \|x\|_p \|y\|_q$ for $1/p + 1/q = 1$ and all $x, y \in \mathbb{R}^n$.

Also, for $x \in \mathbb{R}^n$ each norm can be bounded above and below by a multiple of any of the other norms.

- For $A \in \mathbb{R}^{n \times m}$, the induced matrix norms are defined by vector norms: $\|A\| = \max_{x \neq 0} \|Ax\|/\|x\|$ for any p-norm. It is therefore consistent with vector norms and satisfies $\|A\|\,\|x\| \ge \|Ax\|$. Examples include

  $\|A\|_1 = \max_j (\sum_{i=1}^n |\{A\}_{ij}|)$ (max column sum of A),

  $\|A\|_\infty = \max_i (\sum_{j=1}^m |\{A\}_{ij}|)$ (max row sum of A),

  $\|A\|_2 =$ largest singular value of A.

- The Frobenius norm $\|A\|_F = (\sum_i \sum_j (\{A\}_{ij})^2)^{\frac{1}{2}}$ is not consistent with a vector norm but has some useful properties. This norm as well as the induced norms satisfy $\|AB\| \le \|A\|\,\|B\|$ for any two compatible matrices A and B. In addition, the induced and Frobenius matrix norms can be bounded above and below by a multiple of any of the other matrix norms.

- The *condition number* is defined by $\kappa(A) = \|A\|\,\|A^{-1}\|$. Using the induced 2-norm the condition number is given by $\sigma_{\max}/\sigma_{\min}$. If A is a symmetric matrix, then this is equivalent to using the eigenvalues of A with $\lambda_{\max}/\lambda_{\min}$.

2.2.2 Quadratic Forms

Before considering the nonlinear unconstrained problem, we briefly consider the special case of the quadratic form, given by

$$f(x) := c + a^T x + \frac{1}{2} x^T B x \tag{2.8}$$

with the scalar function $f : \mathbb{R}^n \to \mathbb{R}$ and constant c, $a, x \in \mathbb{R}^n$, and $B \in \mathbb{R}^{n \times n}$. To find extreme points of (2.8) and to characterize the neighborhood around the extremum in n dimensions, we use the concepts of eigenvalues and eigenvectors to "see" into these higher dimensions. Here, the matrix B can be further decomposed into eigenvalues and eigenvectors, i.e., $B = V \Lambda V^T$. Defining a new coordinate system with $z = V^T x$, we can rewrite the quadratic function in the space of the z variables:

$$f(z) := c + \bar{a}^T z + \frac{1}{2} z^T \Lambda z, \tag{2.9}$$

where $\bar{a} = V^T a$. We further define a point z^*, where

$$z_j^* = \begin{cases} 0 & \text{if } j \in J_0 = \{j \,|\, \lambda_j = 0\}, \\ -\bar{a}_j/\lambda_j & \text{otherwise.} \end{cases}$$

Note that z^* is a stationary point, i.e., $\nabla f(z^*) := \bar{a} + \Lambda z^* = 0$ if there are no zero eigenvalues; otherwise, if $\lambda_j = 0$, there is no stationary point unless $\bar{a}_j = 0, j \in J_0$. Using z^* we can represent (2.9) as

$$\begin{aligned}
f(z) &= c + \bar{a}^T z + \frac{1}{2} z^T \Lambda z \\
&= c + \bar{a}^T z + z^{*T} \Lambda z - \frac{1}{2} z^{*T} \Lambda z^* + \frac{1}{2}(z - z^*)^T \Lambda (z - z^*) \\
&= \bar{c} + (\bar{a} + \Lambda z^*)^T z + \frac{1}{2}(z - z^*)^T \Lambda (z - z^*) \\
&= \bar{c} + \sum_{j \in J_0} \bar{a}_j z_j + \frac{1}{2} \sum_{j=1}^{n} \lambda_j (z_j - z_j^*)^2
\end{aligned} \tag{2.10}$$

with the constant $\bar{c} = c - \frac{1}{2}(z^*)^T \Lambda z^*$, and the last equation follows from $z_j^* = -\bar{a}_j/\lambda_j$, $j \notin J_0$. Centered about z^*, (2.10) allows us to see the locations of extreme points by inspection, for the following cases:

- *B is singular with no stationary point:* If some of the eigenvalues are zero and $\bar{a}_j \neq 0$ for some $j \in J_0$, then $f(z)$ becomes unbounded below by setting $z_j = -\alpha \bar{a}_j, j \in J_0$, and $z_j = z_j^*, j \notin J_0$, and allowing $\alpha \to \infty$. Moreover, a direction that is a linear combination of the eigenvectors, $v_j, j \in J_0$, is called a direction of *zero curvature*.

- *B is positive definite:* If all of the eigenvalues are positive, then (2.10) clearly shows that z^* is the minimum point for $f(z)$, which is unbounded above. In two dimensions, this can be seen through the top contour plot in Figure 2.2.

- *B is negative definite:* If all of the eigenvalues are negative, then (2.10) clearly shows that z^* is the maximum point for $f(z)$, which is unbounded below. In two dimensions, this is similar to the top contour plot in Figure 2.2, but with the signs of the objective function reversed.

- *B is semidefinite with a stationary point:* If $\bar{a}_j = 0, j \in J_0$, and $\lambda_j > 0$ (or < 0), $j \notin J_0$, then z^* is the minimum (or maximum) point. In two dimensions, this can be seen in the middle contour plot in Figure 2.2.

- *B is indefinite:* If some of the eigenvalues are negative and some are positive, then $f(z)$ is unbounded above and below. In this case, if there are no zero eigenvalues, then the stationary point, z^*, is a saddle point. In two dimensions, this can be seen in the bottom contour plot in Figure 2.2. Also, a linear combination of the eigenvectors corresponding to the positive (or negative) eigenvalues is called a *direction of positive (or negative) curvature*.

Example 2.8 Consider the quadratic function given by $f(x) = a^T x + \frac{1}{2} x^T B x$ with $a^T = [1\ 1]$ and $B = \begin{bmatrix} 2 & 1 \\ 1 & 2 \end{bmatrix}$. The eigenvalues can be found by solving the characteristic polynomial $\det(B - \lambda I) = 3 - 4\lambda + \lambda^2 = 0 \implies \lambda = 1, 3$. Thus we have the eigenvalue and eigenvector matrices $\Lambda = \begin{bmatrix} 1 & 0 \\ 0 & 3 \end{bmatrix}$ and $V = \begin{bmatrix} 2^{-\frac{1}{2}} & 2^{-\frac{1}{2}} \\ -2^{-\frac{1}{2}} & 2^{-\frac{1}{2}} \end{bmatrix}$, and because the eigenvalues are both positive, we have the minimum $z^* = -\Lambda^{-1} V^T a = [0\ -\frac{2^{\frac{1}{2}}}{3}]^T$ or $x^* = -V\Lambda^{-1}V^T a = [-\frac{1}{3}\ -\frac{1}{3}]^T$. The contour plots for this quadratic function are given in Figure 2.3, where x^*

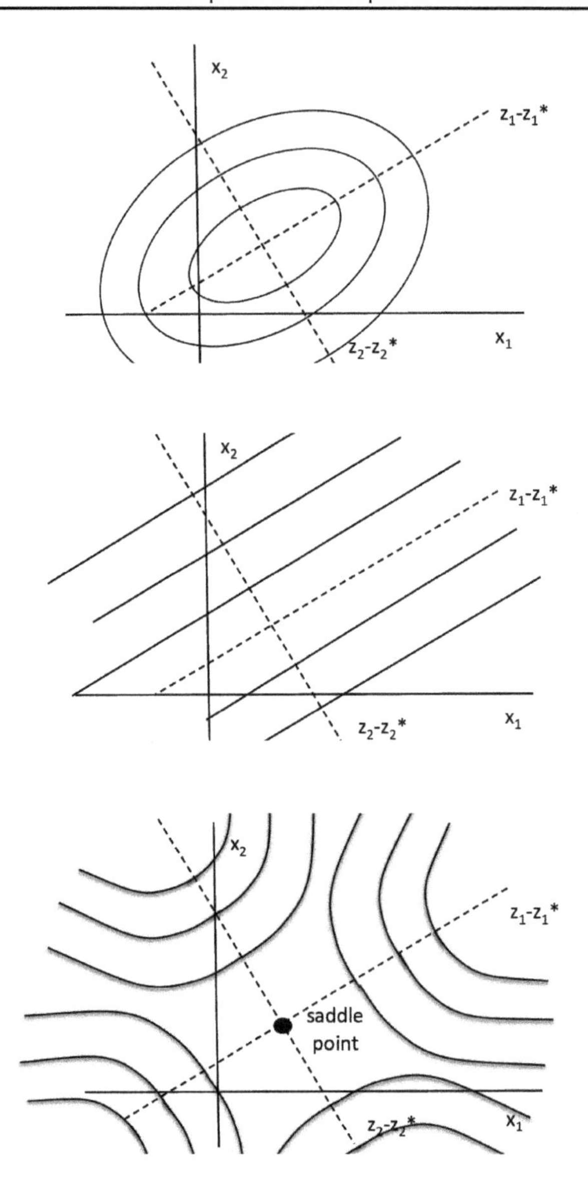

Figure 2.2. *Contour plots for quadratic objective functions: B is positive definite (top), B is singular (middle), B is indefinite (bottom).*

and the eigenvectors v_1 and v_2 are shown. Note that the contours show ellipsoidal shapes with condition number $\kappa(B) = \lambda_{\max}/\lambda_{\min} = 3$, and with $\kappa^{\frac{1}{2}}$ that gives the ratio of major and minor axes in these contours. Note that for large values of κ the contours become more "needle like" and, as κ approaches infinity, the contours approach those of the middle plot in Figure 2.2. ∎

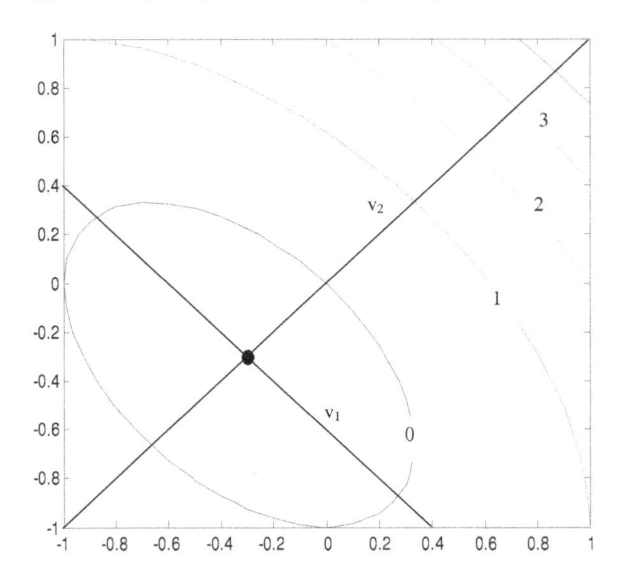

Figure 2.3. *Contour plot for quadratic example.*

2.2.3 Classification of Functions

Figure 2.4 considers a number of contour plots for different types of objective functions. Some basic definitions of these functions are given as follows.

Definition 2.9 A function $\phi(x) : \mathbb{R}^n \to \mathbb{R}$ is *continuous* in $\mathbb{R}^n$ if for every point $\bar{x}$ and all $\epsilon > 0$ there is a value $\delta > 0$ such that

$$\|x - \bar{x}\| < \delta \implies \|\phi(x) - \phi(\bar{x})\| < \epsilon, \tag{2.11}$$

and therefore $\lim_{x \to \bar{x}} \phi(x) \to \phi(\bar{x})$. Note that the bottom left contour plot in Figure 2.4 shows $\phi(x)$ to be discontinuous.

Definition 2.10 A continuous function $\phi(x) : \mathbb{R} \to \mathbb{R}^n$ is *Lipschitz continuous* in $\mathbb{R}^n$ if for any two points $x, y \in \mathbb{R}^n$ there exists a finite $L > 0$ such that

$$\|\phi(x) - \phi(y)\| < L\|x - y\|. \tag{2.12}$$

Definition 2.11 Let $\gamma(\xi)$ be a scalar function of a scalar variable, ξ. Assuming that the limit exists, the first derivative is defined by

$$\frac{d\gamma}{d\xi} = \gamma'(\xi) := \lim_{\epsilon \to 0} \frac{\gamma(\xi + \epsilon) - \gamma(\xi)}{\epsilon} \tag{2.13}$$

and the second derivative is defined by

$$\frac{d^2\gamma}{d\xi^2} = \gamma''(\xi) := \lim_{\epsilon \to 0} \frac{\gamma'(\xi + \epsilon) - \gamma'(\xi)}{\epsilon}. \tag{2.14}$$

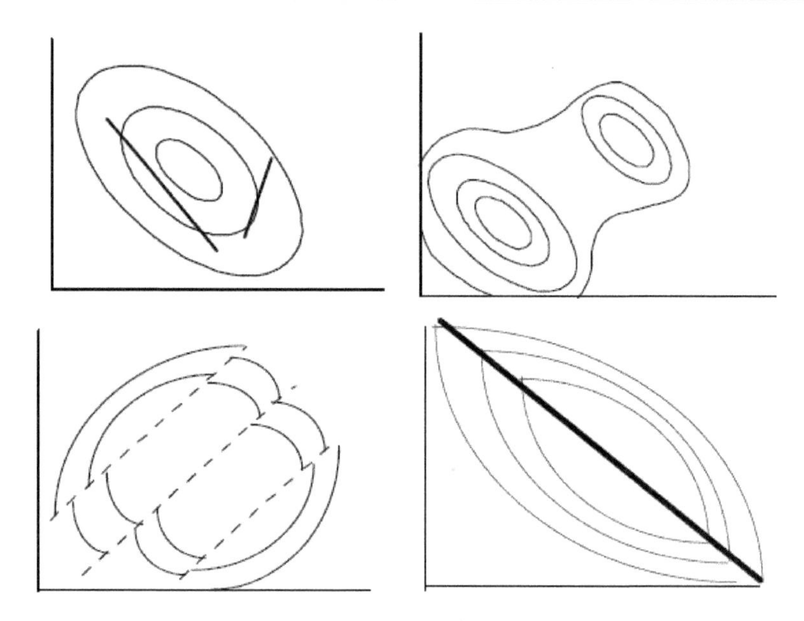

Figure 2.4. *Contour plots for different objective functions: Convex function (top left), nonconvex function (top right), discontinuous function (bottom left), nondifferentiable (but convex) function (bottom right).*

Definition 2.12 Define the unit vector $e_i = [0,0,\dots,1,\dots,0]$ (i.e., a "1" in the ith element of e_i; the other elements are zero). Assuming that the limits exist, the first partial derivative of the multivariable $\phi(x)$ is defined by

$$\frac{\partial \phi}{\partial x_i} := \lim_{\epsilon \to 0} \frac{\phi(x+\epsilon e_i) - \phi(x)}{\epsilon} \tag{2.15}$$

and the second partial derivative is defined by

$$\frac{\partial^2 \phi}{\partial x_i x_j} := \lim_{\epsilon \to 0} \frac{\frac{\partial \phi(x+\epsilon e_i)}{\partial x_j} - \frac{\partial \phi(x)}{\partial x_j}}{\epsilon}. \tag{2.16}$$

We define $\phi(x)$ as *(twice) differentiable* if all of the (second) partial derivatives exist and as *(twice) continuously differentiable* if these (second) partial derivatives are continuous. Second derivatives that are continuous have the property $\frac{\partial^2 \phi}{\partial x_i x_j} = \frac{\partial^2 \phi}{\partial x_j x_i}$. Note that the bottom right contour plot in Figure 2.4 shows $\phi(x)$ to be nondifferentiable.

Assembling these partial derivatives leads to the *gradient* vector

$$\nabla_x \phi(x) = \begin{bmatrix} \frac{\partial \phi(x)}{\partial x_1} \\ \vdots \\ \frac{\partial \phi(x)}{\partial x_n} \end{bmatrix} \tag{2.17}$$

and the *Hessian* matrix

$$\nabla_{xx}\phi(x) := \begin{bmatrix} \frac{\partial^2\phi(x)}{\partial x_1^2} & \frac{\partial^2\phi(x)}{\partial x_1 x_2} & \cdots & \frac{\partial^2\phi(x)}{\partial x_1 x_n} \\ \frac{\partial^2\phi(x)}{\partial x_2 x_1} & \frac{\partial^2\phi(x)}{\partial x_2^2} & \cdots & \frac{\partial^2\phi(x)}{\partial x_2 x_n} \\ \vdots & \vdots & \vdots & \vdots \\ \frac{\partial^2\phi(x)}{\partial x_n x_1} & \frac{\partial^2\phi(x)}{\partial x_n x_2} & \cdots & \frac{\partial^2\phi(x)}{\partial x_n^2} \end{bmatrix}. \tag{2.18}$$

Also, when a function has only one argument, we suppress the subscript on "∇" and simply write $\nabla\phi(x)$ and $\nabla^2\phi(x)$ for the gradient and Hessian, respectively.

In addition, for equality constraints given by $h(x) = 0, h : \mathbb{R}^n \to \mathbb{R}^m$, which we will consider in Chapter 4, we define the *Jacobian* matrix as follows:

$$\nabla h(x)^T := \begin{bmatrix} \frac{\partial h_1(x)}{\partial x_1} & \frac{\partial h_1(x)}{\partial x_2} & \cdots & \frac{\partial h_1(x)}{\partial x_n} \\ \frac{\partial h_2(x)}{\partial x_1} & \frac{\partial h_2(x)}{\partial x_2} & \cdots & \frac{\partial h_2(x)}{\partial x_n} \\ \vdots & \vdots & \vdots & \vdots \\ \frac{\partial h_m(x)}{\partial x_1} & \frac{\partial h_m(x)}{\partial x_2} & \cdots & \frac{\partial h_m(x)}{\partial x_n} \end{bmatrix}. \tag{2.19}$$

Finally we establish properties for convex functions. As already mentioned in Chapter 1, a convex function is defined as follows.

Definition 2.13 A function $\phi(x)$, $x \in \mathbb{R}^n$, is convex if and only if

$$\alpha\phi(x^a) + (1-\alpha)\phi(x^b) \geq \phi(\alpha x^a + (1-\alpha)x^b) \tag{2.20}$$

holds for all $\alpha \in (0,1)$ and all points $x^a, x^b \in \mathbb{R}^n$. Strict convexity requires that inequality (2.20) be strict. For differentiable functions this can be extended to the following statements:

- A continuously differentiable function $\phi(x)$ is convex if and only if

$$\phi(x+p) \geq \phi(x) + \nabla\phi(x)^T p \tag{2.21}$$

  holds for all $x, p \in \mathbb{R}^n$. Strict convexity requires that inequality (2.21) be strict.

- A twice continuously differentiable function $\phi(x)$ is convex if and only if $\nabla^2\phi(x)$ is positive semidefinite for all $x \in \mathbb{R}^n$.

- If $\nabla^2\phi(x)$ is positive definite for all $x \in \mathbb{R}^n$, then the function $\phi(x)$ is defined as *strongly convex*. A strongly convex function is always strictly convex, but the converse is not true. For instance, the function $\phi(x) = x^4$ is strictly convex but not strongly convex.

2.3 Optimality Conditions

The previous section provides the background needed to characterize optimum points. Moreover, for the case of quadratic functions, these points could be identified from the stationary points if particular curvature conditions are satisfied. For general nonlinear objective functions, we first consider a well-known theorem (found in any calculus book).

Theorem 2.14 (Taylor's Theorem [294]). Suppose that $f(x)$ is continuously differentiable, then we have for all $x, p \in \mathbb{R}^n$,

$$f(x+p) = f(x) + \nabla f(x+tp)^T p \quad \text{for some } t \in (0,1). \tag{2.22}$$

Moreover, if $f(x)$ is twice continuously differentiable, then we have for all $x, p \in \mathbb{R}^n$,

$$f(x+p) = f(x) + \nabla f(x)^T p + \frac{1}{2} p^T \nabla^2 f(x+tp)^T p \quad \text{for some } t \in (0,1) \tag{2.23}$$

and

$$\nabla f(x+p) = \nabla f(x) + \int_0^1 \nabla^2 f(x+tp)\, p dt. \tag{2.24}$$

Definition 2.15 Unconstrained Optimal Solutions.

- A point x^* is a *global minimizer* if $f(x^*) \le f(x)$ for all $x \in \mathbb{R}^n$.
- A point x^* is a *local minimizer* if $f(x^*) \le f(x)$ for all x in a neighborhood around x^*. This neighborhood can be defined as $\mathcal{N}(x^*) = \|x - x^*\| < \epsilon$, $\epsilon > 0$.
- A point x^* is a *strict local minimizer* if $f(x^*) < f(x)$ for all x in a neighborhood around x^*.
- A point x^* is an *isolated local minimizer* if there is a neighborhood around x^* that contains no other local minimizers. For instance, the solution $x^* = 0$ is a strict, but not isolated, minimum for the function $f(x) = x^4(2+\cos(1/x))$.

As we will see, finding global minimizers is generally much harder than finding local ones. However, if $f(x)$ is a convex function, then we can invoke the following theorem.

Theorem 2.16 If $f(x)$ is convex, then every local minimum is a global minimum. If $f(x)$ is strictly convex, then a local minimum is the unique global minimum. If $f(x)$ is convex and differentiable, then a stationary point is a global minimum.

Proof: For the first statement, we assume that there are two local minima, x^a and x^b with $f(x^a) > f(x^b)$, and establish a contradiction. Note that x^a is not a global minimum and we have $f(x^a) \le f(x)$, $x \in \mathcal{N}(x^a)$, and $f(x^b) \le f(x)$, $x \in \mathcal{N}(x^b)$. By convexity,

$$f((1-\alpha)x^a + \alpha x^b) \le (1-\alpha) f(x^a) + \alpha f(x^b) \quad \text{for all } \alpha \in (0,1). \tag{2.25}$$

Now choosing α so that $\bar{x} = (1-\alpha)x^a + \alpha x^b \in \mathcal{N}(x^a)$, we can write

$$f(\bar{x}) \le f(x^a) + \alpha(f(x^b) - f(x^a)) < f(x^a) \tag{2.26}$$

which shows a contradiction.

For the second statement, we assume that, again, there are two local minima, x^a and x^b with $f(x^a) \ge f(x^b)$, and establish a contradiction. Here we have $f(x^a) \le f(x)$, $x \in \mathcal{N}(x^a)$, $f(x^b) \le f(x)$, $x \in \mathcal{N}(x^b)$. By strict convexity,

$$f((1-\alpha)x^a + \alpha x^b) < (1-\alpha) f(x^a) + \alpha f(x^b) \quad \text{for all } \alpha \in (0,1), \tag{2.27}$$

and by choosing α so that $\bar{x} = (1-\alpha)x^a + \alpha x^b \in \mathcal{N}(x^a)$, we can write

$$f(\bar{x}) < f(x^a) + \alpha(f(x^b) - f(x^a)) \leq f(x^a) \tag{2.28}$$

which again shows a contradiction.

For the third statement, we note that the stationary point satisfies $\nabla f(x^*) = 0$, and from (2.21), we have

$$f(x^* + p) \geq f(x^*) + \nabla f(x^*)^T p = f(x^*) \tag{2.29}$$

for all $p \in \mathbb{R}^n$. □

Convexity is sufficient to show that a local solution is a global solution. On the other hand, in the absence of convexity, showing that a particular local solution is also a global solution often requires application of rigorous search methods. A comprehensive treatment of global optimization algorithms and their properties can be found in [144, 203, 379]. Instead, this text will focus on methods that guarantee only local solutions.

To identify locally optimal solutions, we consider the following properties.

Theorem 2.17 (Necessary Conditions for Local Optimality). If $f(x)$ is twice continuously differentiable and there exists a point x^* that is a local minimum, then $\nabla f(x^*) = 0$ and $\nabla^2 f(x^*)$ must be positive semidefinite.

Proof: To show $\nabla f(x^*) = 0$, we assume that it is not and apply Taylor's theorem (2.22) to get

$$f(x^* + tp) = f(x^*) + t\nabla f(x^* + \tau p)^T p \quad \text{for some } \tau \in (0,t) \text{ and for all } t \in (0,1). \tag{2.30}$$

Choosing $p = -\nabla f(x^*)$ and, because of continuity of the first derivatives, we can choose t sufficiently small so that $-t\nabla f(x^* + \tau p)^T \nabla f(x^*) < 0$, and $x^* + tp \in \mathcal{N}(x^*)$ is in a sufficiently small neighborhood. Then $f(x^* + tp) = f(x^*) + t\nabla f(x^* + \tau p)^T p < f(x^*)$, which is a contradiction.

To show that $\nabla^2 f(x^*)$ must be positive semidefinite, we assume it is not and show a contradiction. Applying Taylor's theorem (2.23) leads to

$$\begin{aligned} f(x^* + tp) &= f(x^*) + t\nabla f(x^*)^T p + \frac{t^2}{2} p^T \nabla^2 f(x^* + \tau p)^T p \quad (\text{for some } \tau \in (0,t)) \\ &= f(x^*) + \frac{t^2}{2} p^T \nabla^2 f(x^* + \tau p)^T p. \end{aligned}$$

Assuming $p^T \nabla^2 f(x^*)^T p < 0$ and choosing the neighborhood $\mathcal{N}(x^* + tp)$ sufficiently small leads to $\frac{t^2}{2} p^T \nabla^2 f(x^* + \tau p) p < 0$ by continuity of the second derivatives. This also leads to the contradiction, $f(x^* + tp) < f(x^*)$. □

Theorem 2.18 (Sufficient Conditions for Local Optimality). If $f(x)$ is continuously twice differentiable and there exists a point x^* where $\nabla f(x^*) = 0$ and $\nabla^2 f(x^*)$ is positive definite, then x^* is a strict, isolated local minimum.

Proof: Applying these conditions with Taylor's theorem (2.23), we have

$$\begin{aligned} f(x^*+p) &= f(x^*)+\nabla f(x^*)^T p+\frac{1}{2}p^T\nabla^2 f(x^*+tp)^T p \quad \text{(for some } t \in (0,1)) \\ &= f(x^*)+\frac{1}{2}p^T\nabla^2 f(x^*+tp)^T p \end{aligned} \tag{2.31}$$

for any $p \in \mathbb{R}^n$. Choosing p so that $x^* + p \in \mathcal{N}(x^*)$ is in a sufficiently small neighborhood, we have, from continuity of the second derivatives, that $\frac{1}{2}p^T\nabla^2 f(x^*+tp)^T p > 0$ for all $t \in (0,1)$ and $f(x^*+p) > f(x^*)$ for all $x^*+p \in \mathcal{N}(x^*)$.

To show that x^* must be an isolated minimum, we choose a neighborhood $\mathcal{N}(x^*)$ sufficiently small so that $\nabla^2 f(x)$ is positive definite for all $x \in \mathcal{N}(x^*)$. We now assume that x^* is not isolated and show a contradiction. Consider two local solutions $x^*, x^+ \in \mathcal{N}(x^*)$. From Theorem 2.17 and Taylor's theorem (2.24) we have

$$\int_0^1 \nabla^2 f(x^*+t(x^+-x^*))\,(x^+-x^*)dt = \nabla f(x^+) - \nabla f(x^*) = 0, \tag{2.32}$$

which contradicts positive definiteness of $\nabla^2 f(x)$ for all $x \in \mathcal{N}(x^*)$. □

Example 2.19 To illustrate these optimality conditions, we consider the following two-variable unconstrained optimization problem [205]:

$$\min\ f(x) = \alpha\ \exp(-\beta) \tag{2.33}$$

$$\text{with } u = x_1 - 0.8, \tag{2.34}$$

$$v = x_2 - (a_1 + a_2u^2(1-u)^{\frac{1}{2}} - a_3u), \tag{2.35}$$

$$\alpha = -b_1 + b_2u^2(1+u)^{\frac{1}{2}} + b_3u, \tag{2.36}$$

$$\beta = c_1v^2(1-c_2v)/(1+c_3u^2) \tag{2.37}$$

with $a^T = [0.3, 0.6, 0.2]$, $b^T = [5, 26, 3]$, and $c^T = [40, 1, 10]$. The solution to this problem is given by $x^* = [0.7395, 0.3144]$ with $f(x^*) = -5.0893$. At this solution, $\nabla f(x^*) = 0$ and the Hessian is given by $\nabla^2 f(x^*) = \begin{bmatrix} 77.012 & 108.334 \\ 108.334 & 392.767 \end{bmatrix}$, which has eigenvalues $\lambda = 43.417$ and 426.362. Therefore this solution is a strict local minimum. The contour plot for this problem is given in the top of Figure 2.5. From this plot, the careful reader will note that $f(x)$ is a nonconvex function. Moreover, the bottom of Figure 2.5 shows regions of positive and negative eigenvalues. ■

2.4 Algorithms

From the conditions that identify locally optimal solutions, we now consider methods that will locate them efficiently. In Section 2.2, we defined a number of function types. Our main focus will be to develop fast methods for continuously differentiable functions. Before developing these methods, we first provide a brief survey of methods that do not require differentiability of the objective function.

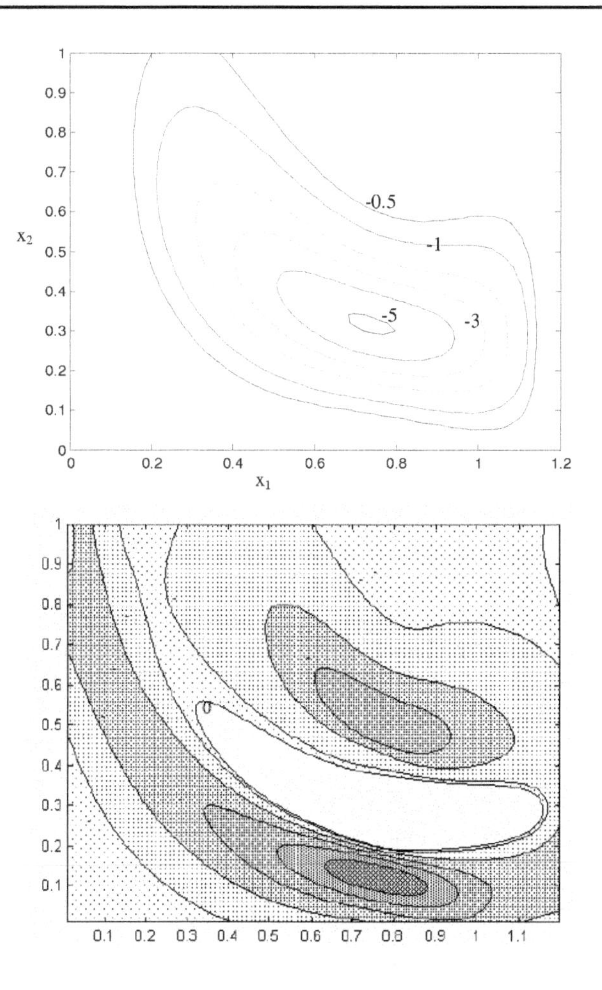

Figure 2.5. *Contours for nonlinear unconstrained example (top) and regions of minimum eigenvalues (bottom). The clear region has nonnegative eigenvalues, and the regions are decremented by values of* 50 *and with darker shading.*

2.4.1 Direct Search Methods

A broad class of optimization strategies does not require derivative information; we call this category *direct search methods*. These methods have the advantage of easy implementation and require little prior knowledge of the optimization problem. In particular, such methods are well suited for optimization studies that can explore the scope of optimization for new applications, prior to investing effort for more sophisticated model reformulation and solution strategies. Many of these methods are derived from heuristics that naturally lead to numerous variations, and a very broad literature describes these methods. Here we discuss only a few of the many important trends in this area.

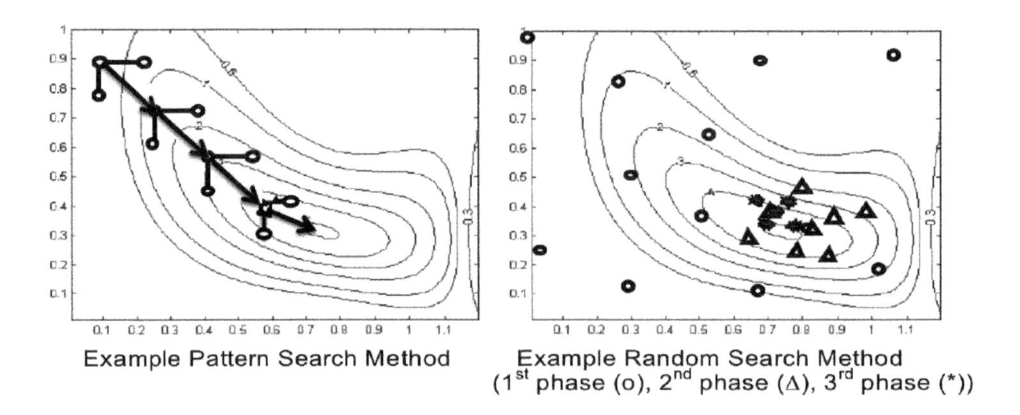

Figure 2.6. *Examples of optimization methods without derivatives.*

Classical Direct Search Methods

Developed in the 1960s and 1970s, these methods include one-at-a-time search and methods based on experimental designs [69]. Methods that also fall into this class include the pattern search of Hooke and Jeeves [202], simplex and complex searches, in particular that of Nelder and Mead [288], and the adaptive random search methods of Luus and Jaakola [268], Goulcher and Cesares Long [168], and Banga and Seider [22]. In addition, the conjugate direction method of Powell [312] takes a systematic approach by constructing and minimizing an approximation based on quadratic forms. All of these methods require only objective function values for unconstrained minimization. Associated with these methods are numerous studies on a wide range of process problems. Moreover, many of these methods include heuristics that prevent premature termination (e.g., directional flexibility in the complex search as well as random restarts and direction generation). To illustrate these methods, Figure 2.6 shows the performance of a pattern search method as well as a random search method on an unconstrained problem.

Simulated Annealing

This strategy is related to random search methods and is derived from a class of heuristics with analogies to the motion of molecules in the cooling and solidification of metals [239]. Here a "temperature" parameter, θ, can be raised or lowered to influence the probability of accepting points that do not improve the objective function. The method starts with a base point, x, and objective value, $f(x)$. The next point x' is chosen at random from a distribution. If $f(x') < f(x)$, the move is accepted with x' as the new point. Otherwise, x' is accepted with probability $p(\theta, x', x)$. Options include the Metropolis distribution,

$$p(\theta, x, x') = \exp(-(f(x') - f(x))/\theta),$$

and the Glauber distribution,

$$p(\theta, x, x') = \exp(-(f(x') - f(x))/\theta)/(1 + \exp(-(f(x') - f(x))/\theta)).$$

The θ parameter is then reduced and the method continues until no further progress is made.

Genetic Algorithms

This approach, first proposed by Holland [200], is based on the analogy of improving a population of solutions through modifying its gene pool. It also has performance characteristics similar as random search methods and simulated annealing. Two forms of genetic modification, crossover or mutation, are used, and the elements of the optimization vector, x, are represented as binary strings. Crossover deals with random swapping of vector elements (among parents with highest objective function values or other rankings of population) or any linear combinations of two parents. Mutation deals with the addition of a random variable to elements of the vector. Genetic algorithms (GAs) have seen widespread use in process engineering and a number of codes are available. Edgar, Himmelblau, and Lasdon [122] described a related GA that is available in Microsoft Excel™.

Derivative-Free Optimization (DFO)

In the past decade, the availability of parallel computers and faster computing hardware, and the need to incorporate complex simulation models within optimization studies, have led a number of optimization researchers to reconsider classical direct search approaches. Dennis and Torczon [111] developed a multidimensional search algorithm that extends the simplex approach of Nelder and Mead [288]. They noted that the Nelder–Mead algorithm fails as the number of variables increases, even for very simple problems. To overcome this, their multidimensional pattern search approach combines reflection, expansion, and contraction steps that act as line search algorithms for a number of linear independent search directions. This approach is easily adapted to parallel computation and can be tailored to the number of processors available. Moreover, this approach converges to locally optimal solutions for unconstrained problems and enjoys an unexpected performance synergy when multiple processors are used. The work of Dennis and Torczon has spawned considerable research on the analysis and code development for DFO methods. Powell [315] recently developed a family of methods for unconstrained and bound constrained optimization through quadratic approximation. In addition, Conn, Scheinberg, and Toint [101] constructed a multivariable DFO algorithm that uses a surrogate model for the objective function within a trust region method. Here, points are sampled to obtain a well-defined quadratic interpolation model, and descent conditions from trust region methods enforce convergence properties. A number of trust region methods that rely on this approach are reviewed in Conn, Gould, and Toint [100] and more recently in Conn, Scheinberg, and Vicente [102].

Finally, a number of DFO codes has been developed that lead to "black box" optimization implementations for large, complex simulation models. These include UOBYQA and NEWUOA by Powell [315, 316], the DAKOTA package at Sandia National Lab [124], and FOCUS developed at Boeing Corporation [66].

Direct search methods are easy to apply to a wide variety of problem types and optimization models. Moreover, because their termination criteria are not based on gradient information and stationary points, they are more likely to favor the search for globally optimal rather than locally optimal solutions. Some of these methods can also be adapted to include integer variables. However, no rigorous convergence properties to globally optimal solutions have yet been discovered. Finally, the performance of direct search methods often scales poorly with the number of decision variables. While performance can be improved with the use of parallel computing, these methods are rarely applied to problems with more than a few dozen decision variables.

2.4.2 Methods That Require Derivatives

For smooth objective functions, i.e., those that are at least three times continuously differentiable, we can focus on much faster methods to locate local optima. Many derivative-based methods are available for unconstrained optimization. In this text we will focus on Newton-type methods because they are fast local solvers and because many of the concepts for unconstrained optimization extend to fast constrained optimization algorithms as well. In addition to these methods, it is also important to mention the matrix-free conjugate gradient method. More detail on this method can be found in [294].

To derive Newton's method for unconstrained optimization, we consider the smooth function $f(x)$ and apply Taylor's theorem at a fixed value of x and a direction p:

$$f(x+p) = f(x) + \nabla f(x)^T p + \frac{1}{2} p^T \nabla^2 f(x) p + O(\|p\|^3). \tag{2.38}$$

If we differentiate this function with respect to p, we obtain

$$\nabla f(x+p) = \nabla f(x) + \nabla^2 f(x) p + O(\|p\|^2). \tag{2.39}$$

At the given point x we would like to determine the vector p that locates the stationary point, i.e., $\nabla f(x+p) = 0$. Taking only the first two terms on the right-hand side, we can define the search direction by

$$\nabla f(x) + \nabla^2 f(x) p = 0 \implies p = -(\nabla^2 f(x))^{-1} \nabla f(x) \tag{2.40}$$

if $\nabla^2 f(x)$ is nonsingular. In fact, for x in the neighborhood of a strict local optimum, with $\|p\|$ small, we would expect the third order term in (2.38) to be negligible; $f(x)$ would then behave like a quadratic function. Moreover, $\nabla^2 f(x)$ would need to be positive definite in order to compute the Newton step in (2.40), and from Theorem 2.18 this would be consistent with conditions for a local minimum.

This immediately leads to a number of observations:

- If $f(x)$ is a quadratic function where $\nabla^2 f(x)$ is positive definite, then the Newton step $p = -(\nabla^2 f(x))^{-1} \nabla f(x)$ applied at any point $x \in \mathbb{R}^n$ immediately finds the global solution at $x + p$.

- Nonsingularity of $\nabla^2 f(x)$ is required at any point x where the Newton step p is computed from (2.40). This is especially important for $\nabla^2 f(x^*)$, as a strict local solution is required for this method. On the other hand, if $\nabla^2 f(x)$ is singular, some corrections need to be made to obtain good search directions and develop a convergent method. More detail on this issue will be given in Chapter 3.

- To promote convergence to a local solution, it is important that the Newton step be a *descent step*, i.e., $\nabla f(x)^T p < 0$, so that $f(x^k)$ can decrease from one iteration k to the next. If $x^k \in \mathcal{N}(x^*)$ and $x^{k+1} = x^k + \alpha p$, where the step size $\alpha \in (0,1]$, then from (2.23) we require that

$$0 > f(x^k + \alpha p) - f(x^k) = \alpha \nabla f(x^k)^T p + \frac{\alpha^2}{2} p^T \nabla^2 f(x + tp) p \quad \text{for some } t \in (0,1). \tag{2.41}$$

Since $p^T \nabla^2 f(x+tp)p > 0$ if x^* is a strict local solution and $\mathcal{N}(x^*)$ is sufficiently small, then we must have $\nabla f(x^k)^T p < 0$. This is a key property for the algorithmic development in the next chapter.

The resulting Newton algorithm with full steps, i.e., $\alpha = 1$, is given by the following algorithm.

ALGORITHM 2.1.
Choose a starting point x^0.
For $k \geq 0$ while $\|p^k\| > \epsilon_1$ and $\|\nabla f(x^k)\| > \epsilon_2$:

1. At x^k, evaluate $\nabla f(x^k)$ and $\nabla^2 f(x^k)$. If $\nabla^2 f(x^k)$ is singular, STOP.
2. Solve the linear system $\nabla^2 f(x^k)p^k = -\nabla f(x^k)$.
3. Set $x^{k+1} = x^k + p^k$ and $k = k+1$.

This basic algorithm has the desirable property of fast convergence, which can be quantified by the following well-known property.

Theorem 2.20 Assume that $f(x)$ is twice differentiable and $\nabla^2 f(x)$ is Lipschitz continuous in a neighborhood of the solution x^*, which satisfies the sufficient second order conditions. Then, by applying Algorithm 2.1 and with x^0 sufficiently close to x^*, there exists a constant $\hat{L} > 0$ such that

- $\|x^{k+1} - x^*\| \leq \hat{L}\|x^k - x^*\|^2$, i.e., the convergence rate for $\{x^k\}$ is quadratic;
- the convergence rate for $\{\nabla f(x^k)\}$ is also quadratic.

Proof: For the Newton step, $p^k = -(\nabla^2 f(x^k))^{-1}\nabla f(x^k)$, we note that, by continuity of the second derivatives, $\nabla^2 f(x^k)$ is nonsingular and satisfies $\|\nabla^2 f(x^k)^{-1}\| < C$ for some $C > 0$ if x^k is sufficiently close to x^*. Thus the Newton step is well defined and we can then write

$$\begin{aligned}
x^k + p^k - x^* &= x^k - x^* - (\nabla^2 f(x^k))^{-1}\nabla f(x^k) \\
&= \nabla^2 f(x^k)^{-1}(\nabla^2 f(x^k)(x^k - x^*) - \nabla f(x^k)) \\
&= \nabla^2 f(x^k)^{-1}(\nabla^2 f(x^k)(x^k - x^*) - (\nabla f(x^k) - \nabla f(x^*))) \\
&\quad \text{and from (2.24)} \\
&= \nabla^2 f(x^k)^{-1}\int_0^1 (\nabla^2 f(x^k) - \nabla^2 f(x^k + t(x^* - x^k)))(x^k - x^*)dt. \qquad (2.42)
\end{aligned}$$

Taking norms of both sides of (2.42) leads to

$$\begin{aligned}
\|x^{k+1} - x^*\| &\leq \|\nabla^2 f(x^k)^{-1}\| \left\| \int_0^1 (\nabla^2 f(x^k) - \nabla^2 f(x^k + t(x^* - x^k)))dt \right\| \|(x^k - x^*)\| \\
&\leq C\bar{L}\|x^k - x^*\|^2 = \hat{L}\|x^k - x^*\|^2. \qquad (2.43)
\end{aligned}$$

Similarly, from $\nabla f(x^k) + \nabla^2 f(x^k) p^k = 0$ we can write

$$\begin{aligned} \|\nabla f(x^{k+1})\| &= \|\nabla f(x^{k+1}) - \nabla f(x^k) - \nabla^2 f(x^k) p^k\| \\ &= \left\| \int_0^1 (\nabla^2 f(x^k + tp^k) - \nabla^2 f(x^k)) p^k dt \right\| \quad \text{from (2.24)} \\ &\leq \bar{L} \|p^k\|^2 \\ &\leq \bar{L} \|(\nabla^2 f(x^k))^{-1}\|^2 \|\nabla f(x^k)\|^2 \\ &\leq \bar{L} C^2 \|\nabla f(x^k)\|^2. \end{aligned} \tag{2.44}$$

□

Example 2.21 Consider the problem described in Example 2.19. Applying the basic Newton algorithm from a starting point close to the solution leads to the iteration sequence given in Table 2.2. Here it is clear that the solution can be found very quickly and, based on the errors $\|x^k - x^*\|$ and $\|\nabla f(x^k)\|$, we observe that both $\{x^k\}$ and $\{f(x^k)\}$ have quadratic convergence rates, as predicted by Theorem 2.20. Moreover, from Figure 2.5 we note that the convergence path remains in the region where the Hessian is positive definite.

On the other hand, if we choose a starting point farther away from the minimum, then we obtain the iteration sequence in Table 2.3. Note that the starting point, which is not much

Table 2.2. *Iteration sequence with basic Newton method with starting point close to solution.*

Iteration (k)	x_1^k	x_2^k	$f(x^k)$	$\|\nabla f(x^k)\|$	$\|x^k - x^*\|$
0	0.8000	0.3000	−5.000	3.000	6.2175×10^{-2}
1	0.7423	0.3115	−5.0882	0.8163	3.9767×10^{-3}
2	0.7395	0.3143	−5.0892	6.8524×10^{-3}	9.4099×10^{-5}
3	0.7395	0.3143	−5.0892	2.6847×10^{-6}	2.6473×10^{-8}
4	0.7395	0.3143	−5.0892	1.1483×10^{-13}	2.9894×10^{-16}

Table 2.3. *Iteration sequence with basic Newton method with starting point far from solution.*

Iteration (k)	x_1^k	x_2^k	$f(x^k)$	$\|\nabla f(x^k)\|$
0	1.0000	0.5000	−1.1226	9.5731
1	1.0748	0.6113	−0.3778	3.3651
2	1.0917	0.7272	−0.1521	1.2696
3	1.0950	0.8398	-8.2079×10^{-2}	0.4894
4	1.0879	0.9342	-6.6589×10^{-2}	0.1525
5	1.0781	0.9515	-6.6956×10^{-2}	1.5904×10^{-2}
6	1.0769	0.9504	-6.6964×10^{-2}	2.0633×10^{-4}
7	1.0768	0.9505	-6.6965×10^{-2}	3.7430×10^{-8}
8	1.0768	0.9504	-6.6964×10^{-2}	9.9983×10^{-15}

farther away, lies in a region where the Hessian is indefinite, and the method terminates at a saddle point where the eigenvalues of the Hessian are 11.752 and −3.034. Moreover, other starting points can also lead to iterates where the objective function is undefined and the method fails. ∎

Example 2.21 illustrates that the optimum solution can be found very accurately if a good starting point is found. On the other hand, it also shows that more reliable algorithms are needed to find the unconstrained optimum. While Newton's method converges very quickly in the neighborhood of the solution, it can fail if

- the Hessian matrix is not positive definite at a given iterate x^k,
- the Hessian matrix is not continuous,
- the starting point x^0 is not sufficiently close to the solution.

In the next chapter we will develop unconstrained algorithms that overcome these difficulties and still retain the fast convergence properties of Newton's method.

2.5 Summary and Notes for Further Reading

This chapter develops concepts and background for unconstrained, multivariable optimization. Beginning with basic concepts of linear algebra, we also explore quadratic forms and further properties of convex functions. This material allows us to characterize global and local optima and develop necessary and sufficient conditions for locally optimal solutions. These conditions then motivate the development of search methods to find optimal solutions. Both derivative-free and gradient-based methods are discussed and strong emphasis is placed on Newton's method and its properties.

More detail on the background concepts for optimality conditions and algorithm development can be found in [110, 134, 166, 226, 227, 294]. In particular, the characterization of local solutions and properties of Newton's methods relies heavily on Taylor's theorem. The presentation above follows the development in [294], where extensions of these properties can also be found.

2.6 Exercises

1. While searching for the minimum of $f(x) = [x_1^2 + (x_2+1)^2][x_1^2 + (x_2-1)^2]$, we terminate at the following points:
   (a) $x = [0,0]^T$,
   (b) $x = [0,1]^T$,
   (c) $x = [0,-1]^T$,
   (d) $x = [1,1]^T$.
   Classify each point.

2. Consider the quadratic function
   $$f(x) = 3x_1 + x_2 + 2x_3 + 4x_1^2 + 3x_2^2 + 2x_3^2 + 2x_1x_2 + 2x_2x_3.$$
   Find the eigenvalues and eigenvectors and any stationary points. Are the stationary points local optima? Are they global optima?

3. Repeat the previous exercise with the quadratic function

$$f(x) = 3x_1 + x_2 + 2x_3 + 4x_1^2 + 3x_2^2 + 2x_3^2 + (M-2)x_1x_2 + 2x_2x_3.$$

For $M = 0$ find all stationary points. Are they optimal? Find the path of optimal solutions as M increases.

4. Apply Newton's method to the optimization problem in Example 2.1, starting from $k_1 = 0.5$, $k_2 = 0.4$ and also from $k_1 = 2.0$, $k_2 = 2.0$. Explain the performance of this method from these points.

5. Download and apply the NEWUOA algorithm to Example 2.19. Use the two starting points from Example 2.21. How does this method compare to the results in Example 2.21?

Chapter 3

Newton-Type Methods for Unconstrained Optimization

Newton-type methods are presented and analyzed for the solution of unconstrained optimization problems. In addition to covering the basic derivation and local convergence properties, both line search and trust region methods are described as globalization strategies, and key convergence properties are presented. The chapter also describes quasi-Newton methods and focuses on derivation of symmetric rank one (SR1) and Broyden–Fletcher–Goldfarb–Shanno (BFGS) methods, using simple variational approaches. The chapter includes a small example that illustrates the characteristics of these methods.

3.1 Introduction

Chapter 2 concluded with the derivation of Newton's method for the unconstrained optimization problem

$$\min_{x \in \mathbb{R}^n} \quad f(x). \tag{3.1}$$

For unconstrained optimization, Newton's method forms the basis for the most efficient algorithms. Derived from Taylor's theorem, this method is distinguished by its fast performance. As seen in Theorem 2.20, this method has a quadratic convergence rate that can lead, in practice, to inexpensive solutions of optimization problems. Moreover, extensions to constrained optimization rely heavily on this method; this is especially true in chemical process engineering applications. As a result, concepts of Newton's method form the core of all of the algorithms discussed in this book.

On the other hand, given a solution to (3.1) that satisfies first and second order sufficient conditions, the basic Newton method in Algorithm 2.1 may still have difficulties and may be unsuccessful. Newton's method can fail on problem (3.1) for the following reasons:

1. The objective function is not smooth. Here, first and second derivatives are needed to evaluate the Newton step, and Lipschitz continuity of the second derivatives is needed to keep them bounded.

2. The Newton step does not generate a descent direction. This is associated with Hessian matrices that are not positive definite. A singular matrix produces Newton steps that are unbounded, while Newton steps with *ascent directions* lead to an

increase in the objective function. These arise from Hessian matrices with negative curvature.

3. The starting point is not sufficiently close to solution. For general unconstrained problems, this property is the hardest to check. While estimates of regions of attraction for Newton's method have been developed in [113], they are not easy to apply when the solution, and its relation to the initial point, is unknown.

These three challenges raise some important questions on how to develop reliable and efficient optimization algorithms, based on Newton's method. This chapter deals with these questions in the following way:

1. In the application of Newton's method throughout this book, we will focus only on problems with smooth functions. Nevertheless, there is a rich literature on optimization with nonsmooth functions. These include development of nonsmooth Newton methods [97] and the growing field of nonsmooth optimization algorithms.

2. In Section 3.2, we describe a number of ways to modify the Hessian matrix to ensure that the modified matrix at iteration k, B^k, has a bounded condition number and remains positive definite. This is followed by Section 3.3 that develops the concept of quasi-Newton methods, which do not require the calculation of the Hessian matrix. Instead, a symmetric, positive definite B^k matrix is constructed from differences of the objective function gradient at successive iterations.

3. To avoid the problem of finding a good starting point, globalization strategies are required that ensure sufficient decrease of the objective function at each step and lead the algorithm to converge to locally optimal solutions, even from distant starting points. In Section 3.4, this *global convergence property* will be effected by line search methods that are simple modifications of Algorithm 2.1 and require that B^k be positive definite with bounded condition numbers. Moreover, these positive definiteness assumptions can be relaxed if we apply trust region methods instead. These strategies are developed and analyzed in Section 3.5.

3.2 Modification of the Hessian Matrix

To generate directions that lead to improvement of the objective function, we first consider modifications of the Hessian matrix. We begin with a minor modification of Algorithm 2.1 and state this as follows.

ALGORITHM 3.1.
Choose a starting point x^0 and tolerances $\epsilon_1, \epsilon_2 > 0$.

For $k \geq 0$ while $\|p^k\| > \epsilon_1$ and $\|\nabla f(x^k)\| > \epsilon_2$:

1. At x^k, evaluate $\nabla f(x^k)$ and the matrix B^k, which is positive definite and bounded in condition number.

2. Solve the linear system $B^k p^k = -\nabla f(x^k)$.

3. Set $x^{k+1} = x^k + p^k$ and $k = k + 1$.

The modified Hessian, B^k, satisfies $v^T(B^k)^{-1}v > \epsilon\|v\|^2$, for all vectors $v \neq 0$ and for some $\epsilon > 0$. The step p^k determined from B^k leads to the descent property:

$$\nabla f(x^k)^T p^k = -\nabla f(x^k)^T (B^k)^{-1} \nabla f(x^k) < -\epsilon\|\nabla f(x^k)\|^2 < 0. \tag{3.2}$$

Moreover, if p^k is sufficiently small, then from Taylor's theorem and continuity of $\nabla f(x^k)$ we have that

$$f(x^k + p^k) - f(x^k) = \nabla f(x^k + tp^k)^T p^k \quad \text{for some } t \in (0,1) \tag{3.3}$$
$$= -\nabla f(x^k + tp^k)^T (B^k)^{-1} \nabla f(x^k) \tag{3.4}$$
$$\leq -\frac{1}{2}\epsilon\|\nabla f(x^k)\|^2 < 0. \tag{3.5}$$

In the next section, a line search method will be used to modify this property in order to enforce a sufficient decrease of $f(x)$ at each iteration. With this decrease, we will then be able to show a *global convergence* property, i.e., convergence to a local solution from poor starting points.

Global convergence is a useful but not sufficient condition to develop an efficient and reliable unconstrained optimization algorithm. For example, an arbitrary choice of B^k could lead to poor steps and small decreases in the objective function. On the other hand, in Chapter 2 we saw from Theorem 2.20 that Newton steps with the actual Hessian, if positive definite, lead to very fast convergence to the optimum.

If second derivatives of the objective function are available, then a modified Hessian can be determined that maintains positive definiteness with a bounded condition number, i.e., $B^k = \nabla^2 f(x^k) + E^k$, where E^k is some correction term. A number of methods are described in [294] and two of the most popular are the *Levenberg–Marquardt correction* and the *modified Cholesky factorization*.

Modified Cholesky Factorization: A positive definite matrix B can be represented by $B = LDL^T$ through a Cholesky factorization where L is a lower triangular matrix with "1" on the diagonal and D is a diagonal matrix with positive diagonal elements. This factorization is determined by pivoting operations and is widely used for the calculation of the Newton step: $\nabla^2 f(x^k)p^k = -\nabla f(x^k)$. However, when the symmetric matrix $\nabla^2 f(x^k)$ is not sufficiently positive definite, the Cholesky factors may not exist or the factorization procedure may be unstable. Under these circumstances, the factorization can be modified while the elements of L and D are calculated. The result is a Cholesky factor of $B^k = \nabla^2 f(x^k) + E^k = \tilde{L}\tilde{D}\tilde{L}^T$, where E^k is determined implicitly, simply by ensuring that the elements of $\tilde{D}$ are chosen to be sufficiently positive. The modified Cholesky factorization described in [162, 294] also ensures that the elements of $\tilde{D}$ and $\tilde{D}^{\frac{1}{2}}\tilde{L}$ are bounded below and above, respectively, by prespecified values. This method therefore leads to a positive definite modified Hessian with a bounded condition number.

Levenberg–Marquardt Correction: Here we can choose a simple correction $E_k = \delta I$ where $\delta > 0$ is selected to ensure positive definiteness. The value of δ can be determined by representing the Hessian with an eigenvalue decomposition (see Definition 2.6) as follows:

$$\begin{aligned} B^k &= \nabla^2 f(x^k) + E^k \\ &= V^k \Lambda^k V^{k,T} + \delta I = V^k(\Lambda^k + \delta I)V^{k,T}, \end{aligned}$$

where the matrices Λ and V incorporate the eigenvalues λ_j and (orthonormal) eigenvectors of $\nabla^2 f(x^k)$, respectively, and the diagonal elements of $\Lambda + \delta I$ are the eigenvalues of B^k. Choosing $\delta = -\min_j(\lambda_j - \epsilon, 0)$ for some tolerance $\epsilon > 0$ leads to eigenvalues of B^k no less than ϵ. If we also assume that the largest eigenvalue is finite, then B^k is a positive definite matrix with a bounded condition number.

While both approaches modify the Hessian to allow the calculation of descent directions, it is not clear how to choose the adjustable parameters that obtain corrections and still lead to fast convergence. In particular, one would like these corrections, E^k, not to interfere with fast convergence to the solution. For instance, if we can set $E^k = 0$ in a neighborhood of the solution and calculate "pure" Newton steps, we obtain quadratic convergence from Theorem 2.20. A weaker condition that leads to *superlinear* convergence is given by the following property.

Theorem 3.1 [294] Assume that $f(x)$ is three times differentiable and that Algorithm 3.1 converges to a point that is a strict local minimum. Then x^k converges at a superlinear rate, i.e.,

$$\lim_{k\to\infty} \frac{\|x^k + p^k - x^*\|}{\|x^k - x^*\|} = 0 \tag{3.6}$$

if and only if

$$\lim_{k\to\infty} \frac{\|(B^k - \nabla^2 f(x^k))p^k\|}{\|p^k\|} = 0. \tag{3.7}$$

In Section 3.5, we will see that such a judicious modification of the Hessian can be performed together with a globalization strategy. In particular, we will consider a systematic strategy for the Levenberg–Marquardt step that is tied to the trust region method.

3.3 Quasi-Newton Methods

The previous section outlined positive definite modifications to the Hessian matrix. However, Algorithm 3.1 also requires calculation of the Hessian as well as solution of a linear system. We now consider an alternative strategy that develops an approximation B^k to $\nabla^2 f(x^k)$ without second derivatives.

For this approach we consider *quasi-Newton* approximations to the B^k matrix that are constructed from gradient information calculated at previous iterations. The basic concept behind the update approximation at a new point x^{k+1} is the *secant* relation given by

$$B^{k+1}(x^{k+1} - x^k) = \nabla f(x^{k+1}) - \nabla f(x^k). \tag{3.8}$$

Here we define $s = x^{k+1} - x^k, y = \nabla f(x^{k+1}) - \nabla f(x^k)$, and this leads to

$$B^{k+1}s = y. \tag{3.9}$$

If $f(x)$ is a quadratic function, the secant relation is exactly satisfied when B^k is the Hessian matrix. Also, from Taylor's theorem (2.23), we see that (3.9) can provide a reasonable approximation to the curvature of $f(x)$ along the direction s. Therefore we consider this relation to motivate a formula to describe B^k. Finally, because $\nabla^2 f(x)$ is a symmetric matrix, we also want B^k to be symmetric as well.

The simplest way to develop an update formula for B^k is to postulate the rank-one update: $B^{k+1} = B^k + ww^T$. Applying (3.9) to this update (see Exercise 1) leads to

$$B^{k+1} = B^k + \frac{(y - B^k s)(y - B^k s)^T}{(y - B^k s)^T s} \tag{3.10}$$

which is the symmetric rank 1 (SR1) update formulation. The SR1 update asymptotically converges to the (positive definite) Hessian of the objective function as long as the steps s are linearly independent. On the other hand, the update for B^{k+1} can be adversely affected by regions of negative or zero curvature and can become ill-conditioned, singular, or unbounded in norm. In particular, care must be taken so that the denominator in (3.10) is bounded away from zero, e.g., $|(y - B^k s)^T s| \geq C_1 \|s\|^2$ for some $C_1 > 0$. So, while this update can work well, it is not guaranteed to be positive definite and may not lead to descent directions.

Instead, we also consider a rank-two quasi-Newton update formula that allows B^k to remain symmetric and positive definite as well. To do this, we define the current Hessian approximation as $B^k = JJ^T$, where J is a square, nonsingular matrix. Note that this definition implies that B^k is positive definite. To preserve symmetry, the update to B^k can be given as $B^{k+1} = J^+(J^+)^T$, where J^+ is also expected to remain square and nonsingular. By working with the matrices J and J^+, it will also be easier to monitor the symmetry and positive definiteness properties of B^k.

Using the matrix J^+, the secant relation (3.9) can be split into two parts. From

$$B^{k+1}s = J^+(J^+)^T s = y, \tag{3.11}$$

we introduce an additional variable vector v and obtain

$$J^+ v = y \quad \text{and} \quad (J^+)^T s = v. \tag{3.12}$$

The derived update satisfies the secant relation and symmetry. In order to develop a unique update formula, we also assume the update has the least change in some norm. Here we obtain an update formula by invoking a least change strategy for J^+, leading to

$$\min \|J^+ - J\|_F \tag{3.13}$$

$$\text{s.t. } J^+ v = y, \tag{3.14}$$

where $\|J\|_F$ is the Frobenius norm of matrix J. Solving (3.13) (see Exercise 8 in Chapter 4) leads to the so-called *Broyden update* used to solve nonlinear equations:

$$J^+ = J + \frac{(y - Jv)v^T}{v^T v}. \tag{3.15}$$

Using (3.15) we can recover an update formula in terms of s, y, and B^k by using the following identities for v. From (3.12), we have $v^T v = (y^T (J^+)^{-T})(J^+)^T s = s^T y$. Also, postmultiplying $(J^+)^T$ by s and using (3.15) leads to

$$v = (J^+)^T s = J^T s + \frac{v(y - Jv)^T s}{v^T v} \tag{3.16}$$

$$= J^T s + v - \frac{v^T J^T s}{s^T y} v \tag{3.17}$$

and $v = \frac{s^T y}{v^T J^T s} J^T s$. Premultiplying v by $s^T J$ and simplifying the expression leads to

$$v = \left(\frac{s^T y}{s^T B^k s}\right)^{1/2} J^T s.$$

Finally, from the definition of v, B^k, and B^{k+1} as well as (3.15), we have

$$\begin{aligned} B^{k+1} &= \left(J + \frac{(y - Jv)v^T}{v^T v}\right)\left(J + \frac{(y - Jv)v^T}{v^T v}\right)^T \\ &= JJ^T + \frac{yy^T - Jvv^T J^T}{v^T v} \\ &= B^k + \frac{yy^T}{s^T y} - \frac{Jvv^T J^T}{v^T v} \\ &= B^k + \frac{yy^T}{s^T y} - \frac{B^k s s^T B^k}{s^T B^k s}. \end{aligned} \tag{3.18}$$

From this derivation, we have assumed B^k to be a symmetric matrix, and therefore B^{k+1} remains symmetric, as seen from (3.18). Moreover, it can be shown that if B^k is positive definite and $s^T y > 0$, then the update, B^{k+1}, is also positive definite. In fact, the condition that $s^T y$ be sufficiently positive at each iteration, i.e.,

$$s^T y \geq C_2 \|s\|^2 \quad \text{for some } C_2 > 0, \tag{3.19}$$

is important in order to maintain a bounded update. As a result, condition (3.19) is checked at each iteration, and if it cannot be satisfied, the update (3.18) is skipped. Another alternative to skipping is known as *Powell damping* [313]. As described in Exercise 2, this approach maintains positive definiteness when (3.19) fails by redefining $y := \theta y + (1-\theta)B^k s$ for a calculated $\theta \in [0, 1]$.

The update formula (3.18) is known as the Broyden–Fletcher–Goldfarb–Shanno (BFGS) update, and the derivation above is due to Dennis and Schnabel [110]. As a result of this updating formula, we have a reasonable approximation to the Hessian matrix that is also positive definite.

Moreover, the BFGS update has a fast rate of convergence as summarized by the following property.

Theorem 3.2 [294] If the BFGS algorithm converges to a strict local solution x^* with $\sum_{k=0}^{\infty} \|x^k - x^*\| < \infty$, and the Hessian $\nabla^2 f(x)$ is Lipschitz continuous at x^*, then (3.7) holds and x^k converges at a superlinear rate.

Finally, while the BFGS update can be applied directly in step 2 of Algorithm 3.1, calculation of the search direction p^k can be made more efficient by implementing the quasi-Newton update through the following options.

- Solution of the linear system can be performed with a Cholesky factorization of $B^k = L^k (L^k)^T$. On the other hand, L^k can be updated directly by applying formula (3.15) with $J := L^k$ and $v = \frac{s^T y}{s^T L^k (L^k)^T s} (L^k)^T s$, i.e.,

$$J^+ = L^k + \frac{y s^T L^k}{(s^T y)^{1/2} (s^T L^k (L^k)^T s)^{1/2}} - \frac{(L^k)^T s s^T L^k}{s^T (L^k)(L^k)^T s}, \tag{3.20}$$

followed by a cheap QR factorization of $J^+ \to L^+$ (via Givens rotations) to obtain the BFGS update of the Cholesky factor $L^{k+1} = L^+$ (see [110]).

- The inverse of B^k can be updated directly, and this inverse can be used directly in the calculation of the search direction, i.e., $p^k = -(B^k)^{-1}\nabla f(x^k) = -H^k \nabla f(x^k)$, where the inverse BFGS update is given by

$$H^{k+1} = \left(I - \frac{sy^T}{s^T y}\right) H^k \left(I - \frac{ys^T}{s^T y}\right) + \frac{ss^T}{s^T y}. \tag{3.21}$$

Therefore, starting with the initial inverse H^0, the step can be calculated with a simple matrix multiplication.

- The BFGS update (3.18) can also be written as

$$B^{k+1} = B^k + \left[v^k (v^k)^T - u^k (u^k)^T\right] \tag{3.22}$$

where

$$u^k = \frac{B^k s^k}{[(s^k)^T B^k s^k]^{1/2}}, \text{ and } v^k = \frac{y^k}{((y^k)^T s^k)^{1/2}}. \tag{3.23}$$

For large-scale problems (say, $n > 1000$), it is advantageous to store only the last m updates and to develop the so-called *limited memory update*:

$$B^{k+1} = B^0 + \sum_{i=\max(0,k-m+1)}^{k} \left[v^i (v^i)^T - u^i (u^i)^T\right]. \tag{3.24}$$

In this way, only the most recent updates are used for B^k, and the older ones are discarded. While the limited memory update has only a linear convergence rate, it greatly reduces the linear algebra cost for large problems. Moreover, by storing only the updates, one can work directly with matrix-vector products instead of B^k, i.e.,

$$B^{k+1} w = B^0 w + \sum_{i=\max(0,k-m+1)}^{k} \left[v^i (v^i)^T w - u^i (u^i)^T w\right]. \tag{3.25}$$

Similar updates have been developed for H^k as well. Moreover, Byrd and Nocedal [83] discovered a particularly efficient compact form of this update as follows:

$$B^{k+1} = B^0 - [B^0 S_k \;\; Y_k] \begin{bmatrix} S_k^T B^0 S_k & L_k \\ L_k^T & -D_k \end{bmatrix}^{-1} \begin{bmatrix} S_k^T B^0 \\ Y_k^T \end{bmatrix}, \tag{3.26}$$

where $D_k = \text{diag}[(s^{k-m+1})^T (y^{k-m+1}), \ldots, (s^k)^T y^k]$, $S_k = [s^{k-m+1}, \ldots, s^k]$, $Y_k = [y^{k-m+1}, \ldots, y^k]$, and

$$(L_k) = \begin{cases} (s^{k-m+i})^T (y^{k-m+j}), & i > j, \\ 0 & \text{otherwise.} \end{cases}$$

The compact limited memory form (3.26) is more efficient to apply than the unrolled form (3.25), particularly when m is large and B^0 is initialized as a diagonal matrix. Similar compact representations have been developed for the inverse BFGS update H^k as well as the SR1 update.

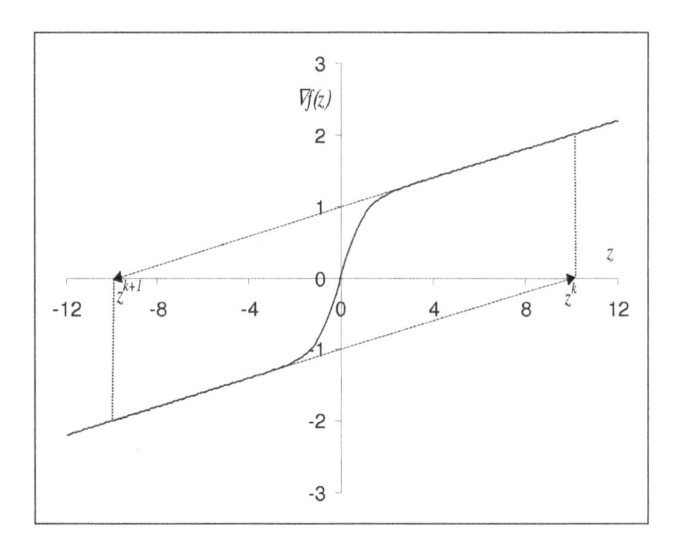

Figure 3.1. *Example that shows cycling of the basic Newton method.*

3.4 Line Search Methods

Choosing a modified Hessian that is positive definite is not sufficient to guarantee convergence from poor starting points. For instance, consider the objective function of the scalar variable z:

$$\begin{aligned} f(z) &= 0.05z^2 + \ln(\cosh(z)) \\ \text{with} \nabla f(z) &= 0.1z + \tanh(z) \\ \text{and } \nabla^2 f(z) &= 1.1 - (\tanh(z))^2. \end{aligned}$$

Note that the Hessian is always positive definite ($f(z)$ is a convex function) and a unique optimum exists at $z^* = 0$. However, as seen in Figure 3.1, applying Algorithm 3.1 with a starting point $z^0 = 10$ leads the algorithm to cycle indefinitely between 10 and -10.

To avoid cycling, convergence requires a sufficient decrease of the objective function. If we choose the step $p^k = -(B^k)^{-1}\nabla f(x^k)$ with B^k positive definite and bounded in condition number, we can modify the selection of the next iterate using a positive step length α with

$$x^{k+1} = x^k + \alpha p^k. \tag{3.27}$$

Using Taylor's theorem, one can show for sufficiently small α that

$$\begin{aligned} f(x^{k+1}) - f(x^k) &= \alpha \nabla f(x^k + t\alpha p^k)^T p^k \quad (\text{for some } t \in (0,1)) &(3.28) \\ &= -\alpha \nabla f(x^k + t\alpha p^k)^T (B^k)^{-1} \nabla f(x^k) &(3.29) \\ &\le -\frac{1}{2}\alpha\epsilon \|\nabla f(x^k)\|^2 < 0. &(3.30) \end{aligned}$$

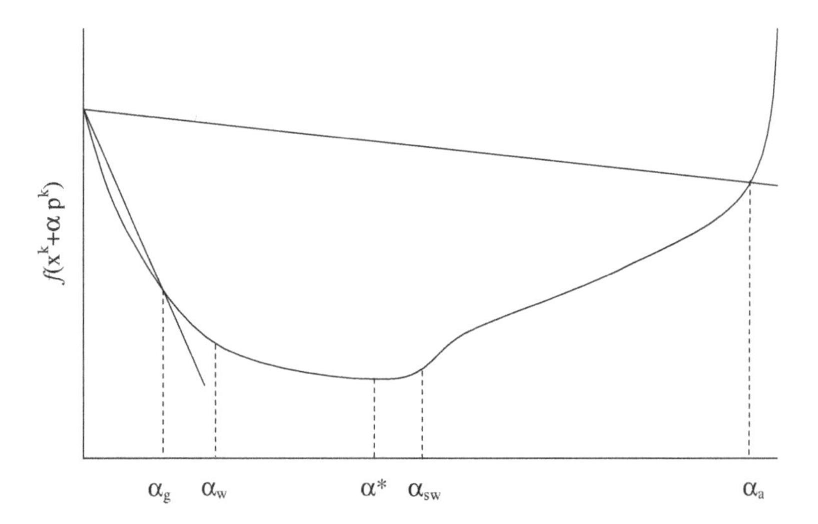

Figure 3.2. *Line search criteria for sufficient decrease.*

On the other hand, restriction to only small α leads to inefficient algorithms. Instead, a systematic *line search* method is needed that allows larger steps to be taken with a sufficient decrease of $f(x)$. The line search method therefore consists of three tasks:

1. At iteration k, start with a sufficiently large value for α. While a number of methods can be applied to determine this initial step length [294, 134], for the purpose of developing this method, we choose an initial α set to one.

2. Set $x^k + \alpha p^k$ and evaluate whether $f(x^k + \alpha p^k) - f(x^k)$ is a *sufficient decrease*. If so, set $\alpha^k := \alpha$ and $x^{k+1} := x^k + \alpha^k p^k$.

3. Otherwise, *reduce* α and repeat the previous step.

Sufficient decrease of $f(x)$ can be seen from the last term in (3.30). Clearly a value of α can be chosen that reduces $f(x)$. As the iterations proceed, one would expect decreases in $f(x)$ to taper off as $\|\nabla f(x)\| \to 0$. On the other hand, convergence would be impeded if $\alpha^k \to 0$ and the algorithm stalls.

An obvious line search option is to perform a single variable optimization in α, i.e., $\min_\alpha f(x^k + \alpha p^k)$, and to choose $\alpha^k := \alpha^*$, as seen in Figure 3.2. However, this option is expensive. And far away from the solution, it is not clear that this additional effort would reduce the number of iterations to converge to x^*. Instead, we consider three popular criteria to determine sufficient decrease during the line search. To illustrate the criteria for sufficient decrease, we consider the plot of $f(x^k + \alpha p^k)$ shown in Figure 3.2.

All of these criteria require the following decrease in the objective function:

$$f(x^k + \alpha^k p^k) \le f(x^k) + \eta \alpha^k \nabla f(x^k)^T p^k, \tag{3.31}$$

where $\eta \in (0, \frac{1}{2}]$. This is also known as the *Armijo condition*. As seen in Figure 3.2, $\alpha \in (0, \alpha_a]$ satisfies this condition. The following additional conditions are also required so that the chosen value of α is not too short:

- The *Wolfe conditions* require that (3.31) be satisfied as well as

$$\nabla f(x^k + \alpha^k p^k)^T p^k \geq \zeta \nabla f(x^k)^T p^k \tag{3.32}$$

  for $\zeta \in (\eta, 1)$. From Figure 3.2, we see that $\alpha \in [\alpha_w, \alpha_a]$ satisfies these conditions.

- The *strong Wolfe conditions* are more restrictive and require satisfaction of

$$|\nabla f(x^k + \alpha^k p^k)^T p^k| \leq \zeta |\nabla f(x^k)^T p^k| \tag{3.33}$$

  for $\zeta \in (\eta, 1)$. From Figure 3.2, we see that $\alpha \in [\alpha_w, \alpha_{sw}]$ satisfies these conditions.

- The *Goldstein* or *Goldstein–Armijo conditions* require that (3.31) be satisfied as well as

$$f(x^k + \alpha^k p^k) \geq f(x^k) + (1 - \eta)\alpha^k \nabla f(x^k)^T p^k. \tag{3.34}$$

  From Figure 3.2, we see that $\alpha \in [\alpha_g, \alpha_a]$ satisfies these conditions. (Note that the relative locations of α_g and α_w change if $(1 - \eta) < \zeta$.)

The Wolfe and strong Wolfe conditions lead to methods that have desirable convergence properties that are analyzed in [294, 110]. The Goldstein conditions are similar but do not require evaluation of the directional derivatives $\nabla f(x^k + \alpha^k p^k)^T p^k$ during the line search. Moreover, in using a backtracking line search, where α is reduced if (3.31) fails, the Goldstein condition (3.34) is easier to check.

We now consider a global convergence proof for the Goldstein conditions. Based on the result by Zoutendijk [422], a corresponding proof is also given for the Wolfe conditions in [294].

Theorem 3.3 (Global Convergence of Line Search Method). Consider an iteration: $x^{k+1} = x^k + \alpha^k p^k$, where $p^k = -(B^k)^{-1} \nabla f(x^k)$ and α^k satisfies the Goldstein–Armijo conditions (3.31), (3.34). Suppose that $f(x)$ is bounded below for $x \in \mathbb{R}^n$, that $f(x)$ is continuously differentiable, and that $\nabla f(x)$ is Lipschitz continuous in an open set containing the level set $\{x | f(x) \leq f(x^0)\}$. Then by defining the angle between p^k and $-\nabla f(x^k)$ as

$$\cos \theta^k = \frac{-\nabla f(x^k)^T p^k}{\|\nabla f(x^k)\| \|p^k\|} \tag{3.35}$$

(see Figure 3.3), we have

$$\sum_{k=0}^{\infty} \cos^2 \theta^k \|\nabla f(x^k)\|^2 < \infty \tag{3.36}$$

and hence, $\lim_{k \to \infty} \cos \theta^k \|\nabla f(x^k)\| \to 0$.

Proof: For $x^{k+1} = x^k + \alpha^k p^k$ satisfying the Goldstein–Armijo conditions, we have

$$-\eta \alpha^k \nabla f(x^k)^T p^k \leq f(x^k) - f(x^{k+1}) \leq -(1 - \eta)\alpha^k \nabla f(x^k)^T p^k. \tag{3.37}$$

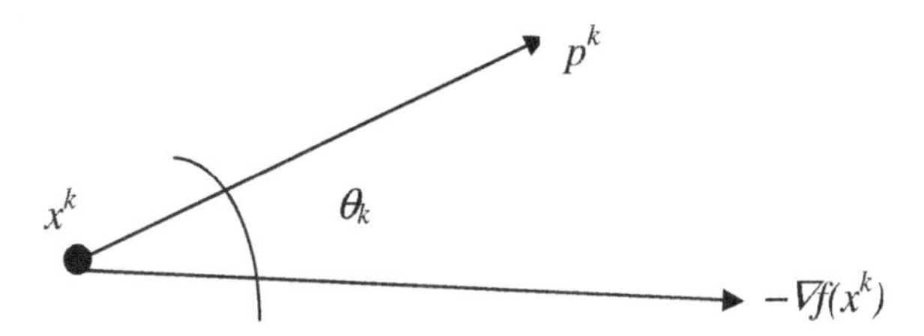

Figure 3.3. *Angle between search direction and negated gradient.*

From Taylor's theorem,

$$f(x^{k+1}) - f(x^k) = \alpha^k \nabla f(x^k + t\alpha^k p^k)^T p^k$$

for some $t \in [0,1]$, thus leading to

$$-\eta\alpha^k \nabla f(x^k)^T p^k \leq -\alpha^k \nabla f(x^k + t\alpha^k p^k) p^k \leq -(1-\eta)\alpha^k \nabla f(x^k)^T p^k. \tag{3.38}$$

From Lipschitz continuity (with a large constant, $L > 0$) and the right-hand side of (3.38), we have

$$L(t\alpha^k)^2 \|p^k\|^2 \geq t\alpha^k (\nabla f(x^k + t\alpha^k p^k)^T p^k - \nabla f(x^k)^T p^k) \geq -(t\alpha^k)\eta \nabla f(x^k)^T p^k, \tag{3.39}$$

and this leads to

$$\alpha^k \geq t\alpha^k \geq -\frac{\eta \nabla f(x^k)^T p^k}{L\|p^k\|^2}. \tag{3.40}$$

Substituting the bound on α^k into the left inequality in (3.37), and noting that $|\nabla f(x^k)^T p^k| = \|\nabla f(x^k)\| \|p^k\| \cos\theta^k$, leads to the inequality

$$f(x^k) - f(x^{k+1}) \geq \frac{(\eta \nabla f(x^k)^T p^k)^2}{L\|p^k\|^2} = \frac{\eta^2}{L} \cos^2\theta^k \|\nabla f(x^k)\|^2. \tag{3.41}$$

Since $f(x)$ is bounded below for $x \in \mathbb{R}^n$, then summing inequality (3.41) over k leads to

$$\infty > f(x^0) - f(x^\infty) \geq \sum_{k=0}^{\infty} f(x^k) - f(x^{k+1}) \geq \frac{\eta^2}{L} \sum_{k=0}^{\infty} \cos^2\theta^k \|\nabla f(x^k)\|^2, \tag{3.42}$$

and this also implies $\lim_{k\to\infty} \cos\theta^k \|\nabla f(x^k)\| = 0$. □

Since θ_k is the angle at x^k between the search direction p^k and the steepest descent direction $-\nabla f(x^k)$, this theorem leads to the result that either $\|\nabla f(x^k)\|$ approaches zero or that $\nabla f(x^k)$ and p^k become orthogonal to each other. However, in the case where we have a positive definite B^k with a bounded condition number, $\kappa(B^k)$, then

$$\cos\theta = \frac{|\nabla f(x^k)^T p^k|}{\|\nabla f(x^k)\| \|p^k\|} = \frac{\nabla f(x^k)^T (B^k)^{-1} \nabla f(x^k)}{\|\nabla f(x^k)\|_2 \|(B^k)^{-1} \nabla f(x^k)\|_2} \geq \frac{1}{\kappa(B^k)}$$

and $\cos\theta^k$ is bounded away from zero so that $\lim_{k\to\infty} \|\nabla f(x^k)\| \to 0$.

With this result, we now state the basic Newton-type algorithm for unconstrained optimization with a backtracking line search.

ALGORITHM 3.2.
Choose a starting point x^0 and tolerances $\epsilon_1, \epsilon_2 > 0$.

For $k \geq 0$ while $\|p^k\| > \epsilon_1$ and $\|\nabla f(x^k)\| > \epsilon_2$:

1. At x^k, evaluate $\nabla f(x^k)$ and evaluate or update B^k, so that it is positive definite and has a bounded condition number.
2. Solve the linear system $B^k p^k = -\nabla f(x^k)$.
3. Set $\alpha = 1$.
4. While (3.31) is *not* satisfied, set $\alpha := \rho\alpha$, where $0 < \rho < 1$.
5. Set $\alpha^k := \alpha$ and $x^{k+1} = x^k + \alpha p^k$.

The value of ρ can be chosen in a number of ways. It can be a fixed fraction (e.g., $\frac{1}{2}$) or it can be determined by minimizing a quadratic (see Exercise 4) or cubic interpolant based on previous line search information. In addition, if $\alpha < 1$, then the Goldstein condition (3.34) should be checked to ensure that α is not too short.

Finally, in addition to global convergence in Theorem 3.3 we would like Algorithm 3.2 to perform well, especially in the neighborhood of the optimum. The following theorem is useful for this feature.

Theorem 3.4 Assume that $f(x)$ is three times differentiable and that Algorithm 3.2 converges to a point that is a strict local minimum x^* and

$$\lim_{k\to\infty} \frac{\|(B^k - \nabla^2 f(x^k))p^k\|}{\|p^k\|} = 0$$

is satisfied. Then there exists a finite k_0, where $\alpha^k = 1$ is admissible for all $k > k_0$ and x^k converges superlinearly to x^*.

Proof: From Taylor's theorem, the calculation of p^k, and (3.7), we have

$$\begin{aligned}
f(x^{k+1}) &= f(x^k) + \nabla f(x^k)^T p^k + \frac{1}{2} p^{k,T} \nabla^2 f(x^k + tp^k) p^k \\
&= f(x^k) + \nabla f(x^k)^T p^k + \frac{1}{2} p^{k,T} (\nabla^2 f(x^k + tp^k) - B^k) p^k + \frac{1}{2} p^{k,T} B^k p^k \\
&= f(x^k) + \frac{1}{2} \nabla f(x^k)^T p^k + \frac{1}{2} p^{k,T} (\nabla^2 f(x^k + tp^k) - B^k) p^k \\
&= f(x^k) + \eta \nabla f(x^k)^T p^k + \left(\frac{1}{2} - \eta\right) \nabla f(x^k)^T p^k + \frac{1}{2} p^{k,T} (\nabla^2 f(x^k + tp^k) - B^k) p^k \\
&\leq f(x^k) + \eta \nabla f(x^k)^T p^k - \left(\frac{1}{2} - \eta\right) |\nabla f(x^k)^T p^k| + o(\|p^k\|^2).
\end{aligned}$$

From the proof of Theorem 3.3 we know that α_k is bounded away from zero, and because we have $\lim_{k\to\infty} \|x^k - x^{k+1}\| \to 0$, we have $\|p^k\| \to 0$. Also, because x^* is a strict local optimum, we have from Taylor's theorem that $|\nabla f(x^k)^T p^k| > \epsilon \|p^k\|^2$ for some $\epsilon > 0$ and k sufficiently large. Consequently, for $\eta < \frac{1}{2}$ there exists a k_0 such that

$$\frac{1}{2} p^{k,T}(\nabla^2 f(x^k + tp^k) - B^k)p^k < \left(\frac{1}{2} - \eta\right) |\nabla f(x^k)^T p^k|$$

leading to

$$f(x^{k+1}) \leq f(x^k) + \eta \nabla f(x^k)^T p^k, \tag{3.43}$$

which satisfies the Armijo condition for $\alpha = 1$. Superlinear convergence then follows from this result and Theorem 3.1. □

Example 3.5 Consider the problem described in Example 2.19:

$$\min\ f(x) = \alpha \exp(-\beta) \tag{3.44}$$
$$\text{with } u = x_1 - 0.8, \tag{3.45}$$
$$v = x_2 - (a_1 + a_2 u^2 (1-u)^{\frac{1}{2}} - a_3 u), \tag{3.46}$$
$$\alpha = -b_1 + b_2 u^2 (1+u)^{\frac{1}{2}} + b_3 u, \tag{3.47}$$
$$\beta = c_1 v^2 (1 - c_2 v)/(1 + c_3 u^2) \tag{3.48}$$

and with $a^T = [0.3, 0.6, 0.2]$, $b^T = [5, 26, 3]$, and $c^T = [40, 1, 10]$. The minimizer occurs at $x^* = [0.73950, 0.31436]$ with $f(x^*) = -5.08926$.[2] As seen from Figure 2.5, this problem has only a small region around the solution where the Hessian is positive definite. As a result, we saw in Example 2.21 that Newton's method has difficulty converging from a starting point far away from the solution. To deal with this issue, let's consider the line search algorithm with BFGS updates, starting with an initial $B^0 = I$ and from a starting point close to the solution. Applying Algorithm 3.2 to this problem, with a termination tolerance of $\|\nabla f(x)\| \leq 10^{-6}$, leads to the iteration sequence given in Table 3.1. Here it is clear that the solution can be found very quickly, although it requires more iterations than Newton's method (see Table 2.3). Also, as predicted by Theorem 3.4, the algorithm chooses step sizes with $\alpha^k = 1$ as convergence proceeds in the neighborhood of the solution. Moreover, based on the error $\|\nabla f(x^k)\|$ we observe superlinear convergence rates, as predicted by Theorem 3.1. Finally, from Figure 2.5 we again note that the convergence path remains in the region where the Hessian is positive definite.

If we now choose a starting point farther away from the minimum, then applying Algorithm 3.2, with a termination tolerance of $\|\nabla f(x)\| \leq 10^{-6}$, leads to the iteration sequence given in Table 3.2. At this starting point, the Hessian is indefinite, and, as seen in Example 2.21, Newton's method was unable to converge to the minimum. On the other hand, the line search method with BFGS updates converges relatively quickly to the optimal solution. The first three iterations show that large search directions are generated, but

[2] To prevent undefined objective and gradient functions, the square root terms are replaced by $f(\xi) = (\max(10^{-6}, \xi))^{1/2}$. While this approximation violates the smoothness assumptions for these methods, it affects only large search steps which are immediately reduced by the line search in the early stages of the algorithm. Hence first and second derivatives are never evaluated at these points.

Table 3.1. *Iteration sequence with BFGS line search method with starting point close to solution.*

Iteration (k)	x_1^k	x_2^k	$f(x^k)$	$\|\nabla f(x^k)\|$	α
0	0.8000	0.3000	−5.0000	3.0000	0.0131
1	0.7606	0.3000	−5.0629	3.5620	0.0043
2	0.7408	0.3159	−5.0884	0.8139	1.000
3	0.7391	0.3144	−5.0892	2.2624×10^{-2}	1.0000
4	0.7394	0.3143	−5.0892	9.7404×10^{-4}	1.0000
5	0.7395	0.3143	−5.0892	1.5950×10^{-5}	1.0000
6	0.7395	0.3143	−5.0892	1.3592×10^{-7}	—

Table 3.2. *Iteration sequence with BFGS line search newton method with starting point far from solution.*

Iteration (k)	x_1	x_2	f	$\|\nabla f(x^k)\|$	α
0	1.0000	0.5000	−1.1226	9.5731	0.0215
1	0.9637	0.2974	−3.7288	12.9460	0.0149*
2	0.8101	0.2476	−4.4641	19.3569	0.0140
3	0.6587	0.3344	−4.9359	4.8700	1.0000
4	0.7398	0.3250	−5.0665	4.3311	1.0000
5	0.7425	0.3137	−5.0890	0.1779	1.0000
6	0.7393	0.3144	−5.0892	8.8269×10^{-3}	1.0000
7	0.7395	0.3143	−5.0892	1.2805×10^{-4}	1.0000
8	0.7395	0.3143	−5.0892	3.1141×10^{-6}	1.0000
9	0.7395	0.3143	−5.0892	5.4122×10^{-12}	—

*BFGS update was reinitialized to I.

the line search leads to very small step sizes. In fact, the first BFGS update generates a poor descent direction, and the matrix B^k had to be reinitialized to I. Nevertheless, the algorithm continues and takes full steps after the third iteration. Once this occurs, we see that the method converges superlinearly toward the optimum. ■

The example demonstrates that the initial difficulties that occurred with Newton's method are overcome by line search methods as long as positive definite Hessian approximations are applied. Also, the performance of the line search method on this example confirms the convergence properties shown in this section. Note, however, that these properties apply only to convergence to *stationary points* and do not guarantee convergence to points that also satisfy second order conditions.

3.5 Trust Region Methods

Line search methods generate steps whose length is adjusted by a step size α, and whose direction is fixed independent of the length. On the other hand, with *trust region methods* the calculated search direction also changes as a function of the step length. This added flexibility leads to methods that have convergence properties superior to line search methods. On the other hand, the computational expense for each trust region iteration may be greater than for line search iterations.

We begin the discussion of this method by defining a trust region for the optimization step, e.g., $\|p\| \leq \Delta$, and a model function $m^k(p)$ that is expected to provide a good approximation to $f(x^k + p)$ within a trust region of size Δ. Any norm can be used to characterize the trust region, although the Euclidean norm is often used for unconstrained optimization. Also, a quadratic model is often chosen for $m^k(p)$, and the optimization step at iteration k is determined by the following optimization problem:

$$\begin{aligned} \min \quad & m^k(p) = \nabla f(x^k)^T p + \frac{1}{2} p^T B^k p \\ \text{s.t.} \quad & \|p\| \leq \Delta, \end{aligned} \tag{3.49}$$

where $B^k = \nabla^2 f(x^k)$ or its quasi-Newton approximation. The basic trust region algorithm can be given as follows.

Algorithm 3.3.
Given parameters $\bar{\Delta}$, $\Delta^0 \in (0, \bar{\Delta}]$, $0 < \kappa_1 < \kappa_2 < 1$, $\gamma \in (0, 1/4)$, and tolerances $\epsilon_1, \epsilon_2 > 0$. Choose a starting point x^0.

For $k \geq 0$ while $\|p^k\| > \epsilon_1$ and $\|\nabla f(x^k)\| > \epsilon_2$:

1. Solve the model problem (3.49) and obtain the solution p^k.
2. Calculate $\rho^k = \frac{ared}{pred}$ where the actual reduction is $ared = f(x^k) - f(x^k + p^k)$ and the predicted reduction from the model is

$$pred = m^k(0) - m^k(p) = -(\nabla f(x^k)^T p^k + \frac{1}{2} p^{k,T} B^k p^k)$$

   - if $\rho^k < \kappa_1$, set $\Delta^{k+1} = \kappa_1 \|p^k\|$;
   - else, if $\rho^k > \kappa_2$ and $\|p^k\| = \Delta^k$, set $\Delta^{k+1} = \min(2\Delta^k, \bar{\Delta})$;
   - otherwise, set $\Delta^{k+1} = \Delta^k$.

3. If $\rho^k > \gamma$, then $x^{k+1} = x^k + p^k$. Else, $x^{k+1} = x^k$.

Typical values of κ_1 and κ_2 are $\frac{1}{4}$ and $\frac{3}{4}$, respectively. Algorithm 3.3 lends itself to a number of variations that will be outlined in the remainder of this section. In particular, if second derivatives are available, then problem (3.49) is exact up to second order, although one may need to deal with an indefinite Hessian and nonconvex model problem. On the other hand, if second derivatives are not used, one may instead use an approximation for B^k such as the BFGS update. Here the model problem is convex, but without second order information it is less accurate.

3.5.1 Convex Model Problems

If B^k is positive semidefinite, then the model problem is convex, and two popular approaches have been developed to solve (3.49).

Levenberg–Marquardt Steps

Levenberg–Marquardt steps were discussed in Section 3.2 as a way to correct Hessian matrices that were not positive definite. For trust region methods, the application of these steps is further motivated by the following property.

Theorem 3.6 Consider the model problem given by (3.49). The solution is given by p^k if and only if there exists a scalar $\lambda \geq 0$ such that the following conditions are satisfied:

$$\begin{aligned} &(B^k + \lambda I)p^k = -\nabla f(x^k), \\ &\lambda(\Delta - \|p^k\|) = 0, \\ &\quad (B^k + \lambda I) \quad \text{is positive semidefinite.} \end{aligned} \tag{3.50}$$

Note that when $\lambda = 0$, we have the same Newton-type step as with a line search method,

$$p^N = -(B^k)^{-1}\nabla f(x^k).$$

On the other hand, as λ becomes large,

$$p(\lambda) = -(B^k + \lambda I)^{-1}\nabla f(x^k)$$

approaches a small step in the steepest descent direction $p^S \approx -\frac{1}{\lambda}\nabla f(x^k)$. As Δ^k is adjusted in Algorithm 3.3 and if $\Delta^k < \|p^N\|$, then one can find a suitable value of λ by solving the equation $\|p(\lambda)\| - \Delta^k = 0$. With this equation, however, p depends nonlinearly on λ, thus leading to difficult convergence with an iterative method. The alternate form

$$\frac{1}{\|p(\lambda)\|} - \frac{1}{\Delta^k} = 0, \tag{3.51}$$

suggested in [281, 294], is therefore preferred because it is nearly linear in λ. As a result, with an iterative solver, such as Newton's method, a suitable λ is often found for (3.51) in just 2–3 iterations. Figure 3.4 shows how the Levenberg–Marquardt step changes with increasing values of Δ. Note that for $\lambda = 0$ we have the Newton step p^N. Once λ increases, the step takes on a length given by the value of Δ from (3.51). The steps $p(\lambda)$ then decrease in size with Δ (and increasing λ) and trace out an arc shown by the dotted lines in the figure. Finally, as Δ vanishes, $p(\lambda)$ points to the steepest descent direction.

Powell Dogleg Steps

The calculation of Levenberg–Marquardt steps requires an iterative process to determine $p(\lambda)$ for a given Δ. However, as seen in Figure 3.4, limiting cases of $p(\lambda)$ are the Newton step and a vanishing step in the steepest descent direction. Therefore, a straightforward modification is to considering a linear combination based on these limiting cases, in order to develop a noniterative trust region method.

For this method, we define a scaled steepest descent step called the *Cauchy step*, given by

$$p^C = \begin{cases} -\dfrac{\nabla f(x^k)^T \nabla f(x^k)}{\nabla f(x^k)^T B^k \nabla f(x^k)} \nabla f(x^k) & \text{if } \nabla f(x^k)^T B^k \nabla f(x^k) > 0, \\ -\dfrac{\Delta^k}{\|\nabla f(x^k)\|} \nabla f(x^k) & \text{otherwise.} \end{cases} \tag{3.52}$$

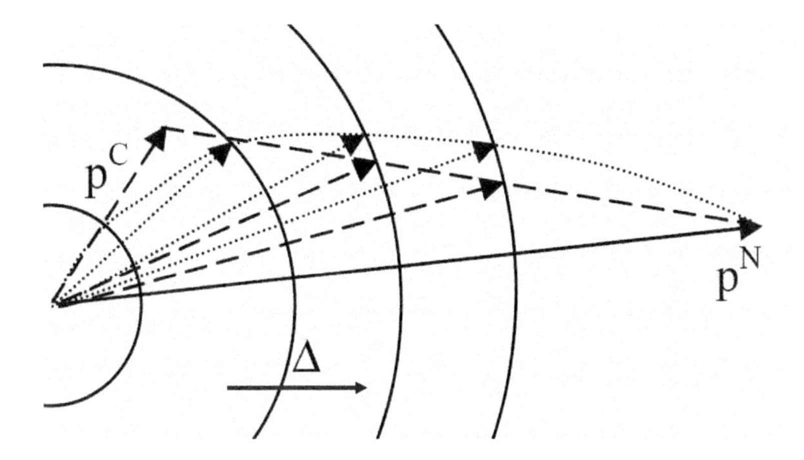

Figure 3.4. *Levenburg–Marquardt (dotted lines) and Powell dogleg (dashed lines) steps for different trust regions.*

In the first case, the Cauchy step can be derived by inserting a trial solution $p = -\tau \nabla f(x^k)$ into the model problem (3.49) and solving for τ with a large value of Δ. Otherwise, if B^k is not positive definite, then the Cauchy step is taken to the trust region bound and has length Δ.

For the dogleg method, we assume that B^k is positive definite and we adopt the Cauchy step for the first case. We also have a well-defined Newton step given by $p^N = -(B^k)^{-1}\nabla f(x^k)$. As a result, the solution to (3.49) is given by the following cases:

- $p^k = p^N$ if $\Delta \geq \|p^N\|$;
- $p^k = \frac{\Delta^k p^C}{\|p^C\|}$ if $\Delta \leq \|p^C\|$;
- otherwise, we select $p^k = \eta p^N + (1-\eta)p^C$ so that $\|p\| = \Delta$, which leads to

$$\eta = \frac{-(p^N - p^C)^T p^C + [((p^N - p^C)^T p^C)^2 + (\Delta^2 - \|p^C\|^2)\|p^N - p^C\|^2]^{\frac{1}{2}}}{\|p^N - p^C\|^2} > 0.$$

Both of these methods provide search directions that address the model problem (3.49). It is also important to note that while the Levenberg–Marquardt steps provide an exact solution to the model problem, the dogleg method solves (3.49) only approximately. In both cases, the following property holds.

Theorem 3.7 [294] Assume that the model problem (3.49) is solved using Levenberg–Marquardt or dogleg steps with $\|p^k\| \leq \Delta^k$; then for some $c_1 > 0$,

$$m^k(0) - m^k(p) \geq c_1 \|\nabla f(x^k)\| \min\left(\Delta, \frac{\|\nabla f(x^k)\|}{\|B^k\|}\right). \tag{3.53}$$

The relation (3.53) can be seen as an analogue to the descent property in line search methods as it relates improvement in the model function back to $\|\nabla f(x)\|$. With this condition, one can show convergence properties similar to (and under weaker conditions than) line search methods [294, 100]. These are summarized by the following theorems.

Theorem 3.8 [294] Let $\gamma \in (0,1/4)$, $\|B^k\| \leq \beta < \infty$, and let $f(x)$ be Lipschitz continuously differentiable and bounded below on a level set $\{x \mid f(x) \leq f(x^0)\}$. Also, in solving (3.49) (approximately), assume that p^k satisfies (3.53) and that $\|p^k\| \leq c_2 \Delta^k$ for some constant $c_2 \geq 1$. Then the algorithm generates a sequence of points with $\lim_{k\to\infty} \nabla f(x^k) = 0$.

The step acceptance condition $f(x^k) - f(x^k + p^k) \geq \gamma(m^k(0) - m^k(p)) > 0$ with $\gamma > 0$ and Lipschitz continuity of ∇f can also be relaxed, but with a weakening of the above property.

Theorem 3.9 [294] Let $\gamma = 0$, $\|B^k\| \leq \beta < \infty$, and let $f(x)$ be continuously differentiable and bounded below on a level set $\{x \mid f(x) \leq f(x^0)\}$. Let p^k satisfy (3.53) and $\|p^k\| \leq c_2 \Delta^k$ for some constant $c_2 \geq 1$. Then the algorithm generates a sequence of points with

$$\lim_{k\to\infty} \inf \nabla f(x^k) = 0.$$

This lim inf property states that without a strict step acceptance criterion (i.e., $\gamma = 0$), $\nabla f(x^k)$ must have a limit point that is *not bounded away from zero*. On the other hand, there is only a subsequence, indexed by k', with $f(x^{k'})$ that converges to zero. A more detailed description of this property can be found in [294].

Finally, note that Theorems 3.8 and 3.9 deal with convergence to stationary points that may not be local optima if $f(x)$ is nonconvex. Here, using B^k forced to be positive definite, the dogleg approach may converge to a point where $\nabla^2 f(x^*)$ is not positive definite. Similarly, the Levenberg–Marquardt method may also converge to such a point if λ remains positive. The stronger property of convergence to a local minimum requires consideration of second order conditions for (3.49), as well as a more general approach with $B^k = \nabla^2 f(x^k)$ for the nonconvex model problem (3.49).

3.5.2 Nonconvex Model Problems

If the exact Hessian is used and it is not positive definite, then a dogleg method that relies on Newton steps may be unsuitable, as they may be ascent steps or unbounded steps. On the other hand, applying the Levenberg–Marquardt method (3.50) to (3.49) will still generate steps for sufficient decrease, but the correspondence between λ and Δ from (3.51) may be lost, along with the flexibility to exploit the size of the trust region. This can lead to poor performance of the method and even premature termination.

To see this effect, we first apply the eigenvalue decomposition $\nabla^2 f(x^k) = V \Lambda V^T$, where the diagonal matrix Λ contains the eigenvalues (λ_i) and the columns of V contain the eigenvectors of $\nabla^2 f(x^k)$. Substituting the decomposition into (3.50) leads to

$$\begin{aligned} p(\lambda) &= -(\nabla^2 f(x^k) + \lambda I)^{-1} \nabla f(x^k) \\ &= -V(\Lambda^k + \lambda I)^{-1} V^T \nabla f(x^k) \\ &= -\sum_{i=1}^{n} \frac{v_i v_i^T \nabla f(x^k)}{\lambda_i + \lambda}. \end{aligned} \tag{3.54}$$

To adjust $p(\lambda)$ to satisfy $\|p(\lambda)\| = \Delta^k$, we see from (3.54) that we can make $p(\lambda)$ small by increasing λ. Also, if $\nabla^2 f(x^k)$ is positive definite, then if $\lambda = 0$ satisfies the acceptance criterion, we can recover the Newton step, and fast convergence is assured.

In the case of negative or zero curvature, we have an eigenvalue $\lambda_{i*} \leq 0$ for a particular index i^*. As long as $v_{i*}^T \nabla f(x^k) \neq 0$, we can still make $p(\lambda)$ large by letting λ approach $|\lambda_{i*}|$. Thus, $p(\lambda)$ could be adjusted so that its length matches Δ^k. However, if we have $v_{i*}^T \nabla f(x^k) = 0$, then no positive value of λ can be found which increases the length $p(\lambda)$ to Δ^k. This case is undesirable, as it precludes significant improvement of $f(x)$ along the direction of negative curvature v_{i*} and could lead to premature termination with large values of λ and very small values of Δ.

This phenomenon is called the *hard case* [282, 100, 294]. For this case, an additional term is needed that includes a direction of negative curvature, z. Here the corresponding eigenvector for a negative eigenvalue, v_{i*}, is an ideal choice. Because it is orthogonal both to the gradient vector and to all of the other eigenvectors, it can independently exploit the negative curvature direction up to the trust region boundary with a step given by

$$p^k = -(\nabla^2 f(x^k) + \lambda I)^{-1} \nabla f(x^k) + \tau z, \tag{3.55}$$

where τ is chosen so that $\|p^k\| = \Delta^k$. Finding the appropriate eigenvector requires an eigenvalue decomposition of $\nabla^2 f(x^k)$ and is only suitable for small problems. Nevertheless, the approach based on (3.55) can be applied in systematic way to find the *global* minimum of the trust region problem (3.49). More details of this method can be found in [282, 100].

For large-scale problems, there is an inexact way to solve (3.49) based on the truncated Newton method. Here we attempt to solve the linear system $B^k p^k = -\nabla f(x^k)$ with the method of conjugate gradients. However, if B^k is indefinite, this method "fails" by generating a large step, which can be shown to be a direction of negative curvature and can be used directly in (3.55). The truncated Newton algorithm for the inexact solution of (3.49) is given as follows.

ALGORITHM 3.4.
Given parameters $\epsilon > 0$, set $p_0 = 0$, $r_0 = \nabla f(x^k)$, and $d_0 = -r_0$. If $\|r_0\| \leq \epsilon$, stop with $p = p_0$.
For $j \geq 0$:

- If $d_j^T B^k d_j \leq 0$ (d_j is a direction of negative curvature), find τ so that $p = p_j + \tau d_j$ minimizes $m(p)$ with $\|p\| = \Delta^k$ and return with the solution $p^k = p$.
- Else, set $\alpha_j = r_j^T r_j / d_j^T B^k d_j$ and $p_{j+1} = p_j + \alpha_j d_j$.
  - If $\|p_{j+1}\| \geq \Delta^k$, find $\tau \geq 0$ so that $\|p\| = \|p_j + \tau d_j\| = \Delta^k$. Return with the solution, $p^k = p$.
  - Else, set $r_{j+1} = r_j + \alpha_j B^k d_j$. If $\|r_{j+1}\| < \epsilon \|r_0\|$, return with the solution, $p^k = p_{j+1}$. Otherwise, set $\beta_{j+1} = r_{j+1}^T r_{j+1} / r_j^T r_j$ and $d_{j+1} = r_{j+1} + \beta_{j+1} d_j$.

The conjugate gradient (CG) steps generated to solve (3.49) for a particular Δ can be seen in Figure 3.5. Note that the first step taken by the method is exactly the Cauchy step.

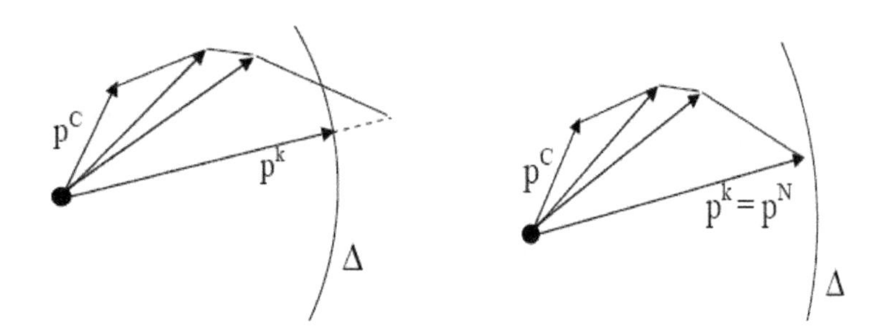

Figure 3.5. *Solution steps for model problem generated by the truncated Newton method: truncated step (left), Newton step (right).*

The subsequently generated CG steps increase in size until they exceed the trust region or converge to the Newton step inside the trust region.

While treatment of negative curvature requires more expensive trust region algorithms, what results are convergence properties that are stronger than those resulting from the solution of convex model problems. In particular, these nonconvex methods can find limit points that are truly local minima, not just stationary points, as stated by the following property.

Theorem 3.10 [294] Let p^* be the exact solution of the model problem (3.49), $\gamma \in (0, 1/4)$, and $B^k = \nabla^2 f(x^k)$. Also, let the approximate solution for Algorithm 3.3 satisfy $\|p^k\| \leq c_2 \Delta^k$ and

$$m^k(0) - m^k(p^k) \geq c_1(m^k(0) - m^k(p^*))$$

for some constants $c_1 > 0$ and $c_2 \geq 1$. Then we have

$$\lim_{k \to \infty} \nabla f(x^k) = 0.$$

Also, if the level set $\{x | f(x) \leq f(x^0)\}$ is closed and bounded, then either the algorithm terminates at a point that satisfies second order *necessary* conditions, or there is a limit point x^* that satisfies second order necessary conditions.

Finally, as with line search methods, global convergence alone does not guarantee efficient algorithms. To ensure fast convergence, we would like to take pure Newton steps at least in the neighborhood of the optimum. For trust region methods, this requires the trust region not to shrink to zero upon convergence, i.e., $\lim_{k \to \infty} \Delta^k \geq \epsilon > 0$. This property is stated by the following theorem.

Theorem 3.11 [294] Let $f(x)$ be twice Lipschitz continuously differentiable and suppose that the sequence $\{x^k\}$ converges to a point x^* that satisfies second order *sufficient* conditions. Also, for sufficiently large k, problem (3.49) is solved asymptotically exactly with $B^k \to \nabla^2 f(x^*)$, and with at least the same reduction as a Cauchy step. Then the trust region bound becomes inactive for k sufficiently large.

Table 3.3. *Iteration sequence with trust region (TR) method and exact Hessian with starting point close to solution.* NFE^k *denotes the number of function evaluations needed to adjust the trust region in iteration k.*

TR Iteration (k)	x_1^k	x_2^k	$f(x^k)$	$\|\nabla f(x^k)\|$	NFE^k
0	0.8000	0.3000	−5.0000	3.0000	3
1	0.7423	0.3115	−5.0882	0.8163	1
2	0.7396	0.3143	−5.0892	6.8524×10^{-3}	1
3	0.73950	0.31436	−5.08926	2.6845×10^{-6}	—

Table 3.4. *Iteration sequence with trust region (TR) method and exact Hessian with starting point far from solution.* NFE^k *denotes the number of function evaluations to adjust the trust region in iteration k.*

TR Iteration (k)	x_1	x_2	f	$\|\nabla f(x^k)\|$	NFE^k
0	1.0000	0.5000	−1.1226	9.5731	11
1	0.9233	0.2634	−4.1492	11.1073	1
2	0.7621	0.3093	−5.0769	1.2263	1
3	0.7397	0.3140	−5.0892	0.1246	1
4	0.7395	0.3143	−5.0892	1.0470×10^{-4}	1
5	0.73950	0.31436	−5.08926	5.8435×10^{-10}	—

Example 3.12 To evaluate trust region methods with exact second derivatives, we again consider the problem described in Example 3.5. This problem has only a small region around the solution where the Hessian is positive definite, and a method that takes full Newton steps has difficulty converging from starting points far away from the solution. Here we apply the trust region algorithm of Gay [156]. This algorithm is based on the exact trust region algorithm [282] described above. More details of this method can also be found in [154, 110]. A termination tolerance of $\|\nabla f(x)\| \leq 10^{-6}$ is chosen and the initial trust region Δ is determined from the initial Cauchy step $\|p^C\|$. Choosing a starting point close to the solution generates the iteration sequence given in Table 3.3. Here it is clear that the solution can be found very quickly, with only three trust region iterations. The first trust region step requires three function evaluations to determine a proper trust region size. After this, pure Newton steps are taken, the trust region becomes inactive, as predicted by Theorem 3.11, and we see quadratic convergence to the optimum solution.

If we choose a starting point farther away from the minimum, then the trust region algorithm, with a termination tolerance of $\|\nabla f(x)\| \leq 10^{-6}$, generates the iteration sequence given in Table 3.4. Here only five trust region iterations are required, but the initial trust region requires 11 function evaluations to determine a proper size. After this, pure Newton steps are taken, the trust region becomes inactive, as predicted by Theorem 3.11, and we can observe quadratic convergence to the optimum solution. ■

This example demonstrates that the initial difficulties that occur with Newton's method are overcome by trust region methods that still use exact Hessians, even if they are indefinite. Also, the performance of the trust region method on this example confirms the trust region convergence properties described in this section.

3.6 Summary and Notes for Further Reading

This chapter develops Newton-type methods for unconstrained optimization. A key concept here is the suitability of the Hessian matrix to take well-defined steps and also achieve fast convergence. In addition, globalization strategies are needed to ensure convergence from poor starting points. To deal with the first issue, we develop ways to modify the Hessian using either modified Cholesky or Levenberg–Marquardt regularization. Moreover, when second derivatives are not available, quasi-Newton methods can be applied. Globalization of Newton-type methods is achieved through line search or trust region strategies. While line search methods merely adjust a step length and are easier to apply, they require a strong assumption on the Hessian, or its approximation. On the other hand, trust region methods converge to stationary points under weaker assumptions. By using exact Hessians and dealing with nonconvex trust region subproblems, it is also possible to show convergence to local solutions, not just stationary points.

Most of the convergence properties for line search and trust region methods are cited or adapted from Nocedal and Wright [294], where more detail can also be found. In addition, unconstrained Newton methods are developed and analyzed in depth in Fletcher [134] and Deuflhard [113]. A detailed treatment of line search methods is given in Dennis and Schnabel [110], which also provides pseudocode for their implementation. Trust region methods are developed, analyzed, and reviewed in encyclopedic detail in [100].

Recent work on trust region methods emphasizes a number of strategies that improve reliability and performance. Advances beyond the pioneering work of [282], in the global solution of nonconvex trust region methods, can be found in [180, 181, 170], as well in the use of cubic model functions and their approximate solution to generate good trust region steps. The recent study [87] shows superior algorithmic properties and performance of the adaptive cubic overestimation (ACO) method over classical trust region methods.

Finally, high-quality software is widely available for the solution of unconstrained problems with Newton-type methods. These include the trust region methods in MINPACK and the TRON code, limited memory BFGS methods in the L-BFGS code, as well as line search and trust region methods in the IMSL and NAg libraries.

3.7 Exercises

1. Using an SR1 update, derive the following formula:

$$B^{k+1} = B^k + \frac{(y - B^k s)(y - B^k s)^T}{(y - B^k s)^T s}.$$

2. In Powell damping, the BFGS update is modified if $s^T y$ is not sufficiently positive by defining $\bar{y} := \theta y + (1-\theta) B^k s$ and substituting for y in (3.18).

   - Show that θ can be found by solving the following one-dimensional NLP:

$$\begin{aligned} \max \quad & \theta \\ \text{s.t.} \quad & \theta s^T y + (1-\theta) s^T B^k s \geq 0.2 s^T B^k s, \\ & 0 \leq \theta \leq 1. \end{aligned}$$

- If $s^T y \geq 0.2 s^T B^k s$, show that Powell damping corresponds to a normal BFGS update.
- If $s^T y \to -\infty$, show that Powell damping corresponds to skipping the BFGS update.

3. Show that if B^k is positive definite, $\cos\theta^k > 1/\kappa(B^k)$ where $\kappa(B^k)$ is the condition number of B^k, based on the 2-norm.

4. Derive a step-size rule for the Armijo line-search that minimizes the quadratic interpolant from the Armijo inequality.

5. Derive the Cauchy step p^C by inserting a trial solution $p = -\tau \nabla f(x^k)$ into the model problem (3.49) and solving for τ with a large value of Δ.

6. Derive the Davidon–Fletcher–Powell (DFP) (complementary BFGS) formula by defining $H^k = (B^k)^{-1}$, $WW^T = H^k$ and invoking a least change strategy for W^+, leading to

$$\begin{aligned} \min \quad & \|W^+ - W\|_F \\ \text{s.t.} \quad & W^+ v = s, \quad (W^+)^T y = v. \end{aligned} \tag{3.56}$$

   Using the inverse update formula (3.21), derive the DFP update for B^k.

7. Download and apply the L-BFGS method to Example 3.5. How does the method perform as a function of the number of updates?

Chapter 4

Concepts of Constrained Optimization

This chapter develops the underlying properties for constrained optimization. It describes concepts of feasible regions and reviews convexity conditions to allow visualization of constrained solutions. Karush–Kuhn–Tucker (KKT) conditions are then presented from two viewpoints. First, an intuitive, geometric interpretation is given to aid understanding of the conditions. Once presented, the KKT conditions are analyzed more rigorously. Several examples are provided to illustrate these conditions along with the role of multipliers and constraint qualifications. Finally, two special cases of nonlinear programming, linear and quadratic, are briefly described, and a case study on portfolio optimization is presented to illustrate their characteristics.

4.1 Introduction

The previous two chapters dealt with unconstrained optimization, where the solution was not restricted in $\mathbb{R}^n$. This chapter now considers the influence of a constrained *feasible region* in the characterization of optimization problems. For continuous variable optimization, we consider the general NLP problem given by

$$\begin{aligned} \min \quad & f(x) \\ \text{s.t.} \quad & g(x) \leq 0, \\ & h(x) = 0, \end{aligned} \tag{4.1}$$

and we assume that the functions $f(x): \mathbb{R}^n \to \mathbb{R}$, $h(x): \mathbb{R}^n \to \mathbb{R}^m$, and $g(x): \mathbb{R}^n \to \mathbb{R}^r$ have continuous first and second derivatives. For ease of notation we will refer to the feasible region as $\mathcal{F} = \{x | g(x) \leq 0, h(x) = 0\}$. As with unconstrained problems, we first characterize the solutions to (4.1) with the following definition.

Definition 4.1 Constrained Optimal Solutions.

- A point x^* is a *global minimizer* if $f(x^*) \leq f(x)$ for all $x \in \mathcal{F}$.
- A point x^* is a *local minimizer* if $f(x^*) \leq f(x)$ for all $x \in \mathcal{N}(x^*) \cap \mathcal{F}$, where we define $\mathcal{N}(x^*) = \|x - x^*\| < \epsilon$, $\epsilon > 0$.

- A point x^* is a *strict local minimizer* if $f(x^*) < f(x)$ for all $x \in \mathcal{N}(x^*) \cap \mathcal{F}$.
- A point x^* is an *isolated local minimizer* if there are no other local minimizers in $\mathcal{N}(x^*) \cap \mathcal{F}$.

As with unconstrained optimization, the following questions need to be considered for problem (4.1):

- If a solution x^* exists to (4.1), is it a global solution in $\mathcal{F}$ or is it only a local solution?
- What conditions characterize solutions to (4.1)?
- Are there special problem classes of (4.1) whose solutions have stronger properties and are easier to solve?
- Are there efficient and reliable methods to solve (4.1)?

This chapter addresses the first three questions, while the next chapter addresses the fourth. The remainder of this section considers *convexity properties* of (4.1), provides a characterization of constrained problems, and presents a small example. Section 4.2 then deals with the presentation of optimality conditions from a kinematic perspective and illustrates these conditions with a small example. Section 4.3 then derives these conditions and provides a more rigorous analysis. Two convex problem classes of (4.1) are considered in Section 4.4, and the chapter concludes by illustrating these problem classes with a case study on portfolio planning.

4.1.1 Constrained Convex Problems

A key advantage occurs when problem (4.1) is convex, i.e., when it has a convex objective function and a convex feasible region. As discussed in Chapter 2, a function $f(x)$ of x in some domain X is convex if and only if for all points $x^a, x^b \in X$, it satisfies

$$\alpha f(x^a) + (1-\alpha) f(x^b) \geq f(\alpha x^a + (1-\alpha) x^b) \tag{4.2}$$

for all $\alpha \in (0,1)$. Strict convexity requires that inequality (4.2) be strict. Similarly, a region Y is convex if and only if for all points $x^a, x^b \in Y$,

$$\alpha x^a + (1-\alpha) x^b \in Y \quad \text{for all } \alpha \in (0,1).$$

We start by characterizing a convex feasible region for (4.1).

Theorem 4.2 If $g(x)$ is convex and $h(x)$ is linear, then the region $\mathcal{F} = \{x \mid g(x) \leq 0, h(x) = 0\}$ is convex, i.e., $\alpha x^a + (1-\alpha)x^b \in \mathcal{F}$ for all $\alpha \in (0,1)$ and $x^a, x^b \in \mathcal{F}$.

Proof: From the definition of convex regions, we consider two points $x^a, x^b \in \mathcal{F}$ and assume there exists a point $\bar{x} = \alpha x^a + (1-\alpha)x^b \notin \mathcal{F}$ for some $\alpha \in (0,1)$. Since $\bar{x} \notin \mathcal{F}$, we can have one or more constraints with $g_i(\bar{x}) > 0$ or $h(\bar{x}) \neq 0$. In the former case, we have by feasibility of x^a and x^b and convexity of $g(x)$,

$$\begin{aligned} 0 &\geq \alpha g_i(x^a) + (1-\alpha) g_i(x^b) \\ &\geq g_i(\alpha x^a + (1-\alpha)x^b) = g_i(\bar{x}) > 0 \end{aligned}$$

which is a contradiction. In the second case, we have from feasibility of x^a and x^b and linearity of $h(x)$,

$$\begin{aligned} 0 &= \alpha h(x^a) + (1-\alpha)h(x^b) \\ &= h(\alpha x^a + (1-\alpha)x^b) = h(\bar{x}) \neq 0, \end{aligned}$$

which again is a contradiction. Since neither case can hold, the statement of the theorem must hold. □

Examples of convex optimization problems are pictured in Figures 4.1 and 4.2. As with unconstrained problems in Chapter 2, we now characterize solutions of convex constrained problems with the following theorem.

Theorem 4.3 If $f(x)$ is convex and $\mathcal{F}$ is convex, then every local minimum in $\mathcal{F}$ is a global minimum. If $f(x)$ is strictly convex in $\mathcal{F}$, then a local minimum is the unique global minimum.

Proof: For the first statement, we assume that there are two local minima x^a, $x^b \in \mathcal{F}$ with $f(x^a) > f(x^b)$ and establish a contradiction. Here we have $f(x^a) \leq f(x)$, $x \in \mathcal{N}(x^a) \cap \mathcal{F}$ and $f(x^b) \leq f(x)$, $x \in \mathcal{N}(x^b) \cap \mathcal{F}$. By convexity, $(1-\alpha)x^a + \alpha x^b \in \mathcal{F}$ and

$$f((1-\alpha)x^a + \alpha x^b) \leq (1-\alpha)f(x^a) + \alpha f(x^b) \quad \text{for all } \alpha \in (0,1). \tag{4.3}$$

Now choosing α so that $\bar{x} = (1-\alpha)x^a + \alpha x^b \in \mathcal{N}(x^a) \cap \mathcal{F}$, we can write

$$f(\bar{x}) \leq f(x^a) + \alpha(f(x^b) - f(x^a)) < f(x^a) \tag{4.4}$$

which shows a contradiction.

For the second statement, we assume that, again, there are two local minima x^a, $x^b \in \mathcal{F}$ with $f(x^a) \geq f(x^b)$ and establish a contradiction. Here we have $f(x^a) \leq f(x)$, $x \in \mathcal{N}(x^a) \cap \mathcal{F}$ and $f(x^b) \leq f(x)$, $x \in \mathcal{N}(x^b) \cap \mathcal{F}$. By convexity of $\mathcal{F}$, $(1-\alpha)x^a + \alpha x^b \in \mathcal{F}$ and strict convexity of $f(x)$ leads to

$$f((1-\alpha)x^a + \alpha x^b) < (1-\alpha)f(x^a) + \alpha f(x^b) \quad \text{for all } \alpha \in (0,1). \tag{4.5}$$

Choosing α so that $\bar{x} = (1-\alpha)x^a + \alpha x^b \in \mathcal{N}(x^a) \cap \mathcal{F}$, we can write

$$f(\bar{x}) < f(x^a) + \alpha(f(x^b) - f(x^a)) \leq f(x^a) \tag{4.6}$$

which again shows a contradiction. □

To obtain an impression of constrained problems, consider the two-dimensional contour plots shown in Figures 4.1–4.3. Figure 4.1 shows contours for linear objective and constraint functions, i.e., for a *linear programming* problem. Not only are all linear programs (LPs) solutions global solutions, but these solutions occur at vertices of the feasible region. The left plot in Figure 4.1 shows a unique solution, while the right plot shows two alternate optima, both with the same objective function value.

Figure 4.2 shows three contour plots of strictly convex objective functions and linear constraints. If the objective function is also quadratic, we have convex quadratic programming problems. In all three plots we see that global solutions are obtained; these solutions

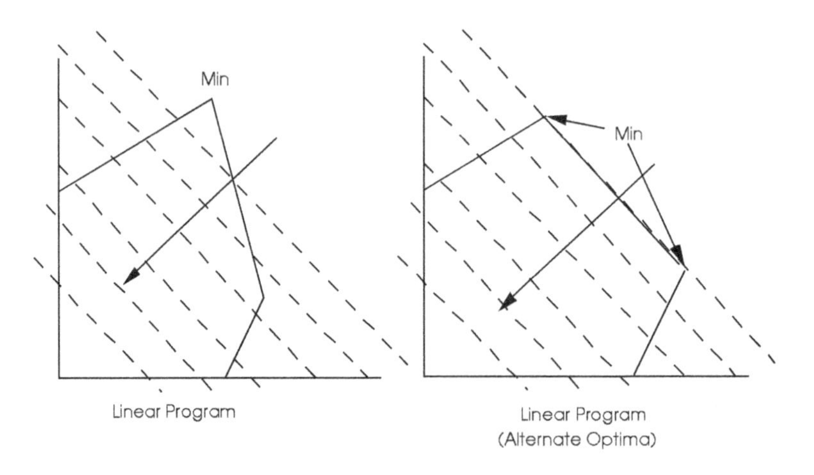

Figure 4.1. *Contours for linear programming problems.*

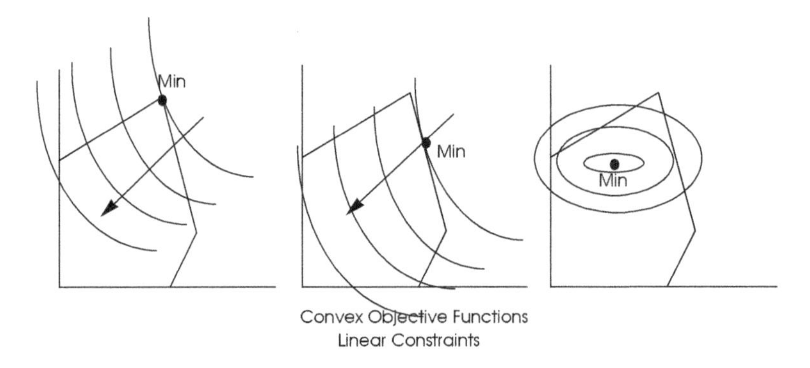

Figure 4.2. *Contours for convex objectives with linear constraints.*

can be observed at vertex points, on a constraint boundary, or in the interior of the feasible region.

Finally, Figure 4.3 shows two contours for nonconvex problems where multiple local solutions are observed. The left plot shows these local solutions as a result of a nonconvex feasible region, even though the objective function is convex, while the right plot has a convex feasible region but shows multiple solutions due to a nonconvex objective function.

In analogy to Chapter 2, we can characterize, calculate, and verify local solutions through well-defined optimality conditions. Without convexity, however, there are generally no well-defined optimality conditions that guarantee global solutions, and a much more expensive enumeration of the search space is required. In this text, we will focus on finding local solutions to (4.1). A comprehensive treatment of global optimization algorithms and their properties can be found in [144, 203, 379].

To motivate the additional complexities of constrained optimization problems, we consider a small geometric example, where the optimal solution is determined by the constraints.

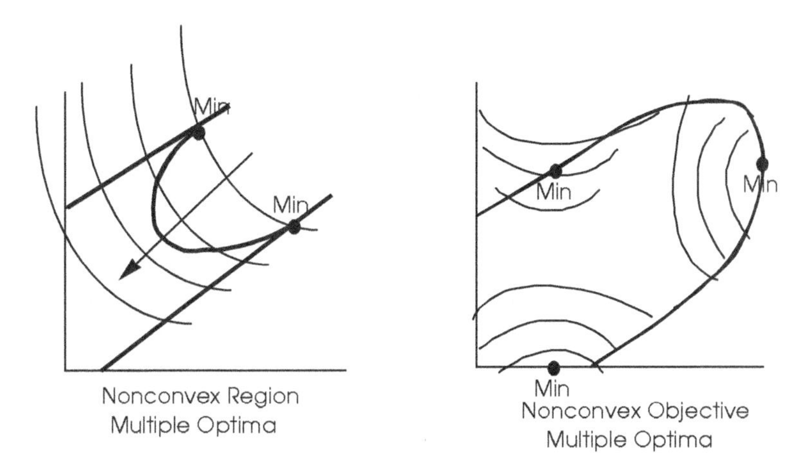

Figure 4.3. *Contours with nonconvex regions or nonconvex objectives.*

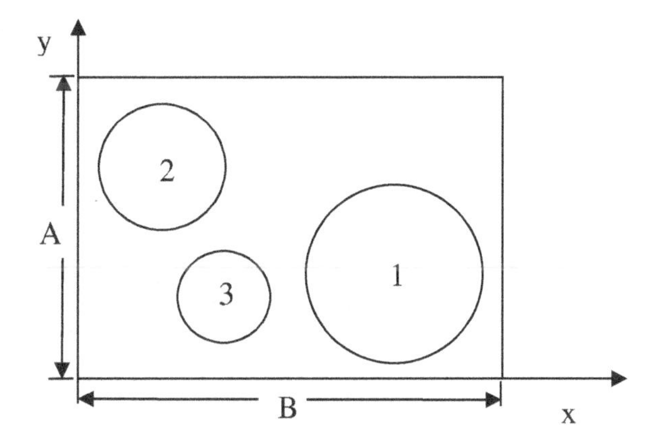

Figure 4.4. *Packing with three circles.*

Example 4.4 We consider three circles of different sizes, as shown in Figure 4.4, and seek the smallest perimeter of a box that encloses these circles.

As decision variables we choose the dimensions of the box, A, B, and the coordinates for the centers of the three circles, $(x_1, y_1), (x_2, y_2), (x_3, y_3)$. As specified parameters we have the radii R_1, R_2, R_3. For this problem we minimize the perimeter $2(A+B)$ and include as constraints the fact that the circles need to remain in the box and cannot overlap with each other. As a result we formulate the following nonlinear program:

$$\begin{aligned} \min \quad & 2(A+B) \\ \text{s.t.} \quad & A \geq 0, B \geq 0, \end{aligned} \tag{4.7}$$

$$\left.\begin{array}{ll} x_1, y_1 \geq R_1,\ x_1 \leq B - R_1, & y_1 \leq A - R_1 \\ x_2, y_2 \geq R_2,\ x_2 \leq B - R_2, & y_2 \leq A - R_2 \\ x_3, y_3 \geq R_3,\ x_3 \leq B - R_3, & y_3 \leq A - R_3 \end{array}\right\} \text{In Box}$$

$$\left.\begin{array}{rcl} (x_1 - x_2)^2 + (y_1 - y_2)^2 & \geq & (R_1 + R_2)^2 \\ (x_1 - x_3)^2 + (y_1 - y_3)^2 & \geq & (R_1 + R_3)^2 \\ (x_2 - x_3)^2 + (y_2 - y_3)^2 & \geq & (R_2 + R_3)^2. \end{array}\right\} \text{No Overlaps}$$

We see that the objective function and the *In Box* constraints are linear, and hence, convex. Similarly, the variable bounds are convex as well. On the other hand, the nonlinear *No Overlap* constraints are nonconvex and there is no guarantee of a global solution. Indeed, we can imagine intuitively the following multiple solutions to this NLP.

- Find a solution and observe an equivalent solution by turning the box by 90°.
- Change the initial arrangement of the circles and manually shrink the walls of the box. The optimal solution will depend on the initial positions of the circles.

This example also shows that the solution is determined entirely by the constraints and needs to be analyzed in a different way than in the previous two chapters. The following two sections provide more detail on this analysis. ∎

4.2 Local Optimality Conditions—A Kinematic Interpretation

Before presenting the formal development of conditions that define a local optimum, we first consider a more intuitive, kinematic illustration. Consider the contour plot of the objective function, $f(x)$, given in Figure 4.5 as a smooth valley in the space of the variables, x_1 and x_2. In the contour plot of the unconstrained problem, min $f(x)$, consider a ball rolling in this valley to the lowest point of $f(x)$, denoted by x^*. This point is at least a local minimum and is defined by a zero gradient and nonnegative curvature in all (nonzero) directions p. In Chapter 2, we stated the first and second order necessary conditions for unconstrained optimality as

$$\nabla f(x^*) = 0 \quad \text{and} \quad p^T \nabla^2 f(x^*) p \geq 0 \quad \text{for all } p \in \mathbb{R}^n, p \neq 0. \tag{4.8}$$

Consider now the imposition of inequality constraint $g(x) \leq 0$ in Figure 4.6. Continuing the kinematic interpretation, the inequality constraints act as "fences" in the valley and we still allow the ball, constrained within fences, to roll to its lowest point. In Figure 4.6, this point x^* is found when the ball is at the fence (at an active inequality, $g(x) = 0$). At x^* the normal force on the ball exerted by the "fence" (in the direction of $-\nabla g(x)$) is balanced by the force of "gravity" (in the direction of $-\nabla f(x)$). Also, if we move the fence farther down and to the left, then the ball will roll to the unconstrained minimum and the influence of the fence will be irrelevant.

Finally, we consider the imposition of both inequality and equality constraints in Figure 4.7, where the inequality constraints $g(x) \leq 0$ again act as "fences" in the valley and the equality constraints $h(x) = 0$ act as "rails." In Figure 4.7, the rolling ball finds its lowest point at x^* where the normal forces exerted on the ball by the "fences" $(-\nabla g(x^*))$ and "rails" $(-\nabla h(x^*))$ are balanced by the force of "gravity" $(-\nabla f(x^*))$.

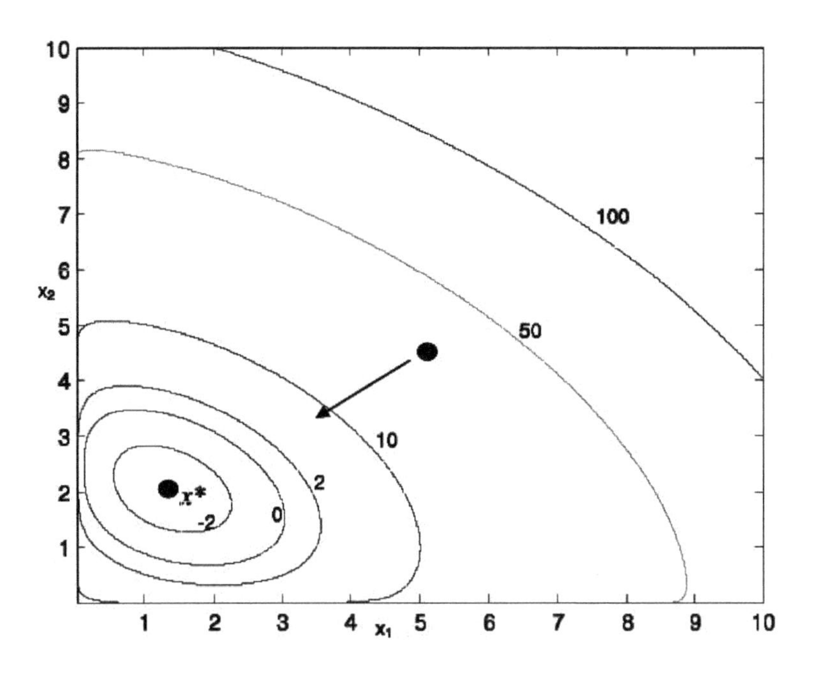

Figure 4.5. *Contours of unconstrained problem.*

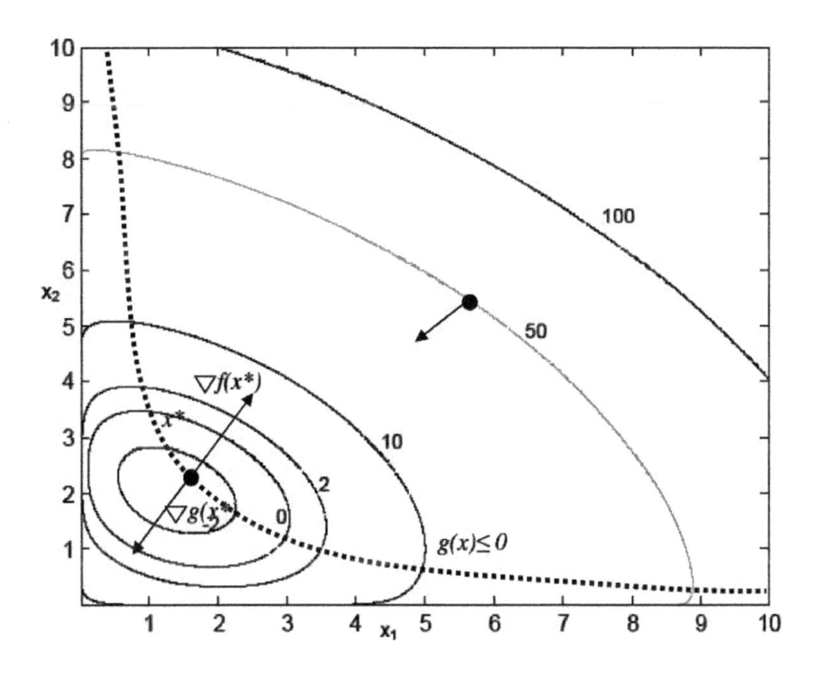

Figure 4.6. *Contours with single inequality.*

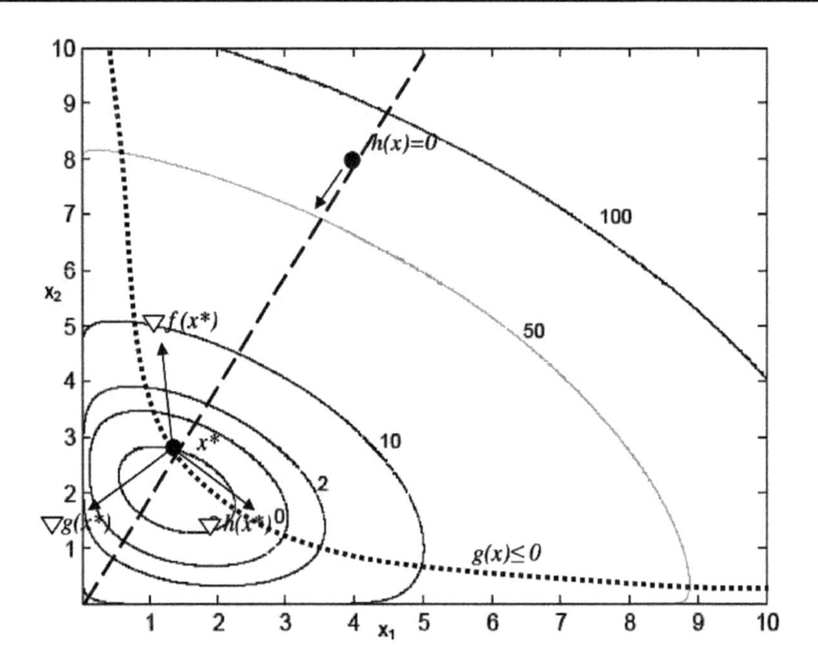

Figure 4.7. *Contours with inequality and equality constraints.*

The conditions that characterize the stationary ball in Figure 4.7 can be stated more precisely by the following Karush–Kuhn–Tucker (KKT) necessary conditions for constrained optimality. They comprise the following elements:

- *Balance of Forces*: It is convenient to define the Lagrangian function $L(x,u,v) = f(x) + g(x)^T u + h(x)^T v$ along with u and v as "weights" or multipliers for the inequality and equality constraints, respectively. At the solution, the stationarity condition (balance of forces acting on the ball) is then given by

$$\nabla_x L(x^*, u^*, v^*) = \nabla f(x^*) + \nabla g(x^*) u^* + \nabla h(x^*) v^* = 0 \tag{4.9}$$

- *Feasibility*: Both inequality and equality constraints must be satisfied; i.e., the ball must lie on the rail and within the fences

$$g(x^*) \leq 0, \quad h(x^*) = 0. \tag{4.10}$$

- *Complementarity*: Inequality constraints are either strictly satisfied (active) or inactive, in which case they are irrelevant to the solution. In the latter case the corresponding KKT multiplier must be zero. Also, while the normal force from the rail can act in either normal direction, the normal force can only "push the ball into the interior." This translates into a sign convention for the inequality constraint multiplier. These conditions can be written as

$$g(x^*)^T u^* = 0, \quad u^* \geq 0. \tag{4.11}$$

- *Constraint Qualification*: The stationarity conditions for a local optimum are based on gradient information for the constrained and the objective functions. Because this linearized information is used to characterize the optimum of a *nonlinear* problem, an additional regularity condition is required on the constraints to ensure that gradients are sufficient for the local characterization of the feasible region at x^*. This concept will be covered in more detail in the next section. A typical condition is that the active constraint gradients at x^* be linearly independent, i.e., the matrix made up of columns of $\nabla h(x^*)$ and $\nabla g_i(x^*)$ with $i \in \{i | g_i(x^*) = 0\}$ is full column rank.

- *Second Order Conditions*: In Figure 4.6 we find that the gradients (i.e., force balance) determine how the ball rolls to x^*. On the other hand, we also want to consider the curvature condition, especially *along* the active constraint. Thus, when the ball is perturbed from x^* in a feasible direction, it needs to "roll back" to x^*. As with unconstrained optimization, nonnegative (positive) curvature is necessary (sufficient) for all of the feasible, i.e., constrained, nonzero directions, p. As will be discussed in the next section, these directions determine the following necessary second order conditions:

$$\begin{aligned} p^T \nabla_{xx} L(x^*, u^*, v^*) p \geq 0&, \\ \text{for all } p \neq 0, \nabla h(x^*)^T p = 0&, \\ \nabla g_i(x^*)^T p = 0, \quad i \in \{i | g_i(x^*) = 0, u_i^* > 0\}&, \\ \nabla g_i(x^*)^T p \leq 0, \quad i \in \{i | g_i(x^*) = 0, u_i^* = 0\}&. \end{aligned} \tag{4.12}$$

  The corresponding sufficient conditions require that the first inequality in (4.12) be strict. Note that for Figure 4.5 the allowable directions p span the entire space. In Figure 4.6 these directions are tangent to the inequality constraint at x^*. Finally, for Figure 4.7, the constraints uniquely define x^* and there are no nonzero directions p that satisfy (4.12).

To close this section we illustrate the KKT conditions with an example drawn from Figures 4.5 and 4.7. An example drawn from Figure 4.6 will be considered in the next section.

Example 4.5 Consider the following unconstrained NLP:

$$\min \quad x_1^2 - 4x_1 + \frac{3}{2}x_2^2 - 7x_2 + x_1 x_2 + 9 - \ln(x_1) - \ln(x_2). \tag{4.13}$$

This function corresponds to the contour plot in Figure 4.5. The optimal solution can be found by solving for the first order conditions (4.8):

$$\nabla f(x^*) = \begin{bmatrix} 2x_1 - 4 + x_2 - \frac{1}{x_1} \\ 3x_2 - 7 + x_1 - \frac{1}{x_2} \end{bmatrix} = 0 \text{ leading to } x^* = \begin{bmatrix} 1.3475 \\ 2.0470 \end{bmatrix} \tag{4.14}$$

and $f(x^*) = -2.8742$. Checking the second order conditions leads to

$$\nabla^2 f(x^*) = \begin{bmatrix} 2 + \frac{1}{(x_1^*)^2} & 1 \\ 1 & 3 + \frac{1}{(x_2^*)^2} \end{bmatrix} \Rightarrow \begin{bmatrix} 2.5507 & 1 \\ 1 & 3.2387 \end{bmatrix}, \tag{4.15}$$

which is positive definite, and x^* is a strict local minimum.

Now consider the constrained NLP:

$$
\begin{aligned}
\min \quad & x_1^2 - 4x_1 + \frac{3}{2}x_2^2 - 7x_2 + x_1x_2 + 9 - \ln(x_1) - \ln(x_2) \\
\text{s.t. } & 4 - x_1x_2 \le 0, \\
& 2x_1 - x_2 = 0,
\end{aligned} \tag{4.16}
$$

which corresponds to the plot in Figure 4.7. The optimal solution can be found by applying the first order KKT conditions (4.9)–(4.10):

$$\nabla L(x^*, u^*, v^*) = \nabla f(x^*) + \nabla g(x^*)u^* + \nabla h(x^*)v^* \tag{4.17}$$

$$= \begin{bmatrix} 2x_1 - 4 + x_2 - \frac{1}{x_1} - x_2u + 2v \\ 3x_2 - 7 + x_1 - \frac{1}{x_2} - x_1u - v \end{bmatrix} = 0, \tag{4.18}$$

$$g(x) = 4 - x_1x_2 \le 0, \qquad h(x) = 2x_1 - x_2 = 0, \tag{4.19}$$

$$u \ge 0, \qquad u^T(4 - x_1x_2) = 0 \tag{4.20}$$

$$\Downarrow \tag{4.21}$$

$$x^* = \begin{bmatrix} 1.4142 \\ 2.8284 \end{bmatrix}, \quad u^* = 1.068, \quad v^* = 1.0355, \tag{4.22}$$

and $f(x^*) = -1.8421$. Checking the second order conditions (4.12) leads to

$$\nabla_{xx} L(x^*, u^*, v^*) = \nabla^2 f(x^*) + \nabla^2 g(x^*)u^* + \nabla^2 h(x^*)v^* \tag{4.23}$$

$$= \begin{bmatrix} 2 + \frac{1}{(x_1^*)^2} & 1 - u^* \\ 1 - u^* & 3 + \frac{1}{(x_2^*)^2} \end{bmatrix} = \begin{bmatrix} 2.5 & 0.068 \\ 0.068 & 3.125 \end{bmatrix} \tag{4.24}$$

with $u^* > 0$ for $g(x^*) = 0$ and

$$\begin{bmatrix} \nabla g(x^*)^T \\ \nabla h(x^*)^T \end{bmatrix} p = \begin{bmatrix} -2.8284 & -1.4142 \\ 2 & -1 \end{bmatrix} p = 0. \tag{4.25}$$

However, because this matrix is nonsingular, there are no nonzero vectors, p, that satisfy the allowable directions. Hence, the *sufficient* second order conditions (4.12) are *vacuously* satisfied for this problem. ■

4.3 Analysis of KKT Conditions

The previous section provides an intuitive explanation of local optimality conditions. This section defines and analyzes these concepts more precisely.

Active Constraints and Multipliers

We define the sets $\mathcal{E} = \{1, \ldots, m\}$ and $\mathcal{I} = \{1, \ldots, r\}$ for the equality and inequality constraints, respectively. At x, a feasible point for (4.1), we define the active set as $\mathcal{A}(x) = \mathcal{E} \cup \{i \mid g_i(x) = 0\}$. Note from (4.9) and (4.11) that the *active* constraints are the only ones needed to define the optimum solution. Inactive constraints at x^* (i.e., $i \in \mathcal{I} \setminus \mathcal{A}(x^*)$) can be discarded.

A more precise definition of the active constraints can be made by considering the multipliers in (4.9). The role of the multipliers can be understood by perturbing the right-hand side of a constraint in (4.1). Here we consider a particular inequality $g_{\hat{i}}(x)+\epsilon \leq 0$ with $\hat{i} \in \mathcal{A}(x^*)$ and consider the solution of the perturbed problem, x^ϵ. For small values of ϵ, one can show that this leads to

$$\begin{aligned} 0 &= h(x^\epsilon)-h(x^*) \approx \nabla h(x^*)^T(x^\epsilon - x^*), \\ 0 &= g_i(x^\epsilon)-g_i(x^*) \approx \nabla g_i(x^*)^T(x^\epsilon - x^*), \quad i \in \mathcal{A}(x^*)\cap \mathcal{I} \setminus \{\hat{i}\}, \text{ and} \\ -\epsilon &= g_{\hat{i}}(x^\epsilon)-g_{\hat{i}}(x^*) \approx \nabla g_{\hat{i}}(x^*)^T(x^\epsilon - x^*). \end{aligned}$$

In addition, from (4.9) we have that

$$\begin{aligned} f(x^\epsilon)-f(x^*) &\approx \nabla f(x^*)^T(x^\epsilon - x^*) \\ &= -\sum_{i\in\mathcal{E}} v_i^* \nabla h_i(x^*)^T(x^\epsilon - x^*) - \sum_{i\in\mathcal{I}} u_i^* \nabla g_i(x^*)^T(x^\epsilon - x^*) \\ &\approx -\sum_{i\in\mathcal{E}} v_i^*(h_i(x^\epsilon)-h_i(x^*)) - \sum_{i\in\mathcal{I}} u_i^*(g_i(x^\epsilon)-g_i(x^*)) \\ &= u_{\hat{i}}^*\epsilon. \end{aligned}$$

Dividing the above expression by ϵ and taking limits as ϵ goes to zero leads to $\frac{df(x^*)}{d\epsilon} = u_{\hat{i}}^*$. Thus we see that the multipliers provide the sensitivity of the optimal objective function value to perturbations in the constraints.

Note that we have not stated any particular properties of the multipliers. In the absence of additional conditions, the multipliers may be nonunique, or may not even exist at an optimal solution. Nevertheless, with values of the multipliers, u^*, v^*, that satisfy (4.9) at the solution to (4.1), we can further refine the definition of the active set. We define a *strongly active set*, $\mathcal{A}_s(x^*) = \{i \,|\, u_i^* > 0, g_i(x^*) = 0\}$, and a *weakly active set*, $\mathcal{A}_w(x^*) = \{i \,|\, u_i^* = 0, g_i(x^*) = 0\}$. This also allows us to state the following definition.

Definition 4.6 Given a local solution of (4.1), x^*, along with multipliers u^*, v^* that satisfy (4.9) and (4.11), the *strict complementarity condition* is given by $u_i^* > 0$ for all $i \in \mathcal{A}(x^*)\cap \mathcal{I}$. This condition holds if and only if $\mathcal{A}_w(x^*)$ is empty.

Example 4.7 To illustrate these concepts, consider the simple example:

$$\min x^2 \text{ s.t. } \frac{1}{2} - x \leq 0, \quad x - 1 \leq 0.$$

The Lagrange function is given by $L(x,u) = x^2 + u_U(x-1) + u_L(\frac{1}{2} - x)$, and the corresponding KKT conditions are given by

$$\begin{gathered} 2x + u_U - u_L = 0, \\ u_U(x-1) = 0, \quad u_L\left(\frac{1}{2} - x\right) = 0, \\ \frac{1}{2} \leq x \leq 1, \quad u_U, u_L \geq 0. \end{gathered}$$

These conditions are satisfied by $x^* = \frac{1}{2}$, $u_U^* = 0$, and $u_L^* = 1$. Thus the lower bound on x is the only active constraint at the solution. This constraint is *strongly active*, and strict complementarity is satisfied. Moreover, if we increase the lower bound by ϵ, the optimal value of the objective function is $f(x^\epsilon) = \frac{1}{4} + \epsilon + \epsilon^2$ and $\frac{df(x^*)}{d\epsilon} = 1 = u_L^*$ as $\epsilon \to 0$.

If we now modify the example to min x^2, $0 \le x \le 1$, with corresponding KKT conditions given by

$$\begin{aligned} 2x + u_U - u_L &= 0, \\ u_U(1-x) = 0, \quad u_L x &= 0, \\ 0 \le x \le 1, \quad u_U, u_L &\ge 0, \end{aligned}$$

we have the solution $x^* = 0$, $u_U^* = 0$, and $u_L^* = 0$. Here the lower bound on x is only *weakly active* at the solution and strict complementarity does not hold. Here, $\frac{df(x^*)}{d\epsilon} = 0 = u_L^*$ and, as verified by the multipliers, we see that reducing the lower bound on x does not change $f(x^*)$, while increasing the lower bound on x increases the optimum objective function value only to $(\epsilon)^2$. ∎

Limiting Directions

To deal with nonlinear objective and constraint functions we need to consider the concept of feasible sequences and their limiting directions. For a given feasible point, $\bar{x}$, we can consider a sequence of points $\{x^k\}$, with each point indexed by k, that satisfies

- $x^k \neq \bar{x}$,
- $\lim_{k \to \infty} x^k = \bar{x}$,
- x^k is feasible for all $k \ge K$, where K is sufficiently large.

Associated with each feasible sequence is a *limiting direction*, d, of unit length defined by

$$\lim_{k \to \infty} \frac{x^k - \bar{x}}{\|x^k - \bar{x}\|} = d. \tag{4.26}$$

In analogy with the kinematic interpretation in Section 4.2, one can interpret this sequence to be points on the "path of a rolling ball" that terminates at or passes through $\bar{x}$. The limiting direction is tangent to this path in the *opposite* direction of the rolling ball.

We now consider *all* possible feasible sequences leading to a feasible point and the limiting directions associated with them. In particular, if the objective function $f(x^k)$ increases monotonically for any feasible sequence to $\bar{x}$, then $\bar{x}$ cannot be optimal. Using the concept of limiting directions, this property can be stated in terms of directional derivatives as follows.

Theorem 4.8 [294] If x^* is a solution of (4.1), then all feasible sequences leading to x^* must satisfy

$$\nabla f(x^*)^T d \ge 0, \tag{4.27}$$

where d is any limiting direction of a feasible sequence.

Proof: We will assume that there exists a feasible sequence $\{x^k\}$ that has a limiting direction $\hat{d}$ with $\nabla f(x^*)^T \hat{d} < 0$ and establish a contradiction. From Taylor's theorem, and the definition of a limiting direction (4.26), we have for sufficiently large k,

$$\begin{aligned} f(x^k) &= f(x^*) + \nabla f(x^*)^T (x^k - x^*) + o(\|x^k - x^*\|) && (4.28)\\ &= f(x^*) + \nabla f(x^*)^T \hat{d} \|x^k - x^*\| + o(\|x^k - x^*\|) && (4.29)\\ &\le f(x^*) + \frac{1}{2} \nabla f(x^*)^T \hat{d} \|x^k - x^*\| && (4.30)\\ &< f(x^*), && (4.31) \end{aligned}$$

which contradicts the assumption that x^* is a solution of (4.1). □

4.3.1 Linearly Constrained Problems

Before we consider the general result for problem (4.1), we first consider the case where the optimization problem is linearly constrained and given by

$$\begin{aligned} \min \quad & f(x) && (4.32)\\ \text{s.t.} \quad & Bx \le b, \\ & Cx = c \end{aligned}$$

with vectors b, c and matrices B, C. For these constraints the limiting directions of *all feasible sequences* to a point $\bar{x}$ can be represented as a system of linear equations and inequalities known as a *cone* [294]. This cone coincides exactly with all of the feasible directions in the neighborhood of $\bar{x}$ and these are derived from the constraints in (4.32). Vectors in this cone will cover limiting directions from all possible feasible sequences for k sufficiently large. As a result, for problem (4.32) we can change the statement of Theorem 4.8 to the following.

Theorem 4.9 If x^* is a solution of (4.32) with active constraints that correspond to appropriate rows of B and elements of b, i.e., $B_i x^* = b_i, i \in \mathcal{A}(x^*)$, then

$$\nabla f(x^*)^T d \ge 0 \tag{4.33}$$

for all d within the cone $Cd = 0$ and $B_i d \le 0, i \in \mathcal{A}(x^*)$.

For systems of linear equations and inequalities, we use Motzkin's theorem of the alternative [273] as a supporting lemma.

Lemma 4.10 Either the linear system

$$a^T z > 0, \quad Az \ge 0, \quad Dz = 0 \tag{4.34}$$

has a solution, or the linear system

$$\begin{aligned} a y_1 + A^T y_2 + D^T y_3 = 0, && (4.35)\\ y_1 > 0, y_2 \ge 0 && (4.36) \end{aligned}$$

has a solution, but not both.

This allows us to prove the following property.

Theorem 4.11 Suppose that x^* is a local solution of (4.32); then

- the following KKT conditions are satisfied:

$$\nabla f(x^*) + B^T u^* + C^T v^* = 0, \tag{4.37a}$$
$$Bx^* \leq b, \quad Cx^* = c, \tag{4.37b}$$
$$(Bx^* - b)^T u^* = 0, \quad u^* \geq 0; \tag{4.37c}$$

- if $f(x)$ is convex and the KKT conditions (4.37) are satisfied at a point x^+, then x^+ is the global solution to (4.32).

Proof: The linear constraints (4.37b) are satisfied as x^* is the solution of (4.32). The complementarity conditions (4.37c) are equivalent to the conditions $u_i^* \geq 0, B_i x^* = b_i, i \in \mathcal{A}(x^*)$, and $u_i^* = 0$ for $i \in \mathcal{I} \setminus \mathcal{A}(x^*)$. By Theorem 4.9, there is no solution d to the linear system

$$\nabla f(x^*)^T d < 0, \quad Cd = 0, \quad B_i d \leq 0, \quad i \in \mathcal{A}(x^*). \tag{4.38}$$

Hence, by Lemma 4.10 and by setting a to $-\nabla f(x^*)$, D to $-C$, and A to the rows of $-B_i, i \in \mathcal{A}(x^*)$, we have (4.37a), $y_3/y_1 = v^*$, and $y_2/y_1 = u^* \geq 0$. (Note that y_1 is a scalar quantity.)

For the second part of the theorem, we know from Lemma 4.10 that (4.37a) implies that there is no solution d to the linear system

$$\nabla f(x^+)^T d < 0, \quad Cd = 0, \quad B_i d \leq 0, \quad i \in \mathcal{A}(x^+). \tag{4.39}$$

Assume now that x^+ is not a solution to (4.32) and there is another feasible point x^* with $f(x^*) < f(x^+)$. We define $d = (x^* - x^+)$, which satisfies $Cd = 0, B_i d \leq 0, i \in \mathcal{A}(x^+)$. By convexity of $f(x)$, $\nabla^2 f(x)$ is positive semidefinite for all x, and by Taylor's theorem,

$$0 > f(x^*) - f(x^+) = \nabla f(x^+)^T d + \frac{1}{2} \int_0^1 d^T \nabla^2 f(x + td) d \, dt \tag{4.40}$$
$$\geq \nabla f(x^+)^T d, \tag{4.41}$$

which contradicts the assumption that there is no solution to (4.39). Therefore, x^+ is a local solution to (4.32), and because the objective function and feasible region are both convex, x^+ is also a global solution by Theorem 4.3. □

4.3.2 Nonlinearly Constrained Problems

In the previous subsection for linearly constrained problems, a crucial condition that links Theorem 4.9 to the KKT conditions (4.37) is the representation of limiting directions to feasible sequences by the cone conditions (4.38).

For nonlinear problems, we also need to ensure a link between Theorem 4.8 and the KKT conditions

$$\nabla f(x^*) + \nabla g(x^*)^T u^* + \nabla h(x^*)^T v^* = 0, \tag{4.42a}$$
$$g(x^*) \leq 0, \quad h(x^*) = 0, \tag{4.42b}$$
$$g(x^*)^T u^* = 0, \quad u^* \geq 0. \tag{4.42c}$$

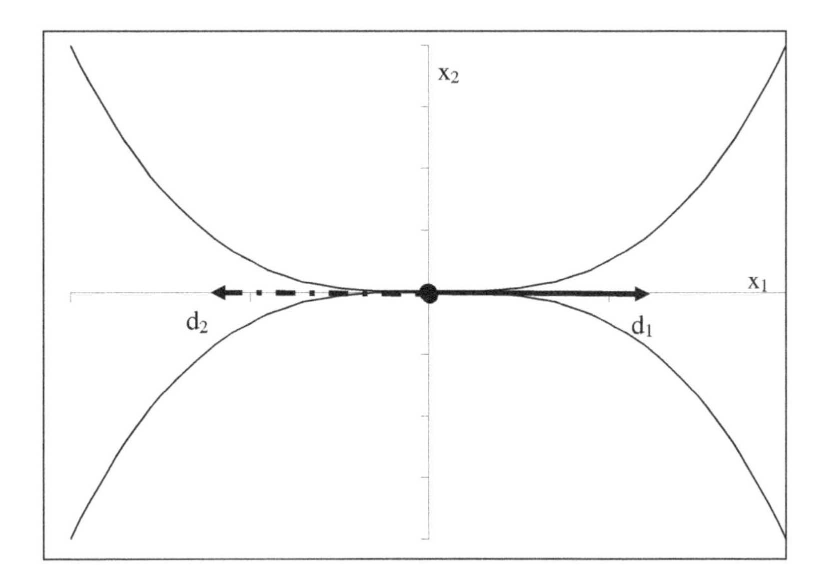

Figure 4.8. *Example that violates constraint qualification.*

However, an additional condition is needed to ensure that limiting directions at the solution can be represented by the following cone conditions derived from linearization of the active constraints:

$$\nabla h_i(x^*)^T d = 0, \quad \nabla g_i(x^*)^T d \le 0, \quad i \in \mathcal{A}(x^*). \tag{4.43}$$

To motivate this point, consider the nonlinear program given by

$$\begin{aligned} \min \quad & x_1 \\ \text{s.t.} \quad & x_2 \le x_1^3, \\ & -x_1^3 \le x_2 \end{aligned} \tag{4.44}$$

and shown in Figure 4.8. The solution of (4.44) is $x_1^* = x_2^* = 0$, and any feasible sequence to the origin will be monotonically decreasing and satisfies the conditions of Theorem 4.8. On the other hand, it is easy to show that the KKT conditions are not satisfied at this point. Moreover, linearization of the active constraints at x^* leads to only two directions, $d_1 = [1,0]^T$ and $d_2 = [-1,0]^T$, as seen in the figure. However, only d_1 is a limiting direction of a feasible sequence. Hence, the essential link between Theorem 4.8 and the KKT conditions (4.42) is broken.

To ensure this link, we need an additional *constraint qualification* that shows that limiting directions of active nonlinear constraints can be represented by their linearizations at the solutions. That is, unlike the constraints in Figure 4.8, the active constraints should not be "too nonlinear." A frequently used condition is the *linear independence constraint qualification* (LICQ) given by the following definition.

Definition 4.12 (LICQ). Given a local solution of (4.1), x^*, and an active set $\mathcal{A}(x^*)$, LICQ is defined by linear independence of the constraint gradients

$$\nabla g_i(x^*), \quad \nabla h_i(x^*), \quad i \in \mathcal{A}(x^*).$$

From Figure 4.8 it is easy to see that (4.44) does not satisfy the LICQ at the solution. To make explicit the need for a constraint qualification, we use the following technical lemma; the proof of this property is given in [294].

Lemma 4.13 [294, Lemma 12.2]. The set of limiting directions from all feasible sequences is a subset of the cone (4.43). Moreover, if LICQ holds for (4.43), then,

- the cone (4.43) is equivalent to the set of limiting directions for all feasible sequences,
- for limiting directions d that satisfy (4.43) with $\|d\| = 1$, a feasible sequence $\{x^k\}$ (with $x^k = x^* + t^k d + o(t^k)$) can always be constructed that satisfies

$$\begin{aligned} h_i(x^k) &= t^k \nabla h_i(x^*)^T d = 0, \\ g_i(x^k) &= t^k \nabla g_i(x^*)^T d \leq 0, \quad i \in \mathcal{A}(x^*), \end{aligned}$$

for some small positive t^k with $\lim t^k \to 0$.

This lemma allows us to extend Theorem 4.11 to nonlinearly constrained problems. Combining the above results, we now prove the main results of this section.

Theorem 4.14 Suppose that x^* is a local solution of (4.1) and the LICQ holds at this solution; then

- the KKT conditions (4.42) are satisfied;
- if $f(x)$ and $g(x)$ are convex, $h(x)$ is linear and the KKT conditions (4.42) are satisfied at a point x^+, then x^+ is the global solution to (4.1).

Proof: The proof follows along similar lines as in Theorem 4.11. The constraints (4.42b) are satisfied as x^* is the solution of (4.1). The complementarity conditions (4.42c) are equivalent to the conditions $u_i^* \geq 0, g_i(x^*) = 0, i \in \mathcal{A}(x^*)$, and $u_i^* = 0$ for $i \in \mathcal{I} \setminus \mathcal{A}(x^*)$. By Theorem 4.8, the LICQ, and Lemma 4.13, there is no solution d to the linear system

$$\nabla f(x^*)^T d < 0, \quad \nabla h(x^*)^T d = 0, \quad \nabla g_i(x^*)^T d \leq 0, \quad i \in \mathcal{A}(x^*). \tag{4.45}$$

Hence, from Lemma 4.10 and by setting a to $-\nabla f(x^*)$, D to $\nabla h(x^*)^T$, and A to the rows of $-\nabla g_i(x^*)$, $i \in \mathcal{A}(x^*)$, we have (4.42a) and $u^* \geq 0$.

For the second part of the theorem, we know from Lemma 4.10 that (4.42a) implies that there is no solution d to the linear system

$$\nabla f(x^+)^T d < 0, \quad \nabla h(x^+)^T d = 0, \quad \nabla g_i(x^+)^T d \leq 0, \quad i \in \mathcal{A}(x^+). \tag{4.46}$$

Assume now that x^+ is not a solution to (4.1) and there is another feasible point x^* with $f(x^*) < f(x^+)$. We define $d = (x^* - x^+)$ and note that, by linearity of $h(x)$ and convexity of $g(x)$, we have $\nabla h(x^+)^T d = 0$ and

$$0 \geq g_i(x^*) \geq g_i(x^+) + \nabla g_i(x^+)^T d = \nabla g_i(x^+)^T d, \quad i \in \mathcal{A}(x^+).$$

By convexity of $f(x)$, $\nabla^2 f(x)$ is positive semidefinite for all x, and by Taylor's theorem,

$$0 > f(x^*) - f(x^+) = \nabla f(x^+)^T d + \frac{1}{2}\int_0^1 d^T \nabla^2 f(x^+ + td) d \, dt \tag{4.47}$$

$$\geq \nabla f(x^+)^T d, \tag{4.48}$$

which contradicts the assumption that there is no solution to (4.46). Therefore, x^+ is a local solution to (4.32), and because the objective function and feasible region are both convex, x^+ is also a global solution by Theorem 4.3. □

The assumption of the LICQ leads to an important additional property on the multipliers, given by the following theorem.

Theorem 4.15 (LICQ and Multipliers). Given a point, x^*, that satisfies the KKT conditions, along with an active set $\mathcal{A}(x^*)$ with multipliers u^*, v^*, if LICQ holds at x^*, then the multipliers are unique.

Proof: We define the vector of multipliers $\lambda^* = [u^{*T}, v^{*T}]^T$ and the matrix $A(x^*)$ made up of the columns of active constraint gradients $\nabla g_i(x^*), \nabla h_i(x^*), i \in \mathcal{A}(x^*)$, and write (4.42a) as

$$\nabla f(x^*) + A(x^*)\lambda^* = 0. \tag{4.49}$$

Now assume that there are two multiplier vectors, λ^a and λ^b, that satisfy (4.42) and (4.49). Substituting both multiplier vectors into (4.49), subtracting the two equations from each other, and premultiplying the difference by $A(x^*)^T$ leads to

$$A(x^*)^T A(x^*)(\lambda^a - \lambda^b) = 0.$$

Since the LICQ holds, $A(x^*)^T A(x^*)$ is nonsingular and we must have $\lambda^a = \lambda^b$, which proves the result. □

Finally, it should be noted that while the LICQ is commonly assumed, there are several other constraint qualifications that link Theorem 4.8 to the KKT conditions (4.42) (see [294, 273]). In particular, the weaker Mangasarian–Fromovitz constraint qualification (MFCQ) has the following definition.

Definition 4.16 (MFCQ). Given a local solution of (4.1), x^*, and an active set $\mathcal{A}(x^*)$, the MFCQ is defined by linear independence of the equality constraint gradients and the existence of a search direction d such that

$$\nabla g_i(x^*)^T d < 0, \quad \nabla h_i(x^*)^T d = 0, \quad i \in \mathcal{A}(x^*).$$

The MFCQ is always satisfied if the LICQ is satisfied. Also, satisfaction of the MFCQ leads to bounded multipliers, u^*, v^*, although they are not necessarily unique.

4.3.3 Second Order Conditions

To show the second order optimality conditions (4.12), we first define two cones from the following linear systems:

$$\mathcal{C}_1(x^*) = \{d \mid \nabla h(x^*)^T d = 0, \nabla g_i(x^*)^T d \le 0, i \in \mathcal{A}(x^*)\},$$
$$\mathcal{C}_2(x^*, u^*) = \{d \mid \nabla h(x^*)^T d = 0, \nabla g_i(x^*)^T d = 0,\ i \in \mathcal{A}_s(x^*), \nabla g_i(x^*)^T d \le 0, i \in \mathcal{A}_w(x^*)\}.$$

It is clear that $\mathcal{C}_2(x^*,u^*) \subseteq \mathcal{C}_1(x^*)$. If we assume that LICQ holds we also know from Lemma 4.13 that $\mathcal{C}_1(x^*)$ represents the space of limiting directions to all feasible sequences. We will see that these limiting directions lead to the *allowable directions* p that were presented in Section 4.2. With these definitions we can show the following properties.

Theorem 4.17 (Second Order Necessary Conditions). Suppose that x^* is a local solution of (4.1), LICQ holds at this solution, and u^*, v^* are the multipliers that satisfy the KKT conditions (4.42). Then

$$d^T \nabla_{xx} L(x^*,u^*,v^*) d \geq 0 \quad \text{for all nonzero } d \in \mathcal{C}_2(x^*,u^*). \tag{4.50}$$

Proof: For $d \in \mathcal{C}_2(x^*,u^*)$ it is easy to see that

$$v_i^* \nabla h_i(x^*)^T d = 0 \quad \text{and} \quad u_i^* \nabla g_i(x^*)^T d = 0, \quad i \in \mathcal{A}(x^*).$$

Since LICQ holds from Lemma 4.13, we can construct for any $d \in \mathcal{C}_1(x^*)$ a feasible sequence $\{x^k\}$ (with $x^k = x^* + t^k d + o(t^k)$) that satisfies

$$h_i(x^k) = t^k \nabla h_i(x^*)^T d = 0, \quad g_i(x^k) = t^k \nabla g_i(x^*)^T d \leq 0, \quad i \in \mathcal{A}(x^*) \tag{4.51}$$

for a sequence of scalars $t^k > 0$ with $\lim t^k \to 0$. For this feasible sequence we then have the result

$$\begin{aligned} L(x^k,u^*,v^*) &= f(x^k) + g(x^k)^T u^* + h(x^k)^T v^* \\ &= f(x^k) + t^k d^T (\nabla g(x^*) u^* + \nabla h(x^*) v^*) \\ &= f(x^k) \end{aligned}$$

and, of course, $L(x^*,u^*,v^*) = f(x^*)$.

Since LICQ holds, the cone $\mathcal{C}_1(x^*)$ contains the limiting directions of all feasible sequences, and, as x^* is a solution to (4.1), we know that there can be no increasing feasible sequence to the solution for k sufficiently large.

We now assume that $d^T \nabla_{xx} L(x^*,u^*,v^*) d < 0$ for some limiting direction $d \in \mathcal{C}_2(x^*,u^*) \subseteq \mathcal{C}_1(x^*)$, and from Lemma 4.13 we can construct a feasible sequence $\{x^k\}$ associated with this limiting direction that satisfies (4.51). Applying Taylor's theorem and noting that $\nabla L(x^*,u^*,v^*) = 0$ leads to the following result:

$$f(x^k) = L(x^k,u^*,v^*) \tag{4.52}$$
$$= L(x^*,u^*,v^*) + t^k \nabla L(x^*,u^*,v^*)^T d \tag{4.53}$$
$$+ \frac{(t^k)^2}{2} d^T \nabla_{xx} L(x^*,u^*,v^*) d + o((t^k)^2) \tag{4.54}$$
$$= f(x^*) + \frac{(t^k)^2}{2} d^T \nabla_{xx} L(x^*,u^*,v^*) d + o((t^k)^2) \tag{4.55}$$
$$< f(x^*) + \frac{(t^k)^2}{4} d^T \nabla_{xx} L(x^*,u^*,v^*) d < f(x^*) \tag{4.56}$$

for k sufficiently large. This contradicts the assumption that x^* is a solution and that there is no increasing feasible sequence to the solution. Hence the result (4.50) must hold. $\square$

Theorem 4.18 (Second Order Sufficient Conditions). Suppose that x^* and the multipliers u^*, v^* satisfy the KKT conditions (4.42) and

$$d^T \nabla_{xx} L(x^*, u^*, v^*) d > 0 \quad \text{for all nonzero } d \in \mathcal{C}_2(x^*, u^*); \tag{4.57}$$

then x^* is a strict local solution of (4.1).

Proof: To show that x^* is a strict local minimum, every feasible sequence must have $f(x^k) > f(x^*)$ for all k sufficiently large. From Lemma 4.13, $\mathcal{C}_1(x^*)$ contains the limiting directions for all feasible sequences. We now consider any feasible sequence associated with these limiting directions and consider the following two cases:

- $d \in \mathcal{C}_2(x^*, u^*) \subseteq \mathcal{C}_1(x^*)$: For any feasible sequence with limiting direction $d \in \mathcal{C}_2(x^*, u^*)$, $\|d\| = 1$, we have $h(x^k)^T v^* = 0, g(x^k)^T u^* \leq 0$, and from (4.42) and (4.26),

$$\begin{aligned}
f(x^k) &\geq f(x^k) + h(x^k)^T v^* + g(x^k)^T u^* \\
&= L(x^k, u^*, v^*) \\
&= L(x^*, u^*, v^*) + \frac{1}{2}(x^k - x^*)^T \nabla_{xx} L(x^*, u^*, v^*)(x^k - x^*) \\
&\quad + o(\|x^k - x^*\|^2) \\
&= f(x^*) + \frac{1}{2} d^T \nabla_{xx} L(x^*, u^*, v^*) d \|x^k - x^*\|^2 + o(\|x^k - x^*\|^2) \\
&\geq f(x^*) + \frac{1}{4} d^T \nabla_{xx} L(x^*, u^*, v^*) d \|x^k - x^*\|^2 > f(x^*),
\end{aligned}$$

  where the higher order term can be absorbed by $\frac{1}{4} d^T \nabla_{xx} L(x^*, u^*, v^*) d \|x^k - x^*\|^2$ for k sufficiently large, and the result follows for the first case.

- $d \in \mathcal{C}_1(x^*) \setminus \mathcal{C}_2(x^*, u^*)$: Unlike the first case, $d^T \nabla_{xx} L(x^*, u^*, v^*) d$ may not be positive for $d \in \mathcal{C}_1(x^*) \setminus \mathcal{C}_2(x^*, u^*)$ so a different approach is needed. Since $d \notin \mathcal{C}_2(x^*, u^*)$, there is at least one constraint with $\nabla g_{\bar{i}}(x^*)^T d < 0, \bar{i} \in \mathcal{A}_s(x^*)$. Also, we have $h(x^k)^T v^* = 0$ and for $i \in \mathcal{A}(x^*) \cap \mathcal{I}$,

$$\begin{aligned}
g_i(x^k) u_i^* &= g_i(x^*) u_i^* + \nabla g_i(x^*)^T (x^k - x^*) u_i^* + o(\|x^k - x^*\|) \\
&= \nabla g_i(x^*)^T d \|x^k - x^*\| u_i^* + o(\|x^k - x^*\|).
\end{aligned}$$

  As a result,

$$\begin{aligned}
L(x^k, u^*, v^*) &= f(x^k) + h(x^k)^T v^* + g(x^k)^T u^* \\
&= f(x^k) + \nabla g_{\bar{i}}(x^*)^T d \|x^k - x^*\| u_{\bar{i}}^* + o(\|x^k - x^*\|).
\end{aligned}$$

  From Taylor's theorem and the KKT conditions (4.42) we have

$$L(x^k, u^*, v^*) = f(x^*) + O(\|x^k - x^*\|^2).$$

  Equating the two relations for $L(x^k, u^*, v^*)$ leads to

$$f(x^k) - f(x^*) + \nabla g_{\bar{i}}(x^*)^T d \|x^k - x^*\| u_{\bar{i}}^* = o(\|x^k - x^*\|).$$

The right-hand side can be absorbed into $-\frac{1}{2}\nabla g_{\bar{i}}(x^*)^T d\|x^k - x^*\|u^*_{\bar{i}}$, for all k sufficiently large, leading to

$$f(x^k) - f(x^*) \geq -\frac{1}{2}\nabla g_{\bar{i}}(x^*)^T d\|x^k - x^*\|u^*_{\bar{i}} > 0$$

which gives the desired result for the second case. □

Properties of Reduced Hessians

The necessary and sufficient conditions developed above are not always easy to check as they require an evaluation of (4.50) or (4.57) for all directions $d \in \mathcal{C}_2(x^*)$. However, these conditions can be simplified through the evaluation of the Hessian matrix projected into a subspace. To see this, we use the following definition.

Definition 4.19 Let A be an $n \times m$ matrix with full column rank; then we can define the null space basis matrix, $Z \in \mathbb{R}^{n \times (n-m)}$, of full column rank that satisfies $A^T Z = 0$.

A straightforward way to determine the null space basis matrix is through a QR factorization of A. This leads to

$$A = Q\begin{bmatrix} R \\ 0 \end{bmatrix} = [Q^R | Q^N]\begin{bmatrix} R \\ 0 \end{bmatrix} = Q^R R, \tag{4.58}$$

where $Q \in \mathbb{R}^{n \times n}$ is an orthonormal matrix, $Q^R \in \mathbb{R}^{n \times m}$, $Q^N \in \mathbb{R}^{n \times (n-m)}$, and $R \in \mathbb{R}^{m \times m}$ is an upper triangular matrix. From this representation it is easy to see that $(Q^R)^T Q^R = I_m$, $(Q^N)^T Q^N = I_{n-m}$, $(Q^R)^T Q^N = 0$, $A^T Q^N = 0$, and the vector d that satisfies $A^T d = 0$ can be represented as $d = Q^N q$ with $q \in \mathbb{R}^{n-m}$. Moreover, for a matrix $W \in \mathbb{R}^{n \times n}$ with $d^T W d > 0$ (≥ 0) for all $A^T d = 0$, we have from $d^T W d = q^T[(Q^N)^T W Q^N]q$ that the *projected matrix* $(Q^N)^T W Q^N$ is positive (semi)definite.

We can derive conditions on the *projected Hessian matrix* by introducing two additional sets that are made up of linear subspaces:

$$\mathcal{C}_3(x^*) = \{d | \nabla h(x^*)^T d = 0, \nabla g_i(x^*)^T d = 0, i \in \mathcal{A}(x^*)\}$$

and

$$\mathcal{C}_4(x^*, u^*) = \{d | \nabla h(x^*)^T d = 0, \nabla g_i(x^*)^T d = 0, i \in \mathcal{A}_s(x^*)\}.$$

With these concepts we can state the following properties:

- The subspaces satisfy $\mathcal{C}_3(x^*) \subseteq \mathcal{C}_4(x^*, u^*)$ and $\mathcal{C}_2(x^*, u^*) \subseteq \mathcal{C}_4(x^*, u^*)$.

- Using $\mathcal{C}_3(x^*)$ we can define the matrix A_3 from the columns of $\nabla h(x^*)$ and $\nabla g_i(x^*)$, $i \in \mathcal{A}(x^*)\}$, and the corresponding null space basis matrix, Q_3^N, with analogous definitions for A_4 and Q_4^N using $\mathcal{C}_4(x^*, u^*)$. All directions $d \in \mathcal{C}_3(x^*)$ ($d \in \mathcal{C}_4(x^*, u^*)$) can be represented by $d = Q_3^N q$ ($d = Q_4^N q$).

- Because $d \in \mathcal{C}_2(x^*, u^*) \subseteq \mathcal{C}_4(x^*, u^*)$, then if $d^T \nabla_{xx} L(x^*, u^*, v^*) d > 0$, $d \in \mathcal{C}_4(x^*, u^*)$, then the sufficient second order condition (4.57) is also satisfied. This is equivalent to positive definiteness of $(Q_4^N)^T \nabla_{xx} L(x^*, u^*, v^*) Q_4^N$. Thus, positive definiteness of this projected Hessian is a (stronger) sufficient second order condition.

- Because $d \in \mathcal{C}_3(x^*) \subseteq \mathcal{C}_2(x^*, u^*)$, then if (4.50) is satisfied, then it must also satisfy $d^T \nabla_{xx} L(x^*, u^*, v^*) d \geq 0$ for $d \in \mathcal{C}_3(x^*) \subseteq \mathcal{C}_2(x^*, u^*)$. This is equivalent to $(Q_3^N)^T \nabla_{xx} L(x^*, u^*, v^*) Q_3^N$ positive semidefinite. Thus positive semidefiniteness of this projected Hessian is a (weaker) necessary second order condition.

- If *strict complementarity* from Definition 4.6 holds at the solution of (4.1), then $\mathcal{A}_s(x^*) = \mathcal{A}(x^*)$, $\mathcal{A}_w(x^*)$ is empty, and $\mathcal{C}_2(x^*, u^*) = \mathcal{C}_3(x^*, u^*) = \mathcal{C}_4(x^*, u^*)$. Under the strict complementarity assumption, the second order conditions are simplified as follows:

  - *Necessary second order condition:* $(Q_3^N)^T \nabla_{xx} L(x^*, u^*, v^*) Q_3^N$ is *positive semidefinite.*
  - *Sufficient second order condition:* $(Q_3^N)^T \nabla_{xx} L(x^*, u^*, v^*) Q_3^N$ is *positive definite.*

To illustrate these properties, we revisit the remaining case related to Example 4.5.

Example 4.20 Consider the constrained NLP

$$\min \quad x_1^2 - 4x_1 + \frac{3}{2}x_2^2 - 7x_2 + x_1 x_2 + 9 - \ln(x_1) - \ln(x_2) \tag{4.59}$$
$$\text{s.t. } 4 - x_1 x_2 \leq 0,$$

which corresponds to the plot in Figure 4.6. The optimal solution can be found by applying the first order KKT conditions (4.42) as follows:

$$\nabla L(x^*, u^*, v^*) = \nabla f(x^*) + \nabla g(x^*) u^* \tag{4.60}$$
$$= \begin{bmatrix} 2x_1 - 4 + x_2 - \frac{1}{x_1} - x_2 u \\ 3x_2 - 7 + x_1 - \frac{1}{x_2} - x_1 u \end{bmatrix} = 0, \tag{4.61}$$
$$g(x) = 4 - x_1 x_2 \leq 0, \quad u^T(4 - x_1 x_2) = 0, \quad u \geq 0 \tag{4.62}$$
$$\Downarrow \tag{4.63}$$
$$x^* = \begin{bmatrix} 1.7981 \\ 2.2245 \end{bmatrix}, \quad u^* = 0.5685, \tag{4.64}$$

$f(x^*) = -2.4944$, and we have strict complementarity at x^*. Checking the second order conditions (4.12) leads to

$$\nabla_{xx} L(x^*, u^*, v^*) = \nabla^2 f(x^*) + \nabla^2 g(x^*) u^* \tag{4.65}$$
$$= \begin{bmatrix} 2 + \frac{1}{(x_1^*)^2} & 1 - u^* \\ 1 - u^* & 3 + \frac{1}{(x_2^*)^2} \end{bmatrix} = \begin{bmatrix} 2.3093 & 0.4315 \\ 0.4315 & 3.2021 \end{bmatrix}. \tag{4.66}$$

Because of strict complementarity,

$$A = \nabla g(x^*) = \begin{bmatrix} -2.2245 \\ -1.7981 \end{bmatrix} \text{ and } Q^N = Q_3^N = Q_4^N = \begin{bmatrix} -0.6286 \\ 0.7777 \end{bmatrix}. \tag{4.67}$$

This leads to $(Q^N)^T \nabla_{xx} L(x^*, u^*, v^*) Q^N = 2.4272$, and the sufficient second order condition holds. ∎

4.4 Special Cases: Linear and Quadratic Programs

Linear programs and quadratic programs are special cases of (4.1) that allow for more efficient solution, based on application of KKT conditions (4.37). This chapter provides only a brief, descriptive coverage of these problem types as there is a vast literature and excellent textbooks in these areas including [108, 122, 134, 195, 287, 294, 332, 411]. The interested reader is referred to these sources for more detailed information on the formulation and solution of these problem types. Because these special cases are convex problems, any locally optimal solution is a global solution.

4.4.1 Description of Linear Programming

If the objective and constraint functions in (4.1) are linear, then the problem can be formulated as a linear program, given in a general way as

$$\min \bar{a}^T y \text{ s.t. } By \leq b. \tag{4.68}$$

Linear programs can be formulated in a number of ways to represent the equivalent problem. In fact, by adding additional variables to (4.68), we can convert the inequalities to equalities and add bounds on the variables. Moreover, by adding or subtracting (possibly large) fixed constants to the variables, one can instead impose simple nonnegativity on *all* the variables and write the LP as

$$\begin{aligned} \min \quad & a^T x \\ \text{s.t.} \quad & Cx = c, \quad x \geq 0, \end{aligned} \tag{4.69}$$

with $a, x \in \mathbb{R}^n$, $C \in \mathbb{R}^{m \times n}$, and $n \geq m$. We will use this form as it is convenient to interpret the solution method for LPs.

In analogy with Section 4.2, the feasible region of (4.69) can be visualized as a box with many "flat" sides; this is called a *polyhedron*. If the box is bounded below, then a ball dropped in this box will find its lowest point at a corner or *vertex*. Thus, an optimal solution of (4.69) is defined by one or more vertices of the polyhedron described by linear constraints. This is shown in Figure 4.1, and in so-called *primal degenerate* cases, multiple vertices can be alternate optimal solutions, with the same values of the objective function. The KKT conditions of (4.69) are

$$a + C^T v^* - u^* = 0, \tag{4.70a}$$

$$Cx^* = c, \quad x^* \geq 0, \tag{4.70b}$$

$$(x^*)^T u^* = 0, \quad u^* \geq 0. \tag{4.70c}$$

Since $\nabla_{xx} L(x^*, u^*, v^*)$ is identically zero, it is easy to see that the necessary second order conditions (4.50) always hold when (4.70) holds. Moreover, at a nondegenerate solution there are no nonzero vectors that satisfy $d \in \mathcal{C}_2(x^*, u^*)$ and the sufficient second order conditions (4.57) hold *vacuously*.

Problem (4.69) can be solved in a finite number of steps. The standard method used to solve (4.69) is the *simplex method*, developed in the late 1940s [108] (although, starting from Karmarkar's discovery in 1984, interior point methods have become quite advanced and competitive for highly constrained problems [411]). The simplex method proceeds by moving successively from vertex to vertex with improved objective function values. At each vertex point, we can repartition the variable vector into $x^T = [x_N^T, x_B^T]$ and C into submatrices $C = [C_N \mid C_B]$ with corresponding columns. Here x_N is the subvector of $n-m$ *nonbasic variables* which are set to zero, and x_B is the subvector of m *basic variables*, which are determined by the square system $C_B x_B = c$. At this vertex, directions to adjacent vertices are identified (with different basic and nonbasic sets) and directional derivatives of the objective are calculated (the so-called *pricing step*). If all of these directional derivatives are positive (nonnegative), then one can show that the KKT conditions (4.70) and the sufficient (necessary) second order conditions are satisfied.

Otherwise, an adjacent vertex associated with a negative directional derivative is selected where a nonbasic variable (the *driving* variable) is increased from zero and a basic variable (the *blocking* variable) is driven to zero. This can be done with an efficient pivoting operation, and a new vertex is obtained by updating the nonbasic and basic sets by swapping the driving and blocking variables in these sets. The sequence of these simplex steps leads to vertices with decreasing objective functions, and the algorithm stops when no adjacent vertices can be found that improve the objective. More details of these simplex steps can be found in [108, 195, 287, 294].

Methods to solve (4.68) are well implemented and widely used, especially in planning and logistical applications. They also form the basis for mixed integer linear programming methods. Currently, state-of-the-art LP solvers can handle millions of variables and constraints and the application of specialized decomposition methods leads to the solution of problems that are even two or three orders of magnitude larger than this.

4.4.2 Description of Quadratic Programming

Quadratic programs (QPs) represent a slight modification of (4.68) and can be given in the following general form:

$$\min\ \bar{a}^T y + \frac{1}{2} y^T \bar{Q} y \text{ s.t. } By \leq b. \tag{4.71}$$

As with LPs, QPs can be formulated in a number of equivalent ways. Thus, by adding additional variables to (4.71), converting inequalities to equalities, and adding bounds on the variables, we can rewrite this QP as

$$\begin{aligned} \min \quad & a^T x + \frac{1}{2} x^T Q x \\ \text{s.t.} \quad & Cx = c, \quad x \geq 0, \end{aligned} \tag{4.72}$$

with $Q \in \mathbb{R}^{n \times n}$. The corresponding KKT conditions for (4.72) are

$$a + Qx^* + C^T v^* - u^* = 0, \tag{4.73a}$$
$$Cx^* = c, \quad x^* \geq 0, \tag{4.73b}$$
$$(x^*)^T u^* = 0, \quad u^* \geq 0. \tag{4.73c}$$

If the matrix Q is positive semidefinite (positive definite) when projected into the null space of the active constraints, then (4.72) is (strictly) convex and (4.73) provides a global (and unique) minimum. Otherwise, multiple local solutions may exist for (4.72) and more extensive global optimization methods are needed to obtain the global solution. Like LPs, convex QPs can be solved in a finite number of steps. However, as seen in Figure 4.2, these optimal solutions can lie on a vertex, on a constraint boundary, or in the interior.

Solution of QPs follows in a manner similar to LPs. For a fixed active set, i.e., nonbasic variables identified and conditions (4.73c) suppressed, the remaining equations constitute a linear system that is often solved directly. The remaining task is to choose among active sets (i.e., nonbasic variables) that satisfy (4.73c), much like the simplex method. Consequently, the QP solution can also be found in a finite number of steps. A number of active set QP strategies have been created that provide efficient updates of active constraints. Popular methods include null-space algorithms, range-space methods, and Schur complement methods. As with LPs, QP problems can also be solved with interior point methods. A thorough discussion of QP algorithms is beyond the scope of this text. For more discussion, the interested reader is referred to [195, 287, 294, 134, 411], as well as the references to individual studies cited therein.

4.4.3 Portfolio Planning Case Study

To illustrate the nature of LP and QP solutions, we consider a simplified financial application. Consider a set of investments $\mathcal{N}$ along with their rates of return $r_i(t)$, $i \in \mathcal{N}$, as a function of time and the average rate of return:

$$\rho_i = \frac{1}{N_t} \sum_{j=1}^{N_t} r_i(t_j), \tag{4.74}$$

where N_t is the number of time periods. We now consider a choice of a portfolio that maximizes the rate of return, and this can be posed as the following LP:

$$\begin{aligned} \max \quad & \sum_{i \in \mathcal{N}} \rho_i x_i \\ \text{s.t.} \quad & \sum_{i \in \mathcal{N}} x_i = 1, \\ & 0 \leq x_i \leq x_{\max}, \end{aligned} \tag{4.75}$$

where x_i is the fraction of the total portfolio invested in investment i and we can choose a maximum amount $x_{\max}$ to allow for some diversity in the portfolio.

However, the LP formulation assumes that there is no risk to these investments, and experience shows that high-yielding investments are likely to be riskier. To deal with risk, one can calculate the variance of investment i as well as the covariance between two investments i and i', given by

$$s_{i,i} = \frac{1}{N_t - 1} \sum_{j=1}^{N_t} (r_i(t_j) - \rho_i)^2, \tag{4.76a}$$

$$s_{i,i'} = \frac{1}{N_t - 1} \sum_{j=1}^{N_t} (r_i(t_j) - \rho_i)(r_{i'}(t_j) - \rho_{i'}), \tag{4.76b}$$

respectively; these quantities form the elements of the matrix S. With this information one can adopt the Markowitz mean/variance portfolio model [275], where the least risky

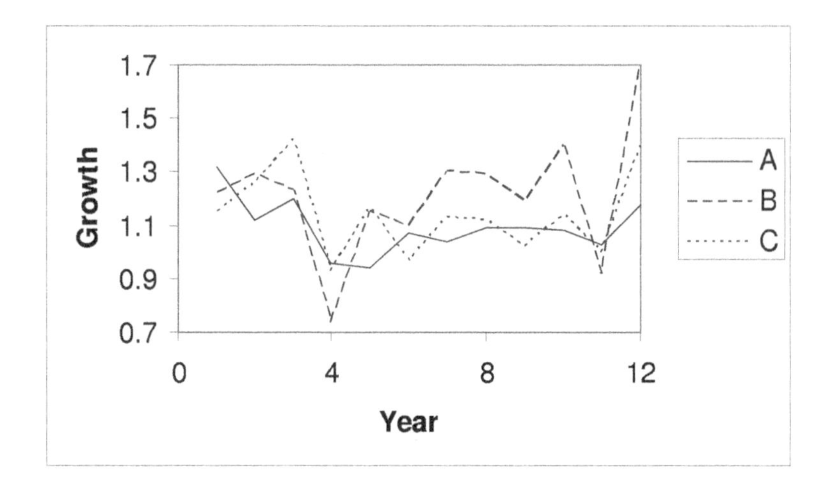

Figure 4.9. *Annual growth in stock prices.*

portfolio is determined that provides (at least) a desired rate of return. This can be written as the following QP:

$$\begin{aligned}
\min \quad obj \equiv \sum_{i,i' \in \mathcal{N}} s_{i,i'} x_i x_{i'} \qquad (4.77)\\
\text{s.t.} \quad \sum_{i \in \mathcal{N}} x_i = 1,\\
\sum_{i \in \mathcal{N}} \rho_i x_i \geq \rho_{\min},\\
0 \leq x_i \leq x_{\max}.
\end{aligned}$$

In addition to these basic formulations there are a number of related LP and QP formulations that allow the incorporation of uncertainties, transaction costs, after-tax returns, as well as differentiation between upside and downside risk. These models constitute important topics in financial planning, and more information on these formulations can be found in [361] and the many references cited therein. The following example illustrates the nature of the solutions to these basic portfolio problems.

Example 4.21 Consider a set with four investments, $\mathcal{N} = \{A, B, C, D\}$. The first three represent stocks issued by large corporations, and the growth over 12 years in their stock prices with dividends, i.e., $(1 + r_i(t))$, is plotted in Figure 4.9. From these data we can apply (4.74) and (4.76) to calculate the mean vector and covariance matrix, respectively, for portfolio planning. Investment D is a risk-free asset which delivers a constant rate of return and does not interact with the other investments. The data for these investments are given in Table 4.1.

Now consider the results of the LP and QP planning strategies. Using the data in Table 4.1, we can solve the LP (4.75) for values of $x_{\max}$ from 0.25 (the lowest value with a feasible solution for (4.75)) to 1.0. This leads to the portfolio plotted in Figure 4.10 with

Table 4.1. *Data for portfolio planning QP.*

Investment (i)	ρ_i	$s_{i,A}$	$s_{i,B}$	$s_{i,C}$	$s_{i,D}$
A	0.0933	0.01090	0.01221	0.00865	0
B	0.2108	0.01221	0.05624	0.02716	0
C	0.1425	0.00865	0.02716	0.02444	0
D	0.03	0	0	0	0

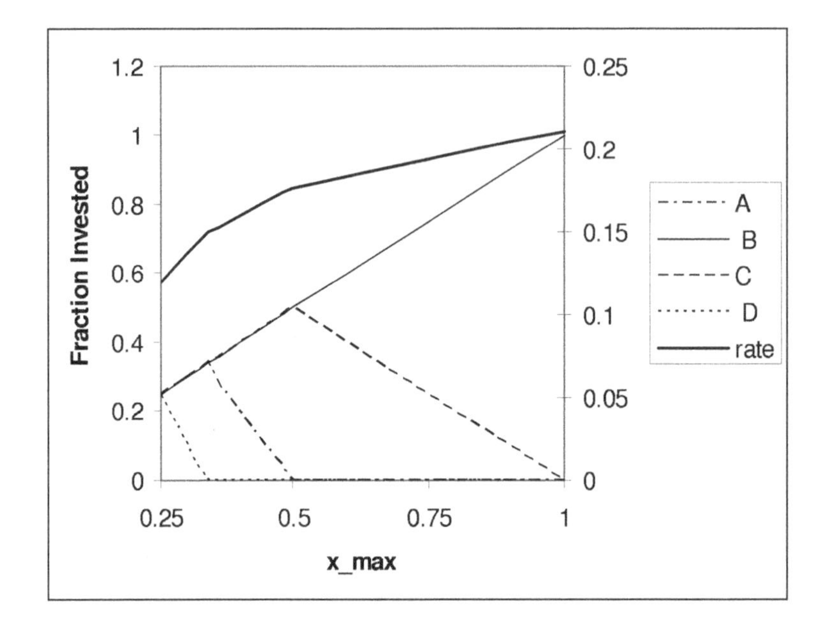

Figure 4.10. *Solutions of the LP portfolio planning model.*

rates of return that vary from 0.1192 to 0.2108. From the figure we see that the optimum portfolio follows a straightforward strategy: put as much as possible ($x_{\max}$) of the fund into the highest yielding investment, then as much of the remainder into the next highest, and so on, until all of the fund is invested. This is consistent with the nature of LPs, where vertices are always optimal solutions. Moreover, as $x_{\max}$ is increased, we see that the fraction of low-yielding investments is reduced and these eventually drop out, while the rate of return on the portfolio increases steadily.

However, B is also the riskiest investment, so the LP solution may not be the best investment portfolio. For the quadratic programming formulation (4.77), one can verify that the S matrix from Table 4.1 is positive semidefinite (with eigenvalues 0, 0.0066, 0.0097, and 0.0753), the QP is convex, and hence its solutions will be global. Here we set $x_{\max} = 1$ and use the data in Table 4.1 to solve this QP for different levels of $\rho_{\min}$. This leads to the solutions shown in Figure 4.11. Unlike the LP solution, the results at intermediate interest rates reflect a mixed or diversified portfolio with minimum risk. As seen in the figure, the amount of risk, obj, increases with the desired rate of return, leading to a trade-off

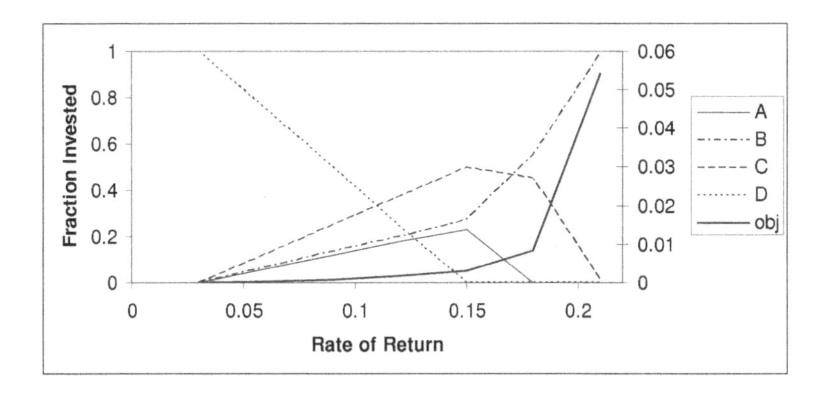

Figure 4.11. *Solutions of the QP portfolio planning model.*

curve between these two measures. One can observe that from $\rho_{\min} = 0.03$ to $\rho_{\min} = 0.15$, the optimal portfolio always has the same ratios of stocks A, B, and C with the fraction invested in D as the only change in the portfolio (see [361] for an interesting discussion on this point). Note also that the risk-free investment disappears at $\rho_{\min} = 0.15$ while the lowest-yielding stock A drops out at $\rho_{\min} = 0.18$, leaving only B and C in the portfolio. At these points, we observe steep rises in the objective. Finally, at $\rho_{\min} = 0.2108$ only B, the highest-yielding stock, remains in the portfolio. ∎

4.5 Summary and Notes for Further Reading

This chapter develops the main concepts that characterize constrained optimal solutions. Unlike the concepts of unconstrained problems developed in Chapter 2, the nature of constrained solutions is considerably more complicated. This was observed in a number of areas. First, the relationship between convexity and global optima is extended to include the nature of the feasible region. Second, stationary points need to be defined in terms of active constraints and the objective function. Third, second order necessary and sufficient directions need to be developed using constrained, i.e., allowable, directions. Local optimality conditions were first described using geometric concepts and then analyzed using the classic development of the KKT conditions. Some care is needed in developing subtle issues including constraint qualifications as well as allowable directions for second order conditions. Special cases of nonlinear programming, i.e., linear and convex quadratic programming, were also considered as they lead to noniterative algorithms that converge to global solutions. Much more information on these problem classes can be found in classic texts on optimization and operations research including [108, 195, 287, 411].

Many of the concepts in this chapter are developed in Nocedal and Wright [294]. Also, they provide additional detail in relating tangent cones to limiting feasible directions for nonlinear problems. A similar approach is given in Fletcher [134] and in Gill, Murray, and Wright [162]. The celebrated KKT conditions were presented by Kuhn and Tucker [236] in 1951, but it was later discovered that these conditions were developed 12 years earlier in an unpublished master's thesis by Karush [224]. An important tool in the development of KKT conditions is the theorem of the alternative (TOA). Moreover, there is a broad set of constraint qualifications (CQ) that lead to weaker assumptions on the nonlinear program to apply optimality conditions. Both TOAs and CQs are explored in depth in the classic text by Mangasarian [273].

4.6 Exercises

1. Show that the "no overlap constraints" in Example 4.4 are not convex.

2. Derive the KKT conditions for Example 4.4.

3. Show that the nonlinear program (4.44) does not satisfy the KKT conditions.

4. Consider the convex problem

$$\min\ x_1 \text{ s.t. } x_2 \le 0, \quad x_1^2 - x_2 \le 0.$$

   Show that this problem does not satisfy the LICQ and does not satisfy the KKT conditions at the optimum.

5. Investigate Lemma 12.2 in [294] and explain why the LICQ is essential to the proof of Lemma 4.13.

6. Apply the second order conditions to both parts of Example 4.7. Define the tangent cones for this problem.

7. Convert (4.68) to (4.69) and compare KKT conditions of both problems.

8. In the derivation of the Broyden update, the following convex equality constrained problem is solved:

$$\begin{aligned} &\min \| J^+ - J \|_F^2 \\ &\text{s.t. } J^+ s = y. \end{aligned}$$

   Using the definition of the Frobenius norm from Section 2.2.1, apply the optimality conditions to the elements of J^+ and derive the Broyden update $J^+ = J + \frac{(y - Js)s^T}{s^T s}$.

9. Convert (4.1) to

$$\begin{aligned} \min \quad & f(x) \\ \text{s.t.} \quad & g(x) + s = 0, \; h(x) = 0, \quad s \ge 0, \end{aligned} \tag{4.78}$$

   and compare KKT conditions of both problems. If (4.1) is a convex problem, what can be said about the global solution of (4.78)?

10. A widely used trick is to convert (4.1) into an equality constrained problem by adding new variables z_j to each inequality constraint to form $g_j(x) - (z_j)^2 = 0$. Compare the KKT conditions for the converted problem with (4.1). Discuss any differences between these conditions as well as the implications of using the converted form within an NLP solver.

Chapter 5

Newton Methods for Equality Constrained Optimization

This chapter extends the Newton-based algorithms in Chapter 3 to deal with equality constrained optimization. It generalizes the concepts of Newton iterations and associated globalization strategies to the optimality conditions for this constrained optimization problem. As with unconstrained optimization we focus on two important aspects: solving for the Newton step and ensuring convergence from poor starting points. For the first aspect, we focus on properties of the KKT system and introduce both full- and reduced-space approaches to deal with equality constraint satisfaction and constrained minimization of the objective function. For the second aspect, the important properties of penalty-based merit functions and filter methods will be explored, both for line search and trust region methods. Several examples are provided to illustrate the concepts developed in this chapter and to set the stage for the nonlinear programming codes discussed in Chapter 6.

5.1 Introduction to Equality Constrained Optimization

The previous chapter deals with optimality conditions for the nonlinear program (4.1). This chapter considers solution strategies for a particular case of this problem. Here we consider the equality constrained NLP problem of (5.1) given by

$$\begin{aligned} \min \quad & f(x) \\ \text{s.t.} \quad & h(x) = 0, \end{aligned} \tag{5.1}$$

and we assume that the functions $f(x) : \mathbb{R}^n \to \mathbb{R}$ and $h(x) : \mathbb{R}^n \to \mathbb{R}^m$ have continuous first and second derivatives. First, we note that if the constraints $h(x) = 0$ are linear, then (5.1) is a convex problem if and only if $f(x)$ is convex. On the other hand, as discussed in Chapter 4, nonlinear equality constraints imply nonconvex problems even if $f(x)$ is convex. As a result, *the presence of nonlinear equalities leads to violation of convexity properties and will not guarantee that local solutions to* (5.1) *are global minima.* Therefore, unless additional information is known about (5.1), we will be content in this chapter to determine only local minima.

In analogy to Chapter 3, we need to consider the following questions:

- When is the Newton iteration well defined, and can we maintain a fast (local) convergence rate to the solution of (5.1)?
- Can we converge to local solutions from distant starting points?

The next section analyzes the Newton iteration through a characterization of the so-called KKT matrix. Here we consider properties that lead to a nonsingular matrix and are also compatible with the first and second order optimality conditions. Moreover, some important remedies are discussed when these properties are not satisfied. Related to these properties, Section 5.3 discusses the solution of the Newton step for full- and reduced-space approaches. Section 5.4 then deals with the extension of these methods if second derivatives are not available. As in Chapter 3, we apply quasi-Newton updates that provide curvature information based on differences in gradient vectors and adapt these updates to constrained optimization.

We then consider convergence from poor starting points using globalization strategies that extend from the line search and trust region strategies discussed in Chapter 3. In contrast to that chapter, progress in solving (5.1) requires satisfaction of the equality constraints along with constrained minimization of $f(x)$. Consequently, Section 5.5 discusses concepts of merit functions and filter methods which measure satisfaction of both goals. Sections 5.6 and 5.7 discuss the application of line search and trust region methods, respectively, that enforce these measures and lead to convergence to local solutions. Finally, reconciliation of these globalization methods with fast Newton steps requires additional analysis. Section 5.8 discusses the Maratos effect which can lead to slow convergence, along with the remedy of second order corrections, which is essential to prevent premature termination of these methods. The treatment of these concepts leads to a family of powerful methods to solve (5.1) and also lays the framework to deal with more general NLP methods in Chapter 6.

5.2 Newton's Method with the KKT Matrix

The first order KKT conditions for local optimality can be simplified for equality constraints. Defining the Lagrange function as $L(x,v) = f(x) + v^T h(x)$, we can write the KKT conditions as

$$\begin{aligned} \nabla_x L(x^*, v^*) = \nabla f(x^*) + \nabla h(x^*) v^* = 0, \\ h(x^*) = 0. \end{aligned} \tag{5.2}$$

As noted in the previous chapter, because linearized information is used to characterize the optimum of a *nonlinearly constrained* problem, a *constraint qualification* must be satisfied at x^* so that (5.2) are necessary conditions for a local optimum for (5.1). For this problem we again require the LICQ: *equality constraint gradients at x^* must be linearly independent, i.e., the matrix $\nabla h(x^*)$ has full column rank.* With this constraint qualification, the KKT multipliers (v^*) are unique at the optimal solution.

The sufficient (necessary) second order conditions state that for x^* and v^* satisfying first order conditions, positive (nonnegative) curvature holds for all constrained directions, p. For NLP (5.1) the sufficient conditions require

$$p^T \nabla_{xx} L(x^*, v^*) p > 0 \quad \text{for all } p \neq 0, \quad \nabla h(x^*)^T p = 0. \tag{5.3}$$

The corresponding necessary condition requires that the inequality in (5.3) be nonstrict.

Finding a local solution of (5.1) can therefore be realized by solving (5.2) and then checking the second order conditions (5.3). In this chapter we develop Newton-based strategies for this task. As with the Newton methods in Chapter 3, a number of important concepts need to be developed.

Solution of (5.2) with Newton's method relies on the generation of Newton steps from the following linear system:

$$\begin{bmatrix} W^k & A^k \\ (A^k)^T & 0 \end{bmatrix} \begin{bmatrix} d_x \\ d_v \end{bmatrix} = - \begin{bmatrix} \nabla L(x^k, v^k) \\ h(x^k) \end{bmatrix}, \tag{5.4}$$

where $W^k = \nabla_{xx} L(x^k, v^k)$ and $A^k = \nabla h(x^k)$. Defining the new multiplier estimate as $\bar{v} = v^k + d_v$ and substituting into (5.4) leads to the equivalent system:

$$\begin{bmatrix} W^k & A^k \\ (A^k)^T & 0 \end{bmatrix} \begin{bmatrix} d_x \\ \bar{v} \end{bmatrix} = - \begin{bmatrix} \nabla f(x^k) \\ h(x^k) \end{bmatrix}. \tag{5.5}$$

Note that (5.5) are also the first order KKT conditions of the following quadratic programming problem:

$$\begin{aligned} \min_{d_x} \quad & \nabla f(x^k)^T d_x + \frac{1}{2} d_x^T W^k d_x \\ \text{s.t.} \quad & h(x^k) + (A^k)^T d_x = 0. \end{aligned} \tag{5.6}$$

Using either (5.4) or (5.5), the basic Newton method can then be stated as follows.

Algorithm 5.1.
Choose a starting point (x^0, v^0).

For $k \geq 0$ while $\|d_x^k\|, \|d_v^k\| > \epsilon_1$ and $\max(\|\nabla L(x^k, v^k)\|, \|h(x^k)\|) > \epsilon_2$:

1. Evaluate $h(x^k)$, $\nabla f(x^k)$, A^k, and W^k at x^k and v^k. If the matrix in (5.4), the so-called KKT matrix, is singular, STOP.
2. Solve the linear system (5.4) (or equivalently, (5.5)).
3. Set $x^{k+1} = x^k + d_x$, $v^{k+1} = v^k + d_v$ (or equivalently, $v^{k+1} = \bar{v}$), and $k = k+1$.

As with unconstrained optimization, we would expect this method to have the following convergence property near the optimal solution.

Theorem 5.1 Consider a solution x^*, v^*, which satisfies the sufficient second order conditions and $\nabla h(x^*)$ is full column rank (LICQ). Moreover, assume that $f(x)$ and $h(x)$ are twice differentiable and $\nabla^2 f(x)$ and $\nabla^2 h(x)$ are Lipschitz continuous in a neighborhood of this solution. Then, by applying Algorithm 5.1, and with x^0 and v^0 sufficiently close to x^* and v^*, there exists a constant $\hat{C} > 0$ such that

$$\left\| \begin{bmatrix} x^{k+1} - x^* \\ v^{k+1} - v^* \end{bmatrix} \right\| \leq \hat{C} \left\| \begin{bmatrix} x^k - x^* \\ v^k - v^* \end{bmatrix} \right\|^2, \tag{5.7}$$

i.e., the convergence rate for $\{x^k, v^k\}$ is quadratic.

The proof of this theorem follows the proof of Theorem 2.20 as long as the KKT matrix in (5.4) is nonsingular at the solution (see Exercise 1). In the remainder of this section we consider properties of the KKT matrix and its application in solving (5.1).

5.2.1 Nonsingularity of KKT Matrix

We begin with the following properties.

Theorem 5.2 (Sylvester's Law of Inertia [169]).

- The inertia of a symmetric matrix B is defined as the triple that contains the numbers of positive, negative, and zero eigenvalues of B, i.e., $In(B) = (n_+, n_-, n_0)$.
- *Sylvester's law of inertia* states that the inertia of a symmetric matrix is invariant under the so-called congruence transformations. That is, given a square nonsingular matrix S, the inertia of $S^T BS$ is the same as the inertia of B.
- It follows that if $S^T BS$ is nonsingular, then so is B.

We now consider a local solution of (5.1), x^*, v^*, that satisfies first and sufficient second order KKT conditions as well as the LICQ. Since $A^* = \nabla h(x^*)$ is full column rank, we can use Definition 4.19 to obtain, $Z^* \in \mathbb{R}^{n\times(n-m)}$, a full rank, null space basis matrix, with $(A^*)^T Z^* = 0$. Moreover, at iteration k we will also assume that A^k is full column rank and similarly define Z^k with $(A^k)^T Z^k = 0$. Finally, we define other basis matrices $Y^*, Y^k \in \mathbb{R}^{n\times m}$ so that the $n \times n$ matrices $[Y^*\ Z^*]$ and $[Y^k\ Z^k]$ are nonsingular.

Partitioning the search direction into two components, $d_x = Y^* p_Y + Z^* p_Z$, we can apply the transformation to (5.4) at the solution of (5.1) to obtain

$$\begin{bmatrix} [Y^*\ Z^*]^T & 0 \\ 0 & I \end{bmatrix} \begin{bmatrix} W^* & A^* \\ (A^*)^T & 0 \end{bmatrix} \begin{bmatrix} [Y^*\ Z^*] & 0 \\ 0 & I \end{bmatrix} \begin{bmatrix} p_Y \\ p_Z \\ d_v \end{bmatrix} = - \begin{bmatrix} (Y^*)^T \nabla L(x^*, v^*) \\ (Z^*)^T \nabla L(x^*, v^*) \\ h(x^*) \end{bmatrix}, \tag{5.8}$$

or simply

$$\begin{bmatrix} (Y^*)^T W^* Y^* & (Y^*)^T W^* Z^* & (Y^*)^T A^* \\ (Z^*)^T W^* Y^* & (Z^*)^T W^* Z^* & 0 \\ (A^*)^T Y^* & 0 & 0 \end{bmatrix} \begin{bmatrix} p_Y \\ p_Z \\ d_v \end{bmatrix} = - \begin{bmatrix} (Y^*)^T \nabla L(x^*, v^*) \\ (Z^*)^T \nabla L(x^*, v^*) \\ h(x^*) \end{bmatrix}. \tag{5.9}$$

This linear system can be characterized by the following properties.

- Because $[Y^*\ Z^*]$ is nonsingular and A^* is full column rank, $(A^*)^T [Y^*\ Z^*] = [(A^*)^T Y^* \mid 0]$ is full row rank, i.e., $(A^*)^T Y^*$ is nonsingular.
- From the sufficient second order conditions (5.3), we can define

  $$p = Z^* \bar{p} \quad \text{and} \quad \bar{p}^T (Z^*)^T \nabla_{xx} L(x^*, v^*) Z^* \bar{p} > 0$$

  for all $\bar{p} \in \mathbb{R}^{n-m}$. This is equivalent to positive definiteness of $(Z^*)^T W^* Z^*$.

- Because the KKT conditions (5.2) are satisfied at x^*, v^*, the right-hand side of (5.9) equals zero.

As a result, we can use the bottom row of (5.9) to solve uniquely for $p_Y = 0$. Then, from the second row of (5.9), we can solve uniquely for $p_Z = 0$. Finally, from the first row of (5.9), we solve uniquely for $d_v = 0$. This unique solution implies that the matrix in (5.9) is nonsingular, and from *Sylvester's law of inertia* we have that the KKT matrix in (5.4) is nonsingular as well at x^* and v^*. Note that nonsingularity is a key property required for the proof of Theorem 5.1.

5.2.2 Inertia of KKT Matrix

We now consider additional information regarding the KKT matrix at the solution of (5.1). If we assume that first order and sufficient second order KKT conditions are satisfied at x^* and v^*, then using the definition in (5.8), we know that the matrices

$$\begin{bmatrix} W^* & A^* \\ (A^*)^T & 0 \end{bmatrix} \quad \text{and} \quad \begin{bmatrix} (Y^*)^T W^* Y^* & (Y^*)^T W^* Z^* & (Y^*)^T A^* \\ (Z^*)^T W^* Y^* & (Z^*)^T W^* Z^* & 0 \\ (A^*)^T Y^* & 0 & 0 \end{bmatrix}$$

have the same inertia. Moreover, by defining the matrix

$$V = \begin{bmatrix} I & 0 & -\frac{1}{2}(Y^*)^T W^* Y^* [(A^*)^T Y^*]^{-1} \\ 0 & I & -(Z^*)^T W^* Y^* [(A^*)^T Y^*]^{-1} \\ 0 & 0 & [(A^*)^T Y^*]^{-1} \end{bmatrix},$$

we have

$$V \begin{bmatrix} (Y^*)^T W^* Y^* & (Y^*)^T W^* Z^* & (Y^*)^T A^* \\ (Z^*)^T W^* Y^* & (Z^*)^T W^* Z^* & 0 \\ (A^*)^T Y^* & 0 & 0 \end{bmatrix} V^T = \begin{bmatrix} 0 & 0 & I \\ 0 & (Z^*)^T W^* Z^* & 0 \\ I & 0 & 0 \end{bmatrix} \tag{5.10}$$

and from Theorem 5.2, it is clear that the right-hand matrix in (5.10) has the same inertia as the KKT matrix in (5.4). A simple permutation of the rows and columns of the matrix on the right-hand side (which leaves the inertia unchanged) leads to

$$\begin{bmatrix} T & 0 \\ 0 & (Z^*)^T W^* Z^* \end{bmatrix},$$

where T is a block diagonal matrix with m blocks of the form $\left[\begin{smallmatrix} 0 & 1 \\ 1 & 0 \end{smallmatrix}\right]$. Because each block has eigenvalues of $+1$ and -1, and because $(Z^*)^T W^* Z^* \in \mathbb{R}^{(n-m)\times(n-m)}$ and is positive definite (it has $n-m$ positive eigenvalues), we see that the KKT matrix at x^* and v^* is nonsingular with an inertia of $(n, m, 0)er, b$. Moreovy Lipschitz continuity of $A(x)$ and $W(x, v)$, the KKT matrix maintains this inertia in a nonzero neighborhood around the optimal solution.

Example 5.3 Consider the following equality constrained quadratic programming problem:

$$\begin{aligned} \min \quad & \frac{1}{2}(x_1^2 + x_2^2) \\ \text{s.t.} \quad & x_1 + x_2 = 1. \end{aligned} \tag{5.11}$$

The solution can be found by solving the first order KKT conditions:

$$x_1 + v = 0,$$
$$x_2 + v = 0,$$
$$x_1 + x_2 = 1$$

with the solution $x_1^* = \frac{1}{2}$, $x_2^* = \frac{1}{2}$, and $v_1^* = -\frac{1}{2}$. We can define a null space basis as $Z^* = [1 \ -1]^T$, so that $A^T Z^* = 0$. The reduced Hessian at the optimum is $(Z^*)^T \nabla_{xx} L(x^*, v^*) Z^* = 2 > 0$ and the sufficient second order conditions are satisfied. Moreover, the KKT matrix at the solution is given by

$$K = \begin{bmatrix} W^* & A^* \\ (A^*)^T & 0 \end{bmatrix} = \begin{bmatrix} 1 & 0 & 1 \\ 0 & 1 & 1 \\ 1 & 1 & 0 \end{bmatrix}$$

with eigenvalues, -1, 1, and 2. The inertia of this system is therefore $In(K) = (2,1,0) = (n,m,0)$ as required. ∎

5.3 Taking Newton Steps

We now consider the solution of the equality constrained problems with the basic Newton method. As seen in Algorithm 5.1, the key step is the solution of the linear system (5.4). Especially for large systems, this linear system can be solved with linear solvers that use one of the following three approaches:

- A *full-space solution* is particularly useful if the linear system (5.4) is sparse and can be represented all at once with no particular decomposition related to structure.
- *Reduced-space decomposition* is particularly useful for (5.4) if $n - m$ is small (i.e., few degrees of freedom) and second derivative information is hard to obtain and must be approximated.
- *Special purpose decompositions* can be applied that exploit other specialized structures, e.g., if the KKT matrix is banded, bordered block diagonal, or preconditioned to solve with iterative linear solvers. Depending on problem size and available architectures, these problems can be treated with either specialized, in-core linear algebra or out of core with a specialized decomposition.

In this section we focus on the first two general approaches and discuss the determination of well-defined Newton steps.

5.3.1 Full-Space Newton Steps

Full-space Newton steps are determined by solution of (5.4), or equivalently, (5.5). If the solution of (5.1) satisfies the LICQ and sufficient second order conditions, we know that the KKT matrix at the solution is nonsingular and has the correct inertia. However, at a given iteration k, particularly far from the solution, there is no expectation that these properties would hold. As a result, some corrections to (5.4) may be required to ensure well-defined Newton steps.

A number of direct solvers is available for symmetric linear systems. In particular, for *positive definite* symmetric matrices, efficient Cholesky factorizations can be applied. These can be represented as LDL^T, with L a lower triangular matrix and D diagonal. On the other hand, because the KKT matrix in (5.4) is indefinite, possibly singular, and frequently sparse, another symmetric factorization (such as the Bunch–Kaufman [74] factorization) needs to be applied. Denoting the KKT matrix as K, and defining P as an orthonormal permutation matrix, the indefinite factorization allows us to represent K as $P^T K P = LBL^T$, where the block diagonal matrix B is determined to have 1×1 or 2×2 blocks. From Sylvester's law of inertia we can obtain the inertia of K cheaply by examining the blocks in B and evaluating the signs of their eigenvalues.

To ensure that the iteration matrix is nonsingular, we first check whether the KKT matrix at iteration k has the correct inertia, $(n,m,0)$, i.e., n positive, m negative, and no zero eigenvalues. If the inertia of this matrix is incorrect, we can modify (5.4) to form the following linear system:

$$\begin{bmatrix} W^k + \delta_W I & A^k \\ (A^k)^T & -\delta_A I \end{bmatrix} \begin{bmatrix} d_x \\ d_v \end{bmatrix} = - \begin{bmatrix} \nabla L(x^k, v^k) \\ h(x^k) \end{bmatrix}. \tag{5.12}$$

Here, different trial values can be selected for the scalars $\delta_W, \delta_A \geq 0$ until the inertia is correct. To see how this works, we first prove that such values of δ_W and δ_A can be found, and then sketch a prototype algorithm.

Theorem 5.4 Assume that matrices W^k and A^k are bounded in norm; then for any $\delta_A > 0$ there exist suitable values of δ_W such that the matrix in (5.12) has the inertia $(n,m,0)$.

Proof: In (5.12) A^k has column rank $r \leq m$. Because A^k may be rank deficient, we represent it by the LU factorization $A^k L^T = [U^T \mid 0]$, with upper triangular $U \in \mathbb{R}^{r \times n}$ of full rank and nonsingular lower triangular $L \in \mathbb{R}^{m \times m}$. This leads to the following factorization of the KKT matrix:

$$\bar{V} \begin{bmatrix} W^k + \delta_W I & A^k \\ (A^k)^T & -\delta_A I \end{bmatrix} \bar{V}^T = \begin{bmatrix} W^k + \delta_W I & [U^T \mid 0] \\ \begin{bmatrix} U \\ 0 \end{bmatrix} & -\delta_A L L^T \end{bmatrix}, \tag{5.13}$$

where

$$\bar{V} = \begin{bmatrix} I & 0 \\ 0 & L \end{bmatrix}.$$

From Theorem 5.2, it is clear that the right-hand matrix has the same inertia as the KKT matrix in (5.12). Moreover, defining the nonsingular matrix

$$\tilde{V} = \begin{bmatrix} I & \delta_A^{-1} [U^T \mid 0] L^{-T} L^{-1} \\ 0 & I \end{bmatrix}$$

allows us to write the factorization

$$\tilde{V} \begin{bmatrix} W^k + \delta_W I & [U^T \mid 0] \\ \begin{bmatrix} U \\ 0 \end{bmatrix} & -\delta_A L L^T \end{bmatrix} \tilde{V}^T = \begin{bmatrix} \hat{W} & 0 \\ 0 & -\delta_A L L^T \end{bmatrix}, \tag{5.14}$$

where

$$\hat{W} = W^k + \delta_W I + \delta_A^{-1}[U^T \mid 0]L^{-T}L^{-1}\begin{bmatrix} U \\ 0 \end{bmatrix} = W^k + \delta_W I + \delta_A^{-1}A^k(A^k)^T.$$

We note that if $\delta_A > 0$, the matrix $-\delta_A LL^T$ has inertia $(0,m,0)$ and we see that the right-hand matrix of (5.14) has an inertia given by $In(\hat{W}) + (0,m,0)$. Again, we note that this matrix has the same inertia as the KKT matrix in (5.12).

To obtain the desired inertia, we now examine $\hat{W}$ more closely. Similar to the decomposition in (5.9), we define full rank range and null-space basis matrices, $Y \in \mathbb{R}^{n\times r}$ and $Z \in \mathbb{R}^{n\times(n-r)}$, respectively, with $Y^T Z = 0$. From $Z^T A^k = 0$, it is clear that $Z^T[U^T \mid 0]L^{-T} = 0$ and also that $(Y)^T A^k \in \mathbb{R}^{r\times m}$ has full column rank. For any vector $p \in \mathbb{R}^n$ with $p = Yp_Y + Zp_Z$, we can choose positive constants $\tilde{a}$ and $\tilde{c}$ that satisfy

$$\tilde{c}\|p_Y\|\|p_Z\| \geq -p_Z^T Z^T(W^k + \delta_W I)Yp_Y = -p_Z^T Z^T W^k Y p_Y,$$
$$p_Y Y^T A^k (A^k)^T Y p_Y \geq \tilde{a}\|p_Y\|^2.$$

For a given $\delta_A > 0$, we can choose δ_W large enough so that for all $p_Z \in \mathbb{R}^{n-r}$ and $p_Y \in \mathbb{R}^r$ with

$$p_Z^T Z^T(W^k + \delta_W I)Zp_Z \geq \tilde{m}\|p_Z\|^2 \quad \text{for } \tilde{m} > 0,$$
$$p_Y^T Y^T(W^k + \delta_W I)Yp_Y \geq -\tilde{w}\|p_Y\|^2 \quad \text{for } \tilde{w} \geq 0,$$

we have $\tilde{a}\delta_A^{-1} > \tilde{w}$ and $\tilde{m} \geq \frac{\tilde{c}^2}{(\tilde{a}\delta_A^{-1} - \tilde{w})}$. With these quantities, we obtain the following relations:

$$\begin{aligned}
p^T\hat{W}p &= \begin{bmatrix} p_Y^T \mid p_Z^T \end{bmatrix}\begin{bmatrix} Y^T(W^k + \delta_W I + \delta_A^{-1}A^k(A^k)^T)Y & Y^T W^k Z \\ Z^T W^k Y & Z^T(W^k + \delta_W I)Z \end{bmatrix}\begin{bmatrix} p_Y \\ p_Z \end{bmatrix} \\
&= p_Y^T Y^T(\hat{W})Yp_Y + 2p_Y^T Y^T W^k Z p_Z + p_Z^T Z^T(W^k + \delta_W I)Zp_Z \\
&\geq (\tilde{a}\delta_A^{-1} - \tilde{w})\|p_Y\|^2 - 2\tilde{c}\|p_Y\|\|p_Z\| + \tilde{m}\|p_Z\|^2 \\
&= \left((\tilde{a}\delta_A^{-1} - \tilde{w})^{1/2}\|p_Y\| - \frac{\tilde{c}}{(\tilde{a}\delta_A^{-1} - \tilde{w})^{1/2}}\|p_z\|\right)^2 \\
&\quad + \left(\tilde{m} - \frac{\tilde{c}^2}{(\tilde{a}\delta_A^{-1} - \tilde{w})}\right)\|p_z\|^2 > 0
\end{aligned}$$

for all values of p. As a result, $\hat{W}$ is positive definite, the inertia of the KKT matrix in (5.12) is $(n,m,0)$, and we have the desired result. □

These observations motivate the following algorithm (simplified from [404]) for choosing δ_A and δ_W for a particular Newton iteration k.

ALGORITHM 5.2.

Given constants $0 < \bar{\delta}_W^{\min} < \bar{\delta}_W^0 < \bar{\delta}_W^{\max}$; $\bar{\delta}_A > 0$; and $0 < \kappa_l < 1 < \kappa_u$. (From [404], recommended values of the scalar parameters are $\bar{\delta}_W^{\min} = 10^{-20}$, $\bar{\delta}_W^0 = 10^{-4}$, $\bar{\delta}_W^{\max} = 10^{40}$, $\bar{\delta}_A = 10^{-8}$, $\kappa_u = 8$, and $\kappa_l = \frac{1}{3}$.) Also, for the first iteration, set $\delta_W^{\text{last}} := 0$.

At each iteration k:

1. Attempt to factorize the matrix in (5.12) with $\delta_W = \delta_A = 0$. For instance, this can be done with an LBL^T factorization; the diagonal (1×1 and 2×2) blocks of B are then used to determine the inertia. If the matrix has correct inertia, then use the resulting search direction as the Newton step. Otherwise, continue with step 2.

2. If the inertia calculation reveals zero eigenvalues, set $\delta_A := \bar{\delta}_A$. Otherwise, set $\delta_A := 0$.

3. If $\delta_W^{last} = 0$, set $\delta_W := \bar{\delta}_W^0$, else set $\delta_W := \max\{\bar{\delta}_W^{min}, \kappa_l \delta_W^{last}\}$.

4. With updated values of δ_W and δ_A, attempt to factorize the matrix in (5.12). If the inertia is now correct, set $\delta_W^{last} := \delta_W$ and use the resulting search direction in the line search. Otherwise, continue with step 5.

5. Continue to increase δ_W and set $\delta_W := \kappa_u \delta_W$.

6. If $\delta_W > \bar{\delta}_W^{max}$, abort the search direction computation. The matrix is severely ill-conditioned.

This correction algorithm provides a systematic way to choose δ_W and δ_A to obtain the required inertia. Step 1 checks to see if (5.4) has the correct inertia so that an uncorrected Newton search direction can be used whenever possible. Otherwise, only a positive (and usually small) δ_A is chosen if the unmodified iteration matrix has a zero eigenvalue, as we assume that the singularity is caused by a rank-deficient constraint Jacobian. This correction helps to maintain the inertia for the negative eigenvalues. On the other hand, if the factorization is still unsuccessful, larger values of δ_W are chosen, but only to find the smallest correction necessary to yield a successful factorization. The algorithm can also be extended (see [404]) to prevent cycling of δ_W when successive iterations and successive iterates are close together.

5.3.2 Reduced-Space Newton Steps

The reduced-space option is well suited to problems where $n - m$ is small and especially when second order information needs to be approximated, as shown in Section 5.4. In contrast to the full-space step (5.12), the reduced-space approach is motivated by the matrix decomposition for the linear system in (5.9) along with the Newton step (5.5). To compute the search direction at iteration k, we consider components of this direction in two separate spaces (see, e.g., [162]). We write the solution d_x of (5.5) as

$$d_x = Y^k p_Y + Z^k p_Z, \tag{5.15}$$

where Z^k is an $n \times (n-m)$ matrix spanning the null space of $(A^k)^T$, and Y^k is any $n \times m$ matrix that allows $[Y^k \; Z^k]$ to be nonsingular. We also refer to the two components $Y^k p_Y$ and $Z^k p_Z$ as *normal* and *tangential* steps, respectively, as shown in Figure 5.1. Applying the decomposition (5.8)–(5.9) with the analogous terms at iteration k leads to

$$\begin{bmatrix} (Y^k)^T W^k Y^k & (Y^k)^T W^k Z^k & (Y^k)^T A^k \\ (Z^k)^T W^k Y^k & (Z^k)^T W^k Z^k & 0 \\ (A^k)^T Y^k & 0 & 0 \end{bmatrix} \begin{bmatrix} p_Y \\ p_Z \\ \bar{v} \end{bmatrix} = - \begin{bmatrix} (Y^k)^T \nabla f(x^k) \\ (Z^k)^T \nabla f(x^k) \\ h(x^k) \end{bmatrix}. \tag{5.16}$$

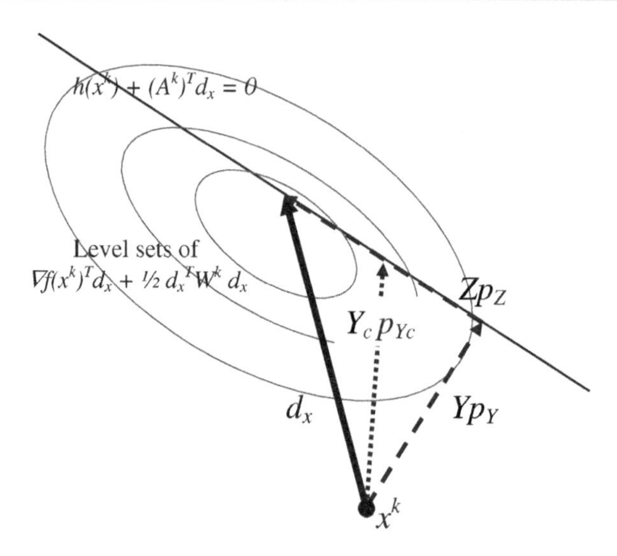

Figure 5.1. *Normal and tangential steps for the solution d_x of* (5.4) *or* (5.6). *Note also that if a coordinate basis (Y_c) is chosen, the normal and tangential steps may not be orthogonal and the steps $Y_c p_{Yc}$ are longer than Yp_Y.*

In contrast to the full-space method, this decomposition requires that A^k have full column rank for all k. From the bottom row in (5.16) we obtain

$$h(x^k)+(A^k)^T Y^k p_Y = 0.$$

Therefore the normal component, p_Y, is given by

$$p_Y = -[(A^k)^T Y^k]^{-1} h(x^k), \tag{5.17}$$

so that d_x^k has the form

$$d_x = -Y^k[(A^k)^T Y^k]^{-1} h(x^k) + Z^k p_Z. \tag{5.18}$$

To determine the tangential component, p_Z, we substitute (5.17) in the second row of (5.16). The following linear system:

$$(Z^k)^T W^k Z^k p_Z = -[(Z^k)^T \nabla f(x^k) + (Z^k)^T W^k Y^k p_Y] \tag{5.19}$$

can then be solved if $(Z^k)^T W^k Z^k$ is positive definite, and this property can be verified through a successful Cholesky factorization. Otherwise, if $(Z^k)^T W^k Z^k$ is not positive definite, this matrix can be modified either by adding a sufficiently large diagonal term, say $\delta_R I$, or by using a modified Cholesky factorization, as discussed in Section 3.2.

Once p_Z is calculated from (5.19) we use (5.18) to obtain d_x. Finally, the top row in (5.16) can then be used to update the multipliers:

$$\begin{aligned}\bar{v} &= -[(Y^k)^T A^k]^{-1}[(Y^k)^T \nabla f(x^k) + (Y^k)^T W^k Z^k p_Z + (Y^k)^T W^k Y^k p_Y] \\ &= -[(Y^k)^T A^k]^{-1}[(Y^k)^T \nabla f(x^k) + (Y^k)^T W^k d_x].\end{aligned} \tag{5.20}$$

Because $d_x^k \to 0$ as the algorithm converges, a first order multiplier calculation may be used instead, i.e.,

$$\bar{v} = -[(Y^k)^T A^k]^{-1}(Y^k)^T \nabla f(x^k), \tag{5.21}$$

and we can avoid the calculation of $(Y^k)^T W^k Y^k$ and $(Y^k)^T W^k Z^k$. Note that the multipliers from (5.21) are still asymptotically correct.

A dominant part of the Newton step is the computation of the null-space and range-space basis matrices, Z and Y, respectively. There are many possible choices for these basis matrices, including the following three options.

- By computing a *QR factorization* of A, we can define Z and Y so that they have orthonormal columns, i.e., $Z^T Z = I_{n-m}$, $Y^T Y = I_m$, and $Z^T Y = 0$. This gives a well-conditioned representation of the null space and range space of A, However, Y and Z are dense matrices, and this can lead to expensive computation when the number of variables (n) is large.

- A more economical alternative for Z and Y can be found through a simple elimination of dependent variables [134, 162, 294]. Here we permute the components of x into m dependent or *basic* variables (without loss of generality, we select the first m variables) and $n-m$ independent or *superbasic* variables. Similarly, the columns of $(A^k)^T$ are permuted and partitioned accordingly to yield

$$(A^k)^T = [A_B^k \mid A_S^k]. \tag{5.22}$$

We assume that the $m \times m$ *basis matrix* A_B^k is nonsingular and we define the *coordinate bases*

$$Z^k = \begin{bmatrix} -(A_B^k)^{-1} A_S^k \\ I \end{bmatrix} \quad \text{and} \quad Y^k = \begin{bmatrix} I \\ 0 \end{bmatrix}. \tag{5.23}$$

Note that in practice Z^k is not formed explicitly; instead we can compute and store the sparse LU factors of A_B^k. Due to the choice (5.23) of Y^k, the normal component determined in (5.17) and multipliers determined in (5.21) simplify to

$$p_Y = -(A_B^k)^{-1} h(x^k), \tag{5.24}$$

$$\bar{v} = -(A_B^k)^{-T}(Y^k)^T \nabla f(x^k). \tag{5.25}$$

- Numerical accuracy and algorithmic performance are often improved if the tangential and normal directions, Zp_Z and Yp_Y, can be maintained orthogonal and the length of the normal step is minimized. This can, of course, be obtained through a QR factorization, but the directions can be obtained more economically by modifying the coordinate basis decomposition and defining the *orthogonal bases*

$$Z^k = \begin{bmatrix} -(A_B^k)^{-1} A_S^k \\ I \end{bmatrix} \quad \text{and} \quad Y^k = \begin{bmatrix} I \\ (A_S^k)^T (A_B^k)^{-T} \end{bmatrix}. \tag{5.26}$$

Note that $(Z^k)^T Y^k = 0$, and from the choice (5.26) of Y^k, the calculation in (5.17) can be written as

$$\begin{aligned} p_Y &= -[I + (A_B^k)^{-1} A_S^k (A_S^k)^T (A_B^k)^{-T}]^{-1} (A_B^k)^{-1} h(x^k) \\ &= -\left[I - (A_B^k)^{-1} A_S^k [I + (A_S^k)^T (A_B^k)^{-T} (A_B^k)^{-1} A_S^k]^{-1} (A_S^k)^T (A_B^k)^{-T}\right] \\ &\quad \times (A_B^k)^{-1} h(x^k), \end{aligned} \tag{5.27}$$

where the second equation follows from the application of the Sherman–Morison–Woodbury formula [294]. An analogous expression can be derived for $\bar{v}$. As with coordinate bases, this calculation requires a factorization of A_B^k, but it also requires a factorization of

$$[I + (A_S^k)^T (A_B^k)^{-T} (A_B^k)^{-1} A_S^k]$$

with an additional cost that is $O((n-m)^3)$.

- The normal and tangential search directions can also be determined implicitly through an appropriate modification of the full-space KKT system (5.4). As shown in Exercise 3, the tangential step $d_t = Z^k p_Z$ can be found from the following linear system:

$$\begin{bmatrix} W^k & A^k \\ (A^k)^T & 0 \end{bmatrix} \begin{bmatrix} d_t \\ v \end{bmatrix} = -\begin{bmatrix} \nabla f(x^k) \\ 0 \end{bmatrix}. \tag{5.28}$$

Implicit calculation of the normal step, $d_n = Y^k p_Y$, requires a substitution of the Hessian matrix by an alternative matrix $\tilde{W}$. Here, the normal step is determined by the following system:

$$\begin{bmatrix} \tilde{W} & A^k \\ (A^k)^T & 0 \end{bmatrix} \begin{bmatrix} d_n \\ v \end{bmatrix} = -\begin{bmatrix} 0 \\ h(x^k) \end{bmatrix}, \tag{5.29}$$

where the condition $(Z^k)^T \tilde{W} Y^k = 0$ must hold. Popular choices for the Hessian term are $\tilde{W} = \left[\begin{smallmatrix} 0 & 0 \\ 0 & I \end{smallmatrix}\right]$ and $\tilde{W} = I$ for the coordinate and orthogonal bases, respectively.

5.4 Quasi-Newton Methods

If exact second derivative information is unavailable or difficult to evaluate, one can apply a quasi-Newton update matrix B^k to substitute for W^k or the reduced Hessian. In either case, the BFGS and SR1 updates developed in Section 3.3 can be applied as before, but with a slight modification for the constrained system. In particular, for the full-space method, we choose the matrix B^k as a quasi-Newton approximation to the Hessian of the Lagrangian, $W(x^k, v^k)$. As in Chapter 3, we apply the secant relation (3.9):

$$B^{k+1} s = y,$$

but with

$$s = x^{k+1} - x^k, \quad y = \nabla L(x^{k+1}, v^{k+1}) - \nabla L(x^k, v^{k+1}). \tag{5.30}$$

Note that because we approximate the Hessian with respect to x, both terms in the definition of y require the multiplier evaluated at v^{k+1}.

With these definitions of s and y, we can directly apply the BFGS update (3.18):

$$B^{k+1} = B^k + \frac{yy^T}{s^T y} - \frac{B^k s s^T B^k}{s^T B^k s} \tag{5.31}$$

or the SR1 update (3.10):

$$B^{k+1} = B^k + \frac{(y - B^k s)(y - B^k s)^T}{(y - B^k s)^T s}. \tag{5.32}$$

However, unlike the Hessian matrix for unconstrained optimization, discussed in Chapter 3, $W(x, v)$ is not required to be positive definite at the optimum. Only its projection, i.e., $Z(x^*)^T W(x^*, v^*) Z(x^*)$, needs to be positive definite to satisfy sufficient second order conditions. As a result, some care is needed in applying the update matrices using (5.31) or (5.32).

As discussed in Chapter 3, these updates are well defined when $s^T y$ is sufficiently positive for BFGS and $(y - B^k s)^T s \neq 0$ for SR1. Otherwise, the updates are skipped for a given iteration, or *Powell damping* (see Section 3.3) can be applied for BFGS.

5.4.1 A Quasi-Newton Full-Space Method

In the full-space method, we substitute B^k for W^k in (5.4), i.e.,

$$\begin{bmatrix} B^k & A^k \\ (A^k)^T & 0 \end{bmatrix} \begin{bmatrix} d_x \\ d_v \end{bmatrix} = - \begin{bmatrix} \nabla L(x^k, v^k) \\ h(x^k) \end{bmatrix}. \tag{5.33}$$

However, because the updated quasi-Newton matrix B^k is dense, handling this matrix directly leads to factorizations of the Newton step that are $O(n^3)$. For large problems, this calculation can be prohibitively expensive. Instead we would like to exploit the fact that A^k is sparse and B^k has the structure given by (5.31) or (5.32). We can therefore store the updates y and s for successive iterations and incorporate these into the solution of (5.33).

For this update, we store only the last q updates and apply the compact limited memory representation from [83] (see Section 3.3), along with a sparse factorization of the initial quasi-Newton matrix, B^0, in (5.33). Also, we assume that B^0 is itself sparse; often it is chosen to be diagonal. For the BFGS update (5.31), this compact representation is given by

$$B^{k+1} = B^0 - [B^0 S_k \;\; Y^k] \begin{bmatrix} S_k^T B^0 S_k & L_k \\ L_k^T & -D_k \end{bmatrix}^{-1} \begin{bmatrix} S_k^T B^0 \\ (Y^k)^T \end{bmatrix}, \tag{5.34}$$

where

$$D_k = \text{diag}[(s^{k-q+1})^T (y^{k-q+1}), \ldots, (s^k)^T y^k], \;\; S_k = [s^{k-q+1}, \ldots, s^k], \;\; Y^k = [y^{k-q+1}, \ldots, y^k],$$

and

$$(L_k) = \begin{cases} (s^{k-q+i})^T (y^{k-q+j}), & i > j, \\ 0 & \text{otherwise.} \end{cases}$$

With this compact form, we represent the quasi-Newton matrix as

$$B^{k+1} = B^0 + V_k V_k^T - U_k U_k^T. \tag{5.35}$$

By writing (5.34) as

$$B^{k+1} = B^0 - [Y^k \;\; B^0 S_k] \begin{bmatrix} -D_k^{1/2} & D_k^{1/2} L_k^T \\ 0 & J_k^T \end{bmatrix}^{-1} \begin{bmatrix} D_k^{1/2} & 0 \\ -L_k^T D_k^{-1/2} & J_k \end{bmatrix}^{-1} \begin{bmatrix} (Y^k)^T \\ S_k^T B^0 \end{bmatrix}, \tag{5.36}$$

where J_k is a lower triangular factorization constructed from the Cholesky factorization that satisfies

$$J_k J_k^T = S_k^T B^0 S_k + L_k D_k^{-1} L_k^T, \tag{5.37}$$

we now define $V_k = Y^k D_k^{-1/2}$, $U_k = (B^0 S_k + Y^k D_k^{-1} L_k^T) J_k^{-T}$, $\tilde{U}^T = [U_k^T \; 0]$, and $\tilde{V}^T = [V_k^T \; 0]$ and consider the matrices

$$K = \begin{bmatrix} B^k & A^k \\ (A^k)^T & 0 \end{bmatrix} = \begin{bmatrix} (B^0 + V_k V_k^T - U_k U_k^T) & A^k \\ (A^k)^T & 0 \end{bmatrix},$$

$$K_0 = \begin{bmatrix} B^0 & A^k \\ (A^k)^T & 0 \end{bmatrix}.$$

Moreover, we assume that K_0 is sparse and can be factorized cheaply using a sparse or structured matrix decomposition. (Usually such a factorization requires $O(n^\beta)$ operations, with the exponent $\beta \in [1,2]$.) Factorization of K can then be made by two applications of the Sherman–Morison–Woodbury formula, yielding

$$K_1^{-1} = K_0^{-1} - K_0^{-1} \tilde{V} [I + \tilde{V}^T K_0^{-1} \tilde{V}]^{-1} \tilde{V}^T K_0^{-1}, \tag{5.38}$$

$$K^{-1} = K_1^{-1} - K_1^{-1} \tilde{U} [I + \tilde{U}^T K_1^{-1} \tilde{U}]^{-1} \tilde{U}^T K_1^{-1}. \tag{5.39}$$

By carefully structuring these calculations so that K_0^{-1}, K_1^{-1}, and K^{-1} are factorized and incorporated into backsolves, these matrices are never explicitly created, and we can obtain the solution to (5.33), i.e.,

$$\begin{bmatrix} d_x \\ d_v \end{bmatrix} = -K^{-1} \begin{bmatrix} \nabla L(x^k, v^k) \\ h(x^k) \end{bmatrix} \tag{5.40}$$

in only $O(n^\beta + q^3 + q^2 n)$ operations. Since only a few updates are stored (q is typically less than 20), this limited memory approach leads to a much more efficient implementation of quasi-Newton methods. An analogous approach can be applied to the SR1 method as well.

As seen in Chapter 3, quasi-Newton methods are slower to converge than Newton's method, and the best that can be expected is a superlinear convergence rate. For equality constrained problems, the full-space quasi-Newton implementation has a local convergence property that is similar to Theorem 3.1, but with additional restrictions.

Theorem 5.5 [63] Assume that $f(x)$ and $h(x)$ are twice differentiable, their second derivatives are Lipschitz continuous, and Algorithm 5.1 modified by a quasi-Newton approximation converges to a KKT point that satisfies the LICQ and the sufficient second order optimality conditions. Then x^k converges to x^* at a superlinear rate, i.e.,

$$\lim_{k \to \infty} \frac{\|x^k + d_x - x^*\|}{\|x^k - x^*\|} = 0 \tag{5.41}$$

if and only if

$$\lim_{k\to\infty} \frac{\|Z(x^k)^T(B^k - W(x^*, v^*))d_x\|}{\|d_x\|} = 0, \tag{5.42}$$

where $Z(x)$ is a representation of the null space of $A(x)^T$ that is Lipschitz continuous in a neighborhood of x^*.

Note that the theorem does not state whether a specific quasi-Newton update, e.g., BFGS or SR1, satisfies (5.42), as this depends on whether B^k remains bounded and well-conditioned, or whether updates need to be modified through damping or skipping. On the other hand, under the more restrictive assumption where $W(x^*, v^*)$ is positive definite, the following property can be shown.

Theorem 5.6 [63] Assume that $f(x)$ and $h(x)$ are twice differentiable, their second derivatives are Lipschitz continuous, and Algorithm 5.1 modified by a BFGS update converges to a KKT point that satisfies the LICQ and the sufficient second order optimality conditions. If, in addition, $W(x^*, v^*)$ and B^0 are positive definite, and $\|B^0 - W(x^*, v^*)\|$ and $\|x^0 - x^*\|$ are sufficiently small, then x^k converges to x^* at a superlinear rate, i.e.,

$$\lim_{k\to\infty} \frac{\|x^k + d_x - x^*\|}{\|x^k - x^*\|} = 0. \tag{5.43}$$

5.4.2 A Quasi-Newton Reduced-Space Method

When n is large and $n - m$ (the number of *degrees of freedom*) is small, it is attractive to simplify the calculation of the Newton step. Using (5.16) and the first order approximation to the Lagrange multipliers, (5.21), the resulting linear system becomes

$$\begin{bmatrix} 0 & 0 & (Y^k)^T A^k \\ (Z^k)^T W^k Y^k & (Z^k)^T W^k Z^k & 0 \\ (A^k)^T Y^k & 0 & 0 \end{bmatrix} \begin{bmatrix} p_Y \\ p_Z \\ \bar{v} \end{bmatrix} = - \begin{bmatrix} (Y^k)^T \nabla f(x^k) \\ (Z^k)^T \nabla f(x^k) \\ h(x^k) \end{bmatrix}. \tag{5.44}$$

This linear system is solved by solving two related subsystems of order m for p_Y and $\bar{v}$, respectively, and a third subsystem of order $n - m$ for p_Z. For this last subsystem, we can approximate the reduced Hessian, i.e., $\bar{B}^k \approx (Z^k)^T W^k Z^k$, by using the BFGS update (5.31). By noting that

$$d_x = Y^k p_Y + Z^k p_Z$$

and assuming that a full Newton step is taken (i.e., $x^{k+1} = x^k + d_x$), we can apply the secant condition based on the following first order approximation to the reduced Hessian:

$$\begin{aligned} (Z^k)^T W^k Z^k p_Z &\approx (Z^k)^T \nabla L(x^k + Z^k p_Z, v^k) - (Z^k)^T \nabla L(x^k, v^k) && (5.45) \\ &= (Z^k)^T \nabla L(x^{k+1} - Y^k p_Y, v^k) - (Z^k)^T \nabla L(x^k, v^k) && (5.46) \\ &\approx (Z^k)^T \nabla L(x^{k+1}, v^k) - (Z^k)^T \nabla L(x^k, v^k) - w^k, && (5.47) \end{aligned}$$

where $w^k = Z^k W^k Y^k p_Y$ or its approximation. Defining

$$s_k = p_Z \quad \text{and} \quad y_k = (Z^k)^T \nabla L(x^{k+1}, v^k) - (Z^k)^T \nabla L(x^k, v^k) - w^k, \tag{5.48}$$

we can write the secant formula $\bar{B}^{k+1} s_k = y_k$ and use this definition for the BFGS update (5.31). However, note that for this update, some care is needed in the choice of $Z(x^k)$ so that it remains continuous with respect to x. Discontinuous changes in $Z(x)$ (due to different variable partitions or QR rotations) could lead to serious problems in the Hessian approximation [162].

In addition, we modify the Newton step from (5.44). Using the normal step p_Y calculated from (5.17), we can write the tangential step as follows:

$$\bar{B}^k p_Z = -[(Z^k)^T \nabla f(x^k) + (Z^k)^T W^k Y^k p_Y] = -[(Z^k)^T \nabla f(x^k) + w^k], \tag{5.49}$$

where the second part follows from writing (or approximating) $(Z^k)^T W^k Y^k p_Y$ as w^k. Note that the BFGS update eliminates the need to evaluate W^k and to form $(Z^k)^T W^k Z^k$ explicitly. In addition, to avoid the formation of $(Z^k)^T W^k Y^k p_Y$, it is appealing to approximate the vectors w^k in (5.49) and (5.47). This can be done in a number of ways.

- Many reduced Hessian methods [294, 100, 134] typically approximate $Z^k W^k Y^k p_Y$ by $w^k = 0$. Neglecting this term often works well, especially when normal steps are much smaller than tangential steps. This approximation is particularly useful when the constraints are linear and p_Y remains zero once a feasible iterate is encountered. Moreover, when Z and Y have orthogonal representations, the length of the normal step $Y^k p_Y$ is minimized. Consequently, neglecting this term in (5.49) and (5.47) still can lead to good performance. On the other hand, coordinate representations (5.23) of Y can lead to larger normal steps (see Figure 5.1), a nonnegligible value for $Z^k W^k Y^k p_Y$, and possibly erratic convergence behavior of the method.

- To ensure that good search directions are generated in (5.44) that are independent of the representations of Z and Y, we can compute a finite difference approximation of $(Z^k)^T W^k Y^k$ along p_Y, for example

$$w^{k+1} = (Z^k)^T \nabla L(x^{k+1}, v^k) - (Z^k)^T \nabla L(x^k + Z^k p_Z, v^k). \tag{5.50}$$

  However, this term requires additional evaluation of reduced gradient $(Z^k)^T \nabla L(x^k + Z^k p_Z, v^k)$ and this may *double the work* per iteration.

- Another approximation, which requires no additional gradient information, uses a *quasi-Newton approximation* S^k for the rectangular matrix $(Z^k)^T W^k$. This can be obtained, for example, with the Broyden update (see (3.15) and the discussion in Chapter 3) and can be used to define $w^k = S^k Y^k p_Y$. This quasi-Newton approximation works well when either the normal steps or the tangential steps dominate the iterations.

- To preserve the economy of computation, we can define a hybrid approach, where w^k is defined via the quasi-Newton approximation S^k whenever conditions on p_Y and p_Z indicate that either step is dominant in iteration k. To develop these conditions, we define a positive constant $\kappa > 0$, a sequence of positive numbers $\{\gamma_k\}$ with

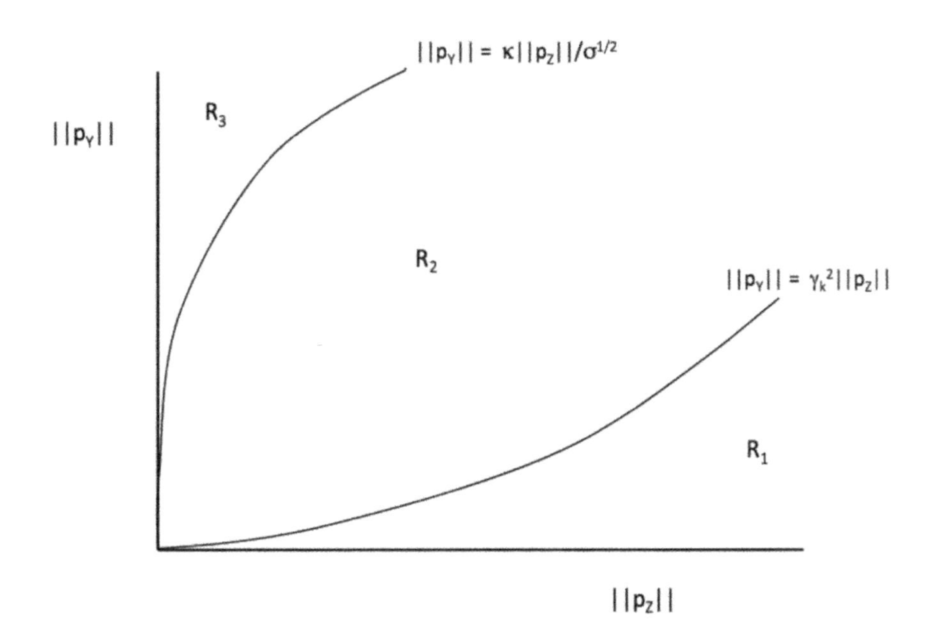

Figure 5.2. *Regions that determine criteria for choosing w^k, the approximation to $(Z^k)^T W^k Y^k p_Y$. Quasi-Newton approximation can be in R_1 or R_3, while R_2 requires a finite difference approximation.*

$\sum_{k=1}^{\infty} \gamma_k < \infty$, and $\sigma_k = \|(Z^k)^T \nabla f(x^k)\| + \|h(x^k)\|$ which is related directly to the distance to the solution, $\|x^k - x^*\|$. In the space of $\|p_Y\|$ and $\|p_Z\|$, we now define three regions given by

$$R_1 : 0 \leq \|p_Y\| \leq \gamma_k^2 \|p_Z\|,$$
$$R_2 : \gamma_k^2 \|p_Z\| < \|p_Y\| \leq \kappa \|p_Z\| / \sigma_k^{1/2},$$
$$R_3 : \|p_Y\| > \kappa \|p_Z\| / \sigma_k^{1/2}$$

and shown in Figure 5.2. In order to obtain a fast convergence rate, we only need to resort to the finite difference approximation (5.50) when the calculated step lies in region R_2. Otherwise, in regions R_1 and R_3, the tangential and the normal steps, respectively, are dominant and the less expensive quasi-Newton approximation can be used for w^k. In [54], these hybrid concepts were incorporated into an algorithm with a line search along with safeguards to ensure that the updates S^k and $\bar{B}^k$ remain uniformly bounded. A detailed presentation of this hybrid algorithm along with analysis of global and local convergence properties and numerical performance is given in [54].

To summarize, the reduced-space quasi-Newton algorithm does not require the computation of the Hessian of the Lagrangian W^k and only makes use of first derivatives of $f(x)$ and $h(x)$. The reduced Hessian matrix $(Z^k)^T W^k Z^k$ is approximated by a positive definite quasi-Newton matrix $\bar{B}^k$, using the BFGS formula, and the $Z^k W^k Y^k p_Y$ term is approximated by a vector w^k, which is computed either by means of a finite difference formula or via a quasi-Newton approximation. The method is therefore well suited for large problems with relatively few degrees of freedom.

The local convergence properties for this reduced-space method are analogous to Theorems 5.5 and 5.6, but they are modified because $\bar{B}^k$ is now a quasi-Newton approximation to the *reduced* Hessian and also because of the choice of approximation w^k. First, we note from (5.15), (5.47), and the definitions for $\bar{B}^k$ and w^k that the condition

$$\lim_{k\to\infty} \frac{\|(\bar{B}^k - (Z^*)^T W(x^*, v^*) Z^*) p_Z\| + \|w^k - (Z^*)^T W(x^*, v^*) Y^* p_Y\|}{\|d_x\|} = 0 \tag{5.51}$$

implies (5.42). Therefore, Theorem 5.7 holds and the algorithm converges superlinearly in x. On the other hand, (5.51) does not hold for all choices of w^k or quasi-Newton updates. For instance, if w^k is a poor approximation to $(Z^k)^T W^k Y^k p_Y$, the convergence rate is modified as follows.

Theorem 5.7 [63] Assume that $f(x)$ and $h(x)$ are twice differentiable, their second derivatives are Lipschitz continuous, and the reduced-space quasi-Newton algorithm converges to a KKT point that satisfies the LICQ and the sufficient second order optimality conditions. If $Z(x)$ and $v(x)$ (obtained from (5.21)) are Lipschitz continuous in a neighborhood of x^*, $Y(x)$ and $Z(x)$ are bounded, and $[Y(x) \mid Z(x)]$ has a bounded inverse, then x^k converges to x^* at a 2-step superlinear rate, i.e.,

$$\lim_{k\to\infty} \frac{\|x^k + d_x^k + d_x^{k+1} - x^*\|}{\|x^k - x^*\|} = 0 \tag{5.52}$$

if and only if

$$\lim_{k\to\infty} \frac{\|(\bar{B}^k - (Z^*)^T W(x^*, v^*) Z^*) p_Z\|}{\|d_x\|} = 0. \tag{5.53}$$

Finally, we consider the hybrid approach that monitors and evaluates an accurate approximation of w^k based on the regions in Figure 5.2. Moreover, since $(Z^*)^T W(x^*, v^*) Z^*$ is positive definite, the BFGS update to the reduced Hessian provides a suitable approximation. As a result, the following, stronger property can be proved.

Theorem 5.8 [54] Assume that $f(x)$ and $h(x)$ are twice differentiable, their second derivatives are Lipschitz continuous, and we apply the hybrid reduced-space algorithm in [54] with the BFGS update (5.31), (5.48). If

- the algorithm converges to a KKT point that satisfies the LICQ and the sufficient second order optimality conditions,
- $Z(x)$ and $v(x)$ (obtained from (5.21)) are Lipschitz continuous,
- $Y(x)$ and $Z(x)$ are bounded and $[Y(x) \mid Z(x)]$ has a bounded inverse in a neighborhood of x^*,

then x^k converges to x^* at a 1-step superlinear rate, i.e.,

$$\lim_{k \to \infty} \frac{\|x^k + d_x - x^*\|}{\|x^k - x^*\|} = 0. \tag{5.54}$$

5.5 Globalization for Constrained Optimization

The previous section discusses the generation of Newton steps that ensure fast convergence to a local solution of problem (5.1). Of course, this requires an initial point that is sufficiently close to the solution. To ensure convergence from poor starting points, the Newton method must be modified to ensure that the iterations make sufficient progress to the solution. This globalization issue was explored in Chapter 3 in the context of unconstrained minimization, where both line search and trust region methods were developed and analyzed. In this chapter, we extend these globalization methods to problem (5.1), but here we need to consider two competing criteria in order to achieve convergence: the reduction of the objective function and satisfaction of the equality constraints.

To balance the reduction in $f(x)$ and $\|h(x)\|$, we explore two popular approaches to deal with this task. First, we consider a composite *merit function*, composed of both $f(x)$ and $\|h(x)\|$, which can be used to determine progress toward the solution to (5.1). Consequently, a step is allowed when it leads to a sufficient reduction in this merit function. There are many ways to pose this function and our goal is to apply a merit function whose minimum corresponds exactly to the solution of (5.1). The second, more recent, approach uses a bicriterion strategy to trade off reductions in the objective function and constraint infeasibility directly. This so-called *filter approach* stores information from previous iterates $(f(x^l), \|h(x^l)\|, l = 1, \ldots, k)$ with the noninferior or nondominated pairs removed; this represents the filter. The filter approach then allows a move to be made to x^{k+1} only if *either* $f(x^{k+1})$ *or* $\|h(x^{k+1})\|$ improves on the points in the current filter. The filter is then updated and the iterations continue. Both approaches lead to robust and efficient extensions to the local Newton-type methods and allow convergence from poor starting points under mild conditions on (5.1).

5.5.1 Concepts of Merit Functions

Merit functions are motivated by a broad literature on methods for the minimization of penalty functions. These methods have the appeal of solving constrained nonlinear problems using unconstrained methods, such as the ones developed in Chapter 3. On the other hand, the direct application of unconstrained minimization to penalty functions is not straightforward and this has implications on the choice of penalty functions as merit functions.

Quadratic Penalty Functions

To explore the use of penalty functions, we first consider the classical quadratic penalty function given by

$$Q(x;\rho) = f(x) + (\rho/2)h(x)^T h(x). \tag{5.55}$$

Minimization of $Q(x;\rho)$ for a fixed value of ρ is typically performed with the unconstrained methods of Chapter 3; if successful, this leads to the local minimizer, $x(\rho)$. Moreover, solving a sequence of these unconstrained minimizations with $\lim_{k\to\infty} \rho^k \to \infty$ leads to a sequence $x(\rho^k)$ that is expected to converge to the solution of (5.1). The quadratic penalty can be characterized by the following properties:

- If the limit point, $\bar{x}$, of the sequence $x(\rho^k)$ is feasible and $\nabla h(\bar{x})$ has linearly independent columns, then $\bar{x}$ is a KKT point of (5.1) with multipliers obtained from

$$v = \lim_{k\to\infty} \rho^k h(x(\rho^k)).$$

- If the limit point, $\bar{x}$, of the sequence $x(\rho^k)$ is infeasible, then it is a stationary point of $h(x)^T h(x)$.
- As ρ increases, care must be taken to deal with ill-conditioning of the Hessian, $\nabla_{xx} Q(x;\rho)$ (or its approximation) when applying an unconstrained method.
- If ρ is not large enough, the quadratic penalty $Q(x,\rho)$ may become unbounded even if a solution exists for (5.1).

Because of these properties, it is clear that the quadratic penalty is not suitable as a merit function, as the constrained solution is obtained only as ρ becomes unbounded. Moreover, for a finite value of ρ, the quadratic penalty is *not exact*; i.e., the minimizer of $Q(x;\rho)$ does not correspond to the solution of (5.1). As a result, $Q(x;\rho)$ can be reduced further by *moving away* from the solution of (5.1). In addition, as discussed in [294], this nonexactness is not limited to quadratic penalties but extends to all smooth penalty functions of the form

$$\phi(x;\rho) = f(x) + \rho \sum_{j=1}^{m} \varphi(h_j(x)) \quad \text{where } \varphi(\xi) \geq 0 \text{ for all } \xi \in \mathbb{R}, \quad \varphi(0) = 0. \tag{5.56}$$

Augmented Lagrangian Functions

The inexact penalty $Q(x;\rho)$ can be modified to form the augmented Lagrange function

$$L_A(x,v;\rho) = f(x) + h(x)^T v + (\rho/2) h(x)^T h(x). \tag{5.57}$$

Because additional variables v of indeterminate sign are included in this function, this function is more general than (5.56). As a result, an unconstrained minimization problem can be formulated that leads to a solution of (5.1) for a *finite* value of ρ. Some properties of the augmented Lagrange function can be summarized as follows:

- Given vectors x^*, v^* that correspond to a local solution of (5.1) that satisfies the LICQ and sufficient second order conditions, for ρ greater than a threshold value, x^* is a strict local minimizer for $L_A(x,v^*;\rho)$.
- $L_A(x,v;\rho)$ has a local minimum $\bar{x}$ that is close to x^* if v is sufficiently close to v^*, *or* ρ is sufficiently large.
- If the multiplier approximation is defined by the least squares estimate

$$v(x) = -[\nabla h(x)^T \nabla h(x)]^{-1} \nabla h(x)^T \nabla f(x) \tag{5.58}$$

in a manner similar to (5.21), then a local solution of (5.1) that satisfies the LICQ and sufficient second order conditions is also a strict local minimizer of $L_A(x,v(x);\rho)$ for ρ greater than a threshold value.

Because the augmented Lagrange function is exact for finite values of ρ and suitable values of v, it forms the basis of a number of NLP methods (see [294, 100, 134]). On the other hand, the suitability of L_A as a merit function is complicated by the following:

- Finding a sufficiently large value of ρ is not straightforward. A suitably large value of ρ needs to be determined iteratively,

- Suitable values for v need to be determined through iterative updates as the algorithm proceeds, or by direct calculation using (5.58). Either approach can be expensive and may lead to ill-conditioning, especially if $\nabla h(x)$ is rank deficient.

Nonsmooth Exact Penalty Functions

Because a smooth function of the form (5.56) cannot be exact, we finally turn to nonsmooth penalty functions. The most common choice for merit functions is the class of ℓ_p penalty functions, given by

$$\phi_p(x;\rho) = f(x) + \rho\, \|h(x)\|_p, \tag{5.59}$$

where, from Section 2.2, we recall the definition $\|y\|_p = (\sum_i |y_{(i)}|^p)^{1/p}$ with $p \geq 1$.

For sufficiently large values of ρ, the minimizer of (5.59), if feasible, corresponds to a local solution of (5.1). On the other hand, because these functions are nonsmooth, their minimizers cannot be found using the unconstrained methods in Chapter 3. Nevertheless, nonsmoothness does not preclude the use of $\phi_p(x;\rho)$ as a merit function. These *exact penalty functions* are characterized by the following properties:

- Let x^* and v^* correspond to a local solution of (5.1) that satisfies the first order KKT conditions and the LICQ. Then for $\rho > \rho^* = \|v^*\|_q$ and $\frac{1}{p} + \frac{1}{q} = 1$, x^* is a strict local minimizer for $\phi_p(x,\rho)$ (see [185]). Note the use of a *dual norm* $\|\cdot\|_q$ to $\|\cdot\|_p$ (see Section 2.2).

- Let $\bar{x}$ be a stationary point of $\phi_p(x;\rho)$ for all values of ρ above a positive threshold; then if $h(\bar{x}) = 0$, $\bar{x}$ is a KKT point for (5.1) (see [294]).

- Let $\bar{x}$ be a stationary point of $\phi_p(x;\rho)$ for all values of ρ above a positive threshold, then if $h(\bar{x}) \neq 0$, $\bar{x}$ is an infeasible stationary point for the penalty function (see [294]).

Compared to (5.55) and (5.57) the nonsmooth exact penalty is easy to apply as a merit function. Moreover, under the threshold and feasibility conditions, the above properties show an equivalence between local minimizers of $\phi_p(x,\rho)$ and local solutions of (5.1). As a result, further improvement of $\phi_p(x;\rho)$ cannot be made from a local optimum of (5.1). Moreover, even if a feasible solution is not found, the minimizer of $\phi_p(x;\rho)$ leads to a "graceful failure" at a local minimum of the infeasibility. This point may be useful to the practitioner to flag incorrect constraint specifications in (5.1) or to reinitialize the algorithm. As a result, $\phi_p(x;\rho)$, with $p = 1$ or 2, is widely used in globalization strategies discussed in the remainder of the chapter and implemented in the algorithms in Chapter 6.

The main concern with this penalty function is to find a reasonable value for ρ. While ρ has a well-defined threshold value, it is unknown a priori, and the resulting algorithm can suffer from poor performance and ill-conditioning if ρ is set too high. Careful attention

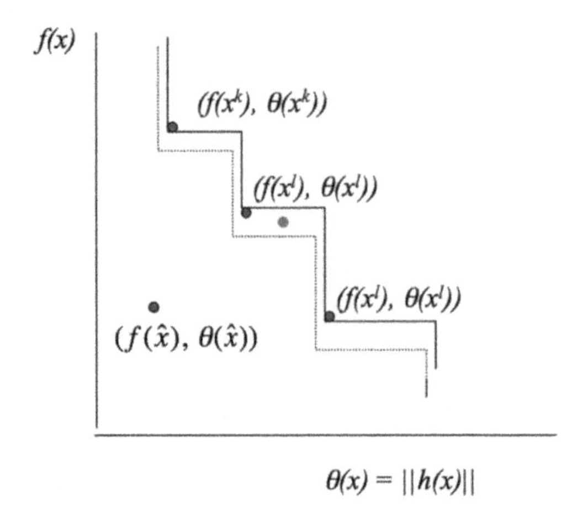

Figure 5.3. *Illustration of filter concepts.*

is needed in updating ρ, which is generally chosen after a step is computed, in order to promote acceptance. For instance, the value of ρ can be adjusted by monitoring the multiplier estimates v^k and ensuring that $\rho > \|v^k\|_q$ as the algorithm proceeds. Recently, more efficient updates for ρ have been developed [78, 79, 294] that are based on allowing ρ to lie within a bounded interval and using the smallest value of ρ in that interval to determine an acceptable step. As described below, this approach provides considerable flexibility in choosing larger steps. Concepts of exact penalty functions will be applied to both the line search and trust region strategies developed in the next section.

5.5.2 Filter Method Concepts

The concept of the filter method was developed by Fletcher and Leyffer [137] as an alternative to merit functions, in order to avoid the determination of an appropriate value of the penalty parameter ρ. Instead, this approach simply requires improvement of *either* $f(x)$ *or* $\|h(x)\|$. The underlying concept of the filter is to interpret the nonlinear program (5.1) as a biobjective optimization problem with two goals: minimizing the constraint infeasibility $\theta(x) = \|h(x)\|$ and minimizing the objective function $f(x)$.[3] Also, more emphasis must be placed on the first measure, since a point has to be feasible in order to be an optimal solution of the nonlinear program. The filter method is illustrated in Figure 5.3. The filter is defined by a set of dominant pairs $(f(x^l),\theta(x^l))$ drawn from previous iterates. As seen in Figure 5.3, a pair $(f(x^l),\theta(x^l))$ is considered dominant if there is no other pair $(f(x^j),\theta(x^j))$, $j \in \{1,\dots,k\}$, that satisfies $f(x^j) \leq f(x^l)$ and $\theta(x^j) \leq \theta(x^l)$. Thus, at iteration k, we would like to improve either $f(x)$ or $\theta(x)$ relative to a filter, defined as a two-dimensional forbidden region $\mathcal{F}_k$ above and to the right of the solid lines in Figure 5.3, defined from the filter pairs.[4]

[3] A scaled norm for $h(x)$ can be used as well.

[4] In [137, 135] $\mathcal{F}_k$ refers to the set of filter pairs, rather than a forbidden region.

A trial point $\hat{x}$ (e.g., generated by a Newton-like step coupled to the line search or trust region approach) is acceptable to the filter if it is *not* in the forbidden region, $\mathcal{F}_k$, *and* a sufficient improvement is realized corresponding to a small fraction of the current infeasibility. In other words, a trial point is accepted if

$$\theta(\hat{x}) \leq (1-\gamma_\theta)\theta(x^k) \text{ or } f(\hat{x}) \leq f(x^k) - \gamma_f \theta(x^k) \text{ and } (f(\hat{x}), \theta(\hat{x})) \notin \mathcal{F}_k \tag{5.60}$$

for some small positive values γ_f, γ_θ. These inequalities correspond to the trial point $(f(\hat{x}),$ $\theta(\hat{x}))$ that lies below the dotted lines in Figure 5.3.

The filter strategy also requires two further enhancements in order to guarantee convergence.

1. Relying solely on criterion (5.60) allows the acceptance of a sequence $\{x^k\}$ that only provides *sufficient reduction* of the constraint violation, but not the objective function. For instance, this can occur if a filter pair is placed at a feasible point with $\theta(x^l) = 0$, and acceptance through (5.60) may lead to convergence to a feasible, but nonoptimal, point. In order to prevent this, we monitor a *switching criterion* to test whenever the constraint infeasibility is too small to promote a sufficient decrease in the objective function. If so, we require that the trial point satisfy a sufficient decrease of the objective function alone.

2. Ill-conditioning at infeasible points can lead to new steps that may be too small to satisfy the filter criteria; i.e., trial points could be "caught" between the dotted and solid lines in Figure 5.3. This problem can be corrected with a feasibility restoration phase which attempts to find a less infeasible point at which to restart the algorithm.

With these elements, the filter strategy, coupled either with a line search or trust region approach, can be shown to converge to either a local solution of (5.1) or a stationary point of the infeasibility, $\theta(x)$. To summarize, the filter strategy is outlined as follows.

- Generate a new trial point, $\hat{x}$. Continue if the trial point is not within the forbidden region, $\mathcal{F}_k$.

- If a switching condition (see (5.85)) is triggered, then accept $\hat{x}$ as x^{k+1} if a sufficient reduction is found in the objective function.

- Else, accept $\hat{x}$ as x^{k+1} if the trial point is acceptable to filter with a margin given by $\theta(x^k)$. Update the filter.

- If the trial point is not acceptable, find another trial point that is closer to x^k.

- If there is insufficient progress in finding trial points, evoke the restoration phase to find a new point, x^R, with a smaller $\theta(x^R)$.

5.5.3 Filter versus Merit Function Strategies

Unlike using the ℓ_p merit function with a fixed penalty parameter, the filter method requires no fixed trade-off function between the objective function and infeasibility. As a result it allows considerable freedom in choosing the next step. A similar advantage can be employed for merit functions as well through the use of flexible penalty parameters, as detailed in

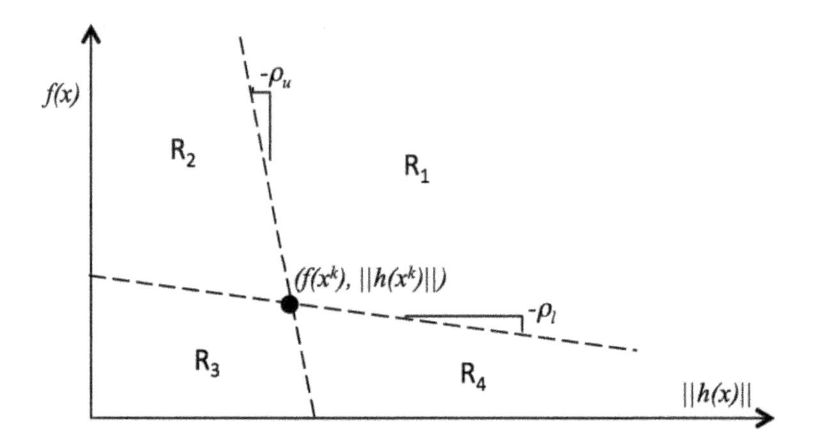

Figure 5.4. *Illustration of merit function concepts with a flexible penalty parameter.*

Curtis and Nocedal [105], and these can be interpreted and compared in the context of filter methods.

Figure 5.4 shows how merit functions can mimic the performance of filter methods in the space of $f(x)$ and $\|h(x)\|$ if we allow the penalty parameter to vary between upper and lower bounds. To see this, consider the necessary condition for acceptance of x^{k+1},

$$f(x^{k+1}) + \rho\|h(x^{k+1})\| < f(x^k) + \rho\|h(x^k)\|, \tag{5.61}$$

$$\text{or if } \|h(x^k)\| \neq \|h(x^{k+1})\|, \quad \rho > \rho_v = \frac{f(x^k) - f(x^{k+1})}{\|h(x^{k+1})\| - \|h(x^k)\|}.$$

Note that the bound on ρ defines the slopes of the lines in Figure 5.4. Clearly, R_1 does not satisfy (5.61), and x^{k+1} is not acceptable in this region. Instead, acceptable regions for x^{k+1} are R_2 and R_3 for $\rho = \rho_u$, while R_3 and R_4 are acceptable regions for $\rho = \rho_l$. If we now allow $\rho \in [\rho_l, \rho_u]$, then all three regions R_2, R_3, and R_4 are acceptable regions for x^{k+1} and these regions behave like one element of the filter method. This is especially true if ρ_u can be kept arbitrarily large and ρ_l can be kept close to zero. Of course, the limits on ρ cannot be set arbitrarily; they are dictated by global convergence properties. As will be seen later, it is necessary to update ρ_u based on a predicted descent property. This property ensures decrease of the exact penalty function as the step size or trust region becomes small. On the other hand, due to nonlinearity, smaller values of ρ may be allowed for satisfaction of (5.61) for larger steps. This relation will guide the choice of ρ_l. More detail on this selection and the resulting convergence properties is given in the next sections. We now consider how both merit function and filter concepts are incorporated within globalization strategies.

5.6 Line Search Methods

Line search strategies were considered in Chapter 3 for unconstrained optimization. In the development of these methods, we were concerned with providing sufficient descent properties as well as step sizes that remain bounded away from zero. These concerns will also be addressed in this section in the context of constrained optimization, for both merit function and filter approaches.

5.6.1 Line Search with Merit Functions

Motivated by the line search algorithm in Chapter 3, this method has the possibility of a straightforward extension, simply through a direct substitution of the merit function for the objective function. As in Section 3.4, we revisit a line search method based on the Armijo criterion (3.31) with backtracking to provide a sufficient decrease in the merit function. However, some care is needed as the exact penalty merit function, $\phi(x;\rho) = f(x)+\rho\|h(x)\|_p$, is nonsmooth. Typically the ℓ_1 penalty (i.e., $p=1$) is used with line search methods.

Translated into this merit function, the Armijo criterion requires satisfaction of

$$\phi_p(x^k+\alpha^k d_x;\rho) \le \phi_p(x^k;\rho)+\eta\alpha^k D_{d_x}\phi_p(x^k;\rho), \tag{5.62}$$

where $x^{k+1}=x^k+\alpha^k d_x$ and $D_{d_x}\phi_p(x^k;\rho)$ is the directional derivative of the merit function along d_x. To evaluate the directional derivative we first consider its definition:

$$D_{d_x}\phi_p(x^k;\rho) = \lim_{\alpha\to 0}\frac{\phi_p(x^k+\alpha d_x;\rho)-\phi_p(x^k;\rho)}{\alpha}. \tag{5.63}$$

For $\alpha \ge 0$ we can apply Taylor's theorem and bound the difference in merit functions by

$$\begin{aligned}\phi_p(x^k+\alpha d_x;\rho)-\phi_p(x^k;\rho) &= (f(x^k+\alpha d_x)-f(x^k))\\ &\quad+\rho(\|h(x^k+\alpha d_x)\|_p-\|h(x^k)\|_p)\\ &\le \alpha\nabla f(x^k)^T d_x+\rho(\|h(x^k)+\alpha\nabla h(x^k)^T d_x\|_p-\|h(x^k)\|_p)\\ &\quad+b_1\alpha^2\|d_x\|^2 \quad \text{for some } b_1>0.\end{aligned} \tag{5.64}$$

Because the Newton step satisfies $\nabla h(x^k)^T d_x = -h(x^k)$, we have

$$\|h(x^k)+\alpha\nabla h(x^k)^T d_x\|_p = (1-\alpha)\|h(x^k)\|_p,$$

thus allowing (5.64) to be simplified to

$$\phi_p(x^k+\alpha d_x;\rho)-\phi_p(x^k;\rho) \le \alpha\nabla f(x^k)^T d_x-\rho\alpha\|h(x^k)\|_p+b_1\alpha^2\|d_x\|^2. \tag{5.65}$$

Using the quadratic term from Taylor's theorem, one can also derive a corresponding lower bound, leading to

$$\alpha\nabla f(x^k)^T d_x-\rho\alpha\|h(x^k)\|_p-b_1\alpha^2\|d_x\|^2 \le \phi_p(x^k+\alpha d_x;\rho)-\phi_p(x^k;\rho). \tag{5.66}$$

Dividing (5.65) and (5.66) by α and applying the definition (5.63) as α goes to zero leads to the following expression for the directional derivative:

$$D_{d_x}\phi_p(x^k;\rho) = \nabla f(x^k)^T d_x-\rho\|h(x^k)\|_p. \tag{5.67}$$

The next step is to tie the directional derivative to the line search strategy to enforce the optimality conditions. In particular, the choices that determine d_x and ρ play important roles. In Sections 5.3 and 5.4 we considered full- and reduced-space representations as well as the use of quasi-Newton approximations to generate d_x. These features also extend to the implementation of the line search strategy, as discussed next.

We start by noting that the Newton steps d_x from (5.5) and (5.16) are equivalent. Substituting (5.15) into the directional derivative (5.67) leads to

$$D_{d_x}\phi_p(x^k;\rho) = \nabla f(x^k)^T(Z^k p_Z + Y^k p_Y) - \rho\|h(x^k)\|_p, \tag{5.68}$$

and from (5.17) we have

$$D_{d_x}\phi_p(x^k;\rho) = \nabla f(x^k)^T Z^k p_Z - \nabla f(x^k)^T Y^k[(A^k)^T Y^k]^{-1}h(x^k) - \rho\|h(x^k)\|_p. \tag{5.69}$$

We now choose $w^k = (Z^k)^T W^k Y^k p_Y$ (or an approximation) and we define $\bar{H}$ as either $(Z^k)^T(W^k + \delta_W I)Z^k$ or the quasi-Newton approximation $\bar{B}^k$. We assume that these matrix choices are bounded, positive definite and that they satisfy

$$s^T\bar{H}^{-1}s \geq c_1\|s\|^2 \quad \text{for all } s \in \mathbb{R}^{n-m}.$$

Using (5.19) and (5.69) leads to

$$\begin{aligned} D_{d_x}\phi_p(x^k;\rho) &= -\nabla f(x^k)^T Z^k \bar{H}^{-1}((Z^k)^T\nabla f(x^k) + w^k) \\ &\quad -\nabla f(x^k)^T Y^k[(A^k)^T Y^k]^{-1}h(x^k) - \rho\|h(x^k)\|_p \\ &\leq -c_1\|(Z^k)^T\nabla f(x^k)\|^2 - \nabla f(x^k)^T Z^k\bar{H}^{-1}w^k \\ &\quad -\nabla f(x^k)^T Y^k[(A^k)^T Y^k]^{-1}h(x^k) - \rho\|h(x^k)\|_p. \end{aligned} \tag{5.70}$$

Finally, using the penalty parameter ρ^k given by

$$\rho^k = \max\left\{\rho^{k-1}, \frac{\nabla f(x^k)^T(Y^k p_Y - Z^k\bar{H}^{-1}w^k)}{\|h(x^k)\|_p} + \epsilon\right\} \tag{5.71}$$

for some $\epsilon > 0$ leads to

$$D_{d_x}\phi_p(x^k;\rho^k) \leq -c_2(\|(Z^k)^T\nabla f(x^k)\|^2 + \|h(x^k)\|_p), \tag{5.72}$$

where $c_2 = \min(c_1, \epsilon)$.

We now consider the following cases:

- In (5.70), the effect of w^k can be modified by multiplying it by a damping factor, ζ, so that

  $$\nabla f(x^k)^T Z^k\bar{H}^{-1}(Z^k f(x^k) + \zeta w^k) \geq \tilde{c}_1\|(Z^k)^T\nabla f(x^k)\|^2$$

  with $c_1 \geq \tilde{c}_1 > 0$. Using

  $$\rho^k = \max\left\{\rho^{k-1}, \frac{\nabla f(x^k)^T Y^k p_Y}{\|h(x^k)\|_p} + \epsilon\right\} \tag{5.73}$$

  allows ρ^k to satisfy (5.72).

- If we set $w^k = 0$ and the multipliers $\bar{v}$ are estimated from (5.21), we have

  $$-\nabla f(x^k)^T Y^k p_Y = \bar{v}^T h(x^k) \leq \|\bar{v}\|_q\|h(x^k)\|_p, \quad \text{where } 1/p + 1/q = 1,$$

and this leads to another bound on ρ^k that satisfies (5.72):

$$\rho^k = \max\{\rho^{k-1}, \|\bar{v}\|_q + \epsilon\}. \tag{5.74}$$

- For the case where $\hat{H} = W^k$ is positive definite for all k or $\hat{H} = B^k$, the full-space BFGS update, we can also choose (5.74) as the update for ρ. From (5.67) and (5.5) we have

$$\begin{aligned} D_{d_x}\phi_p(x^k;\rho) &\le \nabla f(x^k)^T d_x - \rho\|h(x^k)\|_p \\ &\le -d_x^T \hat{H} d_x - d_x^T A^k \bar{v} - \rho\|h(x^k)\|_p \\ &\le -d_x^T \hat{H} d_x + \|h(x^k)\|_p \|\bar{v}\|_q - \rho\|h(x^k)\|_p \\ &\le -d_x^T \hat{H} d_x - \epsilon\|h(x^k)\|_p. \end{aligned} \tag{5.75}$$

  Moreover, from (5.15), nonsingularity of $[Y^k \mid Z^k]$, and linear independence of the columns of A^k (leading to nonsingularity of the KKT matrix in (5.4)), we have for some $c_3, c_4 > 0$,

$$d_x^T \hat{H} d_x \ge c_3(\|p_Z\|^2 + \|p_Y\|^2) \ge c_4(\|h(x^k)\|^2 + \|Z(x^k)^T \nabla f(x^k)\|^2). \tag{5.76}$$

  Substituting this expression into (5.75) and choosing the constants appropriately leads directly to (5.72).

- Finally, for flexible choices of $\rho \in [\rho_l, \rho_u]$ as shown in Figure 5.4, one can choose ρ_u to satisfy (5.72), based on (5.71), (5.73), or (5.74). This approach finds a step size α^k that satisfies the modified Armijo criterion:

$$\phi_p(x^k + \alpha^k d_x; \rho) \le \phi_p(x^k; \rho) + \eta\alpha^k D_{d_x}\phi_p(x^k; \rho_m^k) \tag{5.77}$$

  for some $\rho \in [\rho_l^k, \rho_u^k]$. For this test, ρ_m^k should be chosen large enough to satisfy (5.72). On the other hand, ρ_l^k is chosen as a slowly increasing lower bound given by

$$\rho_l^{k+1} = \min\{\rho_u^k, \rho_l^k + \max\{\epsilon, 0.1(\rho_v - \rho_l^k)\}\}, \tag{5.78}$$

  where ρ_v is as determined in (5.61).

Using any of the above update cases for ρ, we now state the following line search algorithm.

ALGORITHM 5.3.

Choose constants $\eta \in (0, 1/2)$, $\epsilon_1, \epsilon_2 > 0$, and τ, τ' with $0 < \tau < \tau' < 1$. Set $k := 0$ and choose starting points x^0, v^0, and an initial value $\rho_0 > 0$ (or $\rho_l^0, \rho_u^0 > 0$) for the penalty parameter.

For $k \ge 0$, while $\|d_x\| > \epsilon_1$ and $\max(\|\nabla L(x^k, v^k\|, \|h(x^k)\|) > \epsilon_2$:

1. Evaluate $f(x^k), \nabla f(x^k)$, $h(x^k)$, A^k, and the appropriate Hessian terms (or approximations).

2. Calculate the search step d_x and $\bar{v}$ using either the full-space or reduced-space systems, with exact or quasi-Newton Hessian information.

3. Update ρ^k to satisfy (5.71), (5.73), or (5.74), or update ρ_l^k and ρ_u^k for the flexible choice of penalty parameters.

4. Set $\alpha^k = 1$.

5. Test the line search condition (5.62) or (5.77).

6. If the line search condition is not satisfied, choose a new $\alpha^k \in [\tau\alpha^k, \tau'\alpha^k]$ and go to step 5. Otherwise, set

$$x^{k+1} = x^k + \alpha^k d_x, v^{k+1} = \bar{v}. \tag{5.79}$$

Because of the need to update ρ^k at each iteration, we consider only "semiglobal" properties, where stronger assumptions than in Chapter 3 are required on the sequence of iterates.

Assumptions I: The sequence $\{x^k\}$ generated by Algorithm 5.3 is contained in a bounded, convex set D with the following properties:

1. The functions $f : \mathbb{R}^n \to \mathbb{R}$ and $h : \mathbb{R}^n \to \mathbb{R}^m$ are twice differentiable and their first and second derivatives (or approximations) are uniformly bounded in norm over D.

2. The matrix $A(x)$ has full column rank for all $x \in D$, and there exist constants γ_0 and β_0 such that

$$\|Y(x)[A(x)^T Y(x)]^{-1}\| \leq \gamma_0, \quad \|Z(x)\| \leq \beta_0 \tag{5.80}$$

   for all $x \in D$.

3. The Hessian $W(x, v)$ or its approximation is positive definite on the null space of the Jacobian $A(x)^T$ and uniformly bounded.

Lemma 5.9 [54] If Assumptions I hold and if $\rho^k = \rho$ is constant for all sufficiently large k, then there is a positive constant γ_ρ such that for all large k,

$$\phi_p(x^k) - \phi_p(x^{k+1}) \geq \gamma_\rho \left[\|(Z^k)^T \nabla f(x^k)\|^2 + \|h(x^k)\|_p \right]. \tag{5.81}$$

The proof of Lemma 5.9 is a modification from the proof of Lemma 3.2 in [82] and Lemma 4.1 in [54]. Key points in the proof (see Exercise 4) include showing that α^k is always bounded away from zero, using (5.64). Applying (5.72) and the fact that $\|h(x)\|$ has an upper bound in D leads to the desired result.

Finally, Assumptions I and Lemma 5.9 now allow us to prove convergence to a KKT point for a class of line search strategies that use nonsmooth exact penalty merit functions applied to Newton-like steps.

Theorem 5.10 If Assumptions I hold, then the weights $\{\rho^k\}$ are constant for all sufficiently large k, $\lim_{k\to\infty}(\|(Z^k)^T \nabla f(x^k)\| + \|h(x^k)\|) = 0$, and $\lim_{k\to\infty} x^k = x^*$, a point that satisfies the KKT conditions (5.2).

Proof: By Assumptions I, we note that the quantities that define ρ in (5.71), (5.73), or (5.74) are bounded. Therefore, since the procedure increases ρ^k by at least ϵ whenever it changes the penalty parameter, it follows that there is an index k_0 and a value ρ such that for all $k > k_0$, $\rho^k = \rho$ always satisfies (5.71), (5.73), or (5.74).

This argument can be extended to the flexible updates of ρ. Following the proof in [78], we note that ρ_u^k is increased in the same way ρ^k is increased for the other methods, so ρ_u^k eventually approaches a constant value. For ρ_l^k we note that this quantity is bounded above by ρ_u^k and through (5.78) it increases by a finite amount to this value at each iteration. As a result, it also attains a constant value.

Now that ρ^k is constant for $k > k_0$, we have by Lemma 5.9 and the fact that $\phi_p(x^k)$ decreases at each iterate, that

$$\phi_p(x^{k_0};\rho) - \phi_p(x^{k+1};\rho) = \sum_{j=k_0}^{k} (\phi_p(x^j;\rho) - \phi_p(x^{j+1};\rho))$$

$$\geq \gamma_\rho \sum_{j=k_0}^{k} [\|(Z^j)^T \nabla f(x^j)\|^2 + \|h(x^j)\|_p].$$

By Assumptions I, $\phi_p(x,\rho)$ is bounded below for all $x \in D$, so the last sum is finite, and thus

$$\lim_{k\to\infty} [\|(Z^k)^T \nabla f(x^k)\|^2 + \|h(x^k)\|_p] = 0. \tag{5.82}$$

As the limit point, x^*, satisfies $(Z^*)^T \nabla f(x^*) = 0$ and $h(x^*) = 0$, it is easy to see that it satisfies (5.2) as well. □

5.6.2 Line Search Filter Method

For the filter line search method, we again consider a backtracking procedure at iteration k where a decreasing sequence of step sizes $\alpha^{k,l} \in (0,1]$ $(l = 0,1,2,\ldots)$ is tried until some acceptance criterion is satisfied for the corresponding trial point

$$x(\alpha^{k,l}) := x^k + \alpha^{k,l} d_x. \tag{5.83}$$

Here we require a sufficient reduction in one of the measures $\theta = \|h\|$ or f, with a margin related to the infeasibility $\theta(x^k)$. More precisely, for small positive constants $\gamma_\theta, \gamma_f \in (0,1)$, we say that a trial step size $\alpha^{k,l}$ provides sufficient reduction with respect to the current iterate x^k if

$$\theta(x(\alpha^{k,l})) \leq (1-\gamma_\theta)\theta(x^k) \quad \text{or} \quad f(x(\alpha^{k,l})) \leq f(x^k) - \gamma_f \theta(x^k). \tag{5.84}$$

These criteria provide sufficient reduction of the infeasibility, but additional criteria are needed to provide sufficient reduction of the objective function, particularly for near feasible trial points. Here we enforce an *f-type switching condition*

$$m_k(\alpha^{k,l}) < 0 \quad \text{and} \quad [-m_k(\alpha^{k,l})]^{s_f}[\alpha^{k,l}]^{1-s_f} > \delta \left[\theta(x^k)\right]^{s_\theta} \tag{5.85}$$

with constants $\delta > 0$, $s_\theta > 1$, $s_f \geq 1$, where $m_k(\alpha) := \alpha \nabla f(x^k)^T d_x$. If (5.85) holds, then step d_x is a descent direction for the objective function and we require that $\alpha^{k,l}$ satisfies the

Armijo-type condition

$$f(x(\alpha^{k,l})) \leq f(x^k) + \eta_f m_k(\alpha^{k,l}), \tag{5.86}$$

where $\eta_f \in (0, \frac{1}{2})$ is a fixed constant. Note that several trial step sizes may be tried with (5.85) satisfied, but not (5.86). Moreover, for smaller step sizes the f-type switching condition (5.85) may no longer be valid and we revert to the acceptance criterion (5.84).

Note that the second part of (5.85) ensures that the progress for the objective function enforced by the Armijo condition (5.86) is sufficiently large compared to the current constraint violation. Enforcing (5.85) is essential for points that are near the feasible region. Also, the choices of s_f and s_θ allow some flexibility in performance of the algorithm. In particular, if $s_f > 2s_\theta$ (see [403]), both (5.85) and (5.86) will hold for a full step, possibly improved by a second order correction (see Section 5.8) [402], and rapid local convergence is achieved.

During the optimization we make sure that the current iterate x^k is always acceptable to the current filter $\mathcal{F}_k$. At the beginning of the optimization, the forbidden filter region is normally initialized to $\mathcal{F}_0 := \{(\theta, f) \in \mathbb{R}^2 : \theta \geq \theta_{\max}\}$ for some upper bound on infeasibility, $\theta_{\max} > \theta(x^0)$. Throughout the optimization, the filter is then augmented in some iterations after the new iterate x^{k+1} has been accepted. For this, the updating formula

$$\mathcal{F}_{k+1} := \mathcal{F}_k \cup \left\{(\theta, f) \in \mathbb{R}^2 : \theta \geq (1-\gamma_\theta)\theta(x^k) \text{ and } f \geq f(x^k) - \gamma_f \theta(x^k)\right\} \tag{5.87}$$

is used. On the other hand, the filter is augmented only if the current iteration is not an f-type iteration, i.e., if for the accepted trial step size α^k, the f-type switching condition (5.85) does not hold. Otherwise, the filter region is not augmented, i.e., $\mathcal{F}_{k+1} := \mathcal{F}_k$. Instead, the Armijo condition (5.86) must be satisfied, and the value of the objective function is strictly decreased. This is sufficient to prevent cycling.

Finally, finding a trial step size $\alpha^{k,l} > 0$ that provides *sufficient* reduction as defined by criterion (5.84) is not guaranteed. In this situation, the filter method switches to a *feasibility restoration phase*, whose purpose is to find a new iterate x^{k+1} that satisfies (5.84) and is also acceptable to the current filter by trying to decrease the constraint violation. Any iterative algorithm can be applied to find a less infeasible point, and different methods could even be used at different stages of the optimization procedure. To detect the situation where no admissible step size can be found, one can linearize (5.84) (or (5.85) in the case of an f-type iteration) and estimate the smallest step size, $\alpha^{k,\min}$, that allows an acceptable trial point. The algorithm then switches to the feasibility restoration phase when $\alpha^{k,l}$ becomes smaller than $\alpha^{k,\min}$.

If the feasibility restoration phase terminates successfully by delivering an admissible point, the filter is augmented according to (5.87) to avoid cycling back to the problematic point x^k. On the other hand, if the restoration phase is unsuccessful, it should converge to a stationary point of the infeasibility, $\theta(x)$. This "graceful exit" may be useful to the practitioner to flag incorrect constraint specifications in (5.1) or to reinitialize the algorithm.

Combining these elements leads to the following filter line search algorithm.

ALGORITHM 5.4.

Given a starting point x^0; constants $\theta_{\max} \in (\theta(x^0), \infty]$; $\gamma_\theta, \gamma_f \in (0,1)$; $\delta > 0$; $\gamma_\alpha \in (0,1]$; $s_\theta > 1$; $s_f \geq 1$; $\eta_f \in (0, \frac{1}{2})$; $0 < \tau \leq \tau' < 1$.

1. *Initialize* the filter $\mathcal{F}_0 := \{(\theta, f) \in \mathbb{R}^2 : \theta \geq \theta_{\max}\}$ and the iteration counter $k \leftarrow 0$.

2. *Check convergence.* Stop if x^k is a KKT point of the nonlinear program (5.1).

3. *Compute search direction.* Calculate the search step d_x using either the full-space or reduced-space linear systems, with exact or quasi-Newton Hessian information. If this system is detected to be too ill-conditioned, go to feasibility restoration phase in step 8.

4. *Backtracking line search.*

   (a) *Initialize line search.* Set $\alpha^{k,0} = 1$ and $l \leftarrow 0$.

   (b) *Compute new trial point.* If the trial step size becomes too small, i.e., $\alpha^{k,l} < \alpha^{k,\min}$, go to the feasibility restoration phase in step 8. Otherwise, compute the new trial point $x(\alpha^{k,l}) = x^k + \alpha^{k,l} d_x$.

   (c) *Check acceptability to the filter.* If $(f(x(\alpha^{k,l})), \theta(x(\alpha^{k,l}))) \in \mathcal{F}_k$ (i.e., in the forbidden region), then $x(\alpha^{k,l})$ is not acceptable to the filter. Reject the trial step size and go to step 4(e).

   (d) *Check sufficient decrease with respect to current iterate.*

      - *Case* I: $\alpha^{k,l}$ *is an* f*-step size (i.e.,* (5.85) *holds):* If the Armijo condition (5.86) for the objective function holds, accept the trial step and go to step 5. Otherwise, go to step 4(e).
      - *Case* II: $\alpha^{k,l}$ *is not an* f*-step size (i.e.,* (5.85) *is not satisfied):* If (5.84) holds, accept the trial step and go to step 5. Otherwise, go to step 4(e).

   (e) *Choose new trial step size.* Choose $\alpha^{k,l+1} \in [\tau\alpha^{k,l}, \tau'\alpha^{k,l}]$, set $l \leftarrow l+1$, and go back to step 4(b).

5. *Accept trial point.* Set $\alpha^k := \alpha^{k,l}$ and $x^{k+1} := x(\alpha^k)$.

6. *Augment filter if necessary.* If k is not an f-type iteration, augment the filter using (5.87); otherwise leave the filter unchanged, i.e., set $\mathcal{F}_{k+1} := \mathcal{F}_k$.

7. *Continue with next iteration.* Increase the iteration counter $k \leftarrow k+1$ and go back to step 2.

8. *Feasibility restoration phase.* Compute a new iterate x^{k+1} by decreasing the infeasibility measure θ, so that x^{k+1} satisfies the sufficient decrease conditions (5.84) and is acceptable to the filter, i.e., $(\theta(x^{k+1}), f(x^{k+1})) \notin \mathcal{F}_k$. Augment the filter using (5.87) (for x^k) and continue with the regular iteration in step 7.

The filter line search method is clearly more complicated than the merit function approach and is also more difficult to analyze. Nevertheless, the key convergence result for this method can be summarized as follows.

Theorem 5.11 [403] Let Assumption I(1) hold for all iterates and Assumptions I(1,2,3) hold for those iterates that are within a neighborhood of the feasible region ($\|h(x^k)\| \leq \theta_{inc}$ for some $\theta_{inc} > 0$). In this region successful steps can be taken and the restoration phase need not be invoked. Then Algorithm 5.4, with no unsuccessful terminations of the feasibility restoration phase, has the following properties:

$$\lim_{k\to\infty} \|h(x^k)\| = 0 \quad \text{and} \quad \liminf_{k\to\infty} \|\nabla L(x^k, v^k)\| = 0. \tag{5.88}$$

In other words, all limit points are feasible, and if $\{x^k\}$ is bounded, then there exists a limit point x^* of $\{x^k\}$ which is a first order KKT point for the equality constrained NLP (5.1).

The filter line search method avoids the need to monitor and update penalty parameters over the course of the iterations. This also avoids the case where a poor update of the penalty parameter may lead to small step sizes and poor performance of the algorithm. On the other hand, the "lim inf" result in Theorem 5.11 is not as strong as Theorem 5.7, as only convergence of a subsequence can be shown. The reason for this weaker result is that, even for a good trial point, there may be filter information from previous iterates that could invoke the restoration phase. Nevertheless, one should be able to strengthen this property to $\lim_{k\to\infty} \|\nabla_x L(x^k, v^k)\| = 0$ through careful design of the restoration phase algorithm (see [404]).

5.7 Trust Region Methods

In this section we describe trust region globalization strategies with both merit functions and filter methods. As observed for unconstrained problems in Chapter 3, trust region methods, in contrast to line search methods, allow the search direction to be calculated as a function of the step length. This added flexibility leads to methods that have convergence properties superior to line search methods. A key advantage is also a straightforward treatment of indefinite Hessians and negative curvature, so that we can converge to limit points that are truly local minima, not just stationary points. On the other hand, the computational expense for each trust region iteration may be greater than for line search iterations. This remains true for the equality constrained case.

Trust region methods cannot be applied to equality constrained problems in a straightforward way. In Section 5.3.1 Newton steps were developed and were related to the quadratic programming problem (5.6). A simple way to limit the step length would be to extend the QP to

$$\begin{aligned} \min_{d_x} \quad & \nabla f(x^k)^T d_x + \frac{1}{2} d_x^T W^k d_x \\ \text{s.t.} \quad & h(x^k) + (A^k)^T d_x = 0, \\ & \|d_x\| \le \Delta. \end{aligned} \tag{5.89}$$

Here $\|d_x\|$ could be expressed as the *max-norm* leading to simple bound constraints. However, if the trust region is small, there may be no feasible solution for (5.89). As discussed below, this problem can be overcome by using either a merit function or filter approach. With the merit function approach, the search direction d_x is split into two components with separate trust region problems for each, and a composite-step trust region method is developed. In contrast, the filter method applies its restoration phase if the trust region Δ is too small to yield a feasible solution to (5.89).

5.7.1 Trust Regions with Merit Functions

In the Byrd–Omojukun algorithm, the trust region problem is considered through the calculation of *two composite steps*. Because the quadratic program (5.89) may have no solution with adjustment of the trust region, a decomposition is proposed that serves to improve the ℓ_p penalty function:

$$\phi_p(x;\rho) = f(x) + \rho||h(x)||_p.$$

This approach focuses on improving each term separately, and with $p = 2$ quadratic models of each term can be constructed and solved as separate trust region problems. In particular, a trust region problem that improves the quadratic model of the objective function can be shown to give a tangential step, while the trust region problem that reduces a quadratic model of the infeasibility leads to a normal step. Each trust region problem is then solved using the algorithms developed in Chapter 3.

Using the ℓ_2 penalty function, we now postulate a *composite-step trust region* model at iteration k given by

$$m_2(x;\rho) = \nabla f(x^k)^T d_x + \frac{1}{2} d_x^T W^k d_x + \rho \|h(x^k) + (A^k)^T d_x\|_2.$$

Decomposing $d_x = Y^k p_Y + Z^k p_Z$ into normal and tangential components leads to the following subproblems.

Normal Trust Region Subproblem

$$\begin{aligned} \min_{p_Y} \quad & \|h(x^k) + (A^k)^T Y^k p_Y\|_2^2 \\ \text{s.t.} \quad & \|Y^k p_Y\|_2 \le \xi_N \Delta. \end{aligned} \tag{5.90}$$

Typically, $\xi_N = 0.8$. This subproblem is an NLP (with $p_Y \in \mathbb{R}^m$) with a quadratic objective and a single quadratic constraint. Here, we assume A^k to be full column rank so that the problem is strictly convex and can be solved with any of the convex methods in Section 3.5.1 in Chapter 3. Moreover, if m, the number of equations, is large, an inexact solution can be obtained efficiently using Powell's dogleg method.

Tangential Trust Region Subproblem

$$\begin{aligned} & \min_{p_Z} ((Z^k)^T \nabla f(x^k + w^k))^T p_Z + \frac{1}{2} p_Z^T \bar{H}^k p_Z \\ & \text{s.t.} \ ||Y^k p_Y + Z^k p_Z|| \le \Delta, \end{aligned} \tag{5.91}$$

where $\bar{H}^k$ is the reduced Hessian $(Z^k)^T W^k Z^k$, or its approximation $\bar{B}^k$, and where $w^k = (Z^k)^T W^k Y^k p_Y$, or its approximation. This subproblem is an NLP (with $p_Z \in \mathbb{R}^{(n-m)}$) with a quadratic objective and a single quadratic constraint. If $\bar{H}^k$ and w^k are exact, then (5.91) may be a nonconvex model and can be solved with any of the nonconvex methods in Section 3.5.2 in Chapter 3. Moreover, one can obtain an exact, *global* solution to (5.91) using an eigenvalue decomposition and the methods in [282, 100]. Otherwise, if $n - m$ is large, a truncated Newton algorithm is generally applied to solve (5.91) inexactly.

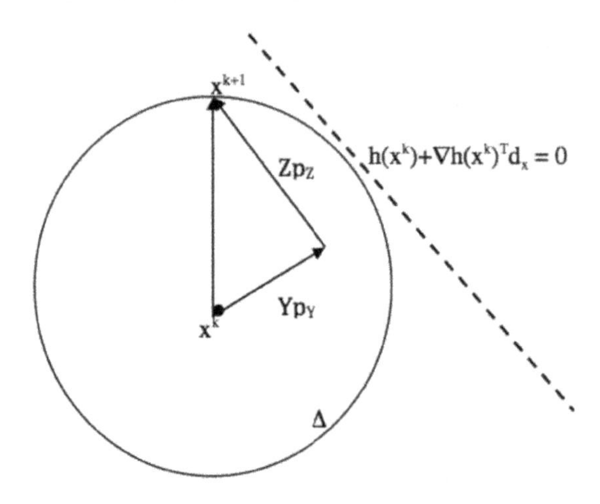

Figure 5.5. *Trust regions for composite steps.*

If acceptable steps have been found that provide sufficient reduction in the merit function, we select our next iterate as

$$x^{k+1} = x^k + Y^k p_Y + Z^k p_Z. \tag{5.92}$$

If not, we reject this step, reduce the size of the trust region, and repeat the calculation of the normal and tangential steps. These steps are shown in Figure 5.5. In addition, we obtain Lagrange multiplier estimates at iteration k using the first order approximation:

$$v^k = -((Y^k)^T A^k)^{-1}(Y^k)^T \nabla f(x^k). \tag{5.93}$$

The criteria for accepting new points and adjusting the trust region are based on predicted and actual reductions of the merit function. The predicted reduction in the quadratic models contains the predicted reduction in the model for the tangential step (q_k),

$$q_k = -((Z^k)^T \nabla f(x^k) + w^k)^T p_Z - \frac{1}{2} p_Z^T \bar{B}^k p_Z, \tag{5.94}$$

and the predicted reduction in the model for the normal step (ϑ_k),

$$\vartheta_k = ||h(x^k)|| - ||h(x^k) + A^k Y^k p_Y||. \tag{5.95}$$

The overall predicted reduction is now

$$pred = q_k + \rho^k \vartheta_k - \nabla f(x^k)^T Y^k p_Y - \frac{1}{2} p_Y^T (Y^k)^T W^k Y^k p_Y, \tag{5.96}$$

where ρ^k is the updated penalty parameter calculated from (5.98). When we update the reduced Hessian with quasi-Newton approximations (and do not know the exact Hessian, W^k), we could omit the term $\frac{1}{2} p_Y^T (Y^k)^T W^k Y^k p_Y$ and obtain a larger value for ρ^k. Otherwise, we use the complete expression (5.96).

The actual reduction is just the difference in the merit function at two successive iterates:

$$ared = f(x^k) - f(x^k + Y^k p_Y + Z^k p_Z) + \rho^k(||h(x^k)|| - ||h(x^k + Y^k p_Y + Z^k p_Z)||). \quad (5.97)$$

The quality of the step is assessed through the ratio $\frac{ared}{pred}$ and, as in Chapter 3, the trust region Δ is adjusted based on how well $ared$ and $pred$ match.

Finally, we choose an update strategy for the penalty parameter ρ that ensures that $pred$ is positive, and ρ is positive and monotonically increasing as we approach the solution. With this in mind, we employ the following updating rule developed in [79]:

$$\rho^k = \max\left[\rho^{k-1}, \frac{-(q^k - \nabla f(x^k)^T Y^k p_Y - \frac{1}{2} p_Y^T (Y^k)^T W^k Y^k p_Y)}{(1-\zeta)\vartheta_k}\right], \quad (5.98)$$

where $\zeta \in (0, 1/2)$. Again, if we use quasi-Newton updates, we would drop the term $\frac{1}{2} p_Y^T (Y^k)^T W^k Y^k p_Y$ in (5.98). A more flexible alternate update for ρ^k is also discussed in [294].

With these elements, the composite-step trust region algorithm follows.

ALGORITHM 5.5.

Set all tolerances and constants; initialize B^0 if using quasi-Newton updates, as well as x^0, v^0, ρ^0, and Δ^0.

For $k \geq 0$ while $\|d_x^k\| > \epsilon_1$ and $\|\nabla L(x^k, v^k)\| > \epsilon_2$:

1. At x^k evaluate $f(x^k)$, $h(x^k)$, $\nabla f(x^k)$, A^k, Y^k, and Z^k.

2. Calculate v^k from (5.93) and if exact second derivatives are available, calculate W^k and $\bar{H}^k = (Z^k)^T W^k Z^k$. Else, calculate $\bar{H}^k = B^k$ from a quasi-Newton update.

3. Calculate $Y^k p_Y$ from (5.90).

4. Set $w^k = (Z^k)^T W^k Y p_Y$ or an approximation.

5. Calculate p_Z from (5.91).

6. Update the penalty parameter (ρ^k) from (5.98).

7. Calculate the merit function $f(x^k) + \rho^k ||h(x^k)||$ as well as $ared$ and $pred$.

   - If $ared/pred < \kappa_1$, set $\Delta^{k+1} = \gamma_1 \Delta^k$,
   - else, if $ared/pred \geq \kappa_2$, set $\Delta^{k+1} = \Delta^k/\gamma_1$,
   - else, if $\kappa_1 \leq ared/pred < \kappa_2$, set $\Delta^{k+1} = \Delta^k$.
   - If $ared/pred > \kappa_0$, then $x^{k+1} = x^k + d_x$. Else, $x^{k+1} = x^k$.

Typical values for this algorithm are $\kappa_0 = 10^{-4}, \kappa_1 = 0.25, \kappa_2 = 0.75, \gamma_1 = 0.5$ and we note that the trust region algorithm is similar to Algorithm 3.3 from Chapter 3.

The trust region algorithm is stated using reduced-space steps. Nevertheless, for large sparse problems, the tangential and normal steps can be calculated more cheaply using only full-space information based on the systems (5.29) and (5.28). Moreover, the tangential problem allows either exact or quasi-Newton Hessian information. The following convergence property holds for all of these cases.

Assumptions II: The sequence $\{x^k\}$ generated by Algorithm 5.5 is contained in a convex set D with the following properties:

1. The functions $f : \mathbb{R}^n \to \mathbb{R}$ and $h : \mathbb{R}^n \to \mathbb{R}^m$ are twice differentiable and their first and second derivatives are Lipschitz continuous and uniformly bounded in norm over D.

2. The matrix $A(x)$ has full column rank for all $x \in D$, and there exist constants γ_0 and β_0 such that
$$\|Y(x)[A(x)^T Y(x)]^{-1}\| \leq \gamma_0, \quad \|Z(x)\| \leq \beta_0, \tag{5.99}$$
for all $x \in D$.

Theorem 5.12 [100] Suppose Assumptions II hold; then Algorithm 5.5 has the following properties:

$$\lim_{k\to\infty} \|h(x^k)\| = 0 \quad \text{and} \quad \lim_{k\to\infty} \|\nabla_x L(x^k, v^k)\| = 0. \tag{5.100}$$

In other words, all limit points are feasible and x^* is a first order stationary point for the equality constrained NLP (5.1).

5.7.2 Filter Trust Region Methods

The filter methods were initially developed in the context of trust region methods. Because these methods incorporate a feasibility restoration step, they can be applied directly to the trust region constrained quadratic programming problem:

$$\begin{aligned} \min_{d_x} \quad & m_f(d_x) = \nabla f(x^k)^T d_x + \frac{1}{2} d_x^T W^k d_x \\ \text{s.t.} \quad & h(x^k) + (A^k)^T d_x = 0, \quad \|d_x\| \leq \Delta^k. \end{aligned} \tag{5.101}$$

The filter trust region method uses similar concepts as in the line search case and can be described by the following algorithm.

ALGORITHM 5.6.
Initialize all constants and choose a starting point (x^0, v^0) and initial trust region Δ^0. Initialize the filter $\mathcal{F}_0 := \{(\theta, f) \in \mathbb{R}^2 : \theta \geq \theta_{\max}\}$ and the iteration counter $k \leftarrow 0$.

For $k \geq 0$ while $\|d_x^k\|, \|d_v^k\| > \epsilon_1$ and $\|\nabla L(x^k, v^k)\| > \epsilon_2$:

1. Evaluate $f(x^k), h(x^k), \nabla f(x^k)$, A^k, and $W(x^k, v^k)$ at x^k and v^k.

2. For the current trust region, check whether a solution exists to (5.101). If so, continue to the next step. If not, add x^k to the filter and invoke the feasibility restoration phase

to find a new point x^r, such that

$$\min_{d_x} \quad m_f(d_x) = \nabla f(x^r)^T d_x + \frac{1}{2} d_x^T W(x^r) d_x \tag{5.102}$$
$$\text{s.t.} \quad h(x^r) + \nabla h(x^r)^T d_x = 0,$$
$$\|d_x\| \leq \Delta^k$$

has a solution and x^r is acceptable to the filter. Set $x^k = x^r$.

3. Determine a trial step d_x that solves (5.101).

4. Check acceptability to the filter.

   - If $x^k + d^k \in \mathcal{F}_k$ or does not satisfy

   $$\theta(x^k + d_x) \leq (1 - \gamma_\theta)\theta(x^k) \text{ or } f(x^k + d_x) \leq f(x^k) - \gamma_f \theta(x^k),$$

     then reject the trial step size, set $x^{k+1} = x^k$ and $\Delta^{k+1} \in [\gamma_0 \Delta^k, \gamma_1 \Delta^k]$ and go to step 1.

   - If

   $$m_f(x^k) - m_f(x^k + d_x) \geq \kappa_\theta \theta(x^k)^2 \tag{5.103}$$

     and

   $$\pi^k = \frac{f(x^k) - f(x^k + d_x)}{m_f(x^k) - m_f(x^k + d_x)} < \eta_1, \tag{5.104}$$

     then we have a rejected f-type step. Set $x^{k+1} = x^k$ and $\Delta^{k+1} \in [\gamma_0 \Delta^k, \gamma_1 \Delta^k]$ and go to step 1.

   - If (5.103) fails, add x^k to the filter. Otherwise, we have an f-type step.

5. Set $x^{k+1} = x^k + d_x$ and choose $\Delta^{k+1} \in [\gamma_1 \Delta^k, \Delta^k]$ if $\pi^k \in [\eta_1, \eta_2)$ or $\Delta^{k+1} \in [\Delta^k, \gamma_2 \Delta^k]$ if $\pi^k \geq \eta_2$.

Typical values for these constants are $\gamma_0 = 0.1, \gamma_1 = 0.5, \gamma_2 = 2, \eta_1 = 0.01, \eta_2 = 0.9$, $\kappa_\theta = 10^{-4}$. Because the filter trust region method shares many of the concepts from the corresponding line search method, it is not surprising that it shares the same convergence property.

Theorem 5.13 [135] Suppose Assumptions I hold; then Algorithm 5.6, with no unsuccessful terminations of the feasibility restoration phase, has the following properties:

$$\lim_{k\to\infty} \|h(x^k)\| = 0 \quad \text{and} \quad \liminf_{k\to\infty} \|\nabla_x L(x^k, v^k)\| = 0. \tag{5.105}$$

5.8 Combining Local and Global Properties

The last part of the convergence analysis requires stitching together the fast local properties of Newton-like steps from Sections 5.3 and 5.4 with global properties of line search or trust region methods in Sections 5.6 and 5.7, respectively. In Chapter 3, we saw that line search and trust region methods applied to unconstrained optimization problems could be designed so that full steps ($\alpha^k = 1$) would be accepted near the solution, thus leading to fast local convergence by Newton-like methods. For constrained optimization problems like (5.1), neither the ℓ_p merit function nor the filter criterion (regardless of a trust region or line search context) can guarantee acceptance of full steps. This failure to allow full steps near the solution is termed the *Maratos effect* and can lead to poor, "creeping" performance near the solution. As a result, a further extension, called a second order correction, needs to be added to the step calculation, whenever the algorithm approaches a neighborhood of the solution.

5.8.1 The Maratos Effect

To appreciate the issues of poor local convergence, we first consider the following classical example, analyzed in [93].

Example 5.14 (Slow Convergence Near Solution). Consider the following equality constrained problem:

$$\begin{aligned} \min \quad & -x_1 + \tau((x_1)^2 + (x_2)^2 - 1) \\ \text{s.t.} \quad & (x_1)^2 + (x_2)^2 = 1 \end{aligned} \tag{5.106}$$

with the constant $\tau > 1$. As shown in Figure 5.6, the solution to this problem is $x^* = [1,\ 0]^T$, $v^* = 1/2 - \tau$ and it can easily be seen that $\nabla_{xx} L(x^*, v^*) = I$ at the solution. To illustrate the effect of slow convergence, we proceed from a feasible point $x^k = [\cos\theta,\ \sin\theta]^T$ and

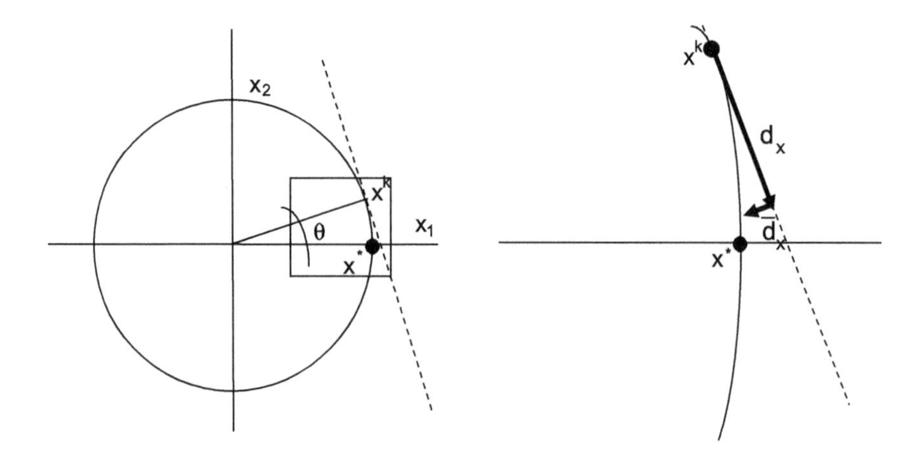

Figure 5.6. *Example of second order correction—detail in right graphic.*

choose θ small enough to start arbitrarily close to the solution. Using the Hessian information at x^* and linearizing (5.5) at x^k allows the search direction to be determined from the linear system

$$\begin{bmatrix} 1 & 0 & 2x_1^k \\ 0 & 1 & 2x_2^k \\ 2x_1^k & 2x_2^k & 0 \end{bmatrix} \begin{bmatrix} d_1 \\ d_2 \\ \bar{v} \end{bmatrix} = \begin{bmatrix} 1-2\tau x_1^k \\ -2\tau x_2^k \\ 1-(x_1^k)^2-(x_2^k)^2 \end{bmatrix} \tag{5.107}$$

yielding $d_1 = \sin^2\theta$, $d_2 = -\sin\theta\cos\theta$ and $x^k + d_x = [\cos\theta + \sin^2\theta,\ \sin\theta(1-\cos\theta)]^T$. As seen in Figure 5.6, $x^k + d_x$ is much closer to the optimum solution, and one can show that the ratio

$$\frac{\|x^k + d_x - x^*\|}{\|x^k - x^*\|^2} = \frac{1}{2} \tag{5.108}$$

is characteristic of quadratic convergence for all values of θ. However, with the ℓ_p merit function, there is no positive value of the penalty parameter ρ that allows this point to be accepted. Moreover, none of the globalization methods described in this chapter accept this point either, because the infeasibility increases:

$$h(x^k) = 0 \rightarrow h(x^k + d_x) = \sin^2\theta$$

and the objective function also increases:

$$f(x^k) = -\cos\theta \rightarrow f(x^k + d_x) = -\cos\theta + (\tau - 1)\sin^2\theta.$$

As a result, $x^k + d_x$ leads to an increase in an ℓ_p merit function and is also not acceptable to a filter method at x^k. Consequently, line search or trust region methods will take smaller steps that lead to slower convergence. ■

To avoid small steps that "creep toward the solution," we develop a second order correction at iteration k. The second order correction is an additional step taken in the range space of the constraint gradients in order to reduce the infeasibility from $x^k + d_x$. Here we define the correction by

$$h(x^k + d_x) + A(x^k)^T \bar{d}_x = 0, \tag{5.109}$$

where $\bar{d}_x = Y^k \bar{p}$ and can be determined from

$$\bar{d}_x = Y^k \bar{p} = -Y^k (A(x^k)^T Y^k)^{-1} h(x^k + d_x). \tag{5.110}$$

Example 5.15 (Maratos Example Revisited). To see the influence of the second order correction, we again refer to Figure 5.6. From (5.110) with $Y^k = A(x^k)$, we have

$$\bar{p} = -\sin^2\theta/4 \quad \text{and} \quad \bar{d}_x = -[\sin^2\theta\cos\theta/2,\ \sin^3\theta/2]^T, \tag{5.111}$$

and we can see that $x^k + d_x + \bar{d}_x$ is even closer to the optimum solution with

$$\begin{aligned} \|x^k + d_x - x^*\| &= 1 - \cos\theta \\ &\geq 1 - \cos\theta - (\cos\theta + 5)(\sin^2\theta)/4 \\ &= \|x^k + d_x + \bar{d}_x - x^*\|, \end{aligned}$$

which is also characteristic of quadratic convergence from (5.108). Moreover, while the infeasibility still increases from $h(x^k) = 0$ to $h(x^k + d_x + \bar{d}_x) = \sin^4\theta/4$, the infeasibility is considerably less than at $x^k + d_x$. Moreover, the objective function becomes

$$f(x^k + d_x + \bar{d}_x) = -(\cos\theta + \sin^2\theta(1 - \cos\theta/2)) + \tau\sin^4\theta/4.$$

For $(1 - \cos\theta/2)/\sin^2\theta > (\tau + \rho)/4$, the objective function clearly decreases from $f(x^k) = -\cos\theta$. Moreover, this point also reduces the exact penalty and is acceptable to the filter. Note that as $\theta \to 0$, this relation holds true for any bounded values of τ and ρ. As a result, the second order correction allows full steps to be taken and fast convergence to be obtained. ∎

Implementation of the second order correction is a key part of all of the globalization algorithms discussed in this chapter. These algorithms are easily modified by first detecting whether the current iteration is within a neighborhood of the solution. This can be done by checking whether the optimality conditions, say $\max(\|h(x^k)\|, \|(Z^k)^T\nabla f(x^k)\|)$ or $\max(\|h(x^k)\|, \|\nabla L(x^k, v^k)\|)$, are reasonably small. If x^k is within this neighborhood, then $\bar{d}_x$ is calculated and the trial point $x^k + d_x + \bar{d}_x$ is examined for sufficient improvement. If the trial point is accepted, we set $x^{k+1} = x^k + d_x + \bar{d}_x$. Otherwise, we discard $\bar{d}_x$ and continue either by restricting the trust region or line search backtracking.

5.9 Summary and Conclusions

This chapter extends all of the topics that relate to Newton-type algorithms in Chapter 3 to equality constrained NLPs. The addition of equality constraints adds a number of subtle complications, which are reflected both in the calculation of search steps and globalization of the algorithm. For the calculation of well-defined Newton-type steps, some care is needed to maintain nonsingularity of the KKT matrix as well as the correct inertia; this is a direct extension of maintaining a positive definite Hessian for unconstrained minimization. Moreover, depending on the problem structure, these steps can be calculated either in the full variable space or in reduced spaces, with steps that are tangential and normal to the linearized constraints. In the absence of exact second derivatives, quasi-Newton methods can be applied, and it should be noted that only the *reduced* Hessian requires a positive definite approximation.

Globalization strategies require that the constraint violations vanish and the objective function be minimized in the constrained space. For this, ℓ_p penalty functions or the more recent filter strategy are typically applied to monitor these competing goals. Either strategy can be applied to the line search or trust region approach. For the line search strategy, a step size α can be found that either improves the exact penalty function or is acceptable to the filter. Otherwise, the extension from the unconstrained case is reasonably straightforward. On the other hand, the trust region method needs to be modified to avoid infeasible solutions of (5.89). This is done either with a composite-step trust region method (with exact penalty) or through a restoration phase (filter strategy).

Finally, unlike the unconstrained case, these globalization methods may not allow full Newton-type steps to be taken near the solution, and "creeping steps" can be observed close to x^*. As a result, a correction step needs to be introduced near the solution to compensate for additional constraint curvature and promote fast convergence.

5.10 Notes for Further Reading

The analysis of the KKT matrix is due to Gould [169] and the inertia corrections are discussed in [404]. The development of reduced-space methods is described in [162, 294, 134] and also in [82, 54], and the development of quasi-Newton methods is described in [405, 83] for the full space and [82, 54] for the reduced space. Local convergence properties with exact and quasi-Newton Hessian information are analyzed and developed in [63, 82, 54]. Penalty functions have a rich literature and the classical treatment of their properties is given in Fiacco and McCormick [131]. The development of properties for ℓ_p penalty functions is due to Han and Mangasarian [185], while augmented Lagrange functions are described in [39, 134, 294]. The treatment of line search and trust region globalization with penalty functions follows along the lines of [294, 100]. Filter methods were developed in [137] and the development and analysis of these methods can be found in [135] and [403] for the trust region and line search cases, respectively. Moreover, a summary on the development of filter methods for a broad variety of algorithms can be found in [139]. Finally, the Maratos effect example is due to [93] and also presented in [294]. In addition to second order corrections, an alternative approach is to monitor improvement after executing a series of full Newton steps. Particular implementations of this concept include the Watchdog algorithm [93] and nonmonotone methods (see [174]), which also lead to fast convergence and overcome the Maratos effect.

5.11 Exercises

1. Prove Theorem 5.1 by modifying the proof of Theorem 2.20.

2. Consider the penalty function given by (5.56) with a finite value of ρ, and compare the KKT conditions of (5.1) with a stationary point of (5.56). Argue why these conditions cannot yield the same solution.

3. Show that the tangential step $d_t = Z^k p_Z$ and normal step $d_n = Y^k p_Y$ can be found from (5.28) and (5.29), respectively.

4. Prove Lemma 5.9 by modifying the proof of Lemma 3.2 in [82] and Lemma 4.1 in [54].

5. Fill in the steps used to obtain the results in Example 5.15. In particular, show that for
$$(1-\cos\theta/2)/\sin^2\theta > \tau + \rho,$$
the second order correction allows a full step to be taken for either the ℓ_p penalty or the filter strategy.

6. Consider the problem
$$\min f(x) = \frac{1}{2}(x_1^2 + x_2^2) \tag{5.112}$$
$$\text{s.t. } h(x) = x_1(x_2 - 1) - \theta x_2 = 0, \tag{5.113}$$
where θ is an adjustable parameter. The solution to this problem is $x_1^* = x_2^* = v^* = 0$ and $\nabla_{xx}^2 L(x^*, v^*) = I$. For $\theta = 1$ and $\theta = 100$, perform the following experiments with $x_i^0 = 1/\theta$, $i = 1,2$.

- Setting $B^k = I$, solve this problem with full-space Newton steps and no globalization.
- Setting $\bar{B}^k = (Z^k)^T Z^k$ and $w^k = 0$, solve the above problem with reduced-space Newton steps and no globalization, using orthogonal bases (5.26).
- Setting $\bar{B}^k = (Z^k)^T Z^k$ and $w^k = 0$, solve the above problem with reduced-space Newton steps and no globalization, using coordinate basis (5.23).
- Setting $\bar{B}^k = (Z^k)^T Z^k$ and $w^k = (Z^k)^T Y^k p_Y$, solve the above problem with reduced-space Newton steps and no globalization, using coordinate basis (5.23).

Chapter 6

Numerical Algorithms for Constrained Optimization

This chapter deals with the solution of nonlinear programs with both equality and inequality constraints. It follows directly from the previous chapter and builds directly on these concepts and algorithms for equality constraints only. The key concept of this chapter is the extension for the treatment of inequality constraints. For this, we develop three approaches: sequential quadratic programming (SQP)-type methods, interior point methods, and nested projection methods. Both global and local convergence properties of these methods are presented and discussed. With these fundamental concepts we then describe and classify many of the nonlinear programming codes that are currently available. Features of these codes will be discussed along with their performance characteristics. Several examples are provided and a performance study is presented to illustrate the concepts developed in this chapter.

6.1 Constrained NLP Formulations

We now return to the general NLP problem (4.1) given by

$$\begin{aligned} \min \quad & f(x) \\ \text{s.t.} \quad & g(x) \leq 0, \\ & h(x) = 0, \end{aligned} \tag{6.1}$$

where we assume that the functions $f(x)$, $h(x)$, and $g(x)$ have continuous first and second derivatives. To derive algorithms for nonlinear programming and to exploit particular problem structures, it is often convenient to consider related, but equivalent NLP formulations. For instance, by introducing nonnegative slack variables $s \geq 0$, we can consider the equivalent bound constrained problem:

$$\begin{aligned} \min \quad & f(x) \\ \text{s.t.} \quad & g(x) + s = 0, s \geq 0, \\ & h(x) = 0, \end{aligned} \tag{6.2}$$

where the nonlinearities are now shifted away from the inequalities.

More generally, we can write the NLP with double-sided bounds given as

$$\begin{aligned} \min \quad & f(x) \\ \text{s.t.} \quad & c(x) = 0, \\ & x_L \leq x \leq x_U \end{aligned} \tag{6.3}$$

with $x \in \mathbb{R}^n$ and $c : \mathbb{R}^n \rightarrow \mathbb{R}^m$.

Through redefinition of the constraints and variables, it is easy to show (see Exercise 1) that problems (6.1), (6.2), and (6.3) are equivalent formulations. On the other hand, it is not true that optimization algorithms will perform identically on these equivalent formulations. Consequently, one may prefer to choose an appropriate NLP formulation that is best suited for the structure of a particular algorithm. For instance, bound constrained problems (6.2) and (6.3) have the advantage that trial points generated from constraint linearizations will never violate the bound constraints. Also, it is easier to identify active constraints through active bounds on variables and to partition the variables accordingly. Finally, by using only bound constraints, the structure of the nonlinearities remains constant over the optimization and is not dependent on the choice of the active set. On the other hand, feasibility measures for merit functions (see Chapter 5) are assessed more accurately in (6.1) than in the bound constrained formulations. This can lead to larger steps and better performance in the line search or trust region method.

For the above reasons and to ease development of the methods as a direct extension of Chapter 5, this chapter derives and analyzes NLP algorithms using the framework of problem (6.3). Nevertheless, the reader should feel comfortable in going back and forth between these NLPs in order to take best advantage of a particular optimization application.

To motivate the development of the algorithms in this chapter, we consider the first order KKT conditions for (6.3) written as

$$\nabla_x L(x^*, u^*, v^*) = \nabla f(x^*) + \nabla c(x^*) v^* - u_L^* + u_U^* = 0, \tag{6.4a}$$

$$c(x^*) = 0, \tag{6.4b}$$

$$0 \leq u_L* \perp (x^* - x_L) \geq 0, \tag{6.4c}$$

$$0 \leq u_U* \perp (x_U - x^*) \geq 0. \tag{6.4d}$$

Note that for the complementarity conditions, $y \perp z$ denotes that $y_{(i)} = 0$ (inclusive) or $z_{(i)} = 0$ for all elements i of these vectors. While it is tempting to solve (6.4) directly as a set of nonlinear equations, conditions (6.4c)–(6.4d) can make the resulting KKT conditions ill-conditioned. This issue will be explored in more detail in Chapter 11.

To deal with these complementarity conditions, two approaches are commonly applied in the design of NLP algorithms. In the *active set* strategy, the algorithm sets either the variable $x_{(i)}$ to its bound, or the corresponding bound multiplier $u_{L,(i)}$ or $u_{U,(i)}$ to zero. Once this assignment is determined, the remaining equations (6.4a)–(6.4b) can be solved for the remaining variables.

On the other hand, for the *interior point* (or barrier) strategy, (6.4c)–(6.4d) are relaxed to

$$U_L(x - x_L) = \mu e, \quad U_U(x_U - x) = \mu e, \tag{6.5}$$

where $\mu > 0$, $e^T = [1, 1, \ldots, 1]$, $U_L = \text{diag}\{u_L\}$, and $U_U = \text{diag}\{u_U\}$. By taking care that the variables x stay strictly within bounds and multipliers u_L and u_U remain strictly positive, the relaxed KKT equations are then solved directly with μ fixed. Then, by solving a sequence of these equations with $\mu \to 0$, we obtain the solution to the original nonlinear program.

With either approach, solving the NLP relies on the solution of nonlinear equations, with globalization strategies to promote convergence from poor starting points.

In this chapter, we develop three general NLP strategies derived from (6.4), using either active set or interior point methods that extend Newton's method to KKT systems. The next section develops the SQP method as a direct active set extension of Chapter 5. In addition to the derivation of the SQP step, line search and trust region strategies are developed and convergence properties are discussed that build directly on the results and algorithms in the previous chapter. Section 6.3 then considers the development of interior point strategies with related Newton-based methods that handle inequalities through the use of barrier functions.

In Section 6.4 we consider a nested, active set approach, where the variables are first partitioned in order to deal with the complementarity conditions (6.4c)–(6.4d). The remaining equations are then solved with (6.4b) nested within (6.4a). These methods apply concepts of reduced gradients as well as gradient projection algorithms. Finally, Section 6.5 considers the implementation of these algorithms and classifies several popular NLP codes using elements of these algorithms. A numerical performance study is presented in this section to illustrate the characteristic features of these methods.

6.2 SQP Methods

SQP is one of the most popular NLP algorithm because it inherits fast convergence properties from Newton methods and can be tailored to a wide variety of problem structures. The basic derivation for SQP follows from taking the KKT conditions (6.4) and assuming that the active bounds are known in advance. Here we define the sets $\mathcal{A}_L = \{i_1, i_2, \ldots, i_j, \ldots\}$ where $x^*_{(i_j)} = (x_L)_{(i_j)}, j = 1, \ldots, |\mathcal{A}_L|$, and $\mathcal{A}_U = \{i'_1, i'_2, \ldots, i'_j, \ldots\}$ where $x^*_{(i'_j)} = (x_U)_{(i'_j)}, j = 1, \ldots, |\mathcal{A}_U|$, as well as active multipliers u_{A_L} and u_{A_U} that correspond to these sets, and matrices E_L and E_U that determine the corresponding variables at bounds. For $j = 1, \ldots, |\mathcal{A}_L|$ and $i = 1, \ldots, n$, E_L has the following elements:

$$\{E_L\}_{ij} = \begin{cases} 1 & \text{if } i = i_j, \\ 0 & \text{otherwise.} \end{cases} \tag{6.6}$$

E_U is defined in a similar manner.

This leads to the following KKT conditions:

$$\begin{aligned} \nabla L(x, u_L, u_U, v) = \nabla f(x) + \nabla c(x) v - E_L u_{A_L} + E_U u_{A_U} = 0, \\ c(x) = 0, E_U^T x = E_U^T x_U, E_L^T x = E_L^T x_L. \end{aligned} \tag{6.7}$$

As in Chapter 5, we can apply Newton's method to this system, leading to the linear system at a particular iteration k:

$$\begin{bmatrix} \nabla_{xx} L(x^k, u_A^k, v^k) & \nabla c(x^k) & -E_L & E_U \\ \nabla c(x^k)^T & 0 & 0 & 0 \\ -E_L^T & 0 & 0 & 0 \\ E_U^T & 0 & 0 & 0 \end{bmatrix} \begin{bmatrix} d_x \\ d_v \\ d_{u_{A_L}} \\ d_{u_{A_U}} \end{bmatrix} = - \begin{bmatrix} \nabla L(x^k, u_A^k, v^k) \\ c(x^k) \\ E_L^T (x_L - x^k) \\ E_U^T (x^k - x_U) \end{bmatrix} \tag{6.8}$$

or, equivalently,

$$\begin{bmatrix} \nabla_{xx}L(x^k,u_A^k,v^k) & \nabla c(x^k) & -E_L & E_U \\ \nabla c(x^k)^T & 0 & 0 & 0 \\ -E_L^T & 0 & 0 & 0 \\ E_U^T & 0 & 0 & 0 \end{bmatrix} \begin{bmatrix} d_x \\ \bar{v} \\ \bar{u}_{A_L} \\ \bar{u}_{A_U} \end{bmatrix} = - \begin{bmatrix} \nabla f(x^k) \\ c(x^k) \\ E_L^T(x_L - x^k) \\ E_U^T(x^k - x_U) \end{bmatrix} \tag{6.9}$$

with $\bar{u}_{A_L} = u_{A_L}^k + d_{u_{A_L}}$, $\bar{u}_{A_U} = u_{A_U}^k + d_{u_{A_U}}$, and $\bar{v} = v^k + d_v$. As shown in Chapter 5, these systems correspond to the first order KKT conditions of the following quadratic programming problem:

$$\begin{aligned} \min_{d_x} \quad & \nabla f(x^k)^T d_x + \frac{1}{2} d_x^T W^k d_x \\ \text{s.t.} \quad & c(x^k) + \nabla c(x^k)^T d_x = 0, \\ & E_U^T(x^k + d_x) = E_U^T x_U, \\ & E_L^T(x^k + d_x) = E_L^T x_L, \end{aligned} \tag{6.10}$$

where $W^k = \nabla_{xx} L(x^k, u^k, v^k) = \nabla_{xx} f(x^k) + \sum_{j=1}^m \nabla_{xx} c_{(j)}(x^k) v_{(j)}^k$ (Note that terms for the linear bound constraints are absent). Moreover, by relaxing this problem to allow selection of active bounds (local to x^k), we can form the following quadratic program (QP) for each iteration k

$$\begin{aligned} \min_{d_x} \quad & \nabla f(x^k)^T d_x + \frac{1}{2} d_x^T W^k d_x \\ \text{s.t.} \quad & c(x^k) + \nabla c(x^k)^T d_x = 0, \\ & x_L \le x^k + d_x \le x_U. \end{aligned} \tag{6.11}$$

In order to solve (6.3), solution of the QP leads to the choice of the likely active set with quadratic programming multipliers $\bar{u}_L$, $\bar{u}_U$, and $\bar{v}$; this choice is then updated at every iteration. Moreover, if (6.11) has the solution $d_x = 0$, one can state the following theorem.

Theorem 6.1 Assume that the QP (6.11) has a solution where the active constraint gradients $A(x^k) = [\nabla c(x^k) \mid E_L \mid E_U]$ have linearly independent columns (LICQ is satisfied) and that the projection of W^k into the null space of $A(x^k)^T$ is positive definite. Then $(x^k, \bar{v}, \bar{u}_L, \bar{u}_U)$ is a KKT point of (6.3) if and only if $d_x = 0$.

Proof: The above assumptions imply that the solution of (6.11) is equivalent to (6.9), and that the KKT matrix in (6.9) is nonsingular. If $(x^k, \bar{u}, \bar{v})$ is a KKT point that satisfies (6.7), then the only solution of (6.9) is $d_x = 0$, $\bar{u}_{A_L}$, $\bar{u}_{A_U}$, and $\bar{v}$. For the converse, if $d_x = 0$ is a solution to (6.11) (and (6.9)), then by inspection of (6.9) we see that the vector $(x^k, \bar{u}_L, \bar{u}_U, \bar{v})$ directly satisfies the KKT conditions (6.7). □

Therefore, by substituting the QP (6.11) for the Newton step, we develop the SQP method as an extension of Algorithm 5.2. Moreover, many of the properties developed in Chapter 5 carry over directly to the treatment of inequality constraints, and analogous algorithmic considerations apply directly to SQP. These issues are briefly considered below and include

- formulation of well-posed QPs, in analogy to nonsingular KKT systems in Chapter 5,
- exploiting problem structure in the solution of (6.4),
- applying quasi-Newton updates when second derivatives are not available, and
- globalization strategies (either line search or trust region) to promote convergence from poor starting points.

6.2.1 The Basic, Full-Space SQP Algorithm

A key requirement for SQP is that the QP subproblem (6.11) have a unique solution at each iteration. In Section 5.2 we explored properties of the KKT matrix, where nonsingularity could be guaranteed by linear independence of the active constraints (LICQ) and a positive definite reduced Hessian (i.e., W^k projected into the null space of the active constraint gradients). With the addition of inequality constraints in (6.11) we also require that the feasible region of the QP be nonempty.

However, since (6.11) is formulated from the gradient and Hessian matrices at iteration k, one cannot always expect that nonsingularity can be guaranteed. For the SQP algorithm based on (6.11), these conditions are handled in the following ways:

- In Chapter 5, loss of positive definiteness in the Hessian was treated through suitable modifications or approximations to W^k and this can also be done for (6.11). In particular, if second derivatives are not available, the Hessian matrix W^k can be approximated by the BFGS update (5.30)–(5.31). In addition, all of the stabilization techniques that maintain positive definiteness carry over to the bound constrained case as well.
- Loss of linear independence for the active constraints is usually handled by the QP solver, and dependent constraints are detected and discarded from the active set. For subproblem (6.11), this detection is somewhat simpler. The QP solver first checks whether $\nabla c(x^k)$ is full rank, and then subsequently checks whether additional active bounds lead to linear dependence.
- Empty feasible regions for (6.11) present a serious challenge to the SQP algorithm. These certainly arise when the nonlinear program (6.3) itself has no feasible region, and this is a frequent occurrence when constraints are specified incorrectly. On the other hand, when $c(x)$ is nonlinear, it is also possible that its linearization at an infeasible point leads to a QP (6.11) without a feasible region.

Handling Infeasible QPs

To deal with infeasible QPs, one needs an alternate strategy to proceed toward the NLP solution, or at least converge to a point where the infeasibility is (locally) minimized. For this purpose, two approaches are commonly considered, often within the same code.

- One can invoke a *restoration phase* in order to find a point that is less infeasible and for which (6.4) is likely to be solved. This approach was discussed in Section 5.5.2 in the context of filter trust region methods and will also be considered later in Section 6.2.3.

- One can formulate a QP subproblem with an *elastic mode* where constraints are relaxed, a feasible region is "created," and a solution can be guaranteed. Below we consider two versions of the elastic mode that are embedded in widely used SQP codes.

To motivate the first relaxation we consider the ℓ_1 penalty function (5.59):

$$\phi_1(x;\rho) = f(x) + \rho\|c(x)\|_1. \tag{6.12}$$

With this function we can formulate the QP subproblem

$$\begin{aligned}
\min \quad & \nabla f(x^k)^T d_x + \frac{1}{2} d_x^T W^k d_x \\
& + \rho \sum_j |c_{(j)}(x^k) + \nabla c_{(j)}(x^k)^T d_x| \\
\text{s.t.} \quad & x_L \le x^k + d_x \le x_U.
\end{aligned}$$

Introducing additional nonnegative variables s and t allows us to reformulate this problem as the so-called ℓ_1 QP:

$$\begin{aligned}
\min_{d_x} \quad & \nabla f(x^k)^T d_x + \frac{1}{2} d_x^T W^k d_x + \rho \sum_j (s_j + t_j) \\
\text{s.t.} \quad & c(x^k) + \nabla c(x^k)^T d_x = s - t, \\
& x_L \le x^k + d_x \le x_U, \quad s,\, t \ge 0.
\end{aligned} \tag{6.13}$$

Because $d_x = 0$, $s_j = \max(0, c_j(x^k))$, $t_j = -\min(0, c_j(x^k))$ is a feasible point for (6.13), and if we choose W^k as a bounded, positive definite matrix, one can always generate a QP solution. Moreover, as discussed in Exercise 2, if $d_x \neq 0$, this solution serves as a descent direction for the ℓ_1 merit function

$$\phi_1(x;\rho) = f(x) + \rho\|c(x)\|_1$$

and allows the SQP algorithm to proceed with a reduction in ϕ_1.

An alternative strategy that allows solution of a smaller QP subproblem is motivated by extending (6.3) to the form

$$\begin{aligned}
\min \quad & f(x) + \bar{M}\bar{\xi} \\
\text{s.t.} \quad & c(x)(1 - \bar{\xi}) = 0, \\
& x_L \le x \le x_U, \quad \bar{\xi} \in [0,1]
\end{aligned} \tag{6.14}$$

with $\bar{M}$ as an arbitrarily large penalty weight on the scalar variable $\bar{\xi}$. Clearly, solving (6.14) with $\bar{\xi} = 0$ is equivalent to solving (6.3). Moreover, a solution of (6.14) with $\bar{\xi} < 1$ also solves (6.3). Finally, if there is no feasible solution for (6.3), one obtains a solution to (6.14) with $\bar{\xi} = 1$. Choosing $\bar{M}$ sufficiently large therefore helps to find solutions to (6.14) with $\bar{\xi} = 0$ when feasible solutions exist for (6.3). For (6.14), we can write the corresponding QP subproblem:

$$\begin{aligned}
\min_{d_x} \quad & \nabla f(x^k)^T d_x + \frac{1}{2} d_x^T W^k d_x + M\xi \\
\text{s.t.} \quad & c(x^k)(1 - \xi) + \nabla c(x^k)^T d_x = 0, \\
& x_L \le x^k + d_x \le x_U, \quad \xi \ge 0
\end{aligned} \tag{6.15}$$

with a slight redefinition of variables, $\xi = \frac{\bar{\xi}-\bar{\xi}^k}{1-\bar{\xi}^k}$, and $M = (1-\bar{\xi}_k)\bar{M}$, assuming that $\bar{\xi}_k < 1$. Again, because $d_x = 0$ and $\xi = 1$ is a feasible point for (6.15), a QP solution can be found. To guarantee a unique solution (and nonsingular KKT matrix) when $\xi > 0$, QP (6.15) can be modified to

$$\begin{aligned} \min_{d_x} \quad & \nabla f(x^k)^T d_x + \frac{1}{2} d_x^T W^k d_x + M(\xi + \xi^2/2) \\ \text{s.t.} \quad & c(x^k)(1-\xi) + \nabla c(x^k)^T d_x = 0, \\ & x_L \leq x^k + d_x \leq x_U, \quad \xi \geq 0. \end{aligned} \tag{6.16}$$

As the objective is bounded over the feasible region with a positive definite Hessian, the QP (6.16) has a unique solution. Moreover, if a solution with $d_x \neq 0$ can be found, the SQP algorithm then proceeds with a reduction in $\phi_1(x,\rho)$, as shown in the following theorem.

Theorem 6.2 Consider the QP (6.16) with a positive definite W^k (i.e., the Hessian of the Lagrange function or its quasi-Newton approximation). If the QP has a solution with $\xi < 1$, then d_x is a descent direction for the ℓ_p merit function $\phi_p(x;\rho) = f(x) + \rho\|c(x)\|_p$ (with $p \geq 1$) for all $\rho > \bar{\rho}$ where $\bar{\rho}$ is chosen sufficiently large.

Proof: We apply the definition of the directional derivative (5.63) to the ℓ_p merit function. From the solution of (6.16) we have

$$\begin{aligned} \nabla c(x^k)^T d_x &= (\xi - 1)c(x^k), \\ x_L &\leq x^k + d_x \leq x_U \end{aligned} \tag{6.17}$$

which allows us to write

$$\|c(x^k) + \alpha \nabla c(x^k)^T d_x\|_p - \|c(x^k)\|_p = \alpha(\xi - 1)\|c(x^k)\|_p$$

and

$$\begin{aligned} \|c(x^k + \alpha d_x)\|_p - \|c(x^k)\|_p &= (\|c(x^k + \alpha d_x)\|_p - \|c(x^k) + \alpha \nabla c(x^k)^T d_x\|_p) \\ &\quad + (\|c(x^k) + \alpha \nabla c(x^k)^T d_x\|_p - \|c(x^k)\|_p) \\ &= (\|c(x^k + \alpha d_x)\|_p - \|c(x^k) + \alpha \nabla c(x^k)^T d_x\|_p) \\ &\quad - \alpha(1-\xi)\|c(x^k)\|_p. \end{aligned}$$

From the derivation of the directional derivatives in Section 5.6.1, we have for some constant $\gamma > 0$,

$$-\gamma(\alpha\|d_x\|)^2 \leq \|c(x^k + \alpha d_x)\| - \|c(x^k) + \alpha \nabla c(x^k)^T d_x\|_p \leq \gamma(\alpha\|d_x\|)^2, \tag{6.18}$$

and applying the definition (5.63) as α goes to zero leads to the following expression for the directional derivative:

$$D_{d_x}\phi_p(x^k;\rho) = \nabla f(x^k)^T d_x - \rho(1-\xi)\|c(x^k)\|_p. \tag{6.19}$$

From the solution of (6.16) we know that

$$\bar{u}_L^T(x_k + d_x - x_L) = 0 \implies \bar{u}_L^T d_x = \bar{u}_L^T(x_L - x_k) \leq 0 \tag{6.20}$$

and similarly $\bar{u}_U^T d_x \geq 0$. The solution of (6.16) therefore yields

$$\nabla f(x^k)^T d_x = -d_x^T (W^k d_x + \nabla c(x^k)\bar{v} + \bar{u}_U - \bar{u}_L),$$
$$\nabla f(x^k)^T d_x \leq -d_x^T (W^k d_x + \nabla c(x^k)\bar{v}),$$

and for $\xi < 1$ and the choice of $\rho \geq \|\bar{v}\|_q + \epsilon$ given by (5.74) (with $\epsilon > 0, 1/p + 1/q = 1$), we have

$$D_{d_x}\phi_p(x^k;\rho) \leq d_x^T W^k d_x - \epsilon(1-\xi)\|c(x^k)\|_p < 0. \quad (6.21)$$

□

Note that this property allows the same selection of ρ as in Section 5.6.1 using (5.74), along with some related alternatives. These lead to a strong descent property for $\phi_1(x,\rho)$ when (5.72) holds. It needs to be emphasized that the assumption that ξ is bounded away from one is crucial for convergence of the SQP algorithm. For NLPs (6.3) with nonlinear equations, this assumption may not always hold. The following example shows a failure of SQP because $\xi = 1$.

Example 6.3 Consider the following NLP:

$$\begin{aligned} \min \quad & x_1 + x_2 \\ \text{s.t.} \quad & 1 + x_1 - (x_2)^2 + x_3 = 0, \\ & 1 - x_1 - (x_2)^2 + x_4 = 0, \\ & 0 \leq x_2 \leq 2, \quad x_3, x_4 \geq 0. \end{aligned} \quad (6.22)$$

This problem has a global solution at $x^* = [-3, 2, 6, 0]^T$. However, by starting from $x^0 = [0, 0.1, 0, 0]^T$, we see that the linearization becomes

$$\begin{aligned} 1.01 + d_1 - 0.2d_2 + d_3 = 0, \\ 0.99 - d_1 - 0.2d_2 + d_4 = 0, \\ -0.1 \leq d_2 \leq 1.9, \quad d_3, d_4 \geq 0. \end{aligned}$$

Adding the two equations and noting that $0.4d_2 \geq 2$ and $d_2 \leq 1.9$ cannot be satisfied simultaneously, we observe that this set of constraints is inconsistent. Nevertheless, a modified QP can be formulated with the relaxation

$$\begin{aligned} 1.01(1-\xi) + d_1 - 0.2d_2 + d_3 = 0, \\ 0.99(1-\xi) - d_1 - 0.2d_2 + d_4 = 0, \\ -0.1 \leq d_2 \leq 1.9, \quad d_3, d_4 \geq 0, \end{aligned}$$

and a feasible region can be created with $\xi \geq 0.62$. On the other hand, if we initialize with $x^0 = [0, 0, 0, 0]^T$, we obtain the linearization

$$\begin{aligned} (1-\xi) + d_1 + d_3 = 0, \\ (1-\xi) - d_1 + d_4 = 0, \\ 0 \leq d_2 \leq 2, \quad d_3, d_4 \geq 0, \end{aligned}$$

and a feasible linearization is admitted only with $\xi = 1$. At this stage, $d_x = 0$ is the only QP solution, and the SQP algorithm terminates. Alternately, a *feasibility restoration phase* could be attempted in order to reduce the infeasibility and avoid termination at x^0. ■

Statement of Basic SQP Method

With this characterization of the QP subproblem and properties of the ℓ_p penalty function, we now state a basic SQP algorithm with line search.

Algorithm 6.1.

Choose constants $\eta \in (0,1/2)$, convergence tolerances $\epsilon_1 > 0$, $\epsilon_2 > 0$, and τ, τ' with $0 < \tau \leq \tau' < 1$. Set $k := 0$ and choose starting points x^0, v^0, u_L^0, u_U^0 and an initial value $\rho^{-1} > 0$ for the penalty parameter.

For $k \geq 0$, while $\|d_x\| > \epsilon_1$ and $\max(\|\nabla L(x^k, u_L^k, u_U^k, v^k\|, \|c(x^k)\|) > \epsilon_2$:

1. Evaluate $f(x^k), \nabla f(x^k), c(x^k), \nabla c(x^k)$, and the appropriate Hessian terms (or approximations).

2. Solve the QP (6.16) to calculate the search step d_x. If $\xi = 1$, stop. The constraints are inconsistent and no further progress can be made.

3. Set $\alpha^k = 1$.

4. Update $\rho^k = \max\{\rho^{k-1}, \|\bar{v}\|_q + \epsilon\}$.

5. Test the line search condition
$$\phi_p(x^k + \alpha^k d_x; \rho^k) \leq \phi_p(x^k; \rho^k) + \eta \alpha^k D_{d_x} \phi_p(x^k; \rho^k). \tag{6.23}$$

6. If (6.23) is not satisfied, choose a new $\alpha^k \in [\tau\alpha^k, \tau'\alpha^k]$ and go to step 5. Otherwise, set
$$x^{k+1} = x^k + \alpha^k d_x, v^{k+1} = \bar{v},$$
$$u_L^{k+1} = \bar{u}_L, u_U^{k+1} = \bar{u}_U.$$

With the relaxation of the QP subproblem, we can now invoke many of the properties developed in Chapter 5 for equality constrained optimization. In particular, from Theorem 6.2 the QP solution leads to a strong descent direction for $\phi_p(x, \rho)$. This allows us to modify Theorem 5.10 to show the following "semiglobal" convergence property for SQP.

Theorem 6.4 Assume that the sequences $\{x^k\}$ and $\{x^k + d_x\}$ generated by Algorithm 6.1 are contained in a closed, bounded, convex region with $f(x)$ and $c(x)$ having uniformly continuous first and second derivatives. If W^k is positive definite and uniformly bounded, LICQ holds and all QPs are solvable with $\xi < 1$, and $\rho \geq \|\bar{v}\| + \epsilon$ for some $\epsilon > 0$, then all limit points of x^k are KKT points of (6.3).

Moreover, the following property provides conditions under which active sets do not change near the solution.

Theorem 6.5 [294, 336] Assume that x^* is a local solution of (6.3) which satisfies the LICQ, sufficient second order conditions, and strict complementarity of the bound multipliers (i.e., $u_{L,(i)}^* + (x_{(i)}^* - x_{L,(i)}) > 0$ and $u_{U,(i)}^* + (x_{U,(i)} - x_{(i)}^*) > 0$ for all vector elements i).

Also, assume that the iterates generated by Algorithm 6.1 converge to x^*. Under these conditions, there exists some neighborhood for iterates $x^k \in \mathcal{N}(x^*)$ where the QP (6.16) solution yields $\xi = 0$ and the active set from (6.16) is the same as that of (6.3).

Therefore, under the assumptions of Theorem 6.5, the active set remains unchanged in a neighborhood around the solution and we can apply the same local convergence results as in the equality constrained case in Chapter 5. Redefining $v^T := [u_L^T, u_U^T, v^T]$, these results can be summarized as follows.

- If we set $W^k = \nabla_{xx} L(x^k, v^k)$, then, by applying Algorithm 6.1 and with x^k and v^k sufficiently close to x^* and v^*, there exists a constant $\hat{C} > 0$ such that

$$\left\| \begin{bmatrix} x^k + d_x - x^* \\ v^k + d_v - v^* \end{bmatrix} \right\| \leq \hat{C} \left\| \begin{bmatrix} x^k - x^* \\ v^k - v^* \end{bmatrix} \right\|^2; \tag{6.24}$$

  i.e., the convergence rate for $\{x^k, v^k\}$ is quadratic.

- If W^k is approximated by a quasi-Newton update B^k, then x^k converges to x^* at a superlinear rate, i.e.,

$$\lim_{k \to \infty} \frac{\|x^k + d_x - x^*\|}{\|x^k - x^*\|} = 0 \tag{6.25}$$

  if and only if

$$\lim_{k \to \infty} \frac{\|Z(x^k)^T (B^k - \nabla_{xx} L(x^*, v^*)) d_x\|}{\|d_x\|} = 0, \tag{6.26}$$

  where $Z(x)$ is an orthonormal representation (i.e., $Z^T Z = I$) of the null space of $A(x)^T = [\nabla c(x^k) \mid E_L \mid E_U]^T$ in a neighborhood of x^*.

- If W^k is approximated by a positive definite quasi-Newton update, B^k, with

$$\lim_{k \to \infty} \frac{\|Z(x^k)^T (B^k - \nabla_{xx} L(x^*, v^*)) Z(x^k) Z(x^k)^T d_x\|}{\|d_x\|} = 0, \tag{6.27}$$

  then x^k converges to x^* at a 2-step superlinear rate, (5.53). Because we expect at most $Z(x^*)^T \nabla_{xx} L(x^*, v^*) Z(x^*)$ (and not $\nabla_{xx} L(x^*, v^*)$) to be positive definite, this property provides the more typical convergence rate for Algorithm 6.1 for most NLP applications.

Finally, as noted in Chapter 5, stitching the global and local convergence properties together follows from the need to take full steps in the neighborhood of the solution. Unfortunately, the Maratos effect encountered in Chapter 5 occurs for the inequality constrained problem as well, with poor, "creeping" performance near the solution due to small step sizes. As noted before, second order corrections and related line searches overcome the Maratos effect and lead to full steps near the solution.

For inequality constrained problems solved with SQP, the Watchdog strategy [93] is a particularly popular line search strategy. This approach is based on ignoring the Armijo descent condition (6.23) and taking a sequence of $i_W \geq 1$ steps with $\alpha = 1$. If one of these points leads to a sufficient improvement in the merit function, then the algorithm proceeds. Otherwise, the algorithm restarts at the beginning of the sequence with the normal line search procedure in Algorithm 6.1. A simplified version of the Watchdog strategy is given below as a modification of Algorithm 6.1.

ALGORITHM 6.2.
Choose constants $\eta \in (0, 1/2)$, convergence tolerances $\epsilon_1 > 0$, $\epsilon_2 > 0$, and τ, τ' with $0 < \tau < \tau' < 1$. Set $k := 0$ and choose starting points x^0, v^0, and an initial value $\rho^{-1} > 0$ for the penalty parameter.

For $k \geq 0$, while $\|d_x\| > \epsilon_1$ and $\max(\|\nabla L(x^k, u_L^k, u_U^k, v^k)\|, \|c(x^k)\|) > \epsilon_2$:

1. Evaluate $f(x^k), \nabla f(x^k), c(x^k), \nabla c(x^k)$, and the appropriate Hessian terms (or approximations).

2. Solve the QP (6.16) to calculate the search step d_x. If $\xi = 1$, stop. The constraints are inconsistent and no further progress can be made.

3. Update $\rho^k = \max\{\rho^{k-1}, \|\bar{v}\|_q + \epsilon\}$.

4. For $i \in \{1, i_W\}$:

   - Set $x^{k+i} = x^{k+i-1} + d_x^i$, where d_x^i is obtained from the solution of QP (6.16) at x^{k+i-1}.
   - Evaluate $f(x^{k+i}), \nabla f(x^{k+i}), c(x^{k+i}), \nabla c(x^{k+i})$ and update the Hessian terms (or approximations).
   - If $\phi_p(x^{k+i}; \rho^k) \leq \phi_p(x^k; \rho^k) + \eta D_{d_x} \phi_p(x^k; \rho^k)$, then accept the step, set $k = k + i - 1$, $v^{k+1} = \bar{v}$, $u_L^{k+1} = \bar{u}_L$, and $u_U^{k+1} = \bar{u}_U$, and proceed to step 1.

5. Reset $x^{k+i_W} = x^k$, $k = k + i_W$, and set $\alpha^k = 1$.

6. Test the line search condition

$$\phi_p(x^k + \alpha^k d_x; \rho^k) \leq \phi_p(x^k; \rho^k) + \eta \alpha^k D_{d_x} \phi_p(x^k; \rho^k). \tag{6.28}$$

7. If (6.28) is not satisfied, choose a new $\alpha^k \in [\tau \alpha^k, \tau' \alpha^k]$ and go to step 6.

8. Set $x^{k+1} = x^k + \alpha^k d_x$, $v^{k+1} = \bar{v}$, $u_L^{k+1} = \bar{u}_L$, and $u_U^{k+1} = \bar{u}_U$.

6.2.2 Large-Scale SQP

The SQP method described above is widely used because it requires few function and gradient evaluations to converge. On the other hand, most QP solvers are implemented with dense linear algebra and become expensive for large NLPs. Large NLPs of the form (6.3) are characterized by many equations and variables (and bounds). Corresponding algorithms exploit the structure of the KKT matrices. Also, with many inequalities, the combinatorial task of choosing an active set may be expensive, and interior point methods in Section 6.3 can be considered. In this section we also discriminate between problems with many degrees of freedom ($n-m$ large) and few degrees of freedom. Examples of process optimization problems with these characteristics are also considered in Chapter 7.

Few Degrees of Freedom

NLPs with $n-m$ small (say, ≤ 100) are frequently encountered in steady state flowsheet optimization, real-time optimization with steady state models, and in parameter estimation. For this case, we consider the decomposition approach in Chapter 5 and extend it to consider bound constraints in the tangential step. This procedure is straightforward for line search SQP methods and can lead to considerable savings in the computational cost of the QP step.

In a similar manner as in Chapter 5, we characterize the solution d_x to (6.11) through two components,

$$d_x = Y^k p_Y + Z^k p_Z, \tag{6.29}$$

where Z^k is a full rank $n \times (n-m)$ matrix spanning the null space of $(\nabla c(x^k))^T$ and Y^k is any $n \times m$ matrix that allows $[Y^k \mid Z^k]$ to be nonsingular. A number of choices for Y^k and Z^k are discussed in Section 5.3.2. These lead to the *normal* and *tangential* steps, $Y^k p_Y$ and $Z^k p_Z$, respectively. Substituting (6.29) into the linearized equations $c(x^k)+\nabla c(x^k)^T d_x = 0$ directly leads to the normal step

$$c(x^k)+(\nabla c(x^k))^T Y^k p_Y = 0.$$

Assuming that $\nabla c(x^k)$ has full column rank, we again write this step, p_Y, as

$$p_Y = -[\nabla c(x^k)^T Y^k]^{-1} c(x^k), \tag{6.30}$$

so that d_x has the form

$$d_x = -Y^k[\nabla c(x^k)^T Y^k]^{-1} c(x^k) + Z^k p_Z. \tag{6.31}$$

To determine the tangential component, p_Z, we substitute (6.30) directly into (6.11) to obtain the following QP in the reduced space:

$$\begin{aligned} \min_{p_Z} \quad & ((Z^k)^T \nabla f(x^k) + w^k)^T p_Z + \frac{1}{2} p_Z^T \bar{B}^k p_Z \\ \text{s.t.} \quad & x_L \leq x^k + Y^k p_Y + Z^k p_Z \leq x_U, \end{aligned} \tag{6.32}$$

where $w^k = (Z^k)^T W^k Y^k p_Y$, or an approximation (see Section 5.6.1), and $\bar{B}^k = (Z^k)^T W^k Z^k$, or a quasi-Newton approximation. Note also that we have dropped the constant terms $\nabla f(x^k)^T Y^k p_Y + \frac{1}{2} p_Y^T (Y^k)^T W^k Y^k p_Y$ from the objective function. To ensure a feasible region, (6.32) is easily modified to include the elastic mode

$$\min_{p_Z} \quad ((Z^k)^T \nabla f(x^k) + w^k)^T p_Z + \frac{1}{2} p_Z^T \bar{B}^k p_Z + M(\xi + \xi^2/2) \tag{6.33}$$
$$\text{s.t.} \quad x_L \leq x^k + (1-\xi) Y^k p_Y + Z^k p_Z \leq x_U.$$

The QP (6.33) can then be solved if $\bar{B}^k$ is positive definite, and this property can be checked through a Cholesky factorization. If $\bar{B}^k$ is not positive definite, the matrix can be modified either by adding to it a sufficiently large diagonal term, say $\delta_R I$, applying a modified Cholesky factorization, or through the BFGS update (5.31) and (5.48).

This reduced-space decomposition takes advantage of the sparsity and structure of $\nabla c(x)$. Moreover, through the formation of basis matrices Z^k and Y^k we can often take advantage of a "natural" partitioning of dependent and decision variables, as will be seen in Chapter 7. The tangential QP requires that the projected Hessian have positive curvature, which is a less restrictive assumption than full-space QP. Moreover, if second derivatives are not available, a BFGS update provides an inexpensive approximation, $\bar{B}^k$, as long as $n-m$ is small. On the other hand, it should be noted that reduced-space matrices in (6.33) are dense and expensive to construct and factorize when $n-m$ becomes large.

Moreover, solution of (6.33) can be expensive if there are many bound constraints. For the solution of (6.11), the QP solver usually performs a search in the space of the primal variables. First a (primal) feasible point is found and the iterates of the solver reduce the quadratic objective until the optimality conditions are satisfied (and dual feasible). On the other hand, the feasible regions for (6.32) or (6.33), may be heavily constrained and difficult to navigate if the problem is ill-conditioned. In this case, dual space QP solvers [26, 40] perform more efficiently as they do not remain in the feasible region of (6.33). Instead, these QP solvers find the unconstrained minimum of the quadratic objective function in (6.33) and then update the solution by successively adding violated inequalities. In this way, the QP solution remains *dual feasible* and the QP solver terminates when its solution is primal feasible.

The reduced-space algorithm can be expressed as a modification of Algorithm 6.1 as follows.

ALGORITHM 6.3.

Choose constants $\eta \in (0,1/2)$, convergence tolerances $\epsilon_1 > 0$, $\epsilon_2 > 0$, and τ, τ' with $0 < \tau \leq \tau' < 1$. Set $k := 0$ and choose starting points x^0, v^0 u_L^0, u_U^0 and an initial value $\rho^{-1} > 0$ for the penalty parameter.

For $k \geq 0$, while $\|d_x\| > \epsilon_1$ and $\max(\|\nabla L(x^k, u_L^k, u_U^k, v^k\|, \|c(x^k)\|) > \epsilon_2$:

1. Evaluate $f(x^k), \nabla f(x^k), c(x^k)$, $\nabla c(x^k)$, and the appropriate Hessian terms (or approximations).

2. Calculate the normal step p_Y from (6.30).

3. Calculate the tangential step p_Z as well as the bound multipliers $\bar{u}_L, \bar{u}_U$ from the QP (6.33). If $\xi = 1$, stop. The constraints are inconsistent and no further progress can be made.

4. Calculate the search step $d_x = Z^k p_Z + Y^k p_Y$ along with the multipliers for the equality constraints:

$$\bar{v} = -((Y^k)^T \nabla c(x^k))^{-1} (Y^k)^T (\nabla f(x^k) + \bar{u}_U - \bar{u}_L). \tag{6.34}$$

5. Set $\alpha^k = 1$.

6. Update $\rho^k = \max\{\rho^{k-1}, \|\bar{v}\|_q + \epsilon\}$.

7. Test the line search condition

$$\phi_p(x^k + \alpha^k d_x; \rho^k) \leq \phi_p(x^k; \rho^k) + \eta \alpha^k D_{d_x} \phi_p(x^k; \rho^k). \tag{6.35}$$

8. If (6.35) is not satisfied, choose a new $\alpha^k \in [\tau \alpha^k, \tau' \alpha^k]$ and go to step 7. Otherwise, set $x^{k+1} = x^k + \alpha^k d_x$, $v^{k+1} = \bar{v}$, $u_L^{k+1} = \bar{u}_L$, and $u_U^{k+1} = \bar{u}_U$.

Many Degrees of Freedom

NLPs with $n - m$ large (say, > 100) are frequently encountered in state estimation, data reconciliation, and dynamic optimization problems. For this case, we need to consider a full-space SQP method that exploits the sparsity of both the Jacobian and Hessian matrices. In addition, it is clear that quasi-Newton updates lead to dense matrices that can become prohibitively expensive to store and factorize. While limited memory updates can also be considered (see Section 5.4.1), an efficient full-space method usually incorporates the use of exact second derivatives. For the construction of this large-scale SQP algorithm, two key aspects need to be considered. First, the algorithm needs to perform appropriate modifications to ensure that the KKT matrix can be factorized. Second, an efficient QP strategy needs to be developed that exploits sparsity, especially in the update of active sets.

Factorization of the KKT matrix with a fixed active set leans heavily on the analysis developed in Chapter 5. A symmetric, indefinite system of the form (6.9) can be factorized with an $L^T BL$ decomposition where the inertia of the KKT matrix can be determined readily. If the inertia is incorrect (does not correspond to n positive eigenvalues, no zero eigenvalues, and the remaining eigenvalues less than zero), then diagonal updates can be made to the KKT matrix to correct for rank-deficient Jacobians and for projected Hessians that are not positive definite. Algorithm 5.2 provides a typical algorithm for this.

Efficient solution of a large, sparse QP is more challenging, particularly due to the combinatorial problem of choosing the correct active set. In this approach, linear systems of the form (6.9) are solved and updated based on feasibility of the primal and dual variables, i.e., variables d_x within bounds and bound multipliers u_L and u_U nonnegative. Both primal and dual space concepts have been adapted to large-scale QP solvers (see [26, 165]). For these methods a key element is the addition and dropping of active constraints and the recalculation of the variables. To see this feature, consider the KKT system (6.9), which we represent as $Kz = r$, with $z^T = [d_x^T, \bar{v}^T, \bar{u}^T]$ and r as the corresponding right-hand side in (6.9). If we update the active set by adding an additional linear constraint, denoted by $e_i^T d_x = b_i$ with its corresponding scalar multiplier $\bar{u}_i$, then (6.9) is augmented as the following linear system:

$$\begin{bmatrix} K & \hat{e}_i \\ \hat{e}_i^T & 0 \end{bmatrix} \begin{bmatrix} z \\ \bar{u}_i \end{bmatrix} = \begin{bmatrix} r \\ b_i \end{bmatrix}, \tag{6.36}$$

where $\hat{e}_i^T = [e_i^T, 0, \ldots, 0]$ has the same dimension as z. Using the Schur complement method, we can apply a previous factorization of K and compute the solution of the augmented system from

$$\bar{u}_i = -(\hat{e}_i^T K^{-1} \hat{e}_i)^{-1}(b_i - \hat{e}_i^T K^{-1} r),$$
$$z = K^{-1}(r - \hat{e}_i \bar{u}_i).$$

Similarly, an active constraint $e_i^T d_x = b_i$ incorporated within K can be dropped by augmenting the KKT system with a slack variable, i.e., $e_i^T d_x + s_i = b_i$, and adding the condition that the corresponding multiplier be set to zero, i.e., $\bar{u}_i = 0$. Defining the vector $\bar{e}_i$ with zero elements, except for a "1" corresponding to the location of the multiplier $\bar{u}_i$ in z, we can form the augmented system

$$\begin{bmatrix} K & \bar{e}_i \\ \bar{e}_i^T & 0 \end{bmatrix} \begin{bmatrix} z \\ s_i \end{bmatrix} = \begin{bmatrix} r \\ 0 \end{bmatrix} \tag{6.37}$$

and compute the solution of this augmented system by the Schur complement method as well. This approach also allows a number of sophisticated matrix updates to the Schur complement as well as ways to treat degeneracy in the updated matrix. More details of these large-scale QP solvers can be found in [26, 163].

Local Convergence Properties

For both large-scale extensions of SQP with line search, the convergence properties remain similar to those developed in Chapter 5 and summarized in the previous section. In particular, global convergence follows from Theorem 6.4 and is obtained by maintaining strong descent directions of the merit function $\phi_p(x, \rho)$. Moreover, the update of ρ follows directly from the rules given in Section 5.6.1. Local convergence rates follow directly from Chapter 5 as well, because we can invoke Theorem 6.5, which ensures constant active sets in the neighborhood of the solution. If no bound constraints are active, the convergence rates are the same as those in Chapter 5. Otherwise, with active bounds at the solution the convergence rates are

- quadratic, if exact second derivatives are available for W^k, or
- 2-step superlinear, if a BFGS update is applied to approximate either $(Z^k)^T W^k Z^k$ or W^k and (6.27) holds.

Finally, to maintain full steps near the solution, and to avoid the Maratos effect, line searches should be modified and built into these large-scale algorithms. Both second order corrections and the Watchdog strategy are standard features in a number of large-scale SQP codes [313, 404, 251].

6.2.3 Extensions of SQP Methods

Up to this point, the discussion of SQP methods has considered only line search methods that use merit functions to promote global convergence. On the other hand, from Chapter 5 we know that additional globalization options can be considered for equality constrained problems. To close this section, we provide brief descriptions of these options for SQP-type methods as well.

SQP with Filter Line Search

As discussed in Chapter 5, the filter method does not use merit functions. Rather it seeks to reduce either the infeasibility or the objective function in order to determine the next point. Section 5.6.2 gives a detailed description of the filter method for equality constrained optimization and summarizes global and local convergence properties for this case. The extension of this approach to inequality constrained problems of the form (6.3) is reasonably straightforward. In particular, Algorithm 5.4 requires the following minor changes for this extension.

- In step 3, the step determination is provided by the QP (6.13).
- If the QP (6.13) in step 3 has no solution, or is "insufficiently consistent," then a restoration phase is invoked and the algorithm proceeds from step 8.
- Once the restoration phase is completed in step 8, the new iterate x_{k+1} must also satisfy the bounds in (6.3).

The modified algorithm enjoys the global convergence property of Theorem 5.11, provided the following additional assumptions are satisfied for all successful iterates k:

- the QP (6.13) has a uniformly positive definite projected Hessian,
- multipliers $\bar{v}$, $\bar{u}_L$, and $\bar{u}_U$ are bounded, and
- the smallest step from x^k to a feasible point of (6.13) is bounded above by the infeasibility measure $M_C \|c(x^k)\|$ for some $M_C > 0$.

These conditions are used to define a *sufficiently consistent* QP that does not require a restoration step.

Finally, the local convergence rates are similar to those of Algorithms 6.1 and 6.3. Again, a modified line search feature, such as the Watchdog strategy or a second order correction, is required to avoid the Maratos effect and take full steps near the solution. If full steps are taken and the solution x^*, v^*, u_L^*, u_U^* satisfies sufficient second order conditions, LICQ, and strict complementarity of the active constraints, then the SQP filter algorithm with the above assumptions has the same convergence rates as in Algorithms 6.1 and 6.3.

SQP with Filter Trust Region

The filter trust region method for equality constrained optimization extends directly to the NLP (6.3). In particular, if a "box" trust region is used, the trust region subproblem can be solved as a QP. Extending problem (5.101) to bound constraints leads directly to

$$\begin{aligned}\min_{d_x} \quad & \nabla f(x^k)^T d_x + \frac{1}{2} d_x^T W^k d_x \\ \text{s.t.} \quad & c(x^k) + \nabla c(x^k)^T d_x = 0, \\ & x_L \leq x^k + d_x \leq x_U, \\ & \|d_x\|_\infty \leq \Delta^k.\end{aligned} \tag{6.38}$$

Therefore, aside from the bound constraints in the trust region subproblem, this approach is the same as in Algorithm 5.6. Moreover, global convergence results are analogous to Theorem 5.13 (see [135]).

Local convergence rates can be developed as well, depending on the choices for the Hessian terms. These results are similar to those obtained for line search SQP methods (see [135, 294]). Nevertheless, as with previous SQP strategies, a modified approach, such as a second order correction, is required to avoid the Maratos effect and take full steps near the solution.

Trust Region with Merit Functions

In Chapter 5, composite-step trust region methods were developed to deal with equality constrained NLPs. Because these algorithms also provide directional flexibility for search steps, in addition to step size, they lead to stronger convergence properties than line search methods. Moreover, because trust region subproblems do not require positive definite Hessian matrices, these methods are not dependent on regularization strategies required for line search methods. On the other hand, there are few active set strategies that are based on composite-step trust region methods, as general inequality constraints are distributed between tangential and normal steps and interfere with the adjustment of trust regions in these subproblems. Instead, we consider a restricted class of NLPs, where the bounds appear only on the variables in the tangential problem.

As defined in Section 5.3.2, and seen later in Section 6.4, we can partition the variables $x^T = [y^T \; z^T]$ with dependent (basic) variables $y \in \mathbb{R}^m$ and describe an NLP with bounds only on the independent (superbasic and nonbasic) variables z as

$$\begin{aligned}\min_{\{y,z\}} \quad & f(y,z) \\ \text{s.t.} \; & c(y,z) = 0, \\ & z_L \leq z \leq z_U\end{aligned} \tag{6.39}$$

with $f : \mathbb{R}^n \to \mathbb{R}$, $c : \mathbb{R}^n \to \mathbb{R}^m$, $z \in \mathbb{R}^{(n-m)}$, $y \in \mathbb{R}^m$, and $n > m$. We also partition

$$\nabla c(x^k)^T = [A_B^k | A_S^k] \tag{6.40}$$

and define the matrices

$$Z^k = \begin{bmatrix} -(A_B^k)^{-1} A_S^k \\ I \end{bmatrix}, \quad Y^k = \begin{bmatrix} I \\ 0 \end{bmatrix}. \tag{6.41}$$

This allows us to write

$$d_x = Y^k p_Y + Z^k p_Z. \tag{6.42}$$

Following the reduced-space decomposition, we can modify the *normal* and *tangential* subproblems, (6.30) and (6.32), respectively, and apply the trust region concepts in Chapter 5. The *normal trust region subproblem*

$$\begin{aligned} \min_{p_Y} \quad & \|c^k + \nabla c(x^k) Y^k p_Y\|^2 \\ \text{s.t.} \quad & \|Y^k p_Y\| \le \xi_N \Delta \end{aligned} \tag{6.43}$$

is typically a large NLP (in $\mathbb{R}^m$) with a quadratic objective and a single quadratic constraint, $\nabla c(x^k)$ is full rank so that this problem is strictly convex, and the problem can be solved inexactly with the aid of sparse matrix factorizations within a dogleg method; $\xi_N = 0.8$ is a typical value.

On the other hand, the *tangential trust region subproblem*

$$\min_{p_Z} ((Z^k)^T \nabla f(x^k) + w^k)^T p_Z + \frac{1}{2} p_Z^T B^k p_Z \tag{6.44}$$

$$\begin{aligned} \text{s.t.} \quad & z_L \le z^k + p_Z \le z_U, \\ & \|p_Z\| \le \tilde{\Delta} \end{aligned}$$

is often a small NLP (in $\mathbb{R}^{(n-m)}$) with a quadratic objective, a single quadratic constraint, and bounds. For this problem we can apply a classical trust region algorithm extended to deal with simple bounds. If $(n-m)$ is small, this problem can be solved with an approach that uses direct matrix factorizations [154]; else an inexact approach can be applied based on truncated Newton methods (see, e.g., [100, 261]). A particular advantage of these approaches is their ability to deal with directions of negative curvature to solve the subproblem within the trust region. For a more comprehensive discussion of these methods, see [281, 154].

In addition, we can coordinate the trust regions, $\xi_N \Delta$ and $\tilde{\Delta}$, in order to enforce $\|Y^k p_Y + Z^k p_Z\| \le \Delta$. Provided that we obtain sufficient reduction in the ℓ_2 merit function $\phi_2(x, \rho)$, this leads to the next iterate:

$$x^{k+1} = x^k + Y^k p_Y + Z^k p_Z. \tag{6.45}$$

If not, we reject this step, reduce the sizes of the trust regions, and repeat the calculation of the quasi-normal and tangential steps. Aside from bound constraints in the tangential subproblem, along with a specialized algorithm to handle them, the approach follows directly from the Byrd–Omojukun algorithm in Section 5.7.1. In particular, with the slight modification for the tangential subproblem, Algorithm 5.5 can be applied directly. Finally, similar global convergence results follow from the analysis in [100].

Assumptions TR: The sequence $\{x^k\}$ generated by the trust region algorithm is contained in a convex set D with the following properties:

1. The functions $f : \mathbb{R}^n \to \mathbb{R}$ and $c : \mathbb{R}^n \to \mathbb{R}^m$ and first and second derivatives are Lipschitz continuous and uniformly bounded in norm over D. Also, W^k is uniformly bounded in norm over D.

2. The matrix $\nabla c(x)$ has full column rank for all $x \in D$, and there exist constants γ_0 and β_0 such that

$$\|Y(x)[\nabla c(x)^T Y(x)]^{-1}\| \le \gamma_0, \quad \|Z(x)\| \le \beta_0, \tag{6.46}$$

for all $x \in D$.

Theorem 6.6 [100] Suppose Assumptions TR hold and strict complementarity holds at the limit points. Then Algorithm 5.5 with the modified tangential subproblem (6.44) has the following properties:

$$\lim_{k\to\infty} \|c(x^k)\| = 0 \quad \text{and} \quad \lim_{k\to\infty} \|Z(x^k)^T(\nabla_x f(x^k) + u_U^k - u_L^k)\| = 0. \tag{6.47}$$

In other words, all limit points x^* are feasible and are first order KKT points for the NLP (6.39).

In addition, fast local convergence rates can be proved, depending on the choices for the Hessian terms. These results are similar to those obtained with line search SQP methods (see [100, 294]).

6.3 Interior Point Methods

As an alternative to active set strategies, one can relax the complementarity conditions using (6.5) and solve a set of relaxed problems for (6.3). While the interior point method can be applied to all of the NLP formulations described in Section 6.1, we simplify the derivation by considering a modified form of (6.3) given by

$$\min_x \quad f(x) \quad \text{s.t. } c(x) = 0,\ x \geq 0. \tag{6.48}$$

Note that a suitable reformulation of variables and equations (see Exercise 3) can be used to represent (6.3) as (6.48). To deal with the nonnegative lower bound on x, we form a log-barrier function and consider the solution of a sequence of equality constrained NLPs of the form

$$\begin{aligned} \min \ & \varphi_{\mu_l}(x) = f(x) - \mu_l \sum_{i=1}^{n_x} \ln(x_{(i)}) \\ \text{s.t. } & c(x) = 0,\ x > 0, \end{aligned} \tag{6.49}$$

where the integer l is the sequence counter and $\lim_{l\to\infty} \mu_l = 0$. Note that the logarithmic barrier term becomes unbounded at $x = 0$, and therefore the path generated by the algorithm must lie in a region that consists of strictly positive points, $x > 0$, in order to determine the solution vector of (6.49). As the barrier parameter decreases, the solutions $x(\mu_l)$ approach the solution of (6.48). This can be seen in Figure 6.1. Also, if $c(x) = 0$ satisfies the LICQ, the solution of (6.49) with $\mu > 0$ satisfies the following first order conditions:

$$\begin{aligned} \nabla f(x) + \nabla c(x)v - \mu X^{-1}e &= 0, \\ c(x) &= 0, \end{aligned} \tag{6.50}$$

where $X = \text{diag}\{x\}$, $e = [1, 1, \ldots, 1]^T$, and the solution vector $x(\mu) > 0$, i.e., it lies strictly in the *interior*. Equations (6.50) are known as the *primal* optimality conditions to denote the absence of multipliers for the inequality constraints. A key advantage of the barrier approach is that decisions about active sets are avoided and the methods developed in Chapter 5 can now be applied. Moreover, the following theorem relates a sequence of solutions to (6.49) to the solution of (6.48).

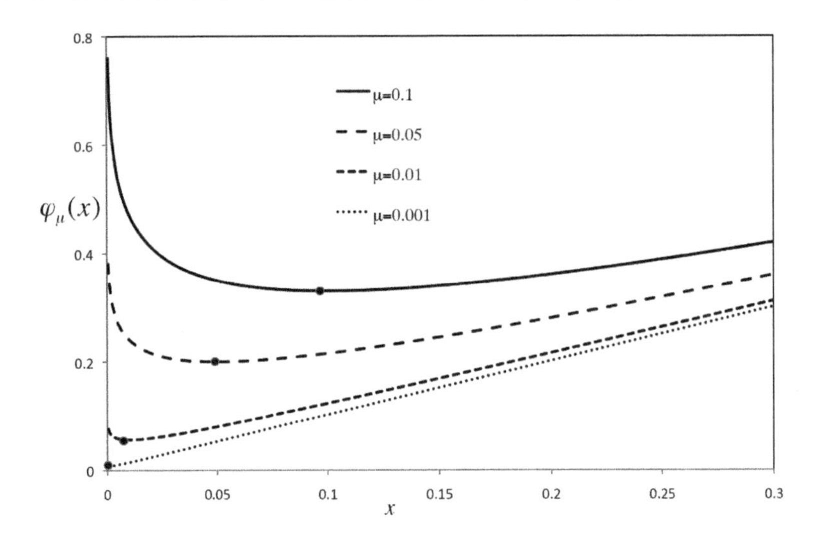

Figure 6.1. *Illustration of barrier solutions for values of* μ *at* 0.1, 0.05, 0.01, 0.001 *for the problem* $\min\ \varphi_\mu(x) = x - \mu\ \ln(x)$. *For this problem* $x(\mu) = \mu$, *and as* $\mu \to 0$ *we approach the solution of the original problem* $\min\ x\ s.t.\ x \geq 0$.

Theorem 6.7 Consider problem (6.48), with $f(x)$ and $c(x)$ at least twice differentiable, and let x^* be a local constrained minimizer of (6.48). Also, assume that the feasible region of (6.48) has a strict interior and that the following sufficient optimality conditions hold at x^*:

1. x^* is a KKT point,
2. the LICQ holds,
3. strict complementarity holds for x^* and for the bound multipliers u^* satisfying the KKT conditions,
4. for $L(x,v) = f(x) + c(x)^T v$, there exists $\omega > 0$ such that $q^T \nabla_{xx} L(x^*, v^*) q \geq \omega \|q\|^2$ for equality constraint multipliers v^* satisfying the KKT conditions and all nonzero $q \in \mathbb{R}^n$ in the null space of the active constraints.

If we now solve a sequence of problems (6.49) with $\mu_l \to 0$, then

- there exists a subsequence of minimizers $(x(\mu_l))$ of the barrier function converging to x^*;
- for every convergent subsequence, the corresponding sequence of barrier multiplier approximations $\mu X^{-1} e$ and $v(\mu)$ is bounded and converges to multipliers u^*, v^* satisfying the KKT conditions for x^*;
- a unique, continuously differentiable vector function $x(\mu)$ of the minimizers of (6.49) exists for $\mu > 0$ in a neighborhood of $\mu = 0$;

- $\lim_{\mu \to 0^+} x(\mu) = x^*$;
- $\|x(\mu_l) - x^*\| = O(\mu_l)$.

The proof of Theorem 6.7 is given through a detailed analysis in [147] (in particular, Theorems 3.12 and Lemma 3.13) and it indicates that nearby solutions of (6.49) provide useful information for bounding solutions of (6.48) for small positive values of μ. On the other hand, note that the gradients of the barrier function in Figure 6.1 are unbounded at the constraint boundary, and this leads to very steep and ill-conditioned response surfaces for $\varphi_\mu(x)$. These can also be observed from the Hessian of problem (6.49):

$$W(x,v) = \nabla^2 f(x) + \sum_{j=1}^{m} \nabla^2 c_{(j)}(x) v_{(j)} + \mu X^{-2}.$$

Consequently, because of the extremely nonlinear behavior of the barrier function, direct solution of barrier problems is often difficult. Moreover, for the subset of variables that are zero at the solution of (6.48), the Hessian becomes unbounded as $\mu \to 0$. On the other hand, if we consider the quantity $q^T W(x^*, v^*) q$, with directions q that lie in the null space of the Jacobian of the active constraints, then the unbounded terms are no longer present and the conditioning of the corresponding reduced Hessian is not affected by the barrier terms.

This motivates an important modification to the primal optimality conditions (6.50) to form the *primal-dual* system. Here we define new "dual" variables along with the equation $Xu = \mu e$ and we replace the barrier contribution to form

$$\nabla f(x) + \nabla c(x) v - u = 0, \tag{6.51a}$$

$$Xu = \mu e, \tag{6.51b}$$

$$c(x) = 0. \tag{6.51c}$$

This substitution and linearization eases the nonlinearity of the barrier terms and now leads to a straightforward relaxation of the KKT conditions for (6.48). Thus, we can view this barrier modification as applying a homotopy method to the primal-dual equations with the homotopy parameter μ, where the multipliers $u \in \mathbb{R}^n$ correspond to the KKT multipliers for the bound constraints (6.48) as $\mu \to 0$. Note that (6.51) for $\mu = 0$, together with "$x, u \geq 0$," are the KKT conditions for the original problem (6.48). This defines the basic step for our interior point method.

We now consider a Newton-based strategy to solve (6.48) via the primal-dual system (6.51). First, using the elements of the primal-dual equations (6.51), we define an error for the optimality conditions of the interior point problem as

$$E_\mu(x,v,u) := \max\{\|\nabla f(x) + \nabla c(x) v - u\|_\infty, \|c(x)\|_\infty, \|Xu - \mu e\|_\infty\}. \tag{6.52}$$

Similarly, $E_0(x,v,z)$ corresponds to (6.52) with $\mu = 0$ and this measures the error in the optimality conditions for (6.48). The overall barrier algorithm terminates if an approximate solution is found that satisfies

$$E_0(\tilde{x}_*, \tilde{v}_*, \tilde{u}_*) \leq \epsilon_{tol}, \tag{6.53}$$

where $\epsilon_{tol} > 0$ is the user-provided error tolerance.

In this description we consider a nested approach to solve (6.49), where the barrier problem is solved (approximately) in an inner loop for a fixed value of μ. Using l as the iteration counter in adjusting μ in the outer loop, we apply the following algorithm.

ALGORITHM 6.4.

Choose constants $\epsilon_{tol} > 0$, $\kappa_\epsilon > 0$, $\kappa_\mu \in (0,1)$, and $\theta_\mu \in (1,2)$. Set $l := 0$ and choose starting points $x^0 > 0$, v^0, u^0, and an initial value $\mu_0 > 0$ for the barrier parameter.

For $l \geq 0$, while $E_0(\tilde{x}(\mu_l), \tilde{v}(\mu_l), \tilde{u}(\mu_l)) \geq \epsilon_{tol}$:

1. Find the *approximate solution* $(\tilde{x}(\mu_l), \tilde{v}(\mu_l), \tilde{u}(\mu_l))$ of the barrier problem (6.49) that satisfies

$$E_{\mu_l}(\tilde{x}(\mu_l, \tilde{v}(\mu_l), \tilde{u}(\mu_l)) \leq \kappa_\epsilon \mu_l. \tag{6.54}$$

2. Set the new barrier parameter

$$\mu_{l+1} = \max\left\{\frac{\epsilon_{tol}}{10}, \min\left\{\kappa_\mu \mu_l, \mu_l^{\theta_\mu}\right\}\right\}. \tag{6.55}$$

Note that the barrier parameter is eventually decreased at a superlinear rate, and this governs the final rate of convergence. On the other hand, it is important that the update (6.55) does not let μ become too small in order to avoid numerical difficulties in the solution of the inner problem (6.51). Typical values for the above constants [404] are $\kappa_\epsilon = 10$, $\kappa_\mu = 0.2$, and $\theta_\mu = 1.5$.

It should also be noted that faster parameter updating strategies are available that vary μ together with the primal and dual variables. These include the predictor-corrector method for the solution of linear and quadratic programs [411] and an adaptive μ strategy for NLPs in [293]. For our nested framework, we fix μ_ℓ and consider the solution of the primal-dual equations in the inner problem.

6.3.1 Solution of the Primal-Dual Equations

In order to solve the barrier problem (6.49) for a given fixed value μ_l, we consider a Newton method with line search applied to the primal-dual equations (6.51) with k chosen for the iteration counter.

Given an iterate (x^k, v^k, u^k) with $x^k, u^k > 0$, search directions (d_x^k, d_v^k, d_u^k) are obtained from the following linearization of (6.51) at (x^k, v^k, u^k):

$$\begin{bmatrix} W^k & \nabla c(x^k) & -I \\ \nabla c(x^k)^T & 0 & 0 \\ U^k & 0 & X^k \end{bmatrix} \begin{bmatrix} d_x^k \\ d_v^k \\ d_u^k \end{bmatrix} = -\begin{bmatrix} \nabla f(x^k) + \nabla c(x^k) v^k - u^k \\ c(x^k) \\ X^k u^k - \mu_l e \end{bmatrix}, \tag{6.56}$$

where $W^k = \nabla_{xx} L(x^k, v^k)$ (or an approximation), $X = \mathrm{diag}\{x\}$, and $U = \mathrm{diag}\{u\}$. Because of the diagonal matrices, (6.56) is easily simplified by first solving the smaller, symmetric linear system

$$\begin{bmatrix} W^k + \Sigma^k & \nabla c(x^k) \\ \nabla c(x^k)^T & 0 \end{bmatrix} \begin{bmatrix} d_x^k \\ d_v^k \end{bmatrix} = -\begin{bmatrix} \nabla \varphi_\mu(x^k) + \nabla c(x^k) v^k \\ c(x^k) \end{bmatrix} \tag{6.57}$$

with $\Sigma^k := (X^k)^{-1}U^k$. This is derived from (6.56) by eliminating the last block row. The vector d_u^k is then obtained from

$$d_u^k = \mu_l (X^k)^{-1} e - u^k - \Sigma^k d_x^k. \tag{6.58}$$

Note that (6.57) has the same structure as the KKT system (5.4) considered in Chapter 5. Consequently, the concepts of (i) maintaining a nonsingular system, (ii) applying quasi-Newton updates efficiently in the absence of second derivatives, and (iii) developing global convergence strategies, carry over directly. In particular, to maintain a nonsingular system, one can modify the matrix in (6.57) and consider the modified linear system

$$\begin{bmatrix} W^k + \Sigma^k + \delta_W I & \nabla c(x^k) \\ \nabla c(x^k)^T & -\delta_A I \end{bmatrix} \begin{bmatrix} d_x^k \\ d_v^k \end{bmatrix} = - \begin{bmatrix} \nabla \varphi_\mu(x^k) + \nabla c(x^k) v^k \\ c(x^k) \end{bmatrix}. \tag{6.59}$$

Again, δ_W and δ_A can be updated using Algorithm 5.2 in order to ensure that (6.59) has the correct inertia.

From the search directions in (6.59) and (6.58), a line search can be applied with step sizes $\alpha^k, \alpha_u^k \in (0,1]$ determined to obtain the next iterate as

$$x^{k+1} := x^k + \alpha^k d_x^k, \tag{6.60a}$$
$$v^{k+1} := v^k + \alpha^k d_v^k, \tag{6.60b}$$
$$u^{k+1} := u^k + \alpha_u^k d_u^k. \tag{6.60c}$$

Since x and u are both positive at an optimal solution of the barrier problem (6.49), and $\varphi_\mu(x)$ is only defined at positive values of x, this property must be maintained for all iterates. For this, we apply the *step-to-the-boundary* rule

$$\alpha_{\max}^k := \max\left\{\alpha \in (0,1] : x^k + \alpha d_x^k \geq (1-\tau_l) x^k\right\}, \tag{6.61a}$$
$$\alpha_u^k := \max\left\{\alpha \in (0,1] : u^k + \alpha d_u^k \geq (1-\tau_l) u^k\right\} \tag{6.61b}$$

with the parameter

$$\tau_l = \max\{\tau_{\min}, 1-\mu_l\}. \tag{6.62}$$

Here, $\tau_{\min}$ is typically set close to 1, say 0.99, and $\tau_l \to 1$ as $\mu_l \to 0$. Note that α_u^k is the actual step size used in (6.60c), while the step size $\alpha^k \in (0, \alpha_{\max}^k]$ for x and v is determined by a backtracking line search procedure that explores a decreasing sequence of trial step sizes $\alpha^{k,j} = 2^{-j}\alpha_{\max}^k$ (with $j = 0,1,2,\ldots$).

6.3.2 A Line Search Filter Method

In order to develop a fast and reliably convergent NLP algorithm, it is tempting to apply the line search strategies from Chapter 5 in a straightforward way. However, the necessity of the "step-to-the-boundary" rule to enforce iterates in the interior, complicates this extension, as seen in the next example.

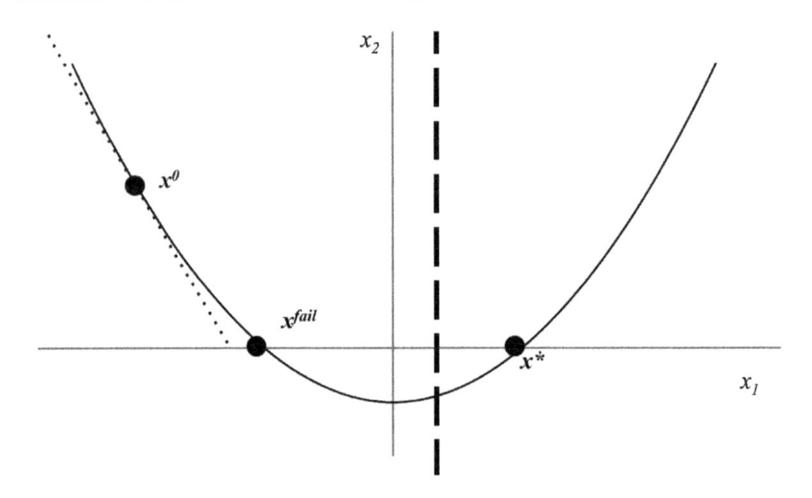

Figure 6.2. *Failure of line search for Newton-based interior point method.*

Example 6.8 Consider the following NLP given by

$$\begin{aligned} \min \quad & x_1 \\ \text{s.t.} \quad & (x_1)^2 - x_2 - 1 = 0, \\ & x_1 - x_3 - 0.5 = 0, \\ & x_2, x_3 \geq 0, \end{aligned} \tag{6.63}$$

and shown in Figure 6.2. For the projection into variables x_1 and x_2, the feasible region lies on the parabola to the right of the dashed line. Also since x_1 is not bounded, the corresponding barrier term can be omitted for this variable. This problem has only one solution at $x^* = [1,0,0.5]^T$ with corresponding multipliers $v^* = [-0.5,0]^T$ and $u_2^* = 0.5, u_3^* = 0$. At the solution the problem satisfies the LICQ and sufficient second order conditions for this problem. However, by starting from an infeasible point, $x^0 = [-2,3,1]^T$, directions calculated from (6.57) and (6.58) are severely truncated by the "step-to-the-boundary" rule (6.61). In less than 5 iterations, $\alpha_{\max} < 10^{-5}$, the variables crash into the bounds for x_2 and x_3 and the algorithm terminates at an arbitrarily infeasible point x^{fail}. Because the *maximum* step size is cut by the variable bounds, this behavior is independent of the particular line search strategy. It is also largely independent of the choice of objective function and the barrier parameter μ. Moreover, the failure occurs for any starting point that lies to the left of the dotted line with $x_2, x_3 > 0$.

Convergence failure in this example is due to the restriction that the Newton step chooses the search direction to satisfy

$$d_x \geq -\tau x^k, c(x^k) + \nabla c(x^k)^T d_x = 0$$

(the dotted line in Figure 6.2). Although the algorithm fails in a different way, this restriction has much in common with the inconsistency of the QP problem seen in Example 6.3. In both cases, the remedy is to produce a step that does not lie in the space of the constraint linearizations. For instance, a Cauchy step exists for an ℓ_p penalty function at x^0 and this does lead to a less infeasible point with improved convergence behavior. Alternately, invoking a *feasibility restoration step* leads to a less infeasible point, from which convergence to the solution is possible. ■

This example motivates the need for a more powerful global convergence approach. In fact, from Chapter 5 the filter line search and trust region methods, as well as the composite-step trust region strategy, overcome the poor behavior of this example.

The filter line search method from Chapter 5 can be applied with few modifications to the solution of the inner problem (6.49) with μ_l fixed. The barrier function $\varphi_\mu(x)$ replaces $f(x)$ in Algorithm 5.4 and the line search proceeds by augmenting the filter with previous iterates or by performing f*-type* iterates which lead to a sufficient reduction in the barrier function. Moreover, if a suitably large step size cannot be found ($\alpha^{k,l} < \alpha^{k,\min}$), then a feasibility restoration phase is invoked. (This occurs in Example 6.8.)

With the application of a reliable globalization strategy, the solution of the Newton step (6.56) can be performed with all of the variations described in Chapter 5. In particular, the symmetric primal-dual system (6.57) can be solved with the following options:

- The Hessian can be calculated directly or a limited memory BFGS update can be performed using the algorithm in section 5.4.1 with $B^0 := B^0 + \Sigma^k$.
- If $n - m$ is small, (6.57) can be decomposed into normal and tangential steps and solved as in Section 5.3.2. Note that the relevant reduced Hessian quantities are modified to $Z^T(W + \Sigma)Z$ and $Z^T(W + \Sigma)Yp_Y$.
- For the decomposed system, quasi-Newton updates for the reduced Hessians can be constructed as described in Section 5.4.2. However, additional terms for $Z^T \Sigma Z$ and $Z^T \Sigma Y p_Y$ must also be evaluated and included in the Newton step calculation.

Convergence Properties

The convergence properties for the filter method follow closely from the results for the equality constrained problem in Chapter 5. In particular, the *global convergence* property from Theorem 5.11 follows directly as long as the iterates generated by the filter line search algorithm for the primal-dual problem remain suitably in the interior of the region defined by the bound constraints, $x \geq 0$. To ensure this condition, we require the following assumptions.

Assumptions F: The sequence $\{x^k\}$ generated by Algorithm 5.4 is contained in a convex set D with the following properties:

1. The functions $f : \mathbb{R}^n \to \mathbb{R}$ and $c : \mathbb{R}^n \to \mathbb{R}^m$, their first and second derivatives, and W^k are uniformly bounded in norm over D.
2. The matrix $\nabla c(x)$ has full column rank for all $x \in D$, and there exist constants γ_0 and β_0 such that

$$\|Y(x)[\nabla c(x)^T Y(x)]^{-1}\| \leq \gamma_0, \quad \|Z(x)\| \leq \beta_0, \tag{6.64}$$

for all $x \in D$.

3. The primal-dual barrier term Σ^k has a bounded deviation from the "primal Hessian" $\mu_l(X^k)^{-2}$. Suitable safeguards have been proposed in [206] to ensure this behavior.

4. The matrices $W^k + \mu(X^k)^{-2}$ are uniformly bounded and uniformly positive definite on the null space of $\nabla c(x^k)^T$.

5. The restoration phase is not invoked arbitrarily close to the feasible region. (In [403] a weaker but more detailed assumption is applied.)

These assumptions lead to the following result.

Theorem 6.9 [403] Suppose Assumptions F hold. Then there exists a constant ϵ_x, so that $x^k \geq \epsilon_x e$ for all k.

This theorem is proved in [403] and can be used directly to establish the following global convergence result.

Theorem 6.10 [403] Suppose Assumptions F(1)-(2) hold for all $x^k \in D$, and the remaining assumptions hold in the neighborhood of the feasible region $\|c(x^k)\| \leq \theta_{inc}$, for some $\theta_{inc} > 0$ (it is assumed that successful steps can be taken here and the restoration phase need not be invoked); then the filter line search algorithm for (6.49) has the following properties:

$$\lim_{k \to \infty} \|c(x^k)\| = 0 \quad \text{and} \quad \liminf_{k \to \infty} \|\nabla \varphi_\mu(x^k) + \nabla c(x^k) v^k\| = 0. \tag{6.65}$$

In other words, all limit points are feasible, and if $\{x^k\}$ is bounded, then there exists a limit point x^* of $\{x^k\}$ which is a first order optimal point for the equality constrained NLP (6.49).

Finally, for the solution of (6.49) with μ_l fixed, one would expect to apply the *local properties* from Chapter 5 as well. Moreover, close to the solution of (6.49) we expect the line search to allow large steps, provided that suitable second order corrections are applied. However, we see that the convergence rate is governed by the value of τ_l in (6.62) and the update of μ_l. While there are a number of updating strategies for μ_l that lead to fast convergence rates (see [294, 206, 293]), these need to be implemented carefully to prevent ill-conditioning of the primal-dual system. Nevertheless, for μ_l updated by (6.55), the convergence rate is at best superlinear for the overall algorithm.

6.3.3 Globalization with Trust Region Methods

The composite-step trust region method developed in the previous chapter has also been applied to the solution of the barrier problem (6.49). By substituting $\varphi_\mu(x)$ for $f(x)$ in Algorithm 5.5, a modified algorithm can be embedded within Algorithm 6.4 to solve the "inner" problem (6.49). On the other hand, as with line search methods, careful attention must be paid to the behavior of the barrier terms in (6.49), and the requirement that x remain positive (i.e., in the interior). As in Chapter 5, we again consider two separate trust region problems to calculate the normal and tangential steps, p_N and p_T, respectively, with $d_x = p_N + p_T$. Both steps are scaled, however, and the trust region is applied to these scaled steps to encourage *larger steps for vector elements away from the boundary*, $x \geq 0$, and smaller steps for elements close to the boundary. The scaled steps are defined as $\tilde{d}_x = (X^k)^{-1} d_x$, $\tilde{p}_N = (X^k)^{-1} p_N$, and $\tilde{p}_T = (X^k)^{-1} p_T$, and the trust region is applied directly to $\tilde{p}_N$ and $\tilde{p}_T$. Using the derivation in Section 5.7.1, we formulate the trust region subproblems as follows:

$$\min_{p_N} \|c(x^k) + \nabla c(x^k)^T p_N\|_2^2$$
$$\text{s.t. } \|(X^k)^{-1} p_N\|_2 \le \xi_N \Delta,$$
$$p_N \ge \frac{-\tau}{2} x^k,$$

where $\xi_N = 0.8$ and $\tau < 1$ is the step to the boundary parameter, as defined in (6.62). The scaled normal problem is given as follows.

Scaled Normal Trust Region Subproblem

$$\min_{p_N} \|c(x^k) + \nabla c(x^k)^T X^k \tilde{p}_N\|_2^2 \tag{6.66a}$$
$$\text{s.t.} \quad \|\tilde{p}_N\|_2 \le \xi_N \Delta, \tag{6.66b}$$
$$\tilde{p}_N \ge \frac{-\tau}{2} e. \tag{6.66c}$$

As in Section 5.7.1, the trust region in problem (6.66) is smaller, in order to leave some additional freedom for the tangential step. Without inequality (6.66c), this convex problem can be solved approximately using the trust region algorithms in Chapter 3. For instance, a dogleg method can be applied, and if (6.66c) is violated at this solution, a backtracking procedure can be performed to satisfy the inequality.

The corresponding tangential trust region subproblem can be formulated from problem (5.91) in Chapter 5 using similar extensions. Alternately, a related subproblem can be formulated that calculates the combined search direction $d_x = p_N + p_T$ directly. In particular, because the normal problem has already reduced the violation of the equality constraint, the equivalent tangential subproblem can be stated as

$$\min_{d_x} (\nabla f(x^k) - \mu (X^k)^{-1} e)^T d_x + \frac{1}{2} d_x^T (W^k + \Sigma^k) d_x$$
$$\text{s.t.} \quad ||(X^k)^{-1} d_x|| \le \Delta,$$
$$\nabla c(x^k)^T d_x = \nabla c(x^k)^T p_N,$$
$$d_x \ge -\tau x^k,$$

where $\nabla c(x^k)^T p_T = 0$ is used. The corresponding scaled problem is given by as follows.

Scaled Tangential Trust Region Subproblem

$$\min_{d_x} (\nabla f(x^k) X^k - \mu e)^T \tilde{d}_x + \frac{1}{2} \tilde{d}_x^T (X^k W^k X^k + X^k \Sigma^k X^k) \tilde{d}_x \tag{6.67a}$$
$$\text{s.t.} \quad ||\tilde{d}_x|| \le \Delta,$$
$$\nabla c(x^k)^T X^k \tilde{d}_x = \nabla c(x^k)^T X^k \tilde{p}_N, \tag{6.67b}$$
$$\tilde{d}_x \ge -\tau e. \tag{6.67c}$$

Without (6.67c), problem (6.67) can be solved with the trust region methods in Chapter 5, even though it may be nonconvex. In particular, for large n the truncated Newton method with conjugate gradient iterations is generally applied to solve (6.67) inexactly. This solution is aided by the scaled barrier term $X^k \Sigma^k X^k$. This is an efficient preconditioner which has eigenvalues clustered close to μ as $\mu \to 0$. As with the normal subproblem, if inequality (6.67c) is violated by the approximate solution, a backtracking procedure can be applied to recover d_x.

Theorem 6.11 Suppose Assumptions TR hold and also assume that $\varphi_\mu(x)$ is bounded below for all $x \in D$, and D is bounded. Then the composite-step trust region algorithm with subproblems (6.66) and (6.67) has the following properties:

$$\lim_{k\to\infty} \|c(x^k)\| = 0 \quad \text{and} \quad \lim_{k\to\infty} \|\nabla_x \varphi_\mu(x^k) + \nabla c(x^k) v^k\| = 0. \tag{6.68}$$

In other words, the limit points x^* are feasible and first order KKT points for problem (6.49).

The above result was originally shown by Byrd, Gilbert, and Nocedal [80] for the barrier trust region method applied to

$$\min f(x) \quad \text{s.t.} \quad g(x) + s = 0, s \geq 0,$$

but modified here with the boundedness assumption for φ_μ and x, so that it applies to problem (6.48). In their study, the resulting algorithm allows a reset of the slack variables to $s^{k+1} = -g(x^{k+1})$, and the assumption on φ_μ is not needed. Moreover, in the absence of an assumption on linear independence of the constraint gradients, they show that one of three situations occurs:

- the iterates x^k approach an infeasible point that is stationary for a measure of infeasibility, and the penalty parameter ρ^k tends to infinity,
- the iterates x^k approach a feasible point but the active constraints are linearly dependent, and the penalty parameter ρ^k tends to infinity, or
- Theorem 6.11 holds, and the active constraints are linearly independent at x^*.

Finally, for the solution of (6.49) with μ_l fixed, one would expect to apply the local convergence properties from Chapter 5 as well. Moreover, close to the solution of (6.49) we expect the trust region to be inactive and allow large steps to be taken, provided that suitable second order corrections are applied. However, the convergence rate is again governed by the value of τ_l in (6.62) and the update of μ_l. While a number of fast updating strategies are available for μ_l (see [294, 206, 293]), some care is needed to prevent ill-conditioning. Again, as in the line search case, for μ updated by (6.55), the convergence rate is at best superlinear for the overall algorithm.

6.4 Nested Strategies

So far, we have developed Newton-based strategies that consider the simultaneous solution of the KKT conditions. In this section, we consider a nested, active set approach instead. Such approaches are especially useful for NLPs with nonlinear objectives and constraints, where it is important for the solver to remain (close to) feasible over the course of iterations. As seen in Example 6.3, linearizations of these functions at infeasible points can lead to poor and misleading information in determining search directions, active constraints, and constraint multipliers. Nested approaches are designed to avoid these problems.

In this approach the variables are first partitioned in order to deal with the constrained portions of (6.3):

$$\begin{aligned} \min \quad & f(x) \\ \text{s.t.} \quad & c(x) = 0, \\ & x_L \leq x \leq x_U. \end{aligned}$$

We partition the variables into three categories: *basic*, *nonbasic*, and *superbasic*. As described in Chapter 4, nonbasic variables are set to their bounds at the optimal solution, while basic variables can be determined from $c(x) = 0$, once the remaining variables are fixed. For linear programs (and NLPs with vertex solutions), a partition for basic and nonbasic variables is sufficient to develop a convergent algorithm. In fact, the LP simplex method proceeds by setting nonbasic variables to their bounds, solving for the basic variables, and then using pricing information (i.e., reduced gradients) to decide which nonbasic variable to relax (the driving variable) and which basic variable to bound (the blocking variable).

On the other hand, if $f(x)$ or $c(x)$ is nonlinear, then superbasic variables are needed as well. Like nonbasic variables, they "drive" the optimization algorithm, but their optimal values are not at their bounds. To develop these methods, we again consider the first order KKT conditions of (6.3):

$$\begin{aligned} \nabla_x L(x^*, u^*, v^*) = \nabla f(x^*) + \nabla c(x^*)^T v^* - u_L^* + u_U^* = 0, \\ c(x^*) = 0, \\ 0 \leq u_L * \perp (x^* - x_L) \geq 0, \\ 0 \leq u_U * \perp (x_U - x^*) \geq 0. \end{aligned}$$

We now partition and reorder the variables x as basic, superbasic, and nonbasic variables: $x = [x_B^T, x_S^T, x_N^T]^T$ with $x_B \in \mathbb{R}^{n_B}$, $x_S \in \mathbb{R}^{n_S}$, $x_N \in \mathbb{R}^{n_N}$, and $n_B + n_S + n_N = n$. This partition is derived from local information and may change over the course of the optimization iterations. Similarly, we partition the constraint Jacobian as

$$\nabla c(x)^T = [A_B(x) \mid A_S(x) \mid A_N(x)]$$

and the gradient of the objective function as

$$\nabla f(x)^T = [f_B(x)^T \mid f_S(x)^T \mid f_N(x)^T].$$

The corresponding KKT conditions can be now written as

$$f_S(x^*) + A_S(x^*)^T v^* = 0, \tag{6.69a}$$

$$f_N(x^*) + A_N(x^*)^T v^* - u_L^* + u_U^* = 0, \tag{6.69b}$$

$$f_B(x^*) + A_B(x^*)^T v^* = 0, \tag{6.69c}$$

$$c(x^*) = 0, \tag{6.69d}$$

$$x^*_{N,(i)} = x_{N,L,(i)} \quad \text{or} \quad x^*_{N,(i)} = x_{N,U,(i)}. \tag{6.69e}$$

Note that (6.69e) replaces the complementarity conditions (6.4c)–(6.4d) with the assumption that $u_L^*, u_U^* \geq 0$. We now consider a strategy of nesting equations (6.69c)–(6.69d) within (6.69a)–(6.69b). At iteration k, for fixed values of x_N^k and x_S^k, we solve for x_B using

(6.69d). We can then solve for $v^k = -(A_B^k)^{-T} f_B^k$ from (6.69c). With x_B^k determined in an inner problem and x_N^k fixed, the superbasic variables can then be updated from (6.69a) and the bound multipliers updated from (6.69b).

Equations (6.69) also provide sensitivity information (i.e., reduced gradients) that indicates how x_B changes with respect to x_S. By considering x_B as an implicit function of x_S, we apply the differential to $c(x) = 0$ to yield

$$0 = dc = A_B\, dx_B + A_S\, dx_S \implies \left(\frac{d\, x_B}{d\, x_S}\right)^T = -(A_B)^{-1} A_S,$$

and we can define the reduced gradient as

$$\begin{aligned}\frac{d\, f(x_B(x_S), x_S)}{d\, x_S} &= f_S + \left(\frac{d\, x_B}{d\, x_S}\right) f_B \\ &= f_S - A_S^T (A_B)^{-T} f_B = f_S + A_S^T v.\end{aligned}$$

Moreover, we define the matrix Z_P that lies in the null space of $\nabla c(x)^T$ as follows:

$$Z_P = \begin{bmatrix} -A_B^{-1} A_S \\ I \\ 0 \end{bmatrix}$$

and the reduced gradient is given by $Z_P^T \nabla f$ in a similar manner as in Sections 5.3.2 and 6.2.3. Following the same reasoning we define the reduced Hessian as $Z_P^T \nabla_{xx} L(x, v) Z_P$. Often, the dimension of the reduced Hessian (n_S) is small and it may be more efficiently approximated by a quasi-Newton update, $\bar{B}$, as in Sections 5.3.2 and 6.2.3.

With these reduced-space concepts, we consider the nested strategy as the solution of bound constrained subproblems in the space of the superbasic variables, i.e.,

$$\begin{aligned} \min_{x_S} \quad & f(x_B(x_S), x_S, x_N) \\ s.t. \quad & x_{S,L} \le x_S \le x_{S,U}. \end{aligned} \tag{6.70}$$

A basic algorithm for this nested strategy, known as the *generalized reduced gradient (GRG)* algorithm, is given as follows.

ALGORITHM 6.5.

Set $l := 0$, select algorithmic parameters including the convergence tolerance ϵ_{tol} and minimum step size $\alpha_{\min}$. Choose a feasible starting point by first partitioning and reordering x into index sets $\mathcal{B}^0, \mathcal{S}^0$, and $\mathcal{N}^0$ for the basic, superbasic, and nonbasic variables, respectively. Then fix x_S^0 and x_N^0 and solve (6.69d) to yield x_B^0. (This may require some trial partitions to ensure that x_B remains within bounds.)

For $l \ge 0$ and the partition sets $\mathcal{B}^l, \mathcal{S}^l$, and $\mathcal{N}^l$, and while $\|\nabla_x L(x^l, v^l, u^l)\| > \epsilon_{tol}$:

1. Determine x^{l+1} from the solution of problem (6.70).

   For $k \ge 0$, while $\|\frac{df(x_B(x_S^{k,l}), x_S^{k,l})}{dx_s}\| \ge \epsilon_{tol}$, at each trial step to solve (6.70), i.e., $x_S(\alpha) = x_S^{k,l} + \alpha d_S$, perform the following steps:

- Attempt to solve $c(x_B, x_S(\alpha), x_N^l) = 0$ to find $x_B(\alpha)$.
- If no solution is obtained because A_B becomes poorly conditioned, exit the solution algorithm and go to step 3.
- If $x_B(\alpha) \notin [x_{B,L}, x_{B,U}]$ or $f(x_B(\alpha), x_S(\alpha), x_N^l)$ is not significantly less than $f(x_B^{k,l}, x_S^{k,l}, x_N^l)$, reduce the step size α. If $\alpha \le \alpha_{\min}$, stop with a line search failure.
- Else, set $x_B^{k+1,l} = x_B(\alpha)$ and set $x_S^{k+1,l} = x_S(\alpha)$.

2. At x^{l+1}, calculate the multiplier estimates v^{l+1} from (6.69c) and u_L^{l+1} and u_U^{l+1} from (6.69b).

3. Update the index sets $\mathcal{B}^{l+1}, \mathcal{S}^{l+1}$, and $\mathcal{N}^{l+1}$ and repartition the variables for the following cases:

   - If x_B cannot be found due to convergence failure, repartition sets $\mathcal{B}$ and $\mathcal{S}$.
   - If $x_B^{k,l} \notin [x_{B,L}, x_{B,U}]$, repartition sets $\mathcal{B}$ and $\mathcal{N}$.
   - If u_L^{l+1} or u_U^{l+1} have negative elements, repartition $\mathcal{S}$ and $\mathcal{N}$. For instance, the nonbasic variable with the most negative multiplier can be added to the superbasic partition.
   - If x_S has elements at their bounds, repartition $\mathcal{S}$ and $\mathcal{N}$. For instance, superbasic variables at upper (lower) bounds with negative (positive) reduced gradients can be added to the nonbasic partition.

Once the algorithm finds the correct variable partition, the algorithm proceeds by updating x_S using reduced gradients in a Newton-type iteration with a line search. Convergence of this step can be guaranteed by Theorem 3.3 presented in Chapter 3.

On the other hand, effective updating of the partition is essential to overcome potential failure cases in step 3 and ensure reliable and efficient performance. The most serious repartitioning case is failure to solve the constraint equations, as can be seen in the example in Figure 6.3. Following Algorithm 6.5, and ignoring nonbasic variables, leads to the horizontal steps taken for the superbasic variable. Each step requires solution to the constraint equation (the dashed line), taken by vertical steps for basic variables. In the left graphic, various values of the superbasic variable fail to solve the constraint and find the basic variable; the step size is therefore reduced. Eventually, the algorithm terminates at a point where the basis matrix A_B is singular (the slope for the equality is infinity at this point). As seen in the right graphic, repartitioning (rotating by 90°) leads to a different basis matrix that is nonsingular, and Algorithm 6.5 terminates successfully. In most implementations, repartitioning sets $\mathcal{B}$ and $\mathcal{S}$ in step 3 is often governed by updating heuristics that can lead to efficient performance, but do not have well-defined convergence properties.

On the other hand, repartitioning between sets $\mathcal{S}$ and $\mathcal{N}$ can usually be handled through efficient solution of problem (6.70), and systematically freeing the bounded variables. Alternately, a slightly larger bound constrained subproblem can be considered as well:

$$\begin{aligned} \min_{x_S, x_N} \quad & f(x_B(x_S, x_N), x_S, x_N) \\ \text{s.t.} \quad & x_{S,L} \le x_S \le x_{S,U}, \\ & x_{N,L} \le x_N \le x_{N,U}, \end{aligned} \tag{6.71}$$

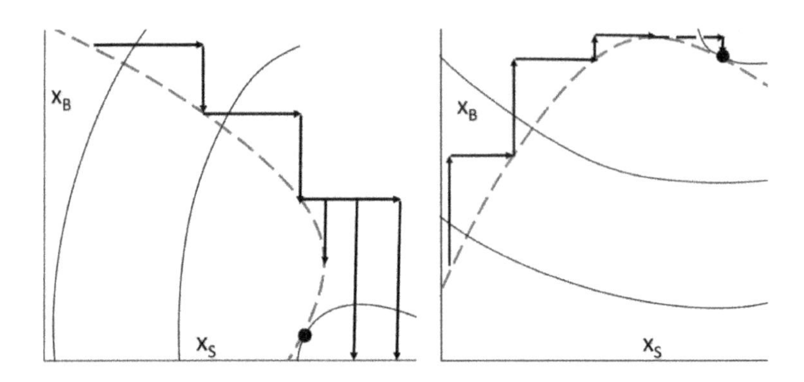

Figure 6.3. *Effect of repartitioning $\mathcal{B}$ and $\mathcal{S}$ to remedy failures to satisfy constraint equations. Here, $c(x) = 0$ is represented by the dashed curve and horizontal and vertical lines represent steps for superbasic and basic variables, respectively.*

where the nonbasic and basic variables are considered together in step 1 of Algorithm 6.5. An effective solution strategy for either (6.70) or (6.71) is the gradient projection method, described next.

6.4.1 Gradient Projection Methods for Bound Constrained Problems

Step 1 of Algorithm 6.5 requires the solution of either problem (6.70) for x_S, or problem (6.71) for x_S and x_N; both of these can be represented as the following bound constrained NLP with variables $z \in \mathbb{R}^{n_z}$:

$$\min_{z} \quad f(z) \quad \text{s.t.} \quad z_L \le z \le z_U. \tag{6.72}$$

To solve (6.72) we first introduce some new concepts. The inequality constraints need to be handled through the concept of gradient projection. We define the projection operator for vector z as the point closest to z in the feasible region Ω, i.e.,

$$z_P = \mathcal{P}_\Omega(z) = \arg\min_{y \in \Omega} \|y - z\|. \tag{6.73}$$

If the feasible region is defined by simple bounds on z, this operator simplifies to

$$\mathcal{P}(z) = \begin{cases} z_{(i)} & \text{if } z_{L,(i)} < z_{(i)} < z_{U,(i)}, \\ z_{L,(i)} & \text{if } z_{(i)} \le z_{L,(i)}, \\ z_{U,(i)} & \text{if } z_{U,(i)} \le z_{(i)}. \end{cases} \tag{6.74}$$

The projected gradient method comprises both projected gradient and projected Newton steps. Using the developments in [39, 227, 294], we define the new iterates based on the projected gradient as

$$z(\alpha) = \mathcal{P}(z - \alpha \nabla f(z)), \tag{6.75}$$

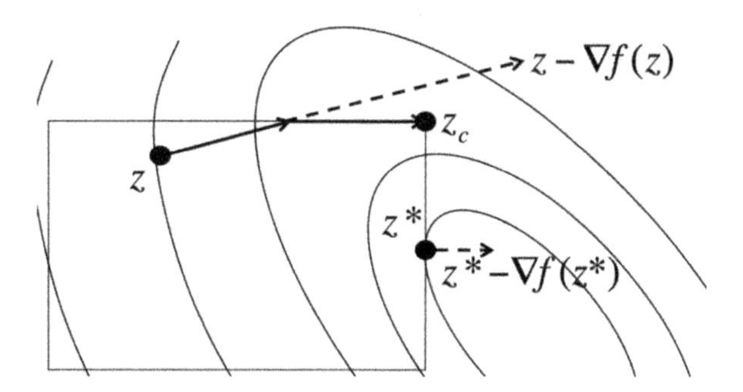

Figure 6.4. *Projected gradient steps and Cauchy points in bound constrained problems. Note that* $z^* = \mathcal{P}(z^* - \nabla f(z^*))$.

where $\alpha \geq 0$ is the step-size parameter. We define the Cauchy point as the first local minimizer of $f(z(\alpha))$ with respect to α, i.e., $z_c = z(\alpha_c)$ where $\alpha_c = \arg\min f(z(\alpha))$, as shown in Figure 6.4. Also, as illustrated in this figure, we note the following termination property.

Theorem 6.12 [227, Theorem 5.2.4] Let $f(z)$ be continuously differentiable. A point $z^* \in \Omega$ is stationary for problem (6.72) if and only if

$$z^* = \mathcal{P}(z^* - \alpha \nabla f(z^*)) \quad \text{for all } \alpha \geq 0. \tag{6.76}$$

These concepts lead to the following simple gradient projection algorithm.

ALGORITHM 6.6.

Choose a starting point z^0, convergence tolerance ϵ_{tol}, and step-size restriction parameter $\beta < 1$.

For $k \geq 0$ with iterate z^k and values $f(z^k)$, and $\nabla f(z^k)$, and while $\|z^* - \mathcal{P}(z^* - \alpha \nabla f(z^*))\| > \epsilon_{tol}$:

1. Find the smallest integer $j \geq 0$ for which $\alpha = (\beta)^j$ and $z(\alpha) = \mathcal{P}(z^k - \alpha \nabla f(z^k))$ satisfies the Armijo inequality

$$f(z(\alpha)) \leq f(z^k) + \delta \nabla f(z^k)^T (z(\alpha) - z^k), \tag{6.77}$$

   where $\delta \in (0, 1/2]$.

2. Set $z^{k+1} = z(\alpha)$ and evaluate $f(z^{k+1})$ and $\nabla f(z^{k+1})$.

This algorithm is globally convergent, but because it is based only on gradient information, it converges at a slow rate. To accelerate the convergence rate, it is tempting to improve the algorithm with a Newton-type step $p_N = -(\bar{B}^k)^{-1} \nabla f(z^k)$, where $\bar{B}^k$ is a positive definite matrix that approximates $\nabla_{zz} f(z^k)$, and then calculate $z(\alpha) = \mathcal{P}(z^k + \alpha p_N)$. However, once projected into a bound, such a step does not necessarily satisfy descent conditions and might lead to a line search failure.

Instead, we note that if the active set is known at the solution, then we can reorder the variables and partition to $z^k = [(z_I^k)^T, (z_A^k)^T]^T$, where $z_{A,(i)}^k$ is at its bound, $z_{I,(i)}^k \in (z_{I,L,(i)}, z_{I,U,(i)})$, and $z_I \in \mathbb{R}^{n_I}$. Defining a projection matrix $Z_I(z^k) = [I_{n_I} \mid 0]^T$ along with the modified Hessian

$$B_R^k = \begin{bmatrix} Z_I^T \bar{B}^k Z_I & 0 \\ 0 & I \end{bmatrix} \tag{6.78}$$

leads to the modified step

$$z(\alpha) = \mathcal{P}(z^k - \alpha(B_R^k)^{-1} \nabla f(z^k)). \tag{6.79}$$

When we are close to z^* and the correct active set is known (see Theorem 6.5), this leads to projected gradient steps for z_A while providing second order information for z_I. On the other hand, far from the solution the partitioning of z must be relaxed and the set of active variables must be overestimated. This can be done by defining an ϵ-active set at iteration k where $\epsilon^k > 0$, z_A^k is given by

$$z_{U,A,(i)} - z_{A,(i)}^k \le \epsilon^k \quad \text{or} \quad z_{A,(i)}^k - z_{L,A,(i)} \le \epsilon^k \tag{6.80}$$

and z_I includes the remaining n_I variables. The quantities B_R and $z(\alpha)$ are now defined with respect to this new variable partition. Moreover, with the following choice of ϵ^k we have the remarkable property that the active set at the solution remains fixed for all $\|z^k - z^*\| \le \delta$ for some $\delta > 0$, if strict complementarity holds at z^*. By defining

$$\epsilon^k = \min[\|z^k - \mathcal{P}(z^k - \nabla f(z^k))\|, \min_i (z_{U,(i)} - z_{L,(i)})/2], \tag{6.81}$$

the variable partition remains unchanged in some neighborhood around the solution. As a result, projected Newton and quasi-Newton methods can be developed that guarantee fast local convergence.

The resulting Newton-type bound constrained algorithm is given as follows.

ALGORITHM 6.7.

Choose a starting point z^0, convergence tolerance ϵ_{tol}, and step size restriction parameter $\beta < 1$.

For $k \ge 0$ with iterate z^k and values $f(z^k)$, $\nabla f(z^k)$, and $\bar{B}^k$, and while $\|z^* - \mathcal{P}(z^* - \alpha \nabla f(z^*))\| > \epsilon_{tol}$:

1. Evaluate ϵ^k from (6.81), determine the ϵ-active set, and reorder and partition the variables z^k according to (6.80). Calculate B_R^k according to (6.78).

2. Find the smallest integer $j \ge 0$ for which $\alpha = (\beta)^j$ and $z(\alpha)$ is evaluated by (6.79) that satisfies the Armijo inequality

$$f(z(\alpha)) \le f(z^k) + \delta \nabla f(z^k)^T (z(\alpha) - z^k), \tag{6.82}$$

   where $\delta \in (0, 1/2]$.

3. Set $z^{k+1} = z(\alpha)$ and evaluate $f(z^{k+1})$, $\nabla f(z^{k+1})$, and $\bar{B}^{k+1}$.

Using the first two parts of Assumptions TR, convergence properties for Algorithm 6.7 can be summarized as follows [227, 39]:

- The algorithm is globally convergent; any limit point is a stationary point.
- If the stationary point z^* satisfies strict complementarity, then the active set and the associated matrix Z_I remains constant for all $k > k_0$, with k_0 sufficiently large.
- If the exact Hessian is used and $Z_I^T \nabla_{zz} f(z^*) Z_I$ is positive definite, then the local convergence rate is quadratic.
- If the BFGS update described in [227] is used and $Z_I^T \nabla_{zz} f(z^*) Z_I$ is positive definite, then the local convergence rate is superlinear.

As noted in [227] some care is needed in constructing the BFGS update, because the partition changes at each iteration and only information about the inactive variables is updated. Nevertheless, Algorithm 6.7 can be applied to a variety of problem types, including bound constrained QPs, and is therefore well suited for step 1 in Algorithm 6.5.

Finally, as detailed in [39, 227, 294], there are a number of related gradient projection methods for bound constrained optimization. A popular alternative algorithm uses limited memory BFGS updates and performs subspace minimizations [294]. Using breakpoints for α defined by changes in the active set (see Figure 6.4), segments are defined in the determination of the Cauchy point, and a quadratic model can be minimized within each segment. This approach forms the basis of the L-BFGS-B code [420]. Also, trust region versions have also been developed for gradient projection; these include the TRON [261] and *TOMS Algorithm 717* [73, 155] codes. As discussed and analyzed in [100], trust region algorithms do not require positive definite Hessians and deal directly with negative curvature in the objective function.

6.4.2 Linearly Constrained Augmented Lagrangian

A significant expense to GRG-type algorithms is the repeated solution of the nonlinear equations. While this is useful for problems with high nonlinearity and poor linearizations, there is also the risk of premature termination due to failure to find feasible points. An alternative to the inner feasibility step is to construct a subproblem with constraints linearized at the current point and a nonlinear objective. Successive linearizations and subproblem solution then lead to iterations that should approach the solution of the nonlinearly constrained problem.

Consider the linearization of $c(x)$ about x^k at iteration k. Using the multiplier estimates v^k, u_L^k, and u_U^k, an augmented Lagrangian function is constructed:

$$L_A(x, v^k, \rho) = f(x) + (c(x) - \bar{c}(x))^T v^k + \frac{\rho}{2}\|c(x) - \bar{c}(x)\|^2, \tag{6.83}$$

where $\bar{c}(x) = c(x^k) + \nabla c(x^k)^T (x - x^k)$. As discussed in Section 5.5.1, augmented Lagrangians have a number of desirable features that make them good candidates for merit functions. In particular, for the appropriate value of v they are *exactly* minimized at the NLP solution with a finite value of ρ, and still remain smooth. The linearly constrained subproblem corresponding to (6.3) is given by

$$\begin{aligned} \min_x \quad & L_A(x, v^k, \rho) \\ \text{s.t.} \quad & c(x^k) + \nabla c(x^k)^T (x - x^k) = 0, \\ & x_L \le x \le x_U. \end{aligned} \tag{6.84}$$

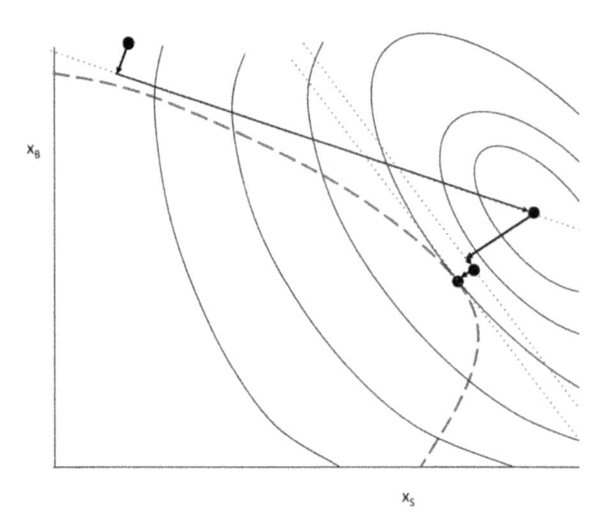

Figure 6.5. *Progress of major iterations for the MINOS algorithm. At each major iteration a linearly constrained subproblem is solved.*

This subproblem can be solved by partitioning x into basic, superbasic, and nonbasic variables and applying Algorithm 6.5. The iterations of this algorithm are known as the *minor iterations*. With linear equalities, the feasibility step is now greatly simplified, because the ∇c matrix does not change. As a result, much more efficient and reliable solutions can be obtained for subproblem (6.84). The solution to (6.84) yields the primal and dual variables $(x^{k+1}, v^{k+1}, u_L^{k+1}, u_U^{k+1})$ for the next *major iteration*. This sequence of the major iterations is illustrated in Figure 6.5.

The MINOS solver has been implemented very efficiently to take care of linearity and sparsity in the Jacobian. In fact, if the problem is totally linear, it becomes the LP simplex method. For nonlinear problems, it uses a reduced-space method and applies quasi-Newton updates to solve (6.84). The major iterations $\|x^{k+1} - x^k\|$ converge at a quadratic rate [162], and because no feasibility step is required, MINOS tends to be more efficient than GRG, especially if the equalities are "mostly linear."

In solving (6.84), the penalty term in $L_A(x)$ controls the infeasibility of the major iterations. Nevertheless, this approach has no other globalization strategies and is not guaranteed to converge on problems with highly nonlinear constraints.

6.5 Nonlinear Programming Codes

Starting from the concepts of Newton-type solvers for equality constrained optimization, this chapter focuses on three approaches to extend these solvers to general NLPs. These concepts have led to a number of implementations and very reliable and efficient software packages for large-scale nonlinear programming. From the algorithmic descriptions, it is clear that NLP solvers are complex algorithms that have required considerable research and development to turn them into reliable and efficient software tools. Practitioners that are confronted with engineering optimization problems should therefore leverage their efforts with these well-developed codes. Much more information on widely available codes can be found on the NEOS server (www-neos.mcs.anl.gov) and in the NEOS Software Guide [1, 283].

We now consider a representative set of NLP codes that build on the concepts in this chapter. This list is necessarily incomplete and represents only a sampling of what is available. Moreover, with ongoing continuous development and improvements for many of these codes, the reader is urged to seek more information from the software developer, along with literature and Internet resources, regarding new developments, software availability, and updates to the codes.

6.5.1 SQP Codes

Modern SQP codes have been developed since the mid 1970s. The first codes used a BFGS update of the full Hessian along with quadratic programming methods based on dense linear algebra. Line searches are generally performed using ℓ_1 merit functions. These codes are quite suitable for NLPs with less than a few hundred variables. Below are listed some of the more popular codes:

- **DONLP2** [367]: Developed by Spellucci [366], this code implements an SQP method with only equality constrained subproblems. It uses quasi-Newton updates and an Armijo-type line search with an ℓ_1 merit function.

- **fmincon** [145]: Coupled to the MATLAB environment, this code applies a primal QP solver, an ℓ_1 merit function line search, and BFGS updates to the Hessian. It also includes more recent features that incorporate trust region globalization and interior point treatment of inequalities.

- **NLPQL** [354]: Developed by Schittkowski [353], the code uses the dual QP code, QL and performs a line search using two alternative merit functions. The Hessian approximation is updated by the modified BFGS formula.

- **NPSOL** [161]: Developed by Gill, Murray, and Wright [162], NPSOL makes use of the QP solver LSSOL, a primal QP method based on QR factorizations. For globalization, it performs a line search using an augmented Lagrangian merit function.

- **VF13AD** [314]: Written by Powell [313], this code is part of the HSL library and is among the earliest SQP codes. This code uses a dense QP solver based on a dual-space algorithm [165] and incorporates the Watchdog technique.

On the other hand, large-scale SQP solvers have also been developed that exploit sparsity of the Jacobian. These take the form of either reduced-space or full-space Newton-type solvers. Representative codes for this algorithm class include the following:

- **filterSQP** [136]: Developed by Fletcher and Leyffer [137], this code implements an SQP trust region algorithm using the filter method to promote global convergence, as described in Section 6.2.3. This code uses the *bqpd* package, a recursive null-space active set method, to solve sparse QP subproblems.

- **MUSCOD-II** [251]: Developed at the IWR at the University of Heidelberg [62, 250, 251], this package employs a reduced-space SQP decomposition strategy with a number of algorithmic options. It includes limited-memory BFGS updates along with a Watchdog line search strategy.

- **MOOCHO/rSQP++** [24]: The "Multifunctional Object-Oriented arCHitecture for Optimization" is part of the Trilinos package from Sandia National Laboratories. Described in [25], this code is an object-oriented framework for reduced-space SQP. It applies a line search with an ℓ_1-type merit function and includes interfaces to a number of QP methods. The framework allows for external configuration of specialized application-specific linear algebra objects and linear solvers.

- **SNOPT** [159]: Also developed by Gill, Murray, and Wright [160], SNOPT performs a full-space, limited-memory BFGS update for the Hessian matrix. It then forms and solves the QP problem using SQOPT, a reduced-Hessian active set method. It performs a line search with an augmented Lagrangian function.

- **SOCS** [41]: Developed at Boeing by Betts and coworkers [40], this code incorporates a primal sparse quadratic programming algorithm that fully exploits exact sparse Jacobians and Hessians. It uses an augmented Lagrangian line search along with specially developed sparse matrix linear algebra.

6.5.2 Interior Point NLP Codes

These NLP codes incorporate Newton-based solvers that are applied to the primal-dual equations (6.51) through the solution of the linear system (6.56). These three codes differ in terms of globalization strategies (line search versus trust region) along with the matrix factorizations needed in (6.56). These codes also include a variety of scaling and regularization strategies that are essential for efficient performance.

- **IPOPT** [206]: Using the algorithm described in [404], this object-oriented code is part of the open-source COIN-OR project and solves the primal-dual equations and the linear system (6.56) through a variety of options for sparse, symmetric, indefinite matrix factorizations. It applies a filter line search and has options for handling exact Hessian or limited-memory full-space BFGS updates. Also included are a number of options for adjustment of the barrier parameter μ.

- **KNITRO** [79]: Using the algorithm described in [81], this code includes two procedures for computing the steps for the primal-dual equations. The Interior/CG version uses the Byrd–Omojukun algorithm and solves the tangential trust region via projected conjugate gradient iterations. The Interior/Direct version solves the primal-dual KKT matrix using direct linear algebra unless poor steps are detected; it then switches to Interior/CG.

- **LOQO** [37]: As described in [392], LOQO is one of the earliest primal-dual NLP codes and incorporates a line search–based Newton solver with a number of options for promoting convergence on difficult problems. Also, through a regularization of the lower right-hand corner of the KKT matrix in (6.56), LOQO formulates and solves quasi-definite primal-dual linear systems through a specialized sparse algorithm.

6.5.3 Nested and Gradient Projection NLP Codes

The following codes consider nested and bound constrained approaches to problem (6.3) and its KKT conditions (6.4). Unlike the previous codes which generally consider the simultaneous solution of the KKT conditions, these consider a decomposition that leads to different nested subproblems, which are solved with Newton-type methods.

- **CONOPT/CONOPT3** [119]: Developed by Drud [120], the latest version of this code (CONOPT3) includes generalized reduced gradient concepts as well as an SQP method with exact second derivatives. The code follows the partitioning and updating concepts in Section 6.4 and incorporates fast sparse linear decompositions for variable reorderings.

- **FSQP/CFSQP** [246]: Developed at the University of Maryland [247], this SQP code is modified to generate feasible iterates through repeated feasibility restoration. It includes a number of options for partially feasible steps as well. A key advantage of this approach is that the objective function is the merit function in the line search step.

- **GRG2** [245]: Developed by Lasdon and coworkers, this code applies the generalized reduced gradient concepts from Section 6.4 and has a long development history. It includes an efficient quasi-Newton algorithm (BFGS in factored form) as well as an optional conjugate gradient method for larger problems.

- **MINOS** [285]: As described in [162] and also in Section 6.4.2, MINOS solves a sequence of linearly constrained subproblems, as illustrated in Figure 6.5. For nonlinear problems, it uses a reduced-space method and applies quasi-Newton updates to solve (6.84).

- **LANCELOT** [99]: As described in [99], this code considers a sequence of bound constrained subproblems formulated from (6.3). Here the objective and equality constraints are combined into an augmented Lagrangian function (5.57). The resulting bound constrained problem is then solved with a trust region–based gradient projection method. Multipliers and the penalty parameter for (5.57) are then updated in an outer loop.

- **PENNON** [231]: As described in [373], this code first converts the constraints by using a transformed penalty/barrier function. It then forms a (generalized) augmented Lagrange function. The NLP is solved through solution of a sequence of unconstrained problems, with multipliers for the augmented Lagrange function updated in an outer loop. The unconstrained minimization (inner loop) is performed by a Newton-type solver with globalization using either line search or trust region options.

6.5.4 Performance Trends for NLP Codes

The above codes have been applied widely to large-scale NLP problems. To provide some insight into their performance characteristics, we recall the algorithmic concepts developed in this chapter and illustrate them with two sets of NLP examples. Readers should be cautioned that this section is not meant to be a performance comparison, as the results are strongly dependent on the hardware environment and operating system, as well as the modeling environment, which provides the problem data to the NLP algorithm. Nevertheless, from the test problem results, we can observe the influence of the Newton step, globalization features, and approximation of second derivatives.

Scalable NLP Test Problem

This problem is derived from an optimal control problem, presented in [194]. After a backward difference formula is applied, the problem can be stated as the following NLP:

$$\begin{aligned}
\min \quad & \sum_{i=1}^{N}[\alpha_1(C_{des}-C_i)^2+\alpha_2(T_{des}-T_i)^2+\alpha_3(u_{des}-u_i)^2]/N \qquad (6.85)\\
\text{s.t.} \quad & C_{i+1}=C_i+\tau\bar{C}_{i+1}/N,\\
& T_{i+1}=T_i+\tau\bar{T}_{i+1}/N,\\
& \bar{C}_i=(1-C_i)/\theta-k_{10}*\exp(w_i)C_i,\\
& \bar{T}_i=(T_f-T_i)/\theta+k_{10}\exp(w_i)C_i-\alpha u_i(T_i-T_c),\\
& w_iT_i+\eta=0,\\
& C_1=C_{init}, \quad T_1=T_{init},\\
& C_i\in[0,\,1], \quad T_i\in[0,\,1], \quad u_i\in[0,\,500],
\end{aligned}$$

where the parameters values are given by $C_{init}=0.1367$, $T_{init}=0.7293$, $C_{des}=0.0944$, $T_{des}=0.7766$, $u_{des}=340$, $\alpha=1.95\times10^{-4}$, $\alpha_1=10^6$, $\alpha_2=2000$, $\alpha_3=0.001$, $k_{10}=300$, $\eta=1$, $\theta=20$, $T_f=0.3947$, $T_c=0.3816$, and $\tau=10$. The variables are initialized by the following infeasible point: $\bar{C}_i=1$, $\bar{T}_i=1$, $C_i=C_{init}+(C_{des}-C_{init})(i-1)/N$, $T_i=T_{init}+(T_{des}-T_{init})(i-1)/N$, $w_i=\eta$, and $u_i=250$.

With a quadratic objective function and strong nonlinear terms in the equality constraints (controlled by the parameter η), this problem can present a challenge to algorithms that are best suited for "mostly linear" constraints. In fact, from this starting point, the MINOS solver (version 5.51, interfaced to GAMS) is unable to find a feasible point, even for small values of N. To assess the performance of Newton-type NLP solvers, we considered a set of codes interfaced to the GAMS (version 23.0.2) modeling environment. Here we apply IPOPT (version 3.5, using the MUMPS sparse linear solver), CONOPT (version 3.14S), KNITRO (version 5.2.0), and SNOPT (version 7.2-4). Problem (6.85) was run for increasing values of N, where the numbers of variables and equations are given by $n=6N-2$ and $m=5N-2$. All codes were run with their default settings on a 2 GB, 2.4 GHz Intel Core2 Duo processor, running Windows XP. The results are given in Table 6.1.

Firm generalizations cannot be drawn from a single case study. Also, for small problems the CPU times are too small to compare algorithmic steps in each of the codes. However, it is interesting to explore a few trends as N increases. First, CONOPT takes the most advantage of problem structure and solves smaller, nested subproblems. It requires the least CPU time, but it requires additional iterations to maintain feasibility. Its performance is also aided by fast SQP steps with exact second derivatives. On the other hand, IPOPT maintains a constant number of iterations, and for N large, the CPU time is roughly linear with N. This performance is characteristic of a Newton-based solution to the primal-dual equations (6.51), aided by a sparse matrix solver and a filter line search to promote larger steps. Since

Table 6.1. *Results for scalable test problem (# iterations [CPU seconds]). Minor iterations are also indicated for SNOPT, and CG iterations are indicated for KNITRO, where applied. The numbers of variables and equations are $n = 6N - 2$ and $m = 5N - 2$ and $n - m = N$.*

N	CONOPT	IPOPT	KNITRO	SNOPT
5	15 [0.11]	9 [0.016]	16/0 [0.046]	13/33 [0.063]
10	15 [0.062]	9 [0.016]	19/0 [0.093]	15/55 [0.063]
50	28 [0.094]	9 [0.032]	20/0 [0.078]	15/357 [0.11]
100	30 [0.047]	9 [0.11]	18/0 [0.109]	13/603 [0.156]
500	32 [0.344]	9 [0.64]	96/337 [4.062]	23/5539 [4.828]
1000	29 [1.375]	9 [1.422]	116/771 [16.343]	31/10093 [13.734]

the direct version of KNITRO also solves (6.51) with a Newton method, its performance is similar to IPOPT for N up to 100, with the difference in iterations probably due to the use of a different line search. For $N = 500$ and $N = 1000$, though, KNITRO automatically switches from the Interior/Direct to the Interior/CG version to provide more careful, but expensive, trust region steps. This can be seen from the added CG iterations in the table. Finally, SNOPT is the only solver considered that does not use exact second derivatives. Table 6.1 lists the major (SQP) iterations and minor (QP) iterations. Note that the number of SQP iterations increases only slightly with N. However, as a result of BFGS updates and heavily constrained QPs, SNOPT requires a much larger increase in CPU time with increasing N.

Mittelmann NLP Benchmarks

We next consider the NLP benchmarks conducted by Prof. Hans Mittelmann [280] for large-scale AMPL-NLP test problems on July 20, 2009. These scalable test problems were drawn from a number of libraries including 12 problems from *COPS*, 6 problems *CUTE*, 4 problems from *Elliptical PDE*, 5 problems from *Parabolic PDE*, 3 problems from the *AMPL* set, 4 problems from *Globallib*, 11 problems from *QCQP*, and 3 problems from *Systems Identification*. They range from 500 to over 260000 variables and from 0 to over 139000 constraints, and are listed in Table 6.2. An important measure is that $n - m \geq 100000$ on four of these examples. For these 48 test problems, we consider five codes: IPOPT (version 3.7.0), KNITRO (version 6.0), PENNON (version 0.9), SNOPT (version 7.2-1), and CONOPT (version 3.14). All algorithms were run in default mode on an 8 GB, 2.67 GHz Intel Core2 Quad processor, with a CPU time limit of 7200 s.

The results of this benchmark are summarized by the Dolan–Morè plot in Figure 6.6. Again, while it is not possible to draw firm generalizations from this problem set alone, one can observe some interesting trends. These mirror some of the trends in Table 6.1 as well. First, both KNITRO and IPOPT have similar performance, with KNITRO slightly faster. PENNON, which performs unconstrained Newton iterations with an augmented Lagrange function (and exact derivatives), performs reasonably well. CONOPT is next, and for these larger problems it is somewhat slower because of the solution of nested subproblems. Finally, SNOPT does not take advantage of second derivatives; this leads to slower performance, especially when $n - m$ is large.

Table 6.2. *Test problems in Mittlemann benchmark study on NLP algorithms.*

Source	Problem	# Variables	# Constraints
COPS	bearing-400	160000	0
	camshape-6400	6400	12800
	dirichlet-120	53881	241
	henon-120	32401	241
	lane-emden-120	7721	241
	elec-400	1200	400
	gasoil-3200	32001	31998
	marine-1600	38415	38392
	pinene-3200	64000	63995
	robot-1600	14399	9601
	rocket-12800	51201	38400
	steering-12800	2000	25601
CUTE	clnlbeam	59999	40000
	corkscrw	44997	35000
	dtoc1nd	6705	4470
	dtoc2	64950	38970
	optmass	60006	50005
	svanberg	50000	50000
Elliptical PDE	ex1-160	50562	25281
	ex4-2-160	51198	25917
	ex1-320	203522	101761
	ex4-2-320	204798	103037
Parabolic PDE	cont5-1-l	90600	90300
	cont5-2-1-l	90600	90300
	cont5-2-2-l	90600	90300
	cont5-2-3-l	90600	90300
	cont5-2-4-l	90600	90300
AMPL	robot-a	1001	52013
	robot-b	1001	52013
	robot-c	1001	52013
Globallib	arki0003	2237	2500
	arki0009	6220	5924
	ex8-2-2	7510	1943
	ex8-2-3	15636	3155
QCQP	qcqp500-3c	500	120
	qcqp750-2c	750	138
	qcqp1000-2c	1000	5107
	qcqp1500-1c	1500	10508
	qcqp500-3nc	500	120
	qcqp750-2nc	750	138
	qcqp1000-1nc	1000	154
	qcqp1000-2nc	1000	5107
	qcqp1500-1nc	1500	10508
	nql180	162001	130080
	qssp180	261365	139141
System Identification	WM-CFy	8520	9826
	Weyl-m0	1680	2049
	NARX-CFy	43973	46744

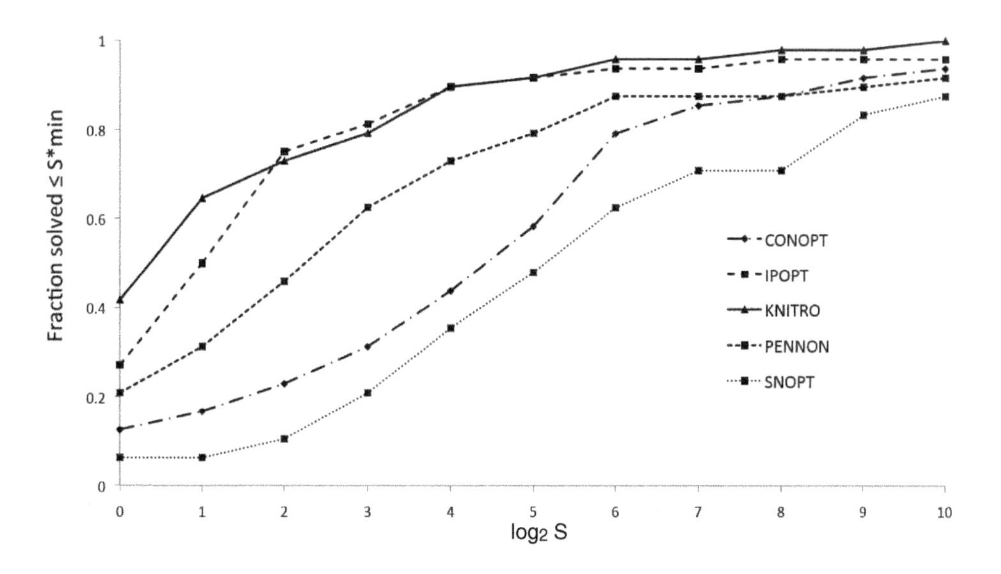

Figure 6.6. *Dolan–Morè plot of Mittelmann NLP benchmark from July 20, 2009. The graph indicates the fraction of test problems solved by an algorithm in* $S\times$ *(minimum CPU time of all* 5 *algorithms); here* S *is a factor that ranges from* 1 *to* 1024. *Comparisons are made on* 48 *large NLPs, ranging from* 500 *to* 261365 *variables.*

6.6 Summary and Conclusions

This chapter completes the development of algorithms that are incorporated into a wide variety of large-scale NLP solvers. Specifically, the focus is on NLP algorithms that extend naturally from the Newton-type solvers developed in Chapter 5. This extension is formulated through the addition of bound constraints in (6.3). While the development of algorithms is more direct for this problem formulation, the reader should be comfortable in extending these derivations to the more general NLP formulation given by (6.1).

The chapter focuses on three classes of NLP solvers: SQP methods, interior point methods, and nested, reduced gradient methods. Key issues for all of these methods include well-posed calculation of Newton-type search directions, application of globalization strategies, using either line searches or trust regions, to ensure convergence from poor starting points, and correction strategies to ensure full steps near the solution and fast local convergence.

SQP methods deal with straightforward extensions of inequality constraints by extending the Newton step to the solution of a quadratic programming subproblem. The efficiency of this method is limited by the expense of the QP solution. This expense is governed by selection of the active set and the linear algebra to solve the KKT conditions for this active set. This chapter has focused on inequality quadratic programming (IQP) methods for SQP, where the QP is solved completely and the active set is recomputed at each SQP iteration. Alternately, one can apply an equality quadratic programming (EQP) method (see [294]), where the active set is fixed for a sequence of SQP iterations, and steps are determined from the linear equations of the form (6.9).

Large-scale SQP methods come in two flavors, full-space and reduced-space methods. Reduced-space methods are best suited for few decision variables, i.e., when $n-m$ is small. This method relies on decomposition of the QP solution into normal and tangential steps. For this case, quasi-Newton updates of the reduced Hessian can also be applied efficiently. Global convergence is promoted through a line search strategy, or, in some cases, a composite-step trust region strategy. On the other hand, full-space methods are especially well suited for problems where $n-m$ is large. A key element of these methods is the provision of second derivatives, the exploitation of sparsity in the QP solver, and treatment of indefinite Hessian matrices. Convergence properties for these methods are described in more detail in [294, 100].

Barrier, or interior point, methods have much in common with SQP methods, as they both depend on Newton-type solvers. On the other hand, rather than considering active set strategies, these methods deal with relaxed KKT conditions, i.e., primal-dual equations. Solving these equations leads to a sequence of relaxed problems solved with Newton-based approaches. These methods provide an attractive alternative to active set strategies in handling problems with large numbers of inequality constraints. However, careful implementation is needed in order to remain in the interior of the inequality constraints. This requires modifications of line search and trust region strategies to prevent the method from "crashing into bounds" and convergence failure. Otherwise, the Newton step can be decomposed or regularized in the same manner as in Chapter 5, with the option of applying quasi-Newton updates as well. Convergence properties for these methods are described in more detail in [294, 100].

Nested NLP solvers based on reduced gradients tend to be applied on problems with highly nonlinear constraints, where linearizations from infeasible points can lead to poor Newton steps and multiplier estimates. These methods partition the NLP problem into basic, superbasic, and nonbasic variables, nest the solution of the KKT conditions (6.69), and update these variables in a sequential manner. There are a number of options and heuristics dealing with repartitioning of these variables as well as relaxation of feasibility. Among these, a particularly useful option is the solution of bound constrained subproblems to deal with superbasic and nonbasic variables (see [227] for algorithmic development and analysis of convergence properties). Nevertheless, these methods require more function evaluations but may be more robust for highly nonlinear problems, as most of them follow feasible paths.

Finally, a set of 19 popular NLP codes is described that incorporate elements of these Newton-type algorithms. In addition, with the help of two brief numerical studies, some trends are observed on characteristics of these methods and their underlying algorithmic components.

6.7 Notes for Further Reading

The SQP method dates back to the work of Wilson in 1963 [409]. This method was developed, extended, and discussed in Beale [35]. However, the original method was not widely adopted. This was apparently due to the need to obtain and deal with an indefinite Hessian matrix, the need to solve QPs efficiently, and the need for a globalization strategy to ensure convergence from poor starting points. The development of quasi-Newton Hessian updates and a line search based on merit functions, particularly in the methods due to Han [184] and Powell [313], led to development of modern SQP methods in the late 1970s.

With the application of SQP on larger problems, both full-space and reduced-space versions were developed, particularly with line search strategies. For applications with few degrees of freedom, reduced-space algorithms were developed using concepts developed in Gill, Murray, and Wright [162, 159] and through composite-step methods analyzed by Coleman and Conn [98] and by Nocedal and coworkers [292, 82, 54]. In particular, early methods with reduced-space QP subproblems are described in [38, 262]. These were followed by developments for both line search [395, 54] and trust region adaptations [11, 380], and also in the development of the SNOPT code [159]. In addition, large-scale, full-space SQP methods were developed by Betts and coworkers [40] and by Lucia and coworkers [265, 266].

Interior point or *barrier* methods for large-scale NLP are grounded in the early work of barrier and other penalty functions developed in Fiacco and McCormick [131]. Moreover, the development of interior point methods for linear programming [223, 411, 385], and the relation of primal-dual equations to stationarity conditions of the barrier function problem, spurred a lot of work in NLP, particularly for convex problems [289]. For general NLP, there has also been a better understanding of the convergence properties of these methods [147], and efficient algorithms have been developed with desirable global and local convergence properties. To allow convergence from poor starting points, interior point methods in both trust region and line search frameworks have been developed that use exact penalty merit functions as well as filter methods [137, 403]. In addition, an interesting survey of filter methods is given in [139].

Convergence properties for line-search-based interior point methods for NLP are developed in [123, 413, 382]. Global and local convergence of an interior point algorithm with a filter line search is analyzed in [403, 402], with less restrictive assumptions. In addition, Benson, Shanno, and Vanderbei [36] discussed numerous options for line search methods based on merit functions and filter methods.

Trust region interior point methods based on exact penalty functions were developed by Byrd, Nocedal, and Waltz [79]. Since the late 1990s these "KNITRO-type" algorithms have seen considerable refinement in the updating of the penalty parameters and solution of the trust region subproblems. In addition, M. Ulbrich, S. Ulbrich, and Vicente [389] considered a trust region filter method that bases the acceptance of trial steps on the norm of the optimality conditions. Also, Ulbrich [390] discussed a filter approach using the Lagrangian function in a trust region setting, including both global and local convergence results. Finally, Dennis, Heinkenschloss, and Vicente [112] proposed a related trust region algorithm for NLPs with bounds on control variables. They applied a reduced-space decomposition and barrier functions for handling the bounds in the tangential step. In addition, they obtained an approximate solution to the tangential problem by using truncated Newton methods.

Reduced gradient methods date back to the work by Rosen [339, 340] and Abadie and Carpentier [5] in the 1960s. Because of their nested structure, these algorithms tend to lead to "monolithic" codes with heuristic steps for repartitioning and decomposition. As a result, they are not easily adapted to NLPs with specific structure. Reduced gradient methods were developed in early work by Sargent [347] and Sargent and Murtagh [348]. Long-term developments by Lasdon and coworkers [245] led to the popular GRG2 code. Similarly, work by Saunders and Murtagh [285] led to the popular MINOS code. More recently, work by Drud has led to the development of CONOPT [119]. Implemented within a number of modeling environments, CONOPT is widely used due to robustness and efficiency with highly nonlinear problems.

The NLP algorithms and associated solvers discussed in this chapter comprise only a sampling of representative codes, based on Newton-type methods. A complete listing is beyond the scope of this book and the reader is referred to the NEOS software guide [1] for a more complete selection and description of NLP codes. Moreover, important issues such as scaling and numerical implementations to improve precision have not been covered here. Readers are referred to [162, 294, 100] for more information on these issues.

Finally, systematic numerical studies are important research elements for NLP algorithm development. A wealth of test problems abound [64, 116, 152, 199] and numerical studies frequently deal with hundreds of test cases. While many benchmarks continue to be available for NLP solvers, it is worth mentioning the frequently updated benchmarks provided by Mittelmann [280]. These serve as extremely useful and impartial evaluations of optimization solvers. Moreover, for numerical comparisons that are geared toward specialized test sets of interest to a particular user, modeling environments like GAMS [71] provide facilities for conducting such solver benchmarks more systematically.

6.8 Exercises

1. Show, by adding and redefining variables and constraints, that problems (6.1), (6.2), and (6.3) have the same optimal solutions.

2. Show that if $d_x \neq 0$, then the solution to (6.13) is a descent direction for the ℓ_1 merit function $\phi_1(x;\rho) = f(x) + \rho\|c(x)\|_1$.

3. Show, by adding and redefining variables and constraints, that problems (6.3) and (6.48) have the same optimal solutions.

4. Rederive the interior point method in Section 6.3 using the double-bounded NLP (6.3).

5. Select solvers from the SQP, interior point, and reduced gradient categories, and apply these to Example 6.3. Use $x^0 = [0, 0.1, 0, 0]^T$ and $x^0 = [0, 0, 0, 0]^T$ as starting points.

6. Extend the derivation of the primal-dual equations (6.51) to the bound constrained NLP (6.3). Derive the resulting Newton step for these equations, analogous to (6.57).

7. Select solvers from the SQP, interior point, and reduced gradient categories, and apply these to Example 6.8. Use $x^0 = [-2, 3, 1]^T$ as the starting point.

8. Consider the placement of three circles of different diameter in a box enclosure of minimum perimeter. As shown in Figure 6.7, this leads to the following nonconvex, constrained NLP:

$$\begin{aligned} \min \quad & (a+b) \\ \text{s.t.} \quad & x_i, y_i \geq R_i, x_i \leq b - R_i, y_i \leq a - R_i, \quad i = 1,\ldots,3, \\ & (x_i - x_{i'})^2 + (y_i - y_{i'})^2 \geq (R_i + R_{i'})^2, \quad 0 < i' < i, i = 1,\ldots,3, \\ & a, b \geq 0, \end{aligned}$$

where R_i is the radius of circle i, (x_i, y_i) represents the center of circle i, and a and b are the horizontal and vertical sides of the box, respectively.

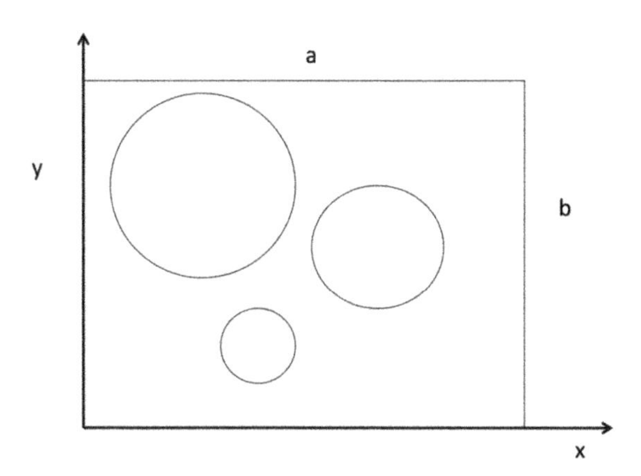

Figure 6.7. *Three circles in enclosure.*

- Let $R_i = 1 + i/10$. Apply an SQP solver to this algorithm with starting points $x_i^0 = i$, $y_i^0 = i$, $a^0 = 6$, and $b^0 = 6$.
- Repeat this study using an interior point NLP solver.

9. Using a modeling environment such as GAMS, repeat the case study for problem (6.85) using $\eta = 5$. Discuss the trends for the NLP algorithms as N is increased.

Chapter 7

Steady State Process Optimization

The chapter deals with the formulation and solution of chemical process optimization problems based on models that describe steady state behavior. Many of these models can be described by algebraic equations, and here we consider models only in this class. In process engineering such models are applied for the optimal design and analysis of chemical processes, optimal process operation, and optimal planning of inventories and products. Strategies for assembling and incorporating these models into an optimization problem are reviewed using popular modular and equation-oriented solution strategies. Optimization strategies are then described for these models and gradient-based NLP methods are explored along with the calculation of accurate derivatives. Moreover, some guidelines are presented for the formulation of equation-oriented optimization models. In addition, four case studies are presented for optimization in chemical process design and in process operations.

7.1 Introduction

The previous six chapters focused on properties of nonlinear programs and algorithms for their efficient solution. Moreover, in the previous chapter, we described a number of NLP solvers and evaluated their performance on a library of test problems. Knowledge of these algorithmic properties and the characteristics of the methods is essential to the solution of *process optimization* models.

As shown in Figure 7.1, chemical process models arise from quantitative knowledge of process behavior based on conservation laws (for mass, energy, and momentum) and constitutive equations that describe phase and chemical equilibrium, transport processes, and reaction kinetics, i.e., the "state of nature." These are coupled with restrictions based on process and product specifications, as well as an objective that is often driven by an economic criterion. Finally, a slate of available decisions, including equipment parameters and operating conditions need to be selected. These items translate to a process model with objective and constraint functions that make up the NLP. Care must be taken that the resulting NLP problem represents the real-world problem accurately and also consists of objective and constraint functions that are well defined and sufficiently smooth.

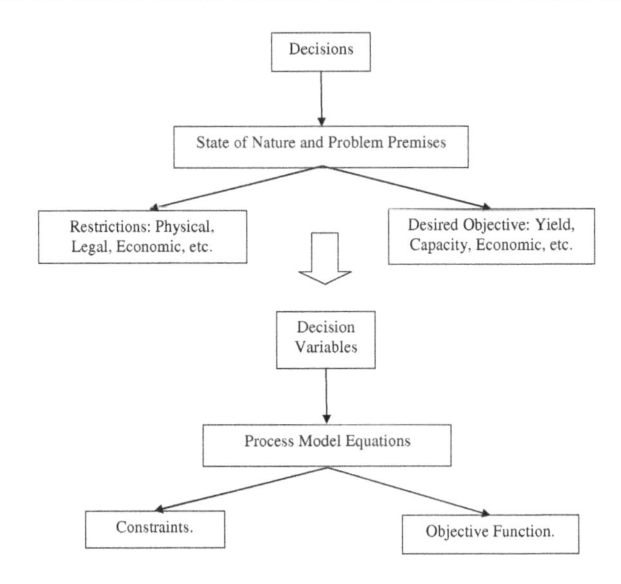

Figure 7.1. *Conceptual formulation for process optimization.*

In this chapter we consider NLP formulations in the areas of design, operation, and planning. Each area is described with optimization examples along with a case study that illustrates the application and performance of appropriate NLP solvers.

Process optimization models arise frequently as off-line studies for design and analysis. These require accurate process *simulation* models to describe steady state behavior of the chemical process. Because of the considerable detail required for these simulation models (particularly in the description of thermodynamic equilibrium), the models need to be structured with specialized solution procedures, along the lines of the individual process units represented in the model. As shown in Section 7.2, this collection of unit models leads to a process flowsheet simulation problem, either for design or analysis, executed in a *modular simulation mode*. Some attention to problem formulation and implementation needs to be observed to adapt this calculation structure to efficient NLP solvers. Nevertheless, the NLP problem still remains relatively small with only a few constraints and few degrees of freedom.

Optimization problems that arise in process operations consist of a problem hierarchy that deals with planning, scheduling, real-time optimization (RTO), and control. In particular, RTO is an online activity that achieves optimal operation in petroleum refineries and chemical plants. Nonlinear programs for RTO are based on process models similar to those used for design and analysis. On the other hand, because these problems need to be solved at regular intervals (at least every few hours), detailed simulation models can be replaced by correlations that are fitted and updated by the process. As shown in Section 7.4, these time-critical nonlinear programs are formulated in the so-called *equation-oriented* mode. As a result, the nonlinear program is generally large, with many equality constraints from the process model, but relatively few degrees of freedom for optimization.

Finally, planning and scheduling problems play a central role in process operations. Optimization models in this area are characterized by both discrete and continuous vari-

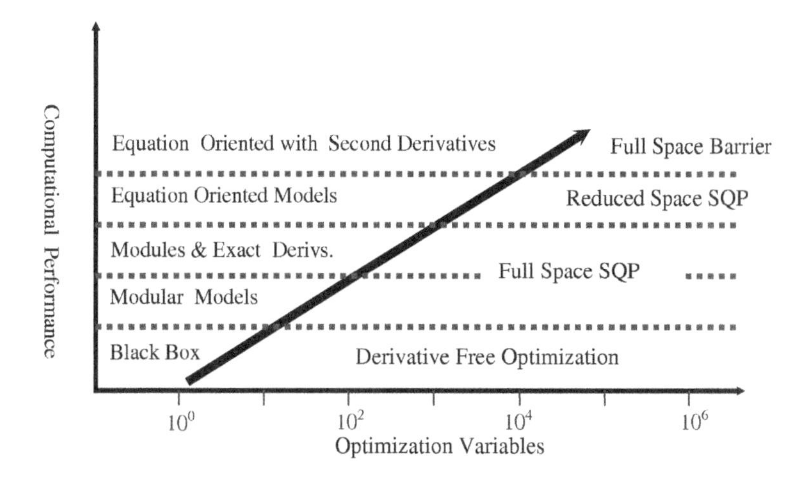

Figure 7.2. *Hierarchy of optimization models and NLP strategies.*

ables, i.e., mixed integer programs, and by simpler nonlinear models. While these models still consist of algebraic equations, they are often spatially distributed over temporal horizons and describe material flows between networks and management of inventories. As seen in Section 7.5 the resulting NLP problems are characterized by many equality constraints (with fewer nonlinearities) and also many degrees of freedom. These applications are illustrated later by gasoline blending models, along with a case study that demonstrates the performance of NLP solvers.

The evolution of these problem formulations is closely tied to the choice of appropriate optimization algorithms. In Chapters 5 and 6, efficient NLP algorithms were developed that assume open, equation-oriented models are available with exact first and second derivatives for all of the constraint and objective functions. These algorithms best apply to the examples in Section 7.5. On the other hand, on problems where function evaluations are expensive, and gradients and Hessians are difficult to obtain, it is not clear that large-scale NLP solvers should be applied. Figure 7.2 suggests a hierarchy of models paired with suitable NLP strategies. For instance, large equation-based models can be solved efficiently with structured barrier NLP solvers. On the other hand, black box optimization models with inexact (or approximated) derivatives and few decision variables are poorly served by large-scale NLP solvers, and derivative-free optimization algorithms should be considered instead. The problem formulations in this chapter also include intermediate levels in Figure 7.2, where SQP and reduced-space SQP methods are expected to perform well.

7.2 Optimization of Process Flowsheets

As shown in Figure 7.3, process flowsheets have a structure represented by collections of equipment units that include the following subprocesses:

- *Feed preparation* consists of equipments including heat exchangers, compressors, pumps, and separators that process the raw material stream to separate unwanted

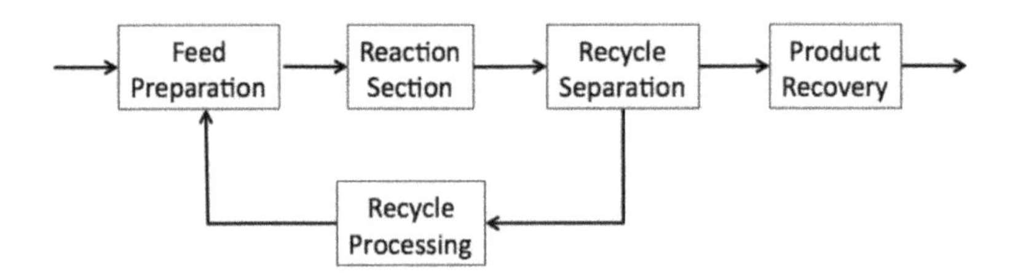

Figure 7.3. *Subprocesses within process flowsheets.*

chemical components and bring the resulting mixture to the desired pressure and temperature for reaction.

- *Reaction* section consists of reactor units that convert feed components to desired products and by-products. These equipment units usually convert only a fraction of the reactants, as dictated by kinetic and equilibrium limits.

- *Recycle separation* section consists of separation units to separate products from reactant components and send their respective streams further downstream.

- *Recycle processing* section consists of pumps, compressors, and heat exchangers that serve to combine the recycled reactants with the process feed.

- *Product recovery* consists of equipment units that provide further temperature, phase, and pressure changes, as well as separation to obtain the product at desired conditions and purity.

Computer models of these tasks are described by a series of unit modules; each software module contains the specific unit model equations as well as specialized procedures for their solution. For modular-based optimization models, we formulate the objective and constraint functions in terms of unit and stream variables in the flowsheet and, through unit modules, these are assumed to be implicit functions of the decision variables. Since we intend to use a gradient-based algorithm, care must be taken so that the objective and constraint functions are continuous and differentiable. Moreover, for the modular approach, derivatives for the implicit module relationships are not directly available, and often not calculated. Instead, they need to be obtained by finite differences (and additional flowsheet evaluations), or by enhancing the unit model calculation to provide exact derivatives directly.

As a result of the above structure, these optimization problems deal with large, arbitrarily complex models but relatively few degrees of freedom. While the number of flowsheet variables could be many thousands, these are "hidden" within the simulation models. On the other hand, even for large flowsheets, there are rarely more than 100 degrees of freedom. The modular mode offers several advantages for flowsheet optimization. First, the flowsheeting problem is relatively easy to construct and to initialize, since numerical procedures are applied that are tailored to each unit. Moreover, the flowsheeting model is relatively easy to debug using process concepts intuitive to the process engineer. On the other hand, calculations in the modular mode are procedural and directed by a rigidly

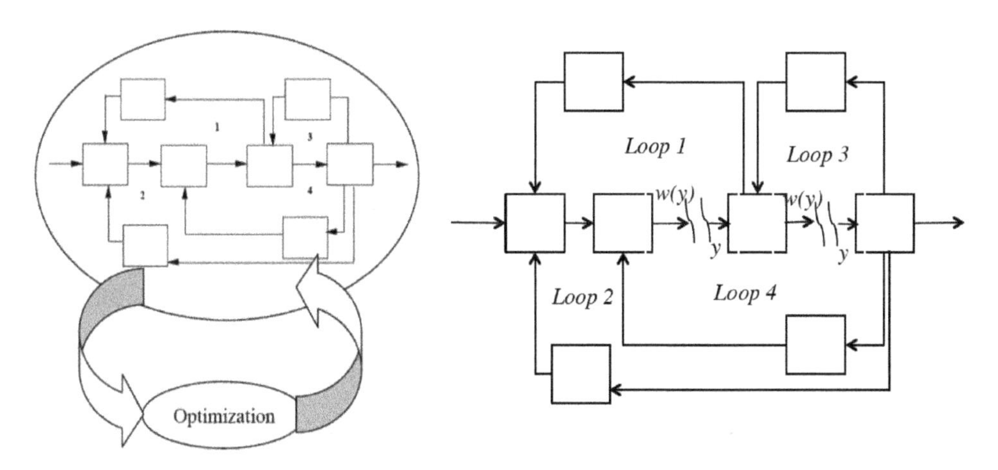

Figure 7.4. *Evolving from black box (left) to modular formulations (right). Here, tear streams are introduced to break all four recycle loops.*

defined information sequence, dictated by the process simulation. Consequently, for optimization one requires that unit models need to be solved repeatedly, and careful problem definition is required to prevent intermediate failure of these process units.

Early attempts at applying gradient-based optimization strategies within the modular mode were based on black box implementations, and these were discouraging. In this simple approach, an optimization algorithm was tied around the process simulator as shown in Figure 7.4. In this "black box" mode, the entire flowsheet needs to be solved repeatedly and failure in flowsheet convergence is detrimental to the optimization. Moreover, as gradients are determined by finite difference, they are often corrupted by errors from loose convergence tolerances in the flowsheet, and this has adverse effects on the optimization strategy. Typically, a flowsheet optimization with 10 degrees of freedom requires the equivalent time of several hundred simulations with the black box implementation.

Since the mid 1980s, however, flowsheet optimization for the modular mode has become a widely used industrial tool. This has been made possible by the following advances in implementation. First, intermediate calculation loops, involving flowsheet recycles and implicit unit specifications, are usually solved in a nested manner with slow fixed-point algorithms. These loops can now be incorporated as equality constraints in the optimization problem, and these additional equality constraints can be handled efficiently with Newton-based NLP solvers. For instance, SQP converges the equality and inequality constraints simultaneously with the optimization problem. This strategy requires relatively few function evaluations and performs very efficiently for process optimization problems, usually in less than the equivalent CPU time of 10 process simulations. Moreover, the NLP solver can be implemented in a nonintrusive way, similar to recycle convergence modules that are already in place. As a result, the structure of the simulation environment and the unit operations blocks does not need to be modified in order to include a simultaneous optimization capability. As seen in Figure 7.4, this approach could be incorporated easily within existing modular simulators and applied directly to flowsheets modeled within these environments.

Given the process optimization problem

$$\begin{aligned} \min \quad & f(z) \\ \text{s.t.} \quad & h(z) = 0, \\ & g(z) \leq 0 \end{aligned} \tag{7.1}$$

with decision variables z, objective function $f(z)$, and constraint functions $h(z), g(z)$, this approach "breaks open" the calculation loops in the simulation problem by considering the so-called *tear variables* y and *tear equations* $y - w(z, y) = 0$ (equations which serve to break every calculation loop at least once) as part of the optimization problem. Adding these variables and equations to (7.1) leads to

$$\begin{aligned} \min \quad & f(z, y) \\ \text{s.t.} \quad & h(z, y) = 0, \\ & y - w(z, y) = 0, \\ & g(z, y) \leq 0. \end{aligned} \tag{7.2}$$

Because slow convergence loops and corresponding convergence errors are eliminated, this strategy is often over an order of magnitude faster than the "black box" approach and converges more reliably, due to more accurate gradients.

Example 7.1 (Formulation of Small Flowsheet Optimization Problem). Consider the simple process flowsheet shown in Figure 7.5. Here the feed stream vector S_1, specified in Table 7.1, has elements j of chemical component flows:

$$j \in \mathcal{C} = \{\text{propane, 1-butene, butane, trans-2-butene, cis-2-butene, pentane}\}$$

as well as temperature and pressure. Stream S_1 is mixed with the "guessed" tear stream vector S_6 to form S_2 which is fed to an adiabatic flash tank, a phase separator operating at vapor-liquid equilibrium with no additional heat input. Exiting from the flash tank is a vapor product stream vector S_3 and a liquid stream vector S_4, which is divided into the bottoms product vector S_7 and recycle vector S_5. The liquid stream S_5 is pumped to 1.02 MPa to form the "calculated" tear stream vector $\bar{S}_6$.

The central unit in the flowsheet is the adiabatic flash tank, where high-boiling components are concentrated in the liquid stream and low-boiling components exit in the vapor stream. This separation module is modeled by the following equations:

$$\text{Overall Mass Balance: } L + V = F, \tag{7.3a}$$

$$\text{Component Mass Balance: } Lx_j + Vy_j = Fz_j, j \in \mathcal{C}, \tag{7.3b}$$

$$\begin{aligned} \text{Energy Balance: } & LH_L(T_f, P_f, x) + VH_V(T_f, P_f, y) \\ & = FH_F(T_{in}, P_{in}, z), \end{aligned} \tag{7.3c}$$

$$\text{Summation: } \sum_{j \in \mathcal{C}} (y_j - x_j) = 0, \tag{7.3d}$$

$$\text{Equilibrium: } y_j - K_j(T_f, P_f, x)x_j = 0, \quad j \in \mathcal{C}, \tag{7.3e}$$

where F, L, V are feed, liquid, and vapor flow rates, x, y, z are liquid, vapor, and feed mole fractions, T_{in}, P_{in} are input temperature and pressure, T_f, P_f are flash temperature

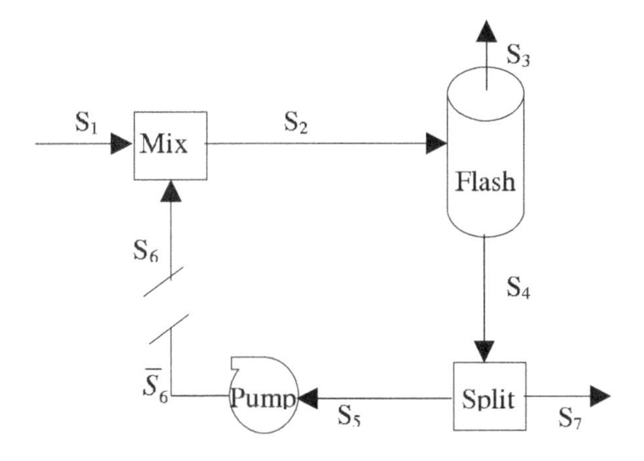

Figure 7.5. *Flowsheet for adiabatic flash loop.*

Table 7.1. *Adiabatic flash problem data and results.*

	Variable	Initial	Optimal	Feed
Tear Stream	Propane (mol/s)	5.682	1.239	10
	1-butene (mol/s)	10.13	2.88	15
	Butane (mol/s)	13.73	4.01	20
	Trans-2-butene (mol/s)	13.78	4.06	20
	Cis-2-butene (mol/s)	13.88	4.13	20
	Pentane (mol/s)	7.30	2.36	10
	Temperature (K)	297.3	277.0	310.9
	Pressure (MPa)	1.02	1.02	1.02
Decision	Flash pressure (MPa)	0.272	0.131	
	Split fraction	0.5	0.8	(upper)
Objective		−1.5711	2.1462	

and pressure, K_j is the equilibrium coefficient, and H_F, H_L, H_V are feed, liquid, and vapor enthalpies, respectively. In particular, K_j and H_F, H_L, H_V are thermodynamic quantities represented by complex nonlinear functions of their arguments, such as cubic equations of state. These quantities are calculated by separate procedures and the model (7.3) is solved with a specialized algorithm, so that vapor and liquid streams S_3 and S_4 are determined, given P_f and feed stream S_2.

The process flowsheet has two decision variables, the flash pressure P_f, and η which is the fraction of stream S_4 which exits as S_7. The individual unit models in Figure 7.5 can be described and linked as follows.

- *Mixer:* Adding stream S_6 to the fixed feed stream S_1 and applying a mass and energy balance leads to the equations $S_2 = S_2(S_6)$.

- *Flash:* By solving (7.3) we separate S_2 adiabatically at a specified pressure, P_f, into vapor and liquid streams leading to $S_3 = S_3(S_2, P_f)$ and $S_4 = S_4(S_2, P_f)$.

- *Splitter:* Dividing S_4 into two streams S_5 and S_7 leads to $S_5 = S_5(S_4, \eta)$ and $S_7 = S_7(S_4, \eta)$.
- *Pump:* Pumping the liquid stream S_5 to 1.02 MPa leads to the equations $\bar{S}_6 = \bar{S}_6(S_5)$.

The tear equations for S_6 can then be formulated by nesting the dependencies in the flowsheet streams:

$$S_6 - \bar{S}_6(S_5(S_4(S_2(S_6), P_f), \eta)) = S_6 - \bar{S}_6[S_6, P_f, \eta] = 0. \tag{7.4}$$

The objective is to maximize a nonlinear function of the elements of the vapor product:

$$S_3(\text{propane})^2 S_3(\text{1-butene}) - S_3(\text{propane})^2 - S_3(\text{butane})^2 + S_3(\text{trans-2-butene}) \\ - S_3(\text{cis-2-butene})^{1/2},$$

and the inequality constraints consist of simple bounds on the decision variables. Written in the form of (7.2), we can pose the following nonlinear program:

$$\begin{aligned} \max \quad & S_3(\text{propane})^2 S_3(\text{1-butene}) - S_3(\text{propane})^2 - S_3(\text{butane})^2 \\ & + S_3(\text{trans-2-butene}) - S_3(\text{cis-2-butene})^{1/2} && (7.5a) \\ \text{s.t.} \quad & S_6 - \bar{S}_6[S_6, P_f, \eta] = 0, && (7.5b) \\ & 0.7 \le P_f \le 3.5, && (7.5c) \\ & 0.2 \le \eta \le 0.8. && (7.5d) \end{aligned}$$

The nonlinear program (7.5) can be solved with a gradient-based algorithm, but derivatives that relate the module outputs to their inputs are usually not provided within simulation codes. Instead, finite difference approximations are obtained through perturbations of the variables y and z in (7.2). Because the accuracy of these derivatives plays a crucial role in the optimization, we consider these more closely before solving this example. ■

7.2.1 Importance of Accurate Derivatives

In the analysis of NLP algorithms we assume that gradient vectors and Hessian matrices could be determined exactly. Errors in these quantities lead to three difficulties. First, inaccurate gradients lead to the calculation of search directions that may not be descent directions for the nonlinear program, leading to premature termination of the algorithm. Second, inaccurate derivative information can lead to Hessians (or approximate Hessians) that are ill-conditioned, also leading to poor search directions. Finally, and most importantly, even if the algorithm finds the solution, accurate derivatives are needed to certify that it is the optimum.

To examine this last issue in more detail, we consider the nonlinear program rewritten as

$$\min \quad f(x) \quad \text{s.t. } c(x) = 0, \; x_L \le x \le x_U. \tag{7.6}$$

As discussed in Chapter 6, one can partition the variables $x = [x_B^T, x_N^T, x_S^T]^T$ into basic, nonbasic, and superbasic variables, respectively, with the first order KKT conditions written as (6.69):

$$c(x_B, x_N, x_S) = 0, \tag{7.7a}$$
$$f_B(x^*) + A_B(x^*)v^* = 0, \tag{7.7b}$$
$$f_N(x^*) + A_N(x^*)v^* - u_L^* + u_U^* = 0, \tag{7.7c}$$
$$f_S(x^*) + A_S(x^*)v^* = 0, \tag{7.7d}$$
$$u_L^*, u_U^* \geq 0, \quad x_L \leq x^* \leq x_U, \tag{7.7e}$$
$$(u_L^*)^T(x_N - x_{N,L}) = 0, \quad (u_U^*)^T(x_N - x_{N,U}) = 0. \tag{7.7f}$$

The nonbasic variables are set to either lower or upper bounds while the basic variables are determined from the equality constraints. For nonbasic variables, the gradients need only be accurate enough so that the signs of u_L^* and u_U^* are correct. Moreover, once x_S and x_N are fixed, x_B is determined from (7.7a), which does not require accurate gradients; multipliers are then determined from (7.7b). Thus, in the absence of superbasic variables (i.e., for vertex optima), correct identification of KKT points may still be possible despite some inaccuracy in the gradients. On the other hand, if superbasic variables are present, accurate identification of x^* relies on accurate gradients for the solution of (7.7b), (7.7a), and (7.7d).

Example 7.2 (Failure Due to Inaccurate Derivatives). To demonstrate the sensitivity of optimization methods and solutions to derivative errors, we consider a simple example proposed by Carter [86]. This unconstrained problem can also be viewed as part of a larger problem where we focus only on the superbasic variables. Consider the unconstrained quadratic problem:

$$\min \quad \frac{1}{2}x^T Ax, \quad \text{where } A = \begin{bmatrix} \epsilon + 1/\epsilon & \epsilon - 1/\epsilon \\ \epsilon - 1/\epsilon & \epsilon + 1/\epsilon \end{bmatrix}, \tag{7.8}$$

with solution $x^* = 0$ and $x^0 = \beta[1, \ 1]^T$ as the starting point with some $\beta \neq 0$. Note that A is positive definite but the condition number of A is $1/\epsilon^2$. The gradient at x^0 is given by

$$\nabla f(x^0) = Ax^0 = 2\epsilon\beta[1\ 1]^T = 2\epsilon x^0,$$

and if the approximated gradient is obtained by numerical perturbations, it is given by:

$$g(x^0) = \nabla f(x^0) + \frac{1}{2}\delta^2(\epsilon + 1/\epsilon)[1\ 1]^T = (2\epsilon\beta + \delta^2(\epsilon + 1/\epsilon)/2)[1\ 1]^T$$

where the second term is the truncation error for the perturbation step δ. If we choose $\beta = -1$ and $\delta^2 = 8\epsilon^2$, we have the following Newton directions:

$$d_{exact} = -A^{-1}\nabla f(x^0) = -x^0 \quad \text{and} \quad d_{actual} = -A^{-1}g(x^0) = (1 + 2\epsilon^2)x^0$$

with directional derivatives

$$\nabla f(x^0)^T d_{exact} = -(x^0)^T Ax^0 < 0 \quad \text{and} \quad \nabla f(x^0)^T d_{actual} = (1 + 2\epsilon^2)(x^0)^T Ax^0 > 0.$$

Note that the inexact gradient leads to *no descent direction* for the Newton step. Consequently, the algorithm fails from a point far from the solution. It is interesting to note that even if we used a "steepest descent direction," $d_{sd} = -g(x^0) = (2\epsilon + 4\epsilon^3)x^0$, similar behavior would be observed for this problem. Moreover, values for δ^2 that lead to failure can actually be quite small. For $\beta < 0$, we require $\delta^2 < -4\beta\epsilon^2/(1+\epsilon^2)$ in order to allow a descent direction for d_{sd}. For example, with $\beta = -1$ and $\epsilon = 10^{-3}$, the algorithm fails for $\delta > 2 \times 10^{-3}$. ∎

This worst-case example shows that gradient error can greatly affect the performance of any derivative-based optimization method. In practice, finite difference approximations may still lead to successful "nearly optimal" NLP solutions, especially for performance or economic optimization, where the optima are usually highly constrained. To minimize the effects of gradient errors, the following guidelines are suggested (see also [162]).

- Modular calculation loops should be converged tightly, so that convergence noise leads to relative errors in module outputs and tear streams, say by $\delta \leq 10^{-8}$.
- Choose a perturbation size so that the relative effect on module outputs and tear streams is approximately $\delta^{1/2}$.
- Finite differenced gradients should be monitored to detect input/output discontinuities, due to failed convergence modules or conditional rules within a module. These are serious problems that must be avoided in the optimization.
- Choose a looser KKT tolerance to include the effect of convergence noise. Alternately, choose a tight tolerance and let the NLP solver terminate with "failure to find a descent direction." Monitoring the final iterates allows the user to assess the effect of inaccurate derivatives on the KKT error, and also to decide whether the final iterate is an acceptable solution.

Example 7.3 (Small Flowsheet Optimization Problem Revisited). The flowsheet given in Figure 7.5 and represented by the optimization problem (7.5) was simulated using the ProSIM process simulator [153] with Soave–Redlich–Kwong (cubic) equations of state for the thermodynamic models. Within this simulation program, gradients for optimization can be obtained either analytically (implemented through successive chain ruling) or by finite difference. Three optimization algorithms discussed in Chapter 6, the reduced gradient method [309], SQP using BFGS updates in the full space, and a reduced-space SQP method, also with BFGS updates, were used to solve this problem with and without analytical derivatives. Using the initial values given in Table 7.1, the optimal solution was obtained for all algorithms considered. At the optimum, also shown in Table 7.1, the split fraction goes to its upper bound while flash pressure is at an intermediate value. Consequently, there is only one allowable direction at this solution (and one superbasic variable) and it is easily verified that the reduced Hessian is positive definite, so that both necessary and sufficient KKT conditions are satisfied at this solution. More detail on this solution can be found in [410].

Performance comparisons (number of flowsheet evaluations (NFE), number of gradient evaluations (NGE), and relative CPU time, normalized to the reduced gradient case with finite difference perturbations) are given for each case in Table 7.2. The reduced gradient case with finite differences corresponds to the equivalent CPU time of about 30 simulations, and the best case finds the optimal solution in about 6 simulation time equivalents. The comparison clearly shows the performance advantages of using exact derivatives. Here

Table 7.2. *Performance comparison for adiabatic flash problem.*

Derivative type	NLP solver	NGE	NFE	Relative effort
Finite	Reduced gradient	20	321	1.00
difference	Full-space SQP	23	228	0.72
	Reduced-space SQP	16	170	0.53
Analytic	Reduced gradient	20	141	0.52
	Full-space SQP	22	25	0.19
	Reduced-space SQP	16	26	0.18

the reduced gradient method appears to be the slowest algorithm, as it devotes considerable effort to an exact line search algorithm. In contrast, the reduced SQP algorithm is slightly faster than the full SQP algorithm when numerical perturbations are used, but with analytical derivatives their computation times are about the same. On the other hand, the use of analytical derivatives leads to important time savings for all three optimization methods. These savings are different for each method (48% for the reduced gradient method, 73% for the full-space SQP, and 66% for reduced-space SQP), since they are realized only for the fraction of time devoted to gradient evaluations. ∎

7.2.2 Ammonia Process Optimization

We consider the optimization of the ammonia process shown in Figure 7.6 with data given in Table 7.3. Production of ammonia (NH_3) from nitrogen (N_2) and hydrogen (H_2) takes place at high temperature and pressure in a catalytic reactor according to the following reaction:

$$N_2 + 3H_2 \rightarrow NH_3.$$

The rest of the flowsheet serves to bring the reactants to the desired reactor conditions and to separate the ammonia product. A feed stream, consisting mostly of hydrogen and nitrogen, is mixed with low-pressure recycle stream S_{15} and compressed in two stages with intercooling. The resulting stream S_4 is combined with the high-pressure recycle stream, S_{14}, heated and sent to the reactor, which has a mass balance model based on a fixed conversion of hydrogen and an effluent temperature that assumes that the reaction is at equilibrium at this conversion. The reactor effluent, S_7, is cooled and separated in two flash stages of decreasing pressure. The overhead vapor streams are S_{10} and S_{15}, which is the low-pressure recycle. The corresponding liquid stream for the low-pressure flash, S_{16}, is the ammonia product. Steam S_{10} is then split to form the high-pressure recycle stream, S_{13}, which is further compressed, and the purge stream, S_{12}. An economic objective function (profit), consisting of net revenue from the product stream S_{16} minus energy costs, is maximized subject to eight equality constraints and three inequalities. Here, seven tear equations for stream S_9 have to be satisfied, and the preheater duty between S_5 and S_6 must match the energy consumption of the reactor. In addition, the product stream S_{16} must contain at least 99% ammonia, the purge stream S_{12} must contain no more than 3.4 mol/s of ammonia, and the pressure drop between S_9 and S_{11} must be at least 0.4 MPa.

Sixteen optimization (seven tear and nine decision) variables have been considered (see Table 7.3) and the Soave–Redlich–Kwong equation of state has been used to calculate

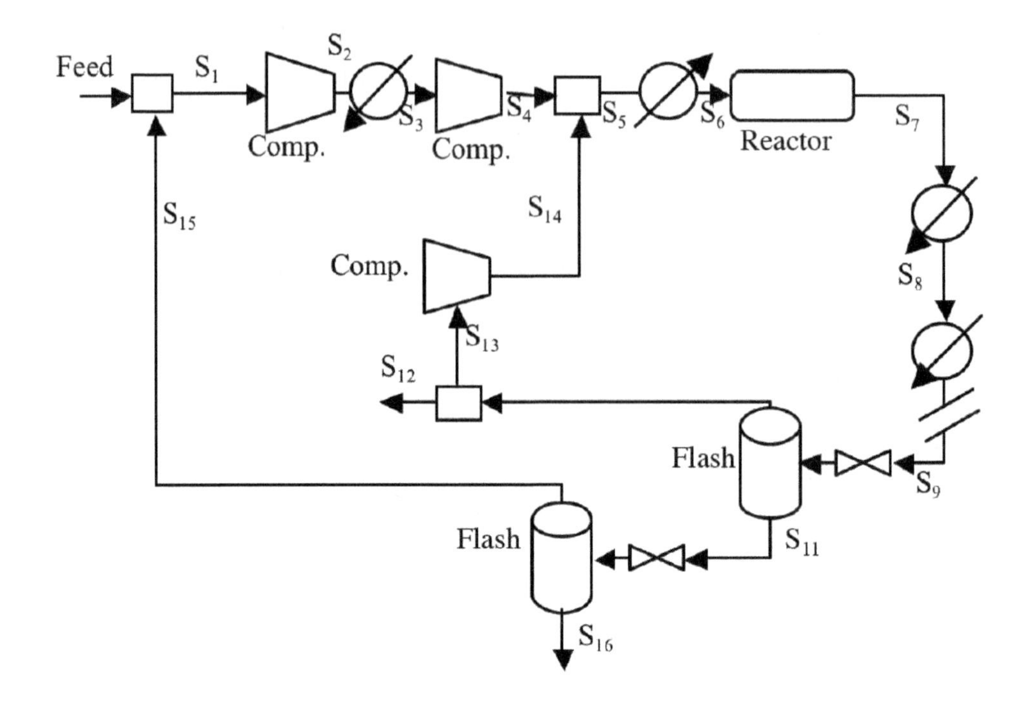

Figure 7.6. *Flowsheet for ammonia synthesis process.*

Table 7.3. *Ammonia synthesis problem data and results.*

Variable Name	Initial	Lower	Upper	Optimal	
Pressure of S_2 (MPa)	4.413	4.137	6.895	4.869	
Pressure of S_4 (MPa)	20.68	19.99	29.72	29.72	Upper Bnd.
Preheater Duty (MW)	3.456	1.46	5.86	3.538	
Reactor Conversion	0.41	0.35	0.45	0.45	Upper Bnd.
Temperature of S_8 (K)	299.8	295.9	310.9	295.9	Lower Bnd.
Temperature of S_9 (K)	242.0	233.1	333.1	244.1	
Pressure of S_{11}, S_{12} (K)	19.99	19.65	29.41	29.31	
Split Ratio (S_{12}/S_{13})	0.1	0.05	0.12	0.053	
Tear Stream (S_9)					Feed
Hydrogen (mol/s)	168.2	0	1000	188.5	110.0
Nitrogen (mol/s)	44.7	0	1000	40.77	35.53
Argon (mol/s)	6.35	0	1000	14.24	0.89
Methane (mol/s)	14.47	0	1000	22.99	1.63
Ammonia (mol/s)	64.76	0	1000	70.23	0.0
Temperature (K)	242	100	1000	244.1	299.82
Pressure (MPa)	20.58	19.89	30	29.62	1.013
Objective	14.27			17.21	

Table 7.4. *Performance comparison for ammonia synthesis problem.*

Derivative type	NLP solver	NGE	NFE	Relative effort
Finite difference	Reduced gradient	87	2092	1.00
	Full-space SQP	49	839	0.40
	Reduced-space SQP	44	760	0.36
Analytic	Reduced gradient	103	927	0.58
	Full-space SQP	49	56	0.10
	Reduced-space SQP	46	51	0.09

the thermodynamic quantities. Three optimization algorithms have also been compared to solve this problem, with and without exact derivatives. All of the algorithms converge to the optimal solution shown in Table 7.3, which improves the profit function by over 20%. Here the purge constraint is active and three decision variables are at their bounds. In particular, the reactor conversion and temperature of stream S_8 are at their upper bounds while the temperature of stream S_9 is at its lower bound. At this solution, the reduced Hessian is a fourth order matrix with eigenvalues $2.87 \times 10^{-2}, 8.36 \times 10^{-10}, 1.83 \times 10^{-4}$, and 7.72×10^{-5}. Consequently, this point satisfies second order necessary conditions, but the extremely low second eigenvalue indicates that the optimum solution is likely nonunique.

Table 7.4 presents a performance comparison among the three algorithms and shows the NFE, the NGE, and the relative CPU time for each case (normalized to the reduced gradient method with finite differenced gradients). The reduced gradient case with finite differences corresponds to the equivalent CPU time of about 30 simulations, while the best case finds the optimal solution in about 3 simulation time equivalents. The comparison shows that the performance results are similar to those obtained for the adiabatic flash loop. The reduced gradient method is slower. When numerical perturbations are used, the reduced-space SQP is faster than full-space SQP, and about the same when analytical derivatives are used. Moreover, significant time savings are observed due to analytical derivatives (42% for reduced gradient, 75% for full-space SQP, and 75% for reduced-space SQP).

7.3 Equation-Oriented Formulation of Optimization Models

The case studies in the previous sections are concerned with application of NLP algorithms for optimization models in modular mode. On the other hand, the hierarchy illustrated in Figure 7.1 shows that fully open, equation-oriented models offer the ability to solve much larger NLPs with many degrees of freedom. Such models allow direct calculation of exact first and second derivatives, along with few restrictions for sparse decomposition strategies. Despite these advantages, successful solution of these optimization models often requires reformulation of the equation-oriented model. While the model must provide a faithful representation of the real-world optimization problem, it should also be compatible with convergence assumptions and particular features of large-scale NLP solvers.

For these Newton-based NLP solvers, linear equalities and variable bounds are satisfied for all iterations, once an iterate x^k is feasible for these constraints. Moreover, the exploitation of sparsity leads to very little additional work for large problems that result from additional variables and equations, particularly if they are linear. As a result, *these algorithms are generally more efficient and reliable in handling large, mostly linear problems than smaller, mostly nonlinear problems*. Using this axiom as a guiding principle, we further motivate the formulation of optimization models by recalling some assumptions required for the convergence analyses in Chapters 5 and 6.

1. The sequence $\{x^k\}$ generated by the algorithm is contained in a convex set D.

2. The objective and constraint functions and their first and second derivatives are Lipschitz continuous and uniformly bounded in norm over D.

3. The matrix of active constraint gradients has full column rank for all $x \in D$.

4. The Hessian matrices are uniformly positive definite and bounded on the null space of the active constraint gradients.

The first two assumptions are relatively easy to enforce for a nonlinear program of the form

$$\min \quad f(x) \quad \text{s.t. } c(x) = 0, \; x_L \leq x \leq x_U.$$

Sensible bounds x_L, x_U can be chosen for many of the variables based on problem-specific information in order to define the convex set D. This is especially important for variables that appear nonlinearly in the objective and constraint functions, although care is needed in bounding nonlinear terms to prevent inconsistent linearizations. In addition, $f(x)$ and $c(x)$ must be bounded, well defined, and smooth for all $x \in D$. This can be addressed in the following ways.

- Nonlinear terms involving inverses, logarithms, trigonometric functions, exponentials, and fractional exponents must be reformulated if they cannot be evaluated over the entire domain of x. For instance, if logarithmic terms in $f(x)$ and $c(x)$ have arguments that take nonpositive values, they can be replaced with additional variables and exponential transformations as in the following:

  $$\ln(g(x)) \implies \exp(y) - g(x) = 0,$$

  where $g(x)$ is the scalar argument and variable y is substituted for $\ln(g(x))$. Upper and lower bounds can also be imposed on y based on limits of x and $g(x)$. The resulting reformulation satisfies the second assumption for any finite bounds on $g(x)$.

- Nonsmooth terms and discrete switches from conditional logic must be avoided in all nonlinear programs. As discussed in Chapter 1, discrete switches are beyond the capabilities of NLP methods; they can be handled through the introduction of integer variables and the formulation of mixed integer nonlinear programs (MINLPs). On the other hand, nonsmooth terms (such as $|x|$) can be handled through smoothing methods or the complementarity reformulations discussed in Chapter 11.

The third assumption requires particular care. Linearly dependent constraint sets are hard to anticipate, as the active set is unknown at the optimum. This assumption can be addressed by first considering the linear constraints and then extending the reformulation to the nonlinear system as follows.

- It is far easier to examine feasible regions made up of simple bounds and linear equalities. Linear independence of linear equalities is relatively easy to check, even for large systems. With the addition of $n-m$ bound constraints (i.e., nonbasic variables), any linear dependence can be detected through elimination of the nonbasic variables in the gradients of the equality constraints. A sufficient condition to satisfy the third assumption is that all combinations of m basic variables lead to nonsingular basis matrices, i.e., A_B in (6.40). This test need only be performed once and is valid for all $x \in D$. Even if the sufficient condition is not satisfied, one can at least identify nonbasic variable sets (and variable bounds) that should be avoided.

- Further difficulties occur with nonlinear equations, especially when elements vanish in the gradient vector. For example, the constraint $x_1^2 + x_2^2 = 1$ is linearly dependent with $x_1 = x_2 = 0$ and most NLP algorithms in Chapter 6 will fail from this starting point. Avoiding this singular point is only possible through the use of additional bounds derived from problem-specific information. A viable approach to handling linear dependencies from poor linearization of nonlinear terms is to reformulate troublesome nonlinear equations by adding new variables that help to *isolate the nonlinear terms*. These terms can then be bounded separately. For instance, if it is known that $x_1 \geq 0.1$, then the constraint $x_1^2 + x_2^2 = 1$ can be rewritten as

$$\begin{aligned} y_1 + y_2 &= 1, \\ y_1 - x_1^2 &= 0, \\ y_2 - x_2^2 &= 0, \\ x_1 &\geq 0.1, \end{aligned}$$

  and linear dependence from these nonlinear terms is avoided.

- An issue related to this assumption is that the objective and constraint functions be well scaled. From the standpoint of Newton-based solvers, good scaling is required for accurate solution of linear systems for determination of Newton steps. A widely used rule of thumb is to scale the objective, constraint functions, and the variables so that magnitudes of the gradient elements are "around 1" [162]. Moreover, many of the NLP algorithms described in Chapter 6 have internal scaling algorithms, or issue warnings if the problem scaling is poor. This usually provides a suitable aid to the user to properly assess the problem scaling.

- Finally, note that linear dependence of constraint gradients is less of a concern in the modular simulation mode because feasible solutions are supplied by internal calculation loops, and because the basis matrix A_B is usually nonsingular at feasible points.

The assumption for the reduced Hessian is the hardest to ensure through reformulation. Positive definiteness is not required for trust region methods; otherwise, the Hessian (or its projection) can be replaced by a quasi-Newton update. However, if the actual projected Hessian has large negative curvature, NLP algorithms can still perform poorly. As a result, some attention needs to be paid to highly nonlinear terms in the objective and constraint functions. For instance, the nonsmooth function $\max\{0, g(x)\}$ is often replaced by a smoothed reformulation $1/2(g(x) + (g(x)^2 + \epsilon)^{1/2})$. For small $\epsilon > 0$, this function provides a reasonable approximation of the max operator, but the higher derivatives become unbounded as $g(x)$, $\epsilon \to 0$; this can adversely affect the curvature of the Hessian. Consequently, a choice of ϵ is required that balances accuracy of the smoothing function with ill-conditioning.

Because process models differ so widely, there are no clear-cut formulation rules that apply to all problems. On the other hand, with the goal of formulating sparse, mostly linear models in mind, along with the above assumptions on convergence properties, one can develop successful optimization models that can be solved efficiently and reliably by modern NLP solvers. To demonstrate the application of the above guidelines, we close this section with a process case study.

7.3.1 Reformulation of the Williams–Otto Optimization Problem

One of the earliest flowsheet simulation problems in the process engineering literature is the Williams–Otto process [408]. Originally formulated as a control problem, this process has been the focus of a number of optimization studies [53, 213, 300, 330, 338]. The process flowsheet is shown in Figure 7.7 with process streams given as mass flows and composition given as mass fractions. Two feed streams (F_A and F_B) containing pure components A and B, respectively, are fed to a stirred tank reactor where the following three reactions occur at temperature T:

$$\begin{aligned} A + B &\xrightarrow{k_1} C, \\ C + B &\xrightarrow{k_2} P + E, \\ P + C &\xrightarrow{k_3} G. \end{aligned} \tag{7.9}$$

Here C is an intermediate component, P is the main product, E is a by-product, and G is a waste product. The effluent stream F_{eff} is cooled in a heat exchanger and is sent to a centrifuge to separate component G from the process in stream F_G. The remaining components are then separated to remove component P in the overhead product stream, F_P. Due to the presence of an azeotrope, some of the product (equal to 10 wt. % of component E) is retained in the bottoms. The bottoms stream is split into purge (F_{purge}) and recycle streams (F_R); the latter is mixed with the feed and sent back to the reactor. More details of this well-known process optimization example can be found in [330, 53]. The objective of the optimization problem is to maximize the return on investment (ROI). Adapted from [330], the optimization model has a compact representation, which leads to a small nonlinear model given by

$$\begin{aligned} \max\ ROI = {} & 100(2207 F_P + 50 F_{purge} - 168 F_A - 252 F_B \\ & - 2.22 F_{eff}^{sum} - 84 F_G - 60 V\rho)/(600 V\rho). \end{aligned}$$

Rate Equations

$$\begin{aligned} k_1 &= a_1 \exp(-120/T), \\ k_2 &= a_2 \exp(-150/T), \\ k_3 &= a_3 \exp(-200/T). \end{aligned}$$

Separator Relation

$$F_{eff}^P = 0.1 F_{eff}^E + F_P.$$

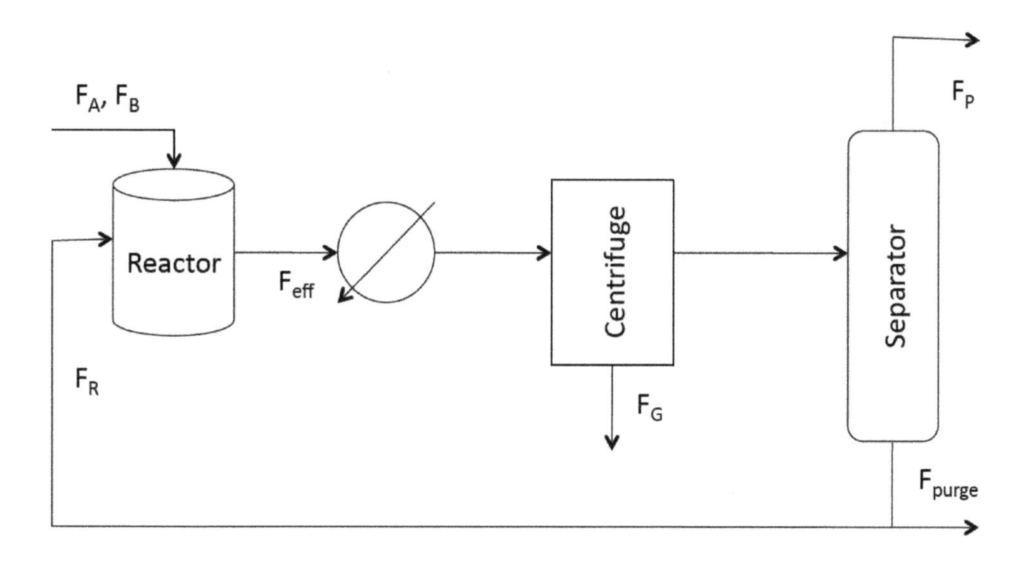

Figure 7.7. *Flowsheet for Williams–Otto optimization problem.*

Overall Total and Component Mass Balances

$$
\begin{aligned}
0 &= F_A + F_B - F_G - F_P - F_{purge},\\
0 &= -k_1 F_{eff}^A F_{eff}^B V\rho/(F_{eff}^{sum})^2\\
&\quad - F_{purge} F_{eff}^A/(F_{eff}^{sum} - F_G - F_P) + F_A,\\
0 &= (-k_1 F_{eff}^A F_{eff}^B - k_2 F_{eff}^B F_{eff}^C) V\rho/(F_{eff}^{sum})^2\\
&\quad - F_{purge} F_{eff}^B/(F_{eff}^{sum} - F_G - F_P) + F_B,\\
0 &= [(2k_1 F_{eff}^A - 2k_2 F_{eff}^C) F_{eff}^B - k_3 F_{eff}^C F_{eff}^P)] V\rho/(F_{eff}^{sum})^2\\
&\quad - F_{purge} F_{eff}^C/(F_{eff}^{sum} - F_G - F_P),\\
0 &= 2k_2 F_{eff}^B F_{eff}^C V\rho/(F_{eff}^{sum})^2\\
&\quad - F_{purge} F_{eff}^E/(F_{eff}^{sum} - F_G - F_P),\\
0 &= (k_2 F_{eff}^B - 0.5k_3 F_{eff}^P) F_{eff}^C V\rho/(F_{eff}^{sum})^2\\
&\quad - F_{purge}(F_{eff}^P - F_P)/(F_{eff}^{sum} - F_G - F_P) - F_P,\\
0 &= 1.5k_3 F_{eff}^C F_{eff}^P V\rho/(F_{eff}^{sum})^2 - F_G,\\
0 &= F_{eff}^A + F_{eff}^B + F_{eff}^C + F_{eff}^E + F_{eff}^P + F_G - F_{eff}^{sum}.
\end{aligned}
$$

Bound Constraints

$$
\begin{aligned}
&V \in [0.03, 0.1], \quad T \in [5.8, 6.8], \quad F_P \in [0, 4.763],\\
&F_{purge}, F_A, F_B, F_G, F_{eff}^j \geq 0,
\end{aligned}
$$

Table 7.5. *Initial and optimal points for the Williams–Otto problem.*

Variable	Initial Point	Optimal Solution
F_{eff}^{sum}	52	366.369
F_{eff}^{A}	10.	46.907
F_{eff}^{B}	30.	145.444
F_{eff}^{C}	3.	7.692
F_{eff}^{E}	3.	144.033
F_{eff}^{P}	5.	19.115
F_P	0.5	4.712
F_G	1.	3.178
F_{purge}	0.	35.910
V	0.06	0.03
F_A	10.	13.357
F_B	20.	30.442
T	5.80	6.744
k_1	6.18	111.7
k_2	15.2	567.6
k_3	10.2	1268.2
ROI	—	121.1

where $\rho = 50$, $a_1 = 5.9755 \times 10^9$, $a_2 = 2.5962 \times 10^{12}$, and $a_3 = 9.6283 \times 10^{15}$. Note that the volume and mass flows are scaled by a factor of 1000, and temperature by a factor of 100.

Table 7.5 shows the solution to this problem that satisfies sufficient second order KKT conditions. Also listed is a suggested starting point. However, from this and other distant starting points, the sparse, large-scale algorithms CONOPT, IPOPT, and KNITRO have difficulty finding the optimal solution. In particular, the mass balances become linearly dependent when flows F_{eff}^{j} tend to zero. Curiously, feasible points can be found for very small values of this stream with essentially zero flow rates for all of the feed and product streams. This leads to an attraction to an unbounded solution *where no production occurs*. Consequently, the above formulation does not allow reliable solution in the equation-oriented mode. To reformulate the model to allow convergence from distant starting points, we apply the guidelines presented above and make the changes below.

- The overall mass balances are reformulated equivalently as mass balances around the reactor, separator, and splitter. This larger model leads to greater sparsity and additional linear equations.

- Additional variables are added to define linear mass balances in the reactor, and the nonlinear reaction terms are defined through these additional variables and isolated as additional equations.

- All nonlinear terms that are undefined at zero are replaced. In particular, the rate constraints are reformulated by adding new variables that render the rate equations linear.

- In addition to existing bound constraints, bounds are placed on all variables that occur nonlinearly. In particular, small lower bounds are placed on F_A and F_B in order to avoid the unbounded solution.

The resulting formulation is given below:

$$\begin{aligned}\max\ ROI = 100(&2207F_P + 50F_{purge} - 168F_A - 252F_B \\ &-2.22F_{eff}^{sum} - 84F_G - 60V\rho)/(600V\rho).\end{aligned}$$

Rate Equations

$$\begin{aligned}
k_1^l &= \ln(a_1) - 120 * T_2, \\
k_2^l &= \ln(a_2) - 150 * T_2, \\
k_3^l &= \ln(a_3) - 200 * T_2, \\
T_2 T &= 1, \\
k_i &= \exp(k_i^l), \quad i = 1, \dots, 3, \\
r_1 &= k_1 x_A x_B V\rho, \\
r_2 &= k_2 x_C x_B V\rho, \\
r_3 &= k_3 x_P x_C V\rho.
\end{aligned}$$

Reactor Balance Equations

$$\begin{aligned}
F_{eff}^A &= F_A + F_R^A - r_1, \\
F_{eff}^B &= F_B + F_R^B - (r_1 + r_2), \\
F_{eff}^C &= F_R^C + 2r_1 - 2r_2 - r_3, \\
F_{eff}^E &= F_R^E + 2r_2, \\
F_{eff}^P &= 0.1F_R^E + r_2 - 0.5r_3, \\
F_{eff}^G &= 1.5r_3, \\
F_{eff}^{sum} &= F_{eff}^A + F_{eff}^B + F_{eff}^C + F_{eff}^E + F_{eff}^P + F_{eff}^G, \\
F_{eff}^j &= F_{eff}^{sum} x_j, \quad j \in \{A, B, C, E, P, G\}, \\
F_{eff}^{sum} &\geq 1.
\end{aligned}$$

Define Waste Stream

$$F_G = F_{eff}^G.$$

Define Product Stream

$$F_P = F_{eff}^P - 0.1F_{eff}^E.$$

Define Purge Stream

$$F_{purge} = \eta(F_{eff}^A + F_{eff}^B + F_{eff}^C + 1.1F_{eff}^E).$$

Define Recycle Stream

$$F_R^j = (1-\eta)F_{eff}^j, \quad j \in \{A,B,C,E,P,G\}.$$

Bound Constraints

$$V \in [0.03, 0.1],\ T \in [5.8, 6.8],\ T_2 \in [0.147, 1.725],\ F_P \in [0, 4.763],$$
$$F_{purge}, F_G, F_{eff}^j \geq 0, F_A, F_B \geq 1.$$

The resulting reformulation increases the optimization model from 17 variables and 13 equations to 37 variables and 33 equations. With this reformulation, all three solvers easily converge to the desired solution from the distant starting point given in Table 7.5. CONOPT, IPOPT, and KNITRO require 72, 43, and 32 iterations, respectively, and less than 0.1 CPUs (with a 2 GB, 2.4 GHz Intel Core2 Duo processor, running Windows XP) for each solver.

7.4 Real-Time Optimization

With appropriate problem formulations, equation-oriented models can be treated directly as equality constraints within the NLP algorithm. This allows ready access to exact derivatives from the model but also places the burden of solving the model equations on the NLP algorithm, rather than specialized solution procedures. A major advantage of this approach is that the models and optimal decision variables are determined simultaneously at the same computational level. Consequently, repeated solutions of individual modules, as seen in the previous section, are avoided. Instead, the cost of an NLP solution is little more than that of the process simulation itself. On the other hand, the NLP algorithm needs to be efficient to handle these larger problems. Equation-oriented optimization models are especially useful in time-critical applications where fast solutions are needed and the nonlinear program can be initialized easily. These characteristics apply to real-time optimization. In this section we deal with real-time optimization problems that rely on NLPs based on large equation-based models (up to 500,000 equations), but with few (about 10–50) degrees of freedom. Because reduced-space curvature information is required in only a few dimensions, second derivatives are rarely provided for these models, and quasi-Newton updates are used instead in the NLP solver.

Real-time optimization (RTO) models have been developed since the early 1980s and are abundant in petrochemical and chemical applications, especially in the production of ethylene. The RTO market is projected to grow to about a billion dollars per year in 2010 (see http://www.arcweb.com). This growth leads to increased agility in plants to execute production plans, eliminate inefficiencies, and profitably capture market opportunities. In particular, manufacturers use RTO to tune existing processes to changes in product demand and price as well as cost and availability of feedstocks. Moreover, RTO is essential to mitigate and reject long-term disturbances and performance losses (for example, through fouling of heat exchangers or deactivation of catalysts).

The structure of an RTO strategy is illustrated in Figure 7.8. Steady state plant data are matched to a plant model in a data reconciliation/parameter estimation (DR-PE) step, and the resulting optimization model is solved to update the steady state operating conditions. These conditions (or setpoints) are then passed to a process control system that steers the plant toward the updated point. This cycle is typically updated every few hours. A key

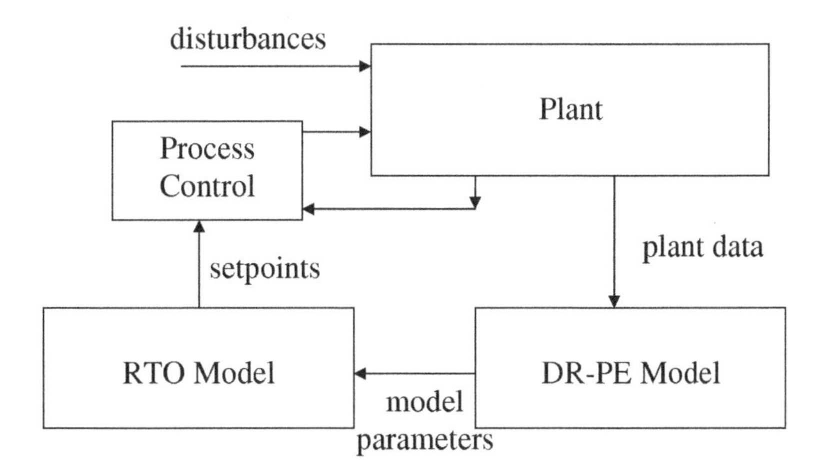

Figure 7.8. *Components of real-time optimization (RTO).*

assumption is that the steady state model is sufficient to describe the plant, and that fast dynamics and disturbances can be handled by the process control system. Along with the benefits of RTO, there are a number of challenges to RTO implementation [276, 146, 415], including formulating models that are consistent with the plant, maintaining stability of the RTO cycle, ensuring useful solutions in the face of disturbances, and solving the NLP problems quickly and reliably.

7.4.1 Equation-Oriented RTO Models

Both the RTO and DR-PE tasks require the solution of large nonlinear programs with process models that have the same components as the flowsheets in the previous section. However, because the model equations now become constraints in the nonlinear program, care is required so that functions and derivatives are well defined. To illustrate the formulation of the RTO model, we consider two standard units in an RTO model, heat exchangers, and trayed separators. Both were discussed in Chapter 1 and they make up the main components of the process shown in Figure 7.9.

Heat Exchanger Models

Heat exchanger models determine temperatures and heat duties from process and utility stream interchanges. Streams participating in the heat exchange are classified as *hot streams*, which need to be cooled, and *cold streams*, which need to be heated. Given two streams a and b, with stream flow rates F_i and heat capacity $C_{p,i}$, inlet and outlet temperatures, T_i^{in} and T_i^{out}, $i \in \{a,b\}$, and exchanger heat duty Q, this unit can be modeled as follows:

- Choosing stream a as the hot stream with $(T_a^{in} - T_a^{out}) > 0$ and b as the cold stream with $(T_b^{in} - T_b^{out}) < 0$, the energy balance for this system is given by

$$Q = F_a C_{p,a}(T_a^{in} - T_a^{out}),$$
$$Q = F_b C_{p,b}(T_b^{out} - T_b^{in}).$$

- Performance of each heat exchanger is based on its available area A for heat exchange along with an overall heat transfer coefficient, U. The resulting area equations are given by

$$\begin{aligned} Q &= UA\Delta T_{lm}, \\ \Delta T_{lm} &= \frac{\Delta T_1 - \Delta T_2}{\ln(\Delta T_1/\Delta T_2)}, \\ \Delta T_1 &= (T_a^{in} - T_b^{out}), \\ \Delta T_2 &= (T_a^{out} - T_b^{in}). \end{aligned} \tag{7.10}$$

- Because the expression for log-mean temperature, ΔT_{lm}, is not well defined when $\Delta T_1 = \Delta T_2$, it is often replaced by an approximation, e.g.,

$$(\Delta T_{lm})^{1/3} = ((\Delta T_1)^{1/3} + (\Delta T_2)^{1/3})/2.$$

Trayed Separation Models

These units include distillation columns as well as absorption and stripping units. A trayed model with chemical species $j \in \mathcal{C}$, trays $i \in 1,\ldots,N_T$, and feed trays $i \in \mathcal{S}$ is constructed using the MESH (Mass-Equilibrium-Summation-Heat) equations given below.

Total Mass Balances

$$L_i + V_i - L_{i+1} - V_{i-1} = 0, \; i \in \{1,\ldots,N_T\}, \quad i \notin \mathcal{S}, \tag{7.11}$$

$$L_i + V_i - L_{i+1} - V_{i-1} - F_i = 0, \quad i \in \mathcal{S}, \tag{7.12}$$

where F_i, L_i, V_i are feed, liquid, and vapor streams. Trays are numbered from bottom to top and streams L_{N+1} and V_0 are external streams that enter the separator.

Component Mass Balances

$$L_i x_{i,j} + V_i y_{i,j} - L_{i+1} x_{i+1,j} - V_{i-1,j} y_{i-1,j} = 0, \quad j \in \mathcal{C}, i \in \{1,\ldots,N_T\}, \quad i \notin \mathcal{S}, \tag{7.13}$$

$$L_i x_{i,j} + V_i y_{i,j} - L_{i+1} x_{i+1,j} - V_{i-1} y_{i-1,j} - F_i z_{F_i,j} = 0, \quad j \in \mathcal{C}, \quad i \in \mathcal{S}, \tag{7.14}$$

where $z_{F_i,j}, x_{i,j}, y_{i,j}$ are mole fractions of the feed, liquid, and vapor streams for component j.

Enthalpy Balances

$$\begin{aligned} &L_i H_{L,i}(T_i, P_i, x_i) + V_i H_{V,i}(T_i, P_i, y_i) - L_{i+1} H_{L,i+1}(T_{i+1}, P_{i+1}, x_{i+1}) \\ &\quad - V_{i-1} H_{V,i-1}(T_{i-1}, P_{i-1}, y_{i-1}) + Q_{i,ext} = 0, \; i \in \{1,\ldots,N_T\}, \quad i \notin \mathcal{S}, \end{aligned} \tag{7.15}$$

$$\begin{aligned} &L_i H_{L,i}(T_i, P_i, x_i) + V_i H_{V,i}(T_i, P_i, y_i) - L_{i+1} H_{L,i+1}(T_{i+1}, P_{i+1}, x_{i+1}) \\ &\quad - V_{i-1} H_{V,i-1}(T_{i-1}, P_{i-1}, y_{i-1}) - F_i H_{F_i}(T_{F,i}, P_{F,i}, z_{F,i}) = 0, \quad i \in \mathcal{S}, \end{aligned}$$

where $H_{F_i}, H_{L,i}, H_{V,i}$ are enthalpies of the feed, liquid, and vapor streams. Also, as shown in Figure 7.9, $Q_{i,ext}$ is the heat added or removed from tray i through external heat exchange.

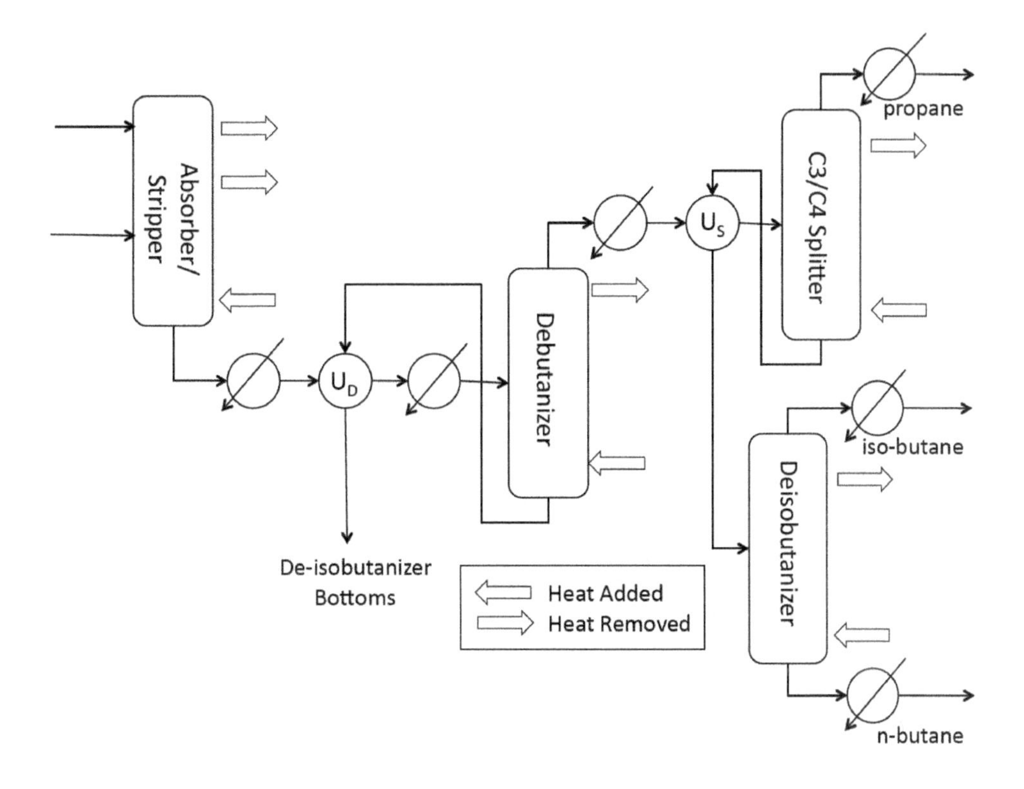

Figure 7.9. *Flowsheet for hydrocracker fractionation plant.*

Summation and Equilibrium Relations

$$\sum_{j \in \mathcal{C}} (y_{i,j} - x_{i,j}) = 0, \quad i \in \{1, \ldots, N_T\},$$

$$y_{i,j} - K_{i,j}(T_i, P_i, x_i) x_{i,j} = 0, \quad j \in \mathcal{C}\; i \in \{1, \ldots, N_T\},$$

where $K_{i,j}, T_i, P_i$ are the equilibrium coefficient, tray temperature, and tray pressure.

Unlike the modular models in the previous section, all of these model equations are declared as constraints in the nonlinear program, from which first derivatives can be evaluated exactly. In particular, equations for the thermodynamic quantities, $C_{p,j}$, $K_{i,j}$, $H_{L,i}$, $H_{V,i}$, $H_{F_i,i}$, are expressed as simplified correlations of their arguments (often polynomials), rather than implicit calculations as procedures. These simplifications are usually determined from regressions against process data or more detailed thermodynamic models.

7.4.2 Case Study of Hydrocracker Fractionation Plant

To illustrate the performance of large-scale NLP solvers on RTO models, we consider the determination of optimal operating conditions for the Sunoco hydrocracker fractionation plant [20, 356]. This model is based on an existing process that was one of the first commissioned for real-time optimization in the early 1980s. Although current RTO models are up to two orders of magnitude larger, the case study is typical of many real-time optimization problems.

The fractionation plant, shown in Figure 7.9, is used to separate the effluent stream from a hydrocracking unit. The process deals with 17 chemical components,

$$\mathcal{C} = \{\text{nitrogen, hydrogen sulfide, hydrogen, methane, ethane, propane, isobutane, n-butane, isopentane, n-pentane, cyclopentane}, C_6, C_7, C_8, C_9, C_{10}, C_{11}\}.$$

The plant includes numerous heat exchangers, including six utility coolers, two interchangers (with heat transfer coefficients U_D and U_S), and additional sources of heating and cooling. It can be described by the following units, each consisting of collections of tray separator and heat exchange models described in the previous section.

- *Absorber/Stripper* separates methane and ethane overhead with the remaining components in the bottom stream. The combined unit is modeled as a single column with 30 trays and feeds on trays 1 and 14. Setpoints include mole fraction of propane in the overhead product, feed cooler duty, and temperature in tray 7.

- *Debutanizer* separates the pentane and heavier components in the bottoms from the butanes and lighter components. This distillation column has 40 trays and feed on tray 20. External heat exchange on the bottom and top trays is provided through a reboiler and condenser, respectively. These are modeled as additional heat exchangers. Setpoints include feed preheater duty, reflux ratio, and mole fraction of butane in the bottom product.

- *C3/C4 splitter* separates propane and lighter components from the butanes. This distillation column has 40 trays, feed on tray 24, and reboiler and condenser (modeled as additional heat exchangers). Setpoints include the mole fractions of butane and propane in the product streams.

- *Deisobutanizer* separates isobutane from butane and has 65 trays with feed on tray 38 and reboiler and condenser (modeled as additional heat exchangers). Setpoints include the mole fractions of butane and isobutane in the product streams.

Note that these 10 unit setpoints represent the decision variables for the RTO. Additional details on the individual units may be found in [20].

As seen in Figure 7.8, a two-step RTO procedure is considered. First, a single-parameter case is solved as the DR-PE step, in order to fit the model to an operating point. The optimization is then performed starting from this "good" starting point. Besides the equality constraints used to represent the individual units, simple bounds are imposed to respect actual physical limits of various variables, bound key variables to prevent the solution from moving too far from the previous point, and fix respective variables (i.e., setpoints and parameters) in the parameter and optimization cases. The objective function which drives the operating conditions of the plant is a profit function consisting of the added value of raw materials as well as energy costs:

$$f(x) = \sum_{k \in G} C_k^G S_k + \sum_{k \in E} C_k^E S_k + \sum_{k \in P} C_k^P S_k - E(x), \tag{7.16}$$

where C_k^G, C_k^E, C_k^P are prices on the feed and product streams (S_k) valued as gasoline, fuel, or pure components, respectively, and $E(x)$ are the utility costs. The resulting RTO problem has 2836 equality constraints and 10 independent variables.

Table 7.6. *Performance results for hydrocracker fractionation problem (with CPU times normalized to MINOS result).*

	Base Case	Case 1 Base Opt.	Case 2 Fouling	Case 3 Fouling	Case 4 Market	Case 5 Market
U_D (base normalized)	1.0	1.0	0.762	0.305	1.0	1.0
U_S (base normalized)	1.0	1.0	0.485	0.194	1.0	1.0
Propane Price ($/$m^3$)	180	180	180	180	300	180
Gasoline Price ($/$m^3$)	300	300	300	300	300	350
Octane Credit ($/(RON-$m^3$))	2.5	2.5	2.5	2.5	2.5	10
Profit ($/day)	230969	239277	239268	236707	258913	370054
Change from Base	—	3.6 %	3.6 %	2.5 %	12.1 %	60.2 %
Poor Initialization						
MINOS Iters. (Major/minor)	5/275	9/788	—	—	—	—
rSQP Iterations	5	20	12	24	17	12
MINOS CPU time (norm.)	0.394	12.48	—	—	—	—
rSQP CPU time (norm.)	0.05	0.173	0.117	0.203	0.151	0.117
Good Initialization						
MINOS Iters. (Major/minor)	—	12/132	14/120	16/156	11/166	11/76
rSQP Iterations	—	13	8	18	11	10
MINOS CPU time (norm.)	—	1.0	0.883	2.212	1.983	0.669
rSQP CPU time (norm.)	—	0.127	0.095	0.161	0.114	0.107

We consider the following RTO cases for this process. Numerical values of the corresponding parameters are included in Table 7.6. In the base optimization case (Case 1), the profit is improved by 3.6% over the base case. This level of performance is typical for RTO implementations. In Cases 2 and 3, the effect of reduced process performance by fouling in heat exchanges is simulated by reducing the heat exchange coefficients for the debutanizer and splitter feed/bottoms exchangers in order to see their effect on the optimal solution. For Case 2 new setpoints are determined so that this process deterioration does not reduce the profit from Case 1. For Case 3, further deterioration of heat transfer leads to lower profit from Case 1, but this is still an improvement over the base case. The effect of changing market prices is seen in Cases 4 and 5. Here, changing market prices are reflected by an increase in the price for propane (Case 4) or an increase in the base price for gasoline, together with an increase in the octane credit (Case 5). In both cases, significant increases in profit are observed, as the RTO determines setpoints that maximize the affected product flows.

All cases were solved to a KKT tolerance of 10^{-8}. Results are reported in Table 7.6, where "poor" initialization indicates initialization at the original initial point, while the "good" initialization results were obtained using the solution to the parameter case as the initial point. Table 7.6 compares RTO performance from two studies [20, 356] using MINOS and reduced-space SQP (rSQP), both described in Chapter 6, as the NLP solvers. For all cases considered, both algorithms terminate with the same optimal solution. From Table 7.6 it is apparent that rSQP is at least as robust and considerably more efficient than MINOS. In particular, for the poor initialization, there is a difference of almost two orders of magnitude. Also, for this problem, rSQP is much less sensitive to a poor initial point than MINOS. Moreover, the results for the good initializations indicate an order of magnitude improvement in CPU times when comparing rSQP to MINOS.

7.5 Equation-Oriented Models with Many Degrees of Freedom

Gasoline blending problems are formulated as large-scale multiperiod nonlinear programs with mass balance constraints, nonlinear blending properties, large-scale structure (particularly across multiperiods), and combinatorial aspects dealing with bounds and possible switching strategies. Gasoline is one of the most important refinery products as it can yield 60–70% of a typical refinery's total revenue. Thus tight control of blending operations can provide a crucial edge to the profitability of a refinery. There are various motivations to blend gasoline online, including variations on regional demands, seasonal variations, particular blends during the summer and winter months, and variations in refining products that must be moderated and improved.

In this section we describe and consider these blending optimization problems, consisting of equation-oriented models with nonlinear mass balances that include concentration, quality, or temperature information. Moreover, their problem sizes can be extended through multiperiod formulations and their degrees of freedom increase proportionally. While these problems are nonconvex and admit to multiple local optima, we consider only local, and not global, optimization solvers here, and rely on a systematic initialization scheme to improve the likelihood of finding global solutions.

The need for blending occurs when product requirements cannot be met by a single source. The classical blending problem arises in refinery processes where feeds with different qualities or properties (e.g., sulfur composition, density or octane number, boiling point temperatures, flow rate) are mixed to obtain final products dispatched to several locations. A generalization of gasoline blending, known as the pooling problem, is used to model systems with intermediate mixing (or pooling) tanks in the blending process. These arise in applications that blend crude or refined petroleum products. Moreover, blending systems are also encountered in other process industries, such as chemical, pharmaceutical, cosmetic, and food. Since there are usually several ways to satisfy product requirements, the problem is posed as maximizing the difference between revenue generated by selling final blends and the cost of purchasing the source streams.

The general gasoline blending formulation is defined for products $k \in \mathcal{K}$, feed sources $i \in \mathcal{I}$, and intermediate tanks $j \in \mathcal{J}$ over a time horizon with N_t time periods, $t \in \{1,\ldots,N_t\}$, as follows:

$$
\begin{aligned}
\max \quad & \sum_{t\in\mathcal{T}} \left(\sum_{k\in\mathcal{K}} c_k s_{t,k} - \sum_{i\in\mathcal{I}} c_i s_{t,i} \right) \qquad (7.17)\\
\text{s.t.} \quad & \sum_{k\in\mathcal{K}} s_{t,jk} - \sum_{i\in\mathcal{I}} s_{t,ij} + v_{t+1,j} = v_{t,j}, \; t \in \{1,\ldots,N_t\},\; j \in \mathcal{J},\\
& \sum_{i\in\mathcal{I}} q_{t,i} s_{t,ij} - \sum_{k\in\mathcal{K}} q_{t,j} s_{t,jk} + q_{t,j} v_{t,j} = q_{t+1,j} v_{t+1,j},\\
& t = 1,\ldots,N_t - 1, j \in \mathcal{J},\\
& q_{t,k} s_{t,k} - \sum_{j\in\mathcal{J}} q_{t,jk} s_{t,jk} = 0, \quad t = 1,\ldots,N_t,\; k \in \mathcal{K},
\end{aligned}
$$

where the indexed variables $s_{t,lm}$ represent a stream flow between tank indices l and m, and $q_{t,l}$ and $v_{t,l}$ are qualities (i.e., blend stream properties) and volumes for index l, respectively,

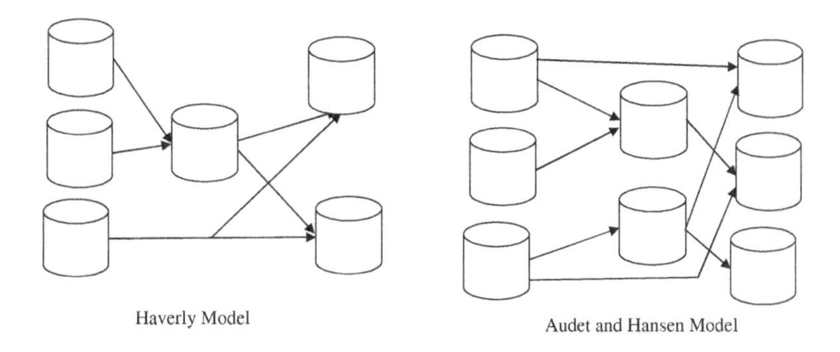

Figure 7.10. *Configuration for blending problems.*

at time t. The classical blending problem mixes feeds directly into blends. The related pooling problem also considers intermediate pools where the feeds are mixed prior to being directed to the final blends. These pools consist of source pools with a single purchased feed, intermediate pools with multiple inputs and outputs, and final pools with a single final blend as output. Also, if intermediate pools have multiple outlets at the same time, then additional equations are added to enforce the same tank qualities on the outlet.

The objective function of the gasoline blending model minimizes cost or maximizes profit of production of blends and remains linear, with the nonlinearities seen only in the constraints. Here we consider three blending models. As shown in Figure 7.10, the first two models were proposed by Haverly [190] (three sources, two products, and one intermediate) and Audet, Hansen, and Brimberg [17] (three sources, three products, and two intermediates) each with only a single quality. The third model applies the blending formulation to a real-world industrial problem [310] (17 sources, 13 intermediate tanks, 7 product tanks, and 5 products) with 48 qualities including octane number, density, and Reid vapor pressure. Among these models we consider 8 examples, $N_t = 1$ and $N_t = 25$ for each of the first two models and $N_t = 1, 5, 10, 15$ for the industrial model. Because all of these problems are nonconvex and may admit locally optimal solutions, we apply the following initialization strategy in an attempt to find globally optimal solutions:

1. Fix quality variables and drop redundant equations in (7.17).
2. Solve this restricted problem as a linear program (LP) for the stream flows.
3. Using the LP solution, fix the streams and solve for the optimal qualities. This provides an upper bound to the solution of (7.17).
4. Using the resulting flows and qualities as a feasible initial guess, solve (7.17) with an NLP solver.

To compare solutions we considered the solvers SNOPT (version 5.3), LOQO (version 4.05), IPOPT (version 2.2.1 using MA27 as the sparse linear solver) with an exact Hessian, and IPOPT with the limited memory BFGS update. As noted in Chapter 6, SNOPT consists of an rSQP method that uses BFGS updates, LOQO uses a full-space barrier method with exact Hessian information and a penalty function line search, and IPOPT,

Table 7.7. *Results of gasoline blending models.*

Model	N_t	n	n_S	Solver	Iterations	CPU s (norm.)
HM	1	13	8	SNOPT	36	0.01
HM	1	13	8	LOQO	30	0.08
HM	1	13	8	IPOPT, Exact	31	0.01
HM	1	13	8	IPOPT, BFGS	199	0.08
AHM	1	21	14	SNOPT	60	0.01
AHM	1	21	14	LOQO	28	0.08
AHM	1	21	14	IPOPT, Exact	28	0.01
AHM	1	21	14	IPOPT, BFGS	44	0.02
HM	25	325	200	SNOPT	739	0.27
HM	25	325	200	LOQO	31	0.33
HM	25	325	200	IPOPT, Exact	47	0.24
HM	25	325	200	IPOPT, BFGS	344	1.99
AHM	25	525	350	SNOPT	1473	0.66
AHM	25	525	350	LOQO	30	0.60
AHM	25	525	350	IPOPT, Exact	44	0.25
AHM	25	525	350	IPOPT, BFGS	76	0.98
IM	1	2003	1449	IPOPT, Exact	21	2.60
IM	1	2003	1449	IPOPT, BFGS	52	8.89
IM	5	10134	7339	IPOPT, Exact	39	1056
IM	5	10134	7339	IPOPT, BFGS	1000	291000
IM	10	20826	15206	IPOPT, Exact	65	11064
IM	15	31743	23073	IPOPT, Exact	110	72460

using a barrier method with a filter line search, is applied in two forms, with exact Hessians and with quasi-Newton updates. Default options were used for all of the solvers; more information on the comparison can be found in [310]. Results for the 8 blending cases were obtained with the NEOS server at Argonne National Laboratory (http://www-neos.mcs.anl.gov) and are presented in Table 7.7. Here n represents the number of variables and n_S is the number of superbasic variables (degrees of freedom) at the solution. Note that for these cases and the initialization above, we always found the same local solutions, although there is no guarantee that these are global solutions.

Table 7.7 shows normalized CPU times as well as iteration counts, which represent the number of linear KKT systems that were solved. First, we consider the results for the Haverly (HM) and the Audet and Hansen (AHM) models. For $N_t = 1$, the objective function values are 400 and 49.2, respectively. These problems have few superbasic variables, all CPU times are small, and there is no significant difference in the solution times for these solvers. Note, however, that solvers that use exact second derivatives (LOQO and IPOPT (exact)) generally require fewer iterations. As a result, this set of results serves as a consistency check that shows the viability of all of the methods. For $N_t = 25$, the objective function values are 10000 and 1229.17, respectively. These models have hundreds of degrees of freedom and the smallest iteration counts are required by both LOQO and IPOPT (exact). Here, methods without exact second derivatives (SNOPT and IPOPT (BFGS)) require at least an order of magnitude more iterations because n_S is large.

Finally, we consider an industrial blending model (IM) with a horizon of one to fifteen days and nonlinear programs with 2003–31743 variables, and 1449–23073 superbasic variables. Most of the solvers have difficulty with this model and only IPOPT was able to solve these cases within 1000 iterations. For $N_t = 1$ (optimum value = 61.35) and $N_t = 5$ (optimum value = 13913.5), IPOPT (exact) was significantly more efficient than IPOPT (BFGS), especially for the larger problem. For $N_t = 10$ and $N_t = 15$ (with optimal objective values of 26388 and 41523.7, respectively), only IPOPT (exact) was able to provide a solution, although quite a long CPU time was required. This problem presents difficulties because redundancies in the constraints also make the KKT matrix ill-conditioned and expensive to factorize with the sparse solver. It is likely that LOQO and SNOPT may have failed for these reasons as well.

7.6 Summary and Notes for Further Reading

This chapter presents a cross section of process optimization problems with algebraic models. These applications lie in the modeling and optimization of steady state processes, and they span the area of off-line design and analysis, online (or real-time) optimization, and planning over multiday periods. Here, we consider the formulation and solution of optimization models in each of these areas. First, detailed models for design and analysis take the form of unit modules that are incorporated in large, comprehensive simulation packages. Each module comprises sets of nonlinear equations and associated solution strategies to calculate output streams from inputs. As a result, the simulation equations are hidden from the NLP solver. This leads to small NLP problems, but care is required to extract accurate gradient information from these models. Second, for real-time optimization, the optimization model usually deals with a simplified form of the process equations that are fully exposed to the NLP solver. Exact derivatives are straightforward to obtain. On the other hand, second derivatives are usually not provided and BFGS updates are used. Finally, planning problems with nonlinear models are spatially and temporally distributed and include many degrees of freedom for optimization, such as inventory levels and intermediate flows between units. These equation-based models are best handled by large-scale NLP solvers which can exploit second derivatives.

In addition to the case studies presented in this chapter, numerous examples of flowsheet simulation can be found in a number of process engineering texts, including [53, 122, 332, 143]. In particular, Seider, Seader, and Lewin [358] and Lewin et al. [254] offer detailed presentations and descriptions of process simulation models and tools. Additional simulation training material with applications is also available from vendor websites, especially for the ASPEN, PRO/II, and HySyS process simulators. There are fewer references in the area of real-time optimization, as most implementations are made by commercial vendors. An excellent analysis of real-time optimization implementations and research issues can be found in [276, 146, 415]. Finally, blending problems represent an important NLP application in the process industries. Additional studies in this area can be found in [228, 114, 17, 6, 190, 362, 171]. More detail on the performance comparisons for blending can also be found in [310].

7.7 Exercises

1. Consider the Williams–Otto optimization problem presented in Section 7.3.1. Reformulate and solve this problem in the modular mode.

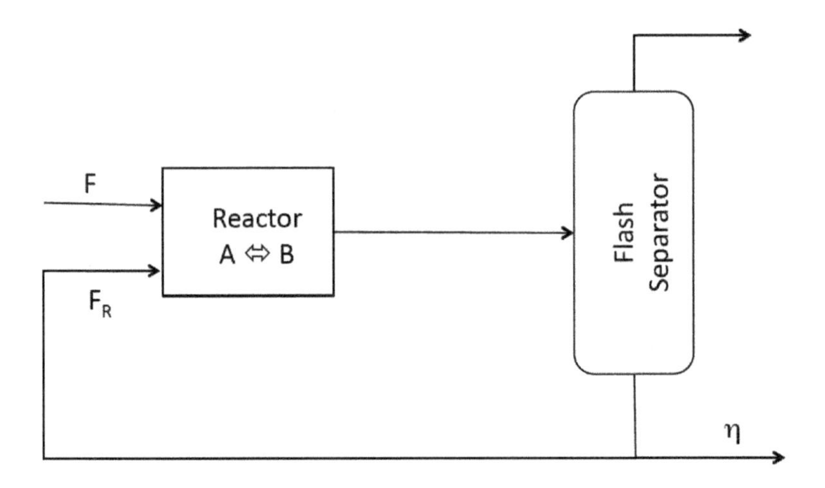

Figure 7.11. *Single-loop reactor-based flowsheet.*

2. Consider the process flowsheet in Figure 7.11. The plug flow reactor is used to convert component A to B in a reversible reaction according to the rate equations

$$(F+F_R)\frac{dC_A}{dV} = -k_1C_A + k_2C_B,$$
$$(F+F_R)\frac{dC_B}{dV} = k_1C_A - k_2C_B,$$

where the feed has $F = 10l/s$ as the volumetric flow rate with concentrations $C_{A,f} = 1$ mol/l and $C_{B,f} = 0$ mol/l. V is the reactor volume, $k_1 = 0.10/s$ and $k_2 = 0.05/s$, the molecular weights of A and B are both 100, and the liquid density is $0.8\,\mathrm{g}/l$. The flash separator operates at 2 atm and temperature T with vapor pressure equations (in atm):

$$\log_{10} P^A_{vap} = 4.665 - 1910/T,$$
$$\log_{10} P^B_{vap} = 4.421 - 1565/T.$$

We assume that the purge fraction is $\eta \in [0.01, 0.99]$, $T \in [380K, 430K]$, $V \in [250l, 1500l]$ and the profit is given by

$$0.5B^{top} - 0.1F_R(500-T) - 10^{-5}V,$$

where F_R is the recycle volumetric flow rate and B^{top} is the mass flow of component B exiting as top product from the flash separator.

(a) Formulate the process optimization model by solving the differential equations analytically and formulating the flash equations.

(b) Set up an equation-oriented optimization model in GAMS, AMPL, or AIMMS and solve. What problems are likely to occur in the solution?

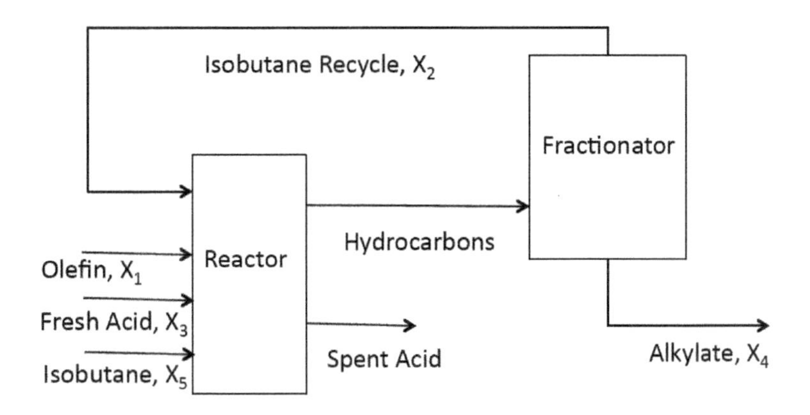

Figure 7.12. *Flowsheet for alkylation problem.*

(c) Comment on satisfaction of the KKT conditions. Calculate the reduced Hessian and comment on the second order conditions.

3. Consider an optimization model for the alkylation process discussed in Liebman et al. [260] and shown in Figure 7.12. The alkylation model is derived from simple mass balance relationships and regression equations determined from operating data. The first four relationships represent characteristics of the alkylation reactor and are given empirically. The alkylate field yield, X_4, is a function of both the olefin feed, X_1, and the external isobutane to olefin ratio, X_8. The following relation is developed from a nonlinear regression for temperature between 80° and 90° F and acid strength between 85 and 93 weight percent:

$$X_4 = X_1(1.12 + .12167X_8 - 0.0067X_8^2).$$

The motor octane number of the alkylate, X_7, is a function of X_8 and the acid strength, X_6. The nonlinear regression under the same conditions as for X_4 yields

$$X_7 = 86.35 + 1.098X_8 - 0.038X_8^2 + 0.325(X_6 - 89).$$

The acid dilution factor, X_9, can be expressed as a linear function of the F-4 performance number, X_{10} and is given by

$$X_9 = 35.82 - 0.222X_{10}.$$

Also, X_{10} is expressed as a linear function of the motor octane number, X_7,

$$X_{10} = 3X_7 - 133.$$

The remaining three constraints represent exact definitions for the remaining variables. The external isobutane to olefin ratio is given by

$$X_8X_1 = X_2 + X_5.$$

The isobutane feed, X_5, is determined by a volume balance on the system. Here olefins are related to alkylated product and there is a constant 22% volume shrinkage, thus giving

$$X_5 = 1.22X_4 - X_1.$$

Finally, the acid dilution strength (X_6) is related to the acid addition rate (X_3), the acid dilution factor (X_9), and the alkylate yield (X_4) by the equation

$$X_6(X_4X_9 + 1000X_3) = 98000X_3.$$

The objective function to be maximized is the profit ($/day)

$$OBJ = 0.063X_4X_7 - 5.04X_1 - 0.035X_2 - 10X_3 - 3.36X_5$$

based on the following prices:

- Alkylate product value = $0.063/octane-barrel
- Olefin feed cost = $5.04/barrel
- Isobutane feed cost = $3.36/barrel
- Isobutane recycle cost = $0.035/barrel
- Acid addition cost = $10.00/barrel.

Use the following variable bounds: $X_1 \in [0, 2000]$, $X_2 \in [0, 16000]$, $X_3 \in [0, 120]$, $X_4 \in [0, 5000]$, $X_5 \in [0, 2000]$, $X_6 \in [85, 93]$, $X_7 \in [90, 95]$, $X_8 \in [3, 12]$, $X_9 \in [1.2, 4]$ for the following exercises:

(a) Set up this NLP problem and solve.

(b) The above regression equations are based on operating data and are only approximations and it is assumed that equally accurate expressions actually lie in a band around these expressions. Therefore, in order to consider the effect of this band, replace the variables X_4, X_7, X_9, and X_{10} with RX_4, RX_7, RX_9, and RX_{10} in the regression equations (only) and impose the constraints

$$0.99X_4 \le RX_4 \le 1.01X_4,$$
$$0.99X_7 \le RX_7 \le 1.01X_7,$$
$$0.99X_9 \le RX_9 \le 1.01X_9,$$
$$0.9X_{10} \le RX_{10} \le 1.11X_{10}$$

to allow for this relaxation. Resolve with this formulation. How would you interpret these results?

(c) Resolve problems 1 and 2 with the following prices:
   - Alkylate product value = $0.06/octane/barrel
   - Olefin feed cost = $5.00/barrel
   - Isobutane feed cost = $3.50/barrel
   - Isobutane recycle cost = $0.04/barrel
   - Acid addition cost = $9.00/barrel.

(d) Calculate the reduced Hessian at optimum of the above three problems and comment on second order conditions.

Chapter 8

Introduction to Dynamic Process Optimization

This chapter provides the necessary background to develop and solve dynamic optimization problems that arise in chemical processes. Such problems arise in a wide variety of areas. Off-line applications range from batch process operation, transition optimization for different product grades, analysis of transients and upsets, and parameter estimation. Online problems include formulations for model predictive control, online process identification, and state estimation. The chapter describes a general multiperiod problem formulation that applies to all of these applications. It also discusses a hierarchy of optimality conditions for these formulations, along with variational strategies to solve them. Finally, the chapter introduces dynamic optimization methods based on NLP which will be covered in subsequent chapters.

8.1 Introduction

With growing application and acceptance of large-scale dynamic simulation in process engineering, recent advances have also been made in the *optimization* of these dynamic systems. Application domains for dynamic optimization cover a wide range of process tasks and include

- design of distributed systems in chemical engineering, including reactors and packed column separators;
- off-line and online problems in process control, particularly for multivariable systems that are nonlinear and output constrained; these are particularly important for nonlinear model predictive control and real-time optimization of dynamic systems;
- trajectory optimization in chemical processes for transitions between operating conditions and to handle load changes;
- optimum batch process operating profiles, particularly for reactors and separators;
- parameter estimation and inverse problems that arise in state estimation for process control as well as in model building applications.

Moreover, in other disciplines, including air traffic management [325] and aerospace applications [40], modern tools for dynamic optimization play an increasingly important role. These applications demand solution strategies that are efficient, reliable, and flexible for different problem formulations and structures.

This chapter introduces dynamic optimization problems related to chemical processes. It provides a general problem statement, examples in process engineering, and optimality conditions for a class of these problems. Since dynamic optimization strategies need to solve (with some reasonable level of approximation) problems in infinite dimensions, they need to determine solutions even for poorly conditioned or unstable systems. Moreover, for online applications, computations are time-limited and efficient optimization formulations and solvers are essential, particularly for large-scale systems. In the next section we describe differential-algebraic models for process engineering and state a general multistage formulation for dynamic optimization problems. Specific cases of this formulation are illustrated with process examples. Section 8.3 then develops the optimality conditions for dynamic optimization problems, based on variational principles. This leads to the examination of a number of cases, which are illustrated with small examples. Two particularly difficult cases merit separate sections. Section 8.4 deals with path constraints, while Section 8.5 deals with singular control problems. Section 8.6 then outlines the need for numerical methods to solve these optimization problems and sets the stage for further developments in Chapters 9 and 10.

Finally, it should be noted that the style of presentation differs somewhat from the previous chapters. As dynamic optimization deals with infinite-dimensional problems, it relies on principles of functional analysis, which are beyond the scope of this book. Although external citations are provided to the relevant theory, the presentation style will rely on informal derivations rather than detailed proofs in order to present the key concepts. Moreover, some notational changes are made that differ from the previous chapters, although the notation shall be clear from the context of the presentation.

8.2 Dynamic Systems and Optimization Problems

Consider the dynamic system given by a fully implicit set of differential-algebraic equations (DAEs). These are expressed with respect to an independent variable, t, which usually represents time or distance. DAEs in process engineering are typically specified as initial value problems with initial conditions at zero:

$$F\left(x, \frac{dx}{dt}, u(t), p, t\right) = 0, \quad h(x(0)) = 0. \tag{8.1}$$

Here $x(t) \in \mathbb{R}^{n_x}$ are state variables that are functions of time, $t \geq 0$, $u(t) \in \mathbb{R}^{n_u}$ are control variables, and $p \in \mathbb{R}^{n_p}$ are variables that are independent of t. Also, most of these systems are *autonomous* and t does not appear explicitly in these equations.

Because the fully implicit DAE (8.1) is more difficult to analyze, we consider a simpler, more structured form that can still be applied to most process applications. Here we partition the state variables $x(t)$ into differential variables $z(t)$ and algebraic variables $y(t)$, leading to the semiexplicit form

$$\frac{dz}{dt} = f(z(t), y(t), u(t), p), \quad z(0) = z_0, \tag{8.2a}$$

$$g(z(t), y(t), u(t), p) = 0, \tag{8.2b}$$

and we assume that $y(t)$ can be solved uniquely from $g(z(t), y(t), u(t), p, t) = 0$, once $z(t), u(t)$, and p are specified. Equivalently, $\frac{\partial g}{\partial y}$ is nonsingular for all values of $z(t), y(t), u(t)$, and p. The invertibility of $g(\cdot, y(t), \cdot, \cdot))$ allows an implicit elimination of the algebraic variables $y(t) = y[z(t), u(t), p]$, which allows us to consider the DAE with the same solution as the related ordinary differential equation (ODE):

$$\frac{dz}{dt} = f(z(t), y[z(t), u(t), p], u(t), p) = \bar{f}(z(t), u(t), p), \quad z(0) = z_0. \tag{8.3}$$

In Section 8.4, we will see that this corresponds to the *index-1* property of the DAE system (8.2). With this property, we can then rely on an important result (the *Picard–Lindelöf theorem*) regarding the solution of initial value ODE problems. This is paraphrased by the following theorem.

Theorem 8.1 [16, 59] For $u(t)$ and p specified, let $\bar{f}(z(t), u(t), p)$ be Lipschitz continuous for all $z(t)$ in a bounded region with $t \in [0, t_f]$. Then the solution of the initial value problem (8.3) exists and is unique, $z(t)$ for $t \in [0, t_f]$.

DAE models of the form (8.2) arise in many areas of process engineering. The differential equations usually arise from conservation laws such as mass, energy, and momentum balances. The algebraic equations are typically derived from constitutive equations and equilibrium conditions. They include equations for physical properties, hydraulics, and rate laws. The decision variables or "degrees of freedom" in dynamic optimization problems are the *control variables* $u(t)$ and the *time-independent variables* p. The former correspond to manipulated variables that determine operating policies over time, while the latter often correspond to equipment parameters, initial conditions, and other steady state decision variables.

Related to the initial value problems (8.2) and (8.3) are boundary value problems (BVPs), where the initial condition $z(0) = z_0$ is replaced by boundary conditions. Much less can be said about existence and uniqueness of solutions for nonlinear boundary value problems of the form

$$\frac{dz}{dt} = \bar{f}(z(t)), \qquad h(z(0), z[t_f; z(0)]) = 0, \tag{8.4}$$

where we suppress the dependence on $u(t)$ and p for the moment, and $z[t_f; z(0)]$ is defined by (8.3) for an unknown $z(0)$ that satisfies the boundary conditions. Solutions to (8.4) may be nonunique, or may not even exist, and only local properties can be considered for this problem. For instance, a key property is that a known solution to (8.4) is *isolated*, i.e., locally unique, as expressed by the following theorem.

Theorem 8.2 [16, pp. 164–165] Consider problem (8.4) with a solution $\hat{z}(t)$. Also, let $\bar{f}(z(t))$ be Lipschitz continuous for all $z(t)$ with $\|z(t) - \hat{z}(t)\| \leq \epsilon$ for some $\epsilon > 0$ and $t \in [0, t_f]$. Then the solution $\hat{z}(t)$ is locally unique if and only if the matrix

$$Q(t) = \frac{\partial h(\hat{z}(0), \hat{z}(t_f))}{\partial z(0)} + Z(t) \frac{\partial h(\hat{z}(0), \hat{z}(t_f))}{\partial z(t_f)}$$

is nonsingular, where the fundamental solution matrix $Z(t) = \frac{dz(t_f)}{dz(0)}$ is evaluated at the solution $\hat{z}(t)$.

We will apply both of these properties in the derivation of optimality conditions for DAE constrained optimization problems.

For the general setting we consider the optimization of dynamic systems over a number of time periods, $l = 1,\dots,N_T$, possibly with different DAE models, states, and decisions in each period, $t \in (t_{l-1}, t_l]$. We formulate this multiperiod dynamic problem in the following form:

$$\min \sum_{l=1}^{N_T} \Phi^l(z^l(t_l), y^l(t_l), u^l(t_l), p^l) \tag{8.5a}$$

$$\text{s.t.} \quad \frac{dz^l}{dt} = f^l(z^l(t), y^l(t), u^l(t), p^l), \quad z^l(t_{l-1}) = z_0^l, \tag{8.5b}$$

$$g^l(z^l(t), y^l(t), u^l(t), p^l) = 0, \tag{8.5c}$$

$$u_L^l \le u^l(t) \le u_U^l, \tag{8.5d}$$

$$p_L^l \le p^l \le p_U^l, \tag{8.5e}$$

$$y_L^l \le y^l(t) \le y_U^l, \tag{8.5f}$$

$$z_L^l \le z^l(t) \le z_U^l, \quad t \in (t_{l-1}, t_l], l = 1,\dots,N_T, \tag{8.5g}$$

$$h(p, z_0^1, z^1(t_1), z_0^2, z^2(t_2),\dots, z_0^{N_T}, z^{N_T}(t_{N_T})) = 0. \tag{8.5h}$$

The dynamic optimization problem (8.5) is defined by separate models in each period l, with initial conditions, z_0^l and inequality constraints represented here as simple bounds (8.5d)–(8.5g) within each period. Note that the state variables are not assumed to be continuous across periods. Instead, a general set of boundary conditions is represented by (8.5h) to link the states of these periods together. The resulting multiperiod formulation (8.5) captures most dynamic optimization problems of interest, including the problem classes in chemical engineering considered below.

Chemical Reactor Design

Optimization of tubular reactors is an especially important task in the chemical industry. For processes operated at steady state, the reactor state variables evolve over distance and not time. The reactor model can be represented by DAE models over multiple periods, and the optimization problem takes the form of problem (8.5), with all periods described by the same model equations and state variables. As shown in Figure 8.1, each period (or zone) can facilitate different feed injections, heating and cooling jackets, and the removal of product streams. Moreover, for catalytic reactions, the packing of the types of catalyst in each zone strongly affects the performance of the reactor. For the degrees of freedom, the decision variables p can include reactor length, diameter, inlet temperature, composition, and pressure for each of the zones. The control profiles $u(t)$ include heat addition along the length of the reactor and packing of different catalyst types.

To illustrate the optimization of tubular reactors, we consider the case of olefin cracking in a furnace reactor. In this system, the cracking of ethane (C_2H_6) to make ethylene (C_2H_4) and hydrogen (H_2) is of particular interest, as the manufacture of many high-volume petrochemicals are based on this product. Moreover, this application is challenging because of the presence of free radicals in the reaction mechanism, and their resulting fast dynamics. For this example, the reaction system includes six molecules, three free radicals

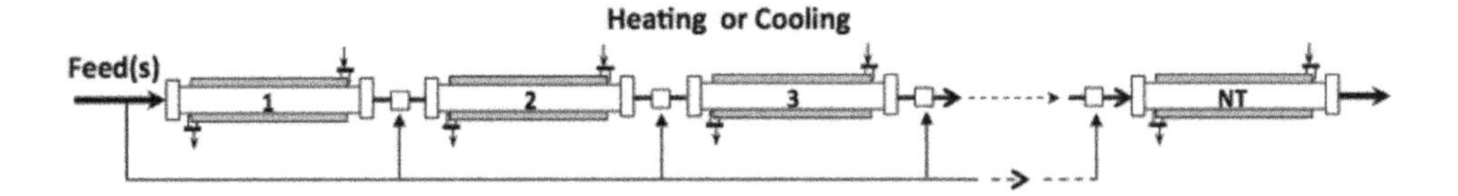

Figure 8.1. *Multizone reactor design.*

($CH_3\cdot$, $C_2H_5\cdot$, $H\cdot$) and the seven reactions shown below:

$$
\begin{aligned}
&C_2H_6 \rightarrow 2\,CH_3\cdot\\
&CH_3\cdot + C_2H_6 \rightarrow CH_4 + C_2H_5\cdot\\
&H\cdot + C_2H_6 \rightarrow H_2 + C_2H_5\cdot\\
&C_2H_5\cdot \rightarrow C_2H_4 + H\cdot\\
&C_2H_5\cdot + C_2H_4 \rightarrow C_3H_6 + \mathrm{CH}_3\cdot\\
&2C_2H_5\cdot \rightarrow C_4H_10\\
&H\cdot + C_2H_4 \rightarrow C_2H_5\cdot
\end{aligned}
$$

As presented in [88], the resulting DAE model contains mass and energy balances for the species in the reactor, mass action reaction kinetics, and a momentum balance that provides the pressure profile in the reactor. The reactor system has only one period (or zone) and there are no time-independent decisions p. The goal of the optimization problem is to find an optimal profile for the heat added along the reactor length that maximizes the production of ethylene. Note that there are also several undesirable by-products that need to be suppressed in the reactor, in order to promote the evolution of product ethylene. The product distribution is therefore determined by the reactor temperature profile, influenced by the heat flux distribution. More details on this dynamic optimization problem and its solution can be found in [88].

Parameter Estimation

Parameter estimation problems arise frequently in the elucidation of kinetic mechanisms in reaction engineering and in model construction for reactive and transport systems. This optimization problem is a crucial aspect in a wide variety of areas in modeling and analysis. It is used in applications ranging from discriminating among candidate mechanisms in fundamental reactive systems to developing predictive models for optimization in chemical plants. Moreover, the resulting NLP solution is subjected to an extensive sensitivity analysis that leads to statistical analysis of the model and its estimated parameters.

For model development, DAEs for parameter estimation arise from a number of dynamic process systems, especially batch reactors. These applications fit the form of problem (8.5), with all periods generally described by the same model equations and state variables. The objective function is usually based on a statistical criterion, typically derived from maximum likelihood assumptions. Depending on knowledge of the error distribution, these assumptions often lead to weighted least squares functions, with a structure that can be exploited by the NLP algorithm. The objective function includes experimental data collected at time periods $t_l, l = 1,\ldots,N_T$, which need to be matched with calculated values

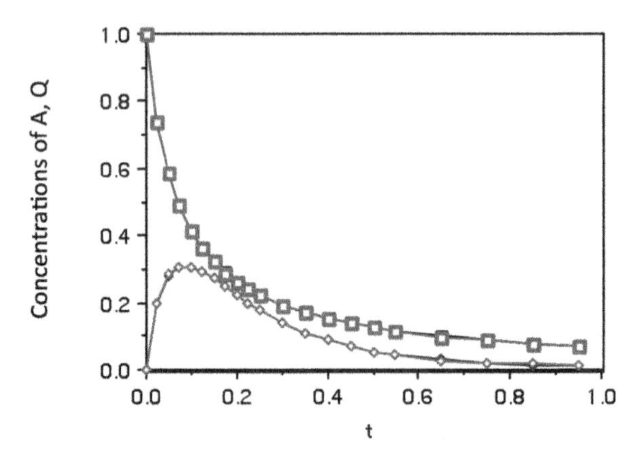

Figure 8.2. *Data and fitted concentration profiles (A, top, and Q, bottom) from parameter estimation of rate constants.*

from the DAE model. These calculated values can be represented in (8.5) as states at the end of each period, $z(t_l), y(t_l)$. Control profiles are rarely included in parameter estimation problems. Instead, the degrees of freedom, represented by p in (8.5), are the model parameters that provide the best fit to the experimental data. This problem is illustrated in Figure 8.2 on a small batch kinetic system. Here three reaction rate parameters need to be estimated for the kinetic model $A \xrightarrow{p_1} Q$, $Q \xrightarrow{p_2} S$, $A \xrightarrow{p_3} S$ to match the concentration data for A and Q. The evolution of these reactions is modeled by two differential equations to calculate concentrations for A and Q. The rate parameters in the DAE model are adjusted by the NLP solver to minimize the squared deviation between the data and calculated concentration values. More details on this application can be found in [383].

Batch Process Optimization

Batch processing constitutes a significant portion of the chemical process industry, and it is the dominant mode of operation for pharmaceutical, food, and specialty chemical manufacturers. These processes allow flexible processing that includes adjustable unit-task assignments, production schedules, storage, and transfer policies. Dynamic optimization problems for batch processes can be described by time periods over different process units or stages. Batch processes operate over *campaigns* which consist of multiple time periods, often for the manufacture of several products. The corresponding optimization problem for this campaign can be represented by problem (8.5), where each period may contain different models, states, and decisions. Batch units either follow a set recipe that can be represented by linear equations with "size factors" [53], or they can be described by detailed DAE models. Processing decisions are then represented by the design and sizing of the batch equipment (through variables p) and dynamic operating profiles (through control variables $u(t)$). In addition, these decisions are usually embedded within a larger problem consisting of discrete decisions that relate to time scheduling of stages to manufacture multiple products. The resulting optimization problem is also accompanied by constraints that dictate transfer policies between units and the possibility of idle times for each unit.

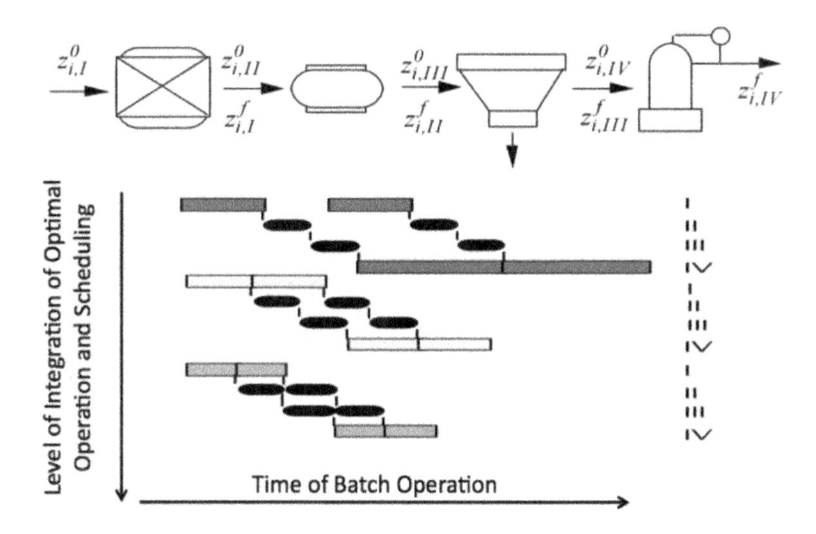

Figure 8.3. *Batch process stages* (I, II, III, IV) *and corresponding process schedules integrated with dynamic optimization.*

The economic objective over the batch campaign is usually motivated by the net present value function that includes product sales, raw material costs, and operating costs. In addition, the *makespan* (the time of the total campaign) is a major consideration in the optimization problem. This dictates the time required for a fixed product slate, or the number of batches that can be produced over a fixed time horizon. Finally, batch processes are often driven by strong interactions between the production schedule, dynamic operation, and equipment design. While optimization models with these interactions are often difficult to solve, they can lead to significant improvements in performance and profitability.

Figure 8.3 illustrates a four-stage batch process with units I (batch reactor), II (heat exchanger), III (centrifugal separator), and IV (batch distillation). Units II and III have no degrees of freedom and are run as "recipe" units, while the reactor can be optimized through a temperature profile and the batch distillation unit can be optimized through the reflux profile. Also shown in Figure 8.3 is a set of production schedules, determined in [46], that result from various levels of integration between the dynamic operation and unit scheduling. As discussed in [46, 142, 317], tighter levels of integration can lead to significant reductions in idle times and makespans.

Dynamic Real-Time Optimization

Real-time optimization was discussed in Chapter 7 in the development of large-scale equation-oriented NLPs, where the resulting optimization models contain equations that describe steady state process operation. As shown in Figure 7.8, the integration of these models with advanced control assumes a decomposition of time scales, with real-time optimization updated more slowly than the dynamics of the control loop. While this implementation is widespread in the process industries, particularly for refineries and petrochemicals, it may not be suitable for a variety of other processes.

In *dynamic real-time optimization* (D-RTO) the goal is to provide a tighter integration of the advanced control and optimization task, so that disturbance rejection, set point tracking, and the optimization of economic objective can be coordinated over similar time scales. Such an approach is particularly useful for processes that are never in steady state. These include batch processes, processes with load changes and grade transitions, such as power plants and polymerization processes, and production units that operate in a periodic manner, such as simulated moving beds (SMBs) [225] and pressure swing adsorption (PSA) [210]. Treating these nonlinear processes requires online optimization with nonlinear DAE models, often over multiple periods of operation, as represented in problem (8.5), where the periods may contain different models, states, and decisions. This formulation includes strategies such as nonlinear model predictive control (NMPC), which was described in Chapter 1 and illustrated in Figure 1.4. In addition to nonlinear DAE models, the dynamic optimization problem (8.5) includes control profiles for optimal operation, state and control profile bounds, and process constraints (8.5h) that link different periods of operation. Finally, the objective function usually deals with minimization of the deviation from a desired steady state or direct maximization of an economic criterion over the time horizon of the process.

An extensive survey of real-time dynamic optimization is given in [177]. More recently, large-scale industrial NMPC and D-RTO applications have been reported [31, 286, 149]. These applications have been aided by powerful large-scale NLP solvers, such as those discussed in Chapter 6. In addition, there is now a much better understanding of stability and robustness properties, and associated dynamic optimization problem formulations that provide them [278, 270]. With these improved optimization formulations and algorithms, the role of online dynamic optimization can be extended to (a) consider economic objectives simultaneously with disturbance rejection and set point tracking, (b) allow longer time horizons with additional constraints and degrees of freedom to improve the objective, and (c) incorporate multiple operating stages over the predictive horizon, including transitions in the predictive horizon due to product changeovers, nonstandard cyclic operations, or anticipated shutdowns [387, 177].

8.3 Optimality Conditions for Optimal Control Problems

We now consider conditions that define locally optimal solutions of dynamic optimization problems. We assume that the state equations are smooth functions in the state and control variables and that, by invertibility of the algebraic equations, $y(t)$ can be solved as implicit functions of $z(t)$, $u(t)$, and p. To develop a compact derivation, we consider only a single-period *optimal control problem*, where the bound constraints are now replaced by general inequality constraints. In addition, we consider an objective function and final time constraints that are functions only of the differential states at final time, $z(t_f)$,

$$\min \quad \Phi(z(t_f)) \tag{8.6a}$$

$$\text{s.t.} \quad \frac{dz}{dt} = f(z(t), y(t), u(t), p), \quad z(0) = z_0, \tag{8.6b}$$

$$g_E(z(t), y(t), u(t), p) = 0, \tag{8.6c}$$

$$g_I(z(t), y(t), u(t), p) \leq 0, \tag{8.6d}$$

$$h_E(z(t_f)) = 0, \quad h_I(z(t_f)) \leq 0. \tag{8.6e}$$

Problem (8.6) with the final time objective function is often called the *Mayer problem*. Replacing this objective by an integral over time leads to the problem of *Lagrange*, and a problem with both integral and final time terms is known as the *Bolza problem*. All of these problems are equivalent and can be formulated interchangeably.

We now assume a local solution to (8.6), $(z^*(t), y^*(t), u^*(t), p^*)$, and derive relations based on perturbations around this solution. Since all constraints are satisfied at the optimal solution, we adjoin these constraints through the use of *costate* or *adjoint variables* as follows:

$$\begin{aligned}\Phi(z(t),u(t),y(t),p) \equiv\ & \Phi(z(t_f)) + \eta_I^T h_I(z(t_f)) + \eta_E^T h_E(z(t_f)) \\ & + \int_0^{t_f} \Bigg[\lambda(t)^T \left(f(z(t),y(t),u(t),p) - \frac{dz}{dt} \right) \\ & + \nu_E^T g_E(z(t),y(t),u(t),p) + \nu_I^T g_I(z(t),y(t),u(t),p) \Bigg] dt, \qquad (8.7)\end{aligned}$$

where $\Phi(z^*(t), u^*(t), y^*(t), p^*) = \Phi(z^*(t_f))$ due to satisfaction of the constraints at the solution. Although we now deal with an infinite-dimensional problem, the development of this adjoined system can be viewed as an extension of the Lagrange function developed in Chapter 4. The adjoint variables λ, ν_E, ν_I are functions of time. Here $\lambda(t)$ serves as a multiplier on the differential equations, while $\nu_E(t)$ and $\nu_I(t)$ serve as multipliers for the corresponding algebraic constraints. In addition η_E and η_I serve as multipliers for the final conditions.

Applying integration by parts to $\int \lambda(t)^T \frac{dz}{dt} dt$ leads to

$$\int_0^{t_f} \lambda(t)^T \frac{dz}{dt} dt = z(t_f)^T \lambda(t_f) - z(0)^T \lambda(0) - \int_0^{t_f} z(t)^T \frac{d\lambda}{dt} dt,$$

and substituting into (8.7) yields the equivalent expression:

$$\begin{aligned}\Phi(z(t),u(t),y(t),p) \equiv\ & \Phi(z(t_f)) + \eta_I^T h_I(z(t_f)) + \eta_E^T h_E(z(t_f)) - z(t_f)^T \lambda(t_f) \\ & + z(0)^T \lambda(0) + \int_0^{t_f} \Bigg[\lambda(t)^T f(z(t),y(t),u(t),p) + z(t)^T \frac{d\lambda}{dt} \\ & + \nu_E(t)^T g_E(z(t),y(t),u(t),p) + \nu_I(t)^T g_I(z(t),y(t),u(t),p) \Bigg] dt. \\ & \qquad (8.8)\end{aligned}$$

We now define perturbations $\delta z(t) = z(t) - z^*(t)$, $\delta y(t) = y(t) - y^*(t)$, $\delta u(t) = u(t) - u^*(t)$, $dp = p - p^*$. We also distinguish between the perturbation $\delta z(t)$ (abbreviated as δz), which applies at a fixed time t, and dp which is independent of time. Because $(z^*(t), y^*(t), u^*(t), p^*)$ is a local optimum, we note that

$$d\Phi^* = \Phi(z^*(t) + \delta z, u^*(t) + \delta u, y^*(t) + \delta y, p^* + dp) - \Phi(z^*(t), u^*(t), y^*(t), p^*) \geq 0$$

for all allowable perturbations (i.e., feasible directions) in a neighborhood around the solution. Using infinitesimal perturbations (where we drop the (t) argument for convenience) allows us to rely on linearizations to assess the change in the objective function $d\Phi^*$.

Applying these perturbations to (8.8) leads to

$$0 \le d\Phi^* = \left[\frac{\partial \Phi(z(t_f))}{\partial z} + \frac{\partial h_I(z(t_f))}{\partial z}\eta_I + \frac{\partial h_E(z(t_f))}{\partial z}\eta_E - \lambda(t_f)\right]^T \delta z(t_f) + \lambda(0)^T \delta z(0)$$
$$+ \int_0^{t_f} \left[\frac{\partial f}{\partial z}\lambda + \frac{d\lambda}{dt} + \frac{\partial g_E}{\partial z}\nu_E + \frac{\partial g_I}{\partial z}\nu_I\right]^T \delta z + \left[\frac{\partial f}{\partial y}\lambda + \frac{\partial g_E}{\partial y}\nu_E + \frac{\partial g_I}{\partial y}\nu_I\right]^T \delta y$$
$$+ \left[\frac{\partial f}{\partial u}\lambda + \frac{\partial g_E}{\partial u}\nu_E + \frac{\partial g_I}{\partial u}\nu_I\right]^T \delta u + \left[\frac{\partial f}{\partial p}\lambda + \frac{\partial g_E}{\partial p}\nu_E + \frac{\partial g_I}{\partial p}\nu_I\right]^T dp\, dt. \quad (8.9)$$

Note that the perturbations of the states and controls are related by the following linearized DAE system:

$$\delta \dot{z} = \frac{\partial f}{\partial z}^T \delta z + \frac{\partial f}{\partial y}^T \delta y + \frac{\partial f}{\partial u}^T \delta u + \frac{\partial f}{\partial p}^T dp, \quad \delta z(0) = 0, \quad (8.10a)$$

$$\frac{\partial g_E}{\partial z}^T \delta z + \frac{\partial g_E}{\partial y}^T \delta y + \frac{\partial g_E}{\partial u}^T \delta u + \frac{\partial g_E}{\partial p}^T dp = 0. \quad (8.10b)$$

By finding the analytical solution of this linearization, one can define δz and δy as implicit functions of δu and dp. This is analogous to our derivation of the KKT multipliers in Section 6.4, where (6.69c) was used to define the multipliers v^*. As shown in [248], one can apply a similar elimination to (8.9) and derive relationships that define the adjoint variables, $\lambda(t)$, $\nu_E(t)$, $\nu_I(t)$, by setting the bracketed quantities for δz, $\delta z(t_f)$, and δy to zero. Note that the perturbations δu and dp will now have a direct influence on $d\Phi^*$ and the perturbations δy and δz satisfy the linearized DAE (8.10).

1. For perturbation of the final state, $\delta z(t_f)$, we have

$$\lambda(t_f) = \frac{\partial \Phi(z(t_f))}{\partial z} + \frac{\partial h_I(z(t_f))}{\partial z}\eta_I + \frac{\partial h_E(z(t_f))}{\partial z}\eta_E. \quad (8.11)$$

   This leads to a boundary or *transversality* condition for $\lambda(t)$.

2. For perturbation of the differential state, δz, we have

$$\frac{d\lambda}{dt} = -\left[\frac{\partial f}{\partial z}\lambda + \frac{\partial g_E}{\partial z}\nu_E + \frac{\partial g_I}{\partial z}\nu_I\right] \quad (8.12)$$

   which gives a differential equation for $\lambda(t)$, the adjoint equations.

3. For perturbation of the algebraic state, δy, we have

$$\frac{\partial f}{\partial y}\lambda + \frac{\partial g_E}{\partial y}\nu_E + \frac{\partial g_I}{\partial y}\nu_I = 0 \quad (8.13)$$

   which leads to an algebraic equation with algebraic multipliers $\nu_E(t)$ and $\nu_I(t)$.

4. For perturbation of the initial state, $\delta z(0)$, we consider three possible transversality cases for $\lambda(0)$:

   - $z(0)$ is fixed to a constant z_0 and $\delta z(0) = 0$ (no feasible perturbation is allowed). In this case there is no initial condition specified for $\lambda(0)$.

- $z(0)$ is not specified and all perturbations are allowed for $\delta z(0)$. In this case $\lambda(0) = 0$.
- The initial condition is specified by variable p in the optimization problem, $z(0) = z_0(p)$. For this case, which subsumes the first two cases, we define

$$\lambda(0)^T \delta z(0) = \lambda(0)^T \frac{\partial z_0^T}{\partial p} dp,$$

and this term is grouped with the terms for dp.

By eliminating the state perturbation terms and suitably defining the adjoint variables above, (8.9) is now simplified, and it is clear that only perturbations in the decisions will continue to influence $d\Phi^*$, as seen in (8.14),

$$\begin{aligned} 0 \le d\Phi^* = & \int_0^{t_f} \left[\frac{\partial f}{\partial u}\lambda + \frac{\partial g_E}{\partial u}\nu_E + \frac{\partial g_I}{\partial u}\nu_I \right]^T \delta u \, dt \\ & + \left\{ [\frac{\partial z_0}{\partial p}\lambda(0)]^T + \int_0^{t_f} \left[\frac{\partial f}{\partial p}\lambda + \frac{\partial g_E}{\partial p}\nu_E + \frac{\partial g_I}{\partial p}\nu_I \right]^T dt \right\} dp. \end{aligned} \tag{8.14}$$

For condition (8.14) we first derive optimality conditions where the inequality constraints are absent. Following this, we modify these conditions to include inequality constraints.

8.3.1 Optimal Control without Inequalities

For the case "unconstrained" by inequalities, we note that if the assumptions of Theorem 8.1 are satisfied for all feasible values of u and p, then the solution of the DAE exists and is unique. Therefore, at the optimal solution *all perturbations δu, dp are allowable* as they lead to feasible state profiles. To satisfy (8.14) for all perturbations, we need to enforce

$$\frac{\partial f}{\partial u}\lambda + \frac{\partial g_E}{\partial u}\nu_E = 0, \tag{8.15a}$$

$$\frac{\partial z_0}{\partial p}\lambda(0) + \int_0^{t_f} \frac{\partial f}{\partial p}\lambda(t) + \frac{\partial g_E}{\partial p}\nu_E \, dt = 0. \tag{8.15b}$$

For convenience we define the Hamiltonian function as

$$H(t) \equiv f(z, y, u, p)^T \lambda + g_E(z, y, u, p)^T \nu_E,$$

and this allows us to concisely represent the optimality conditions (8.12), (8.11), (8.13), and (8.15) as

$$\begin{aligned} & \frac{d\lambda}{dt} = -\frac{\partial H}{\partial z}, \quad \lambda(t_f) = \frac{\partial \Phi(z(t_f))}{\partial z} + \frac{\partial h_E(z(t_f))}{\partial z}\eta_E, \\ & \frac{\partial H(t)}{\partial y} = 0, \quad \frac{\partial H(t)}{\partial u} = 0, \quad \frac{\partial z_0}{\partial p}\lambda(0) + \int_0^{t_f} \frac{\partial H}{\partial p} dt = 0. \end{aligned}$$

These are the *Euler–Lagrange equations* developed for problem (8.6) without inequalities. Adding the state equations to these conditions leads to the following differential-algebraic boundary value problem, with an integral constraint:

$$\frac{dz}{dt} = \frac{\partial H(t)}{\partial \lambda} = f(z(t), y(t), u(t), p), \quad z(0) = z_0, \tag{8.16a}$$

$$\frac{d\lambda}{dt} = -\left[\frac{\partial f}{\partial z}\lambda + \frac{\partial g_E}{\partial z}\nu_E\right] = -\frac{\partial H}{\partial z}, \quad \lambda_f = \frac{\partial \Phi(z_f)}{\partial z} + \frac{\partial h_E(z_f)}{\partial z}\eta_E, \tag{8.16b}$$

$$h_E(z(t_f)) = \frac{\partial H(t)}{\partial \eta_E} = 0, \tag{8.16c}$$

$$g_E(z(t), y(t), u(t), p) = \frac{\partial H(t)}{\partial \nu_E} = 0, \tag{8.16d}$$

$$\frac{\partial f}{\partial y}\lambda + \frac{\partial g_E}{\partial y}\nu_E = \frac{\partial H(t)}{\partial y} = 0, \tag{8.16e}$$

$$\frac{\partial f}{\partial u}\lambda + \frac{\partial g_E}{\partial u}\nu_E = \frac{\partial H(t)}{\partial u} = 0, \tag{8.16f}$$

$$\frac{\partial z_0}{\partial p}\lambda(0) + \int_0^{t_f} \frac{\partial H}{\partial p} dt = 0, \tag{8.16g}$$

where we define $z_f = z(t_f)$ and $\lambda_f = \lambda(t_f)$ for convenience. To illustrate these conditions, we consider a small batch reactor example.

Example 8.3 (Batch Reactor Example without Bounds). Consider the nonisothermal batch reactor with first order series reactions $A \to B \to C$. For optimal reactor operation, we seek a temperature profile that maximizes the final amount of product B. The optimal control problem can be stated as

$$\min \quad -b(t_f) \tag{8.17a}$$

$$\text{s.t.} \quad \frac{da}{dt} = -k_{10}\exp(-E_1/RT)a(t), \tag{8.17b}$$

$$\frac{db}{dt} = k_{10}\exp(-E_1/RT)a(t) - k_{20}\exp(-E_2/RT)b(t), \tag{8.17c}$$

$$a(0) = 1, \quad b(0) = 0. \tag{8.17d}$$

Note that the optimization model has no algebraic equations nor time-independent variables p. To simplify the solution, we redefine the control profile as $u(t) = k_{10}\exp(-E_1/RT)$ and rewrite the problem as

$$\min \quad -b(t_f) \tag{8.18a}$$

$$\text{s.t.} \quad \frac{da}{dt} = -a(t)u(t), \tag{8.18b}$$

$$\frac{db}{dt} = a(t)\,u(t) - k\,b(t)\,u(t)^{\beta}, \tag{8.18c}$$

$$a(0) = 1, \quad b(0) = 0, \tag{8.18d}$$

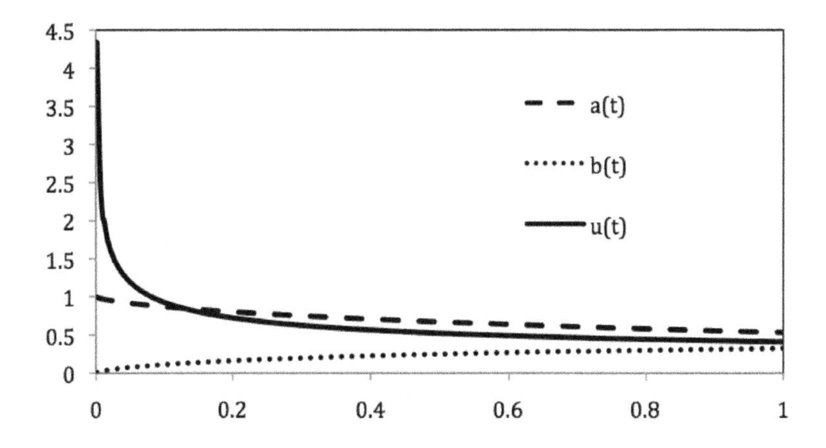

Figure 8.4. *State and control profiles for batch reactor $A \rightarrow B \rightarrow C$.*

where $k = k_{20}/(k_{10}^{\beta})$ and $\beta = E_2/E_1$. To obtain the optimality conditions, we form the Hamiltonian

$$H(t) = (\lambda_2 - \lambda_1)a(t)u(t) - \lambda_2 k u(t)^{\beta} b(t).$$

The adjoint equations are given by

$$\frac{d\lambda_1}{dt} = -(\lambda_2 - \lambda_1)u(t), \tag{8.19a}$$

$$\frac{d\lambda_2}{dt} = \lambda_2 k \, u(t)^{\beta}, \tag{8.19b}$$

$$\lambda_1(t_f) = 0, \quad \lambda_2(t_f) = -1, \tag{8.19c}$$

and the stationarity condition for the Hamiltonian is given by

$$\frac{\partial H}{\partial u} = (\lambda_2 - \lambda_1)a(t) - \beta k \lambda_2 u(t)^{\beta-1} b(t) = 0. \tag{8.20}$$

Equations (8.18)–(8.20) take the form of the optimality conditions (8.16), but without the conditions for the algebraic equations and decisions p. Note that $u(t)$ can be recovered in terms of the state and adjoint variables only if $\beta \neq 1$. Otherwise, the problem is *singular* and the more complex analysis in Section 8.5 is needed. The temperature profile can be found by solving a two-point BVP, which consists of the state equations with initial conditions (8.18b)–(8.18d), adjoint equations with final conditions (8.19a)–(8.19c), and an algebraic equation (8.20), using a DAE BVP solver. The optimal state and control profiles are given in Figure 8.4 for values of $k = 2$, $\beta = 2$, $t_f = 1$. ∎

8.3.2 Optimal Control with Inequality Constraints

With the addition of inequality constraints, we define the Hamiltonian as [72]

$$H = f(z,y,u,p)^T \lambda + g_E(z,y,u,p)^T \nu_E + g_I(z,y,u,p)^T \nu_I$$

and extend the Euler–Lagrange equations as follows:

$$\frac{d\lambda}{dt} = -\frac{\partial H}{\partial z}, \quad \lambda_f = \frac{\partial \Phi(z_f)}{\partial z} + \frac{\partial h_E(z_f)}{\partial z}\eta_E + \frac{\partial h_I(z_f)}{\partial z}\eta_I,$$
$$\frac{\partial H(t)}{\partial y} = 0, \quad \frac{\partial H(t)}{\partial u} = 0, \quad \frac{\partial z_0}{\partial p}\lambda(0) + \int_0^{t_f} \frac{\partial H}{\partial p} dt = 0.$$

For the added inequalities, the multipliers $\nu_I(t)$ and η_I act as switching variables based on activity of the constraints. These can be classified as follows:

1. If $h_I(z_f) < 0$, then $\eta_I = 0$. Similarly, if $g_I(z(t), y(t), u(t), p) < 0$, then $\nu_I(t) = 0$.

2. For $g_I(z(t), y(t), u(t), p) = 0$, $\nu_I(t)$ can be calculated from the Euler–Lagrange conditions. Determination of η_I follows similarly for $h_I(z_f) = 0$.

3. For the case when a single constraint $g_{I,j}$ is active, we note that the allowable perturbation is the feasible direction $(\frac{\partial g_{I,j}}{\partial u})^T \delta u \le 0$. Based on $\frac{\partial H(t)}{\partial u} = 0$,

$$\left[\frac{\partial f}{\partial u}\lambda + \frac{\partial g_E}{\partial u}\nu_E\right]^T \delta u(t) = -\left[\frac{\partial g_{I,j}}{\partial u}\nu_{I,j}\right]^T \delta u(t). \tag{8.21}$$

   Here, the left-hand side indicates the change $d\Phi^*$ if $u(t)$ is perturbed off the constraint. Since $d\Phi^* \ge 0$ is required for optimality, a necessary condition for this is $\nu_{I,j} \ge 0$.

4. For the case when a final constraint $h_{I,j}$ is active, we note that the allowable perturbation is the feasible direction $\frac{\partial h_{I,j}}{\partial u}\delta z_f \le 0$. From the final condition on λ,

$$\left[\frac{\partial \Phi}{\partial z} + \frac{\partial h_E}{\partial z}\eta_E - \lambda_f\right]^T \delta z(t_f) = -\left[\frac{\partial h_{I,j}}{\partial z}(\eta_I)_{(j)}\right]^T \delta z(t_f) \tag{8.22}$$

   and again, the right-hand side indicates the change $d\Phi^*$ when $z(t_f)$ is perturbed off the constraint. Since $d\Phi^* \ge 0$ is required for optimality, a necessary condition for this is $\eta_{I,j} \ge 0$.

These points can be summarized concisely by the complementarity conditions

$$0 \le \nu_I(t) \perp g_I(z(t), y(t), u(t), p) \le 0,$$
$$0 \le \eta_I \perp h_I(z(t_f)) \le 0$$

which can be added to yield the following optimality conditions:

$$\frac{dz}{dt} = \frac{\partial H(t)}{\partial \lambda} = f(z(t), y(t), u(t), p), \quad z(0) = z_0, \tag{8.23a}$$
$$\frac{d\lambda}{dt} = -\left[\frac{\partial f}{\partial z}\lambda + \frac{\partial g_E}{\partial z}\nu_E + \frac{\partial g_I}{\partial z}\nu_I\right] = -\frac{\partial H}{\partial z}, \tag{8.23b}$$
$$\lambda_f = \frac{\partial \Phi(z_f)}{\partial z} + \frac{\partial h_E(z_f)}{\partial z}\eta_E + \frac{\partial h_I(z_f)}{\partial z}\eta_I, \tag{8.23c}$$
$$h_E(z(t_f)) = \frac{\partial H(t)}{\partial \eta_E} = 0, \tag{8.23d}$$
$$0 \le \eta_I \perp h_I(z(t_f)) \le 0, \tag{8.23e}$$

$$g_E(z(t), y(t), u(t), p) = \frac{\partial H(t)}{\partial \nu_E} = 0, \tag{8.23f}$$

$$0 \le \nu_I(t) \perp g_I(z(t), y(t), u(t), p) \le 0, \tag{8.23g}$$

$$\frac{\partial f}{\partial y}\lambda + \frac{\partial g_E}{\partial y}\nu_E + \frac{\partial g_I}{\partial y}\nu_I = \frac{\partial H(t)}{\partial y} = 0, \tag{8.23h}$$

$$\frac{\partial f}{\partial u}\lambda + \frac{\partial g_E}{\partial u}\nu_E + \frac{\partial g_I}{\partial u}\nu_I = \frac{\partial H(t)}{\partial u} = 0, \tag{8.23i}$$

$$\frac{\partial z_0}{\partial p}\lambda(0) + \int_0^{t_f} \frac{\partial H}{\partial p} dt = 0. \tag{8.23j}$$

Solving the above conditions and determining the active constraints is considerably more difficult than solving the BVP (8.16), because the additional complementarity conditions must be considered at each time point. Also, as noted in [72] the junction points, where an inactive inequality constraint becomes active, or vice versa, give rise to "corner conditions" which can lead to nonsmoothness and even discontinuities in $u(t)$. Following the derivations in [72, 248], we note that the optimality conditions (8.23) hold "almost everywhere," with discontinuities in the profiles excluded.

In addition to the optimality conditions (8.23), we consider the following properties:

- For a locally unique solution of (8.23) and the corresponding state and adjoint profiles, we know from Theorem 8.2 that the "algebraic parts" need to be invertible for $u(t)$, $y(t)$, $\nu_E(t)$, and $\nu_I(t)$, and that matrix $Q(t)$ for the associated BVP must be nonsingular.
- The conditions (8.23) represent only first order necessary conditions for optimality. In addition to these, second order conditions are also needed. These are analogous to those developed in Chapter 4 for NLP. In the absence of active constraints, these conditions are known as the *Legendre–Clebsch* conditions and are given as follows:
  - Necessary condition: $\frac{\partial^2 H^*}{\partial u^2}$ is positive semidefinite for $0 \le t \le t_f$.
  - Sufficient condition: $\frac{\partial^2 H^*}{\partial u^2}$ is positive definite for $0 \le t \le t_f$.
- For autonomous problems, the Hamiltonian $H(t)$ is constant over time. This can be seen from

$$\frac{dH}{dt} = \frac{\partial H}{\partial z}\frac{dz}{dt} + \frac{\partial H}{\partial \lambda}\frac{d\lambda}{dt} + \frac{\partial H}{\partial y}\frac{dy}{dt} + \frac{\partial H}{\partial \nu_E}\frac{d\nu_E}{dt} + \frac{\partial H}{\partial \nu_I}\frac{d\nu_I}{dt} + \frac{\partial H}{\partial u}\frac{du}{dt} + \frac{\partial H}{\partial p}\frac{dp}{dt}. \tag{8.24}$$

  From (8.23),

$$\frac{\partial H}{\partial \lambda} = f(z, y, u, p) \quad \text{and} \quad \frac{d\lambda}{dt} = -\frac{\partial H}{\partial z},$$

  we see that the first two terms cancel. Also note that $\frac{dp}{dt} = 0$. Moreover, from (8.23), $\frac{\partial H}{\partial u} = 0$, $\frac{\partial H}{\partial y} = 0$, $\frac{\partial H}{\partial \nu_E} = g_E = 0$, and either $\frac{\partial H}{\partial \nu_I} = g_I = 0$ or $\frac{d\nu_I}{dt} = 0$, because $\nu_I(t) = 0$. As a result, the terms on the right-hand side sum to zero and

$$\frac{dH(t)}{dt} = 0. \tag{8.25}$$

- If final time t_f is not specified, then it can be replaced by a scalar decision variable p_f. In this case, time can be normalized as $t = p_f\ \tau$, $\tau \in [0,1]$, and the DAE system can be rewritten as

$$\frac{dz}{d\tau} = p_f\ f(z(\tau), y(\tau), u(\tau), p)), \quad z(0) = z_0,$$
$$g(z(\tau), y(\tau), u(\tau), p) = 0.$$

The optimality conditions (8.23) can still be applied in the same way.

- Finally, the formulation (8.6) can also accommodate integrals in the objective or final time constraint functions. A particular integral term, $\int_0^{t_f} \phi(z,y,u,p)dt$, can be replaced by a new state variable $\zeta(t_f)$ and a new state equation

$$\frac{d\zeta}{dt} = \phi(z,y,u,p), \quad \zeta(0) = 0.$$

With this substitution, the optimality conditions (8.23) can still be applied as before.

With these properties in hand, we now consider a batch reactor example with control profile inequalities.

Example 8.4 (Batch Reactor Example with Control Bounds). Consider a nonisothermal batch reactor with first order parallel reactions $A \rightarrow B, A \rightarrow C$, where the goal is again to find a temperature profile that maximizes the final amount of product B. However, here the temperature profile has an upper bound. Using the same transformation for temperature as in the previous system, the optimal control problem can be stated as

$$\min \quad -b(t_f) \tag{8.26a}$$
$$\text{s.t.} \quad \frac{da}{dt} = -a(t)(u(t) + ku(t)^{\beta}), \tag{8.26b}$$
$$\frac{db}{dt} = a(t)u(t), \tag{8.26c}$$
$$a(0) = 1, \quad b(0) = 0, \quad u(t) \in [0, U]. \tag{8.26d}$$

We form the Hamiltonian

$$H = -\lambda_1(u(t) + ku(t)^{\beta})a(t) + \lambda_2 u(t)a(t) - \nu_0 u(t) + \nu_U(u(t) - U),$$

and the adjoint equations are given by

$$\frac{d\lambda_1}{dt} = \lambda_1(u(t) + ku(t)^{\beta}) - \lambda_2 u(t), \tag{8.27a}$$
$$\frac{d\lambda_2}{dt} = 0, \tag{8.27b}$$
$$\lambda_1(t_f) = 0, \quad \lambda_2(t_f) = -1. \tag{8.27c}$$

Also, the stationarity conditions for the Hamiltonian are given by

$$\frac{\partial H}{\partial u} = -\lambda_1(1 + k\beta u(t)^{\beta-1})a(t) + \lambda_2 a(t) - \nu_0 + \nu_U = 0,$$
$$0 \le u(t) \perp \nu_0(t) \ge 0, \quad 0 \le (U - u(t)) \perp \nu_U(t) \ge 0. \tag{8.28}$$

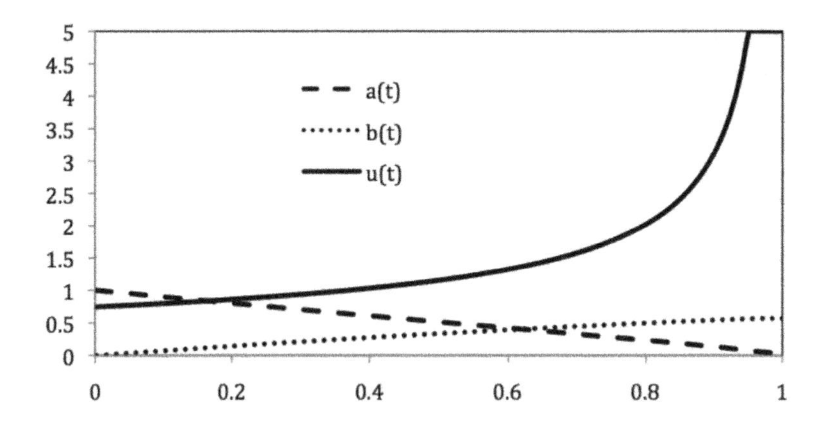

Figure 8.5. *State and control profiles for batch reactor $A \rightarrow B, A \rightarrow C$.*

For $\nu_0(t) = \nu_U(t) = 0$, we see that $u(t)$ can be recovered in terms of the state and adjoint variables only if $\beta \neq 1$. Otherwise, the problem is *singular* and the more complex analysis in Section 8.5 is needed. The optimality conditions now consist of the state equations with initial conditions (8.26b)–(8.26d), adjoint equations with final conditions (8.27a)–(8.27c), and an algebraic equation with complementarity conditions (8.28). We now consider the exact control profile for $k = 0.5$, $\beta = 2$, $t_f = 1$, and $U = 5$. From (8.27c) and (8.29) we have $\lambda_1(1) = 0$, $\lambda_2(1) = -1$, $\nu_0(1) = 0$, $\nu_U(1) > 0$, and therefore $u(1)$ must be at its upper bound. We therefore postulate a two period solution, one where $u(t)$ is unconstrained for $t \in [0, t_s)$ and a second during $t \in [t_s, 1]$ where $u(t) = U = 5$. Starting from $t = 1$ and noting that $\lambda_2(t) = -1$, we solve (8.27a) to determine

$$\lambda_1(t) = \frac{2}{7}[\exp(17.5(t-1)) - 1], t \in [t_s, 1]$$

and use (8.28) to solve for t_s when $\nu_U(t_s) = [\lambda_1(1 + k\beta U^{\beta-1}) + 1]a(t) = 0$. This leads to the kink at $t_s = 0.9500$ where the two periods meet. For the unconstrained portion, we have from (8.28) that $u(t) = -(1 + 1/\lambda_1)$ and after substituting $u(t)$ into (8.27a) and solving for $\lambda_1(t)$ the unconstrained profile is given by:

$$t = (t_s + 2\ \ln(5/6) + 12/5) - 2\ \ln(\lambda_1(t) + 1) - 2/(1 + \lambda_1(t)), \tag{8.29}$$

$$u(t) = -(1 + 1/\lambda_1(t)). \tag{8.30}$$

The analytic optimal control profile $u(t)$ can be seen in Figure 8.5 along with the state profiles, which are determined numerically from (8.26b) and (8.26c). ■

Finally, we consider a special class of optimal control problems where the state variables and the control variables appear only linearly in (8.6); i.e., we have linear autonomous DAEs (assumed index 1) and a linear objective function. Without additional constraints, the solution to the "unconstrained" problem is unbounded. On the other hand, if simple bounds are added to the control profiles, then the optimal control profile is either at its upper bound or its lower bound. This is known as a "bang-bang" solution profile [248, 72, 127]. To close this section, we apply the optimality conditions (8.23) and determine an analytical solution to a linear example.

Example 8.5 (Linear Optimal Control with Control Constraints: Car Problem). Consider the operation of a car over a fixed distance starting at rest and ending at rest. Defining the states as distance $z_1(t)$ and velocity $z_2(t)$ along with the control $u(t)$, we have

$$\begin{aligned}
\min \quad & t_f \\
\text{s.t.} \quad & \frac{dz_1}{dt} = z_2(t), \\
& \frac{dz_2}{dt} = u(t), \\
& z_1(0) = 0, \quad z_1(t_f) = L, \quad u(t) \in [u_L, u_U], \\
& z_2(0) = 0, \quad z_2(t_f) = 0.
\end{aligned}$$

This problem is the classic *double integrator problem*. As it is linear in the state and control variables, a minimum exists only if constraints are specified on either the state or the control variables. Moreover, when constraints are placed on the control variables only, the solution will be on the boundary of the control region. For the case of controls with simple bounds, this leads to a *bang-bang control* profile.

To make this problem autonomous, we define a third state z_3 to represent time and rewrite the problem as

$$\min \quad z_3(t_f) \tag{8.31a}$$

$$\text{s.t.} \quad \frac{dz_1}{dt} = z_2(t), \tag{8.31b}$$

$$\frac{dz_2}{dt} = u(t), \tag{8.31c}$$

$$\frac{dz_3}{dt} = 1, \tag{8.31d}$$

$$z_1(0) = 0, \quad z_1(t_f) = L, \quad u(t) \in [u_L, u_U], \tag{8.31e}$$

$$z_2(0) = 0, \quad z_2(t_f) = 0, \quad z_3(0) = 0. \tag{8.31f}$$

The Hamiltonian can be written as

$$H = \lambda_1 z_2(t) + \lambda_2 u(t) + \lambda_3 + \nu_U(u(t) - u_U) + \nu_L(u_L - u(t)),$$

and the adjoint equations are given by

$$\frac{d\lambda_1}{dt} = 0, \tag{8.32a}$$

$$\frac{d\lambda_2}{dt} = -\lambda_1, \tag{8.32b}$$

$$\frac{d\lambda_3}{dt} = 0, \quad \lambda_3 = 1. \tag{8.32c}$$

Note that because the states $z_1(t)$ and $z_2(t)$ have both initial and final conditions, no conditions are specified for λ_1 and λ_2. Also, the stationarity conditions for the Hamiltonian are given by

$$\frac{\partial H}{\partial u} = \lambda_2 - \nu_L + \nu_U = 0, \tag{8.33a}$$
$$0 \leq (u(t) - u_L) \perp \nu_L \geq 0, \tag{8.33b}$$
$$0 \leq (u_U - u(t)) \perp \nu_U \geq 0. \tag{8.33c}$$

For this linear problem with $\nu_L(t) = \nu_U(t) = 0$, we see that $u(t)$ cannot be recovered in terms of the state and adjoint variables, and no solution can be obtained. Instead, we expect the solution to lie on the bounds. Therefore, to obtain a solution, we need an approach that depends on the sign of $\lambda_2(t)$. Fortunately, this approach is aided by analytic solutions of the state and adjoint equations.

From the adjoint equations (8.32a)–(8.32c), we know that $\lambda_3 = 1$, $\lambda_1 = c_1$, and $\lambda_2 = c_1(t_f - t) + c_2$, where c_1 and c_2 are constants to be determined. We consider the following cases:

- $c_1 = 0$, $c_2 = 0$. This leaves an indeterminate $u(t)$. Moreover, repeated time differentiation of $\frac{\partial H}{\partial u}$ will not yield any additional information to determine $u(t)$.
- $c_1 \geq 0$, $c_2 \geq 0$, or $c_1 \leq 0$, $c_2 \leq 0$. This leads to $u(t) = u_L$ or $u(t) = u_U$, respectively. With $u(t)$ at either bound over the entire time, it is not possible to satisfy the boundary conditions.
- $c_1 \geq 0$, $c_2 \leq 0$. This case leads to a linear profile for $\lambda_2(t)$, with a switching point, that goes from positive to negative as time evolves. The control profile corresponds to full braking at initial time and up to a switching point, and full acceleration from the switching point to final time. Again, this profile does not allow satisfaction of the state boundary conditions.
- $c_1 \leq 0$, $c_2 \geq 0$. This case leads to a linear profile for $\lambda_2(t)$ with a switching point as it goes from negative to positive as time evolves. The control profile corresponds to full acceleration at initial time and up to a switching point, and full braking from the switching point to final time. It is the only case that allows the boundary conditions to be satisfied.

To find t_f and the switching point t_s, we solve the state equations and obtain

$$z_1(t) = 1/2 u_U t^2, \quad z_2(t) = u_U t \quad \text{for } t \in [0, t_s],$$
$$z_1(t) = \frac{u_U t^2}{2} + \left(\frac{u_L - u_U}{2}\right)(t - t_s)^2, \quad z_2(t) = u_U t_s + u_L(t - t_s) \quad \text{for } t \in [t_s, t_f].$$

Substituting the final conditions leads to two equations and two unknowns:

$$z_1(t_f) = \frac{u_U}{2} t_f^2 + \left(\frac{u_L - u_U}{2}\right)(t_f - t_s)^2 = L,$$
$$z_2(t_f) = u_U t_s + u_L(t_f - t_s) = 0$$

with the solution $t_s = (2L/(u_U - u_U^2/u_L))^{1/2}$ and $t_f = (1 - u_L/u_U)t_s$. The solution profiles for this system with $u_a = -2$, $u_b = 1$, and $L = 300$ are shown in Figure 8.6. ∎

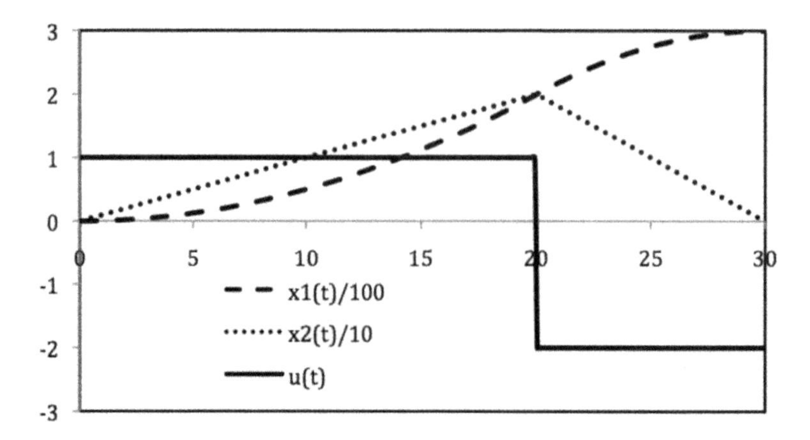

Figure 8.6. *State and control profiles for car problem.*

Finally, there are two classes of optimal control problems where the assumptions made in the derivation of the optimality conditions lead to difficult solution strategies. In the case of state variable path constraints, the adjoint variables can no longer be defined as in (8.14), by shifting the emphasis on the control variables. In the case of singular control problems, the optimality condition (8.15a) is not explicit in the control variable $u(t)$ and the optimal solution may not occur at the control bounds. Both cases require reformulation of the control problem, and considerably more complicated solution strategies. We will explore these challenging cases in the next two sections.

8.4 Handling Path Constraints

Optimality conditions were developed in the previous section for control constraints and mixed constraints of the form $g_E(z, y, u, p) = 0, g_I(z, y, u, p) \leq 0$. This allows definition of the adjoint equations by shifting the perturbations to the control variables. Moreover, by assuming nonsingularity of $(\partial g_E/\partial y)$ we know that the algebraic variables are implicit functions of $z(t)$ and that the algebraic equations can be solved implicitly with the differential equations.

We now consider the case where these properties no longer hold. In this section, inequality and equality constraints are treated separately, but with related methods for constraint reformulation. We first consider the treatment of algebraic equations which are singular in y and require further reformulation. Next we consider path inequality constraints of the form $g_I(z, p) \leq 0$. Case study examples will be presented for both methods to illustrate the reformulation strategies.

8.4.1 Treatment of Equality Path Constraints

Consider the algebraic equality constraints of the form $g_E(z, y, u, p) = 0$. Note that the distinction between the algebraic and the control variable can be left up to the modeler. However, once they are specified, the derivation of the Euler–Lagrange equations requires a nesting of variables and equations, similar to the reduced gradient equations in Section 6.4.

The nesting requires that algebraic equations be paired with algebraic variables so that they can be implicitly eliminated, in analogy to basic variables in Section 6.4. After creating the adjoint equations, the remaining control variables are then used to drive the optimization (in analogy to superbasic variables in Section 6.4).

Consequently, implicit elimination of $y(t)$ from $g_E(z,y,u,p)=0$ is an essential property for algebraic equalities. If this property does not hold, a reformulation strategy must be applied. To develop this strategy, we begin with the definition of the *index* for DAE systems.

Definition 8.6 (Index of Differential-Algebraic Equations [16]). Consider the DAE systems of the form (8.1) or (8.2) with decisions $u(t)$ and p fixed. The index is the integer s that represents the minimum number of differentiations of the DAE system (with respect to time) required to determine an ODE for the variables $z(t)$ and $y(t)$.

Applied to (8.1), this leads to the following *derivative array equations*:

$$0 = F\left(x, \frac{dx}{dt}, t\right), \tag{8.34}$$

$$0 = \frac{dF}{dt}\left(x, \frac{dx}{dt}, \frac{d^2x}{dt^2}, t\right), \tag{8.35}$$

$$\vdots \tag{8.36}$$

$$0 = \frac{d^sF}{dt^s}\left(x, \frac{dx}{dt}, \frac{d^2x}{dt^2}, \ldots, \frac{d^{s+1}x}{dt^{s+1}}, t\right), \tag{8.37}$$

In practice, the index can be determined for semiexplicit systems (8.2) by differentiating only the algebraic equations and substituting the resulting terms $\frac{dz}{dt}$ by their corresponding differential equations. For example, this allows us to determine that the DAE system

$$\frac{dz}{dt} = y(t), \quad z(t) - 5 = 0 \tag{8.38a}$$

$$\Longrightarrow \text{(i) } 0 = \frac{d(z(t)-5)}{dt} = \frac{dz(t)}{dt} = y(t) \tag{8.38b}$$

$$\Longrightarrow \text{(ii) } \frac{dy}{dt} = 0 \tag{8.38c}$$

is index 2. Moreover, we note that for an index-1 semiexplicit DAE (8.2), a single differentiation leads to

$$\frac{dz}{dt} = f(z(t), y(t), u(t), p), \quad z(0) = z_0,$$

$$\frac{\partial g}{\partial y}^T \frac{dy}{dt} + \frac{\partial g}{\partial z}^T f(z(t), y(t), u(t), p) = 0,$$

and because $(\partial g/\partial y)$ is nonsingular, we see that Definition 8.6 applies.

There is a broad literature on the solution of DAEs (see [16, 70, 237]) which deals with (a) regularity properties for existence of solutions for high-index DAEs, (b) stability of high-index DAEs, (c) numerical solution of DAEs, and (d) reformulation of high-index DAEs to index 1. In this section, we deal with only the last issue.

Ascher and Petzold [16] describe a number of ways to deal with high-index systems. These include the development of stabilized DAEs and the application of projection methods. Moreover, for systems with constant rank conditions of $(\partial g_E/\partial y)$ over time [237], it is possible to reformulate a high-index DAE into a lower index DAE.

A naive approach is to modify the semi-explicit high-index DAE by differentiating the algebraic equation s times and substituting for the differentiated variable $z(t)$ along the way. This leads to a set of pure ODEs given by

$$\frac{dz}{dt} = f(z(t), y(t), u(t), p), \quad z(0) = z_0, \tag{8.39a}$$

$$\frac{dy}{dt} = g_s(z(t), y(t), u(t), p), \quad y(0) = y_0, \tag{8.39b}$$

where $g_s(z(t), y(t), u(t), p)$ results from the algebraic equation after s-time differentiations. However, this approach has the following two problems:

- While the reformulated system (8.39) can be solved for any initial conditions (see Theorem 8.1), it may be difficult to determine initial conditions for y_0 that are consistent with the algebraic equations in (8.2). Consequently, the wrong solution may be obtained.

- Even with consistent initial conditions, numerical solution of (8.39) can differ significantly from the solution of (8.2). This follows because the reformulated problem (8.2) also applies to the modified algebraic constraint:

$$g(z(t), y(t), u(t), p) = \sum_{i=0}^{s-1} \beta_i \, t^i$$

  with arbitrary constants β_i for the polynomial in the right-hand side. Consequently, roundoff errors from numerical solution of (8.39) can lead to drift from the algebraic equations in (8.2).

General-purpose solution strategies for high-index DAEs are based on the solution of the derivative array equations (8.34) using specialized algorithms for overdetermined systems [44, 238, 167]. However, another approach to index reduction can be found by creating additional *algebraic* equations, rather than differential equations, and replacing existing differential equations with these new equations. For instance, the index-2 example (8.38),

$$\frac{dz}{dt} = y(t), \quad z(t) - 5 = 0$$

$$\Longrightarrow 0 = \frac{d(z(t)-5)}{dt} = \frac{dz(t)}{dt} = y(t)$$

can be reformulated as the purely algebraic index-1 system:

$$z(t) - 5 = 0, \quad y(t) = 0,$$

which directly yields the solution to (8.38).

These concepts can be extended to the following heuristic algorithm.

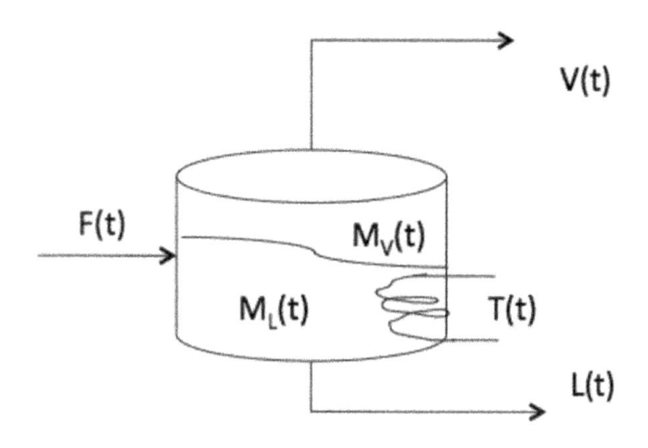

Figure 8.7. *Dynamic flash separation.*

ALGORITHM 8.1. (Reduction of High Index DAEs)
Start with a DAE of the form (8.2).

1. Check if the DAE system is index 1. If yes, stop.

2. Identify a subset of algebraic equations that can be solved for a subset of algebraic variables.

3. Consider the remaining algebraic equations that contain the differential variables z_j. Differentiating these remaining algebraic equations with respect to time leads to terms $\frac{dz_j}{dt}$ in the differentiated equations.

4. For the differential terms $\frac{dz_j}{dt}$, substitute the right-hand sides of the corresponding differential equations $f_j(z,y,u,p)$ into the differentiated algebraic equations, and eliminate (some of) these differential equations. This leads to new algebraic equations that replace the same number of existing differential equations.

5. With this new DAE system, go to step 1.

We now apply this algorithm to the index reduction of a dynamic phase separation.

Example 8.7 (Index Reduction for Dynamic Flash Optimization). Consider the optimal operation of a flash tank with NC components (with index i), a fixed liquid feed $F(t)$ with mole fractions z_i, and vapor and liquid outlet streams, as shown in Figure 8.7. The goal of the optimization is to adjust the tank temperature $T(t)$ to maximize the recovery of an intermediate component subject to purity limits on the recovered product.

The dynamic model of the flash tank can be given by the following DAEs. The differential equations relate to the mass balance:

$$\frac{dM_i}{dt} = F(t)z_i(t) - L(t)x_i(t) - V(t)y_i(t), \quad i = 1,\ldots,NC, \tag{8.40a}$$

while the following algebraic equations describe the equilibrium and hydraulic conditions of the flash:

$$M(t) = \sum_{i=1}^{NC} M_i(t), \tag{8.41a}$$

$$M = M_L + M_V, \tag{8.41b}$$

$$M_i = M_L x_i(t) + M_V y_i(t), \quad i = 1, \dots, NC, \tag{8.41c}$$

$$y_i = K_i(T, P) x_i, \quad i = 1, \dots, NC, \tag{8.41d}$$

$$C_T = M_V / \rho_V(T, P, y) + M_L / \rho_L(T, P, x), \tag{8.41e}$$

$$L = \psi^L(P, M), \tag{8.41f}$$

$$V = \psi^V(P - P_{out}), \tag{8.41g}$$

$$1 = \sum_{i=1}^{NC} x_i. \tag{8.41h}$$

The differential variables are component holdups $M_i(t)$, while algebraic variables are M, the total holdup; M_V, the vapor holdup; M_L, the liquid holdup; x_i, the mole fraction of component i in the liquid phase; y_i, the mole fraction of component i in the vapor phase; P, flash pressure; L, liquid flow rate; and V, vapor flow rate. In addition, P_{out} is the outlet pressure, C_T is the capacity (volume) of the flash tank, ρ_L and ρ_V are molar liquid and vapor densities, $K_i(T, P)$ is the equilibrium expression, ψ^V provides the valve equation, and ψ^L describes the liquid hydraulics of the tank. This model can be verified to be index 1, but as pressure changes much faster than composition, it is also a stiff DAE system.

A simplified form of (8.40)–(8.41) arises by neglecting the vapor holdup $M_V(t) \approx 0$, as this is 2 to 3 orders of magnitude smaller than the liquid holdup. This also allows the pressure P to be specified directly and leads to a less stiff DAE, which can be written as

$$\frac{dM}{dt} = F(t) - L(t) - V(t), \tag{8.42a}$$

$$\frac{dx_i}{dt} = [F(t)(z_i - x_i) - V(t)(y_i - x_i)]/M(t), \quad i = 1, \dots, NC, \tag{8.42b}$$

$$y_i(t) = K_i(T, P) x_i(t), \quad i = 1, \dots, NC, \tag{8.42c}$$

$$L(t) = \psi^L(P, M), \tag{8.42d}$$

$$1 = \sum_{i=1}^{NC} x_i. \tag{8.42e}$$

This model requires much less data than the previous one. With pressure specified, the tank capacity and the vapor valve equations are no longer needed. However, this system is now index 2, and two problems are apparent with (8.42). First, the algebraic variable $V(t)$ cannot be calculated from the algebraic equations; the last equation does not contain the unassigned variable $V(t)$. Second, the initial conditions for $x_i(0)$ cannot be specified independently, since $\sum_i x_i(0) = 1$.

Applying Algorithm 8.1 to (8.42) leads to the following steps:

- Step 1 requires us to consider the last algebraic equation only.
- Step 2 requires differentiating the last equation, leading to $\sum_{i=1}^{NC} \frac{dx_i}{dt} = 0$.

- Step 3 deals with substituting for the differential terms. This leads to the new algebraic equation,

$$\sum_{i=1}^{NC}[F(t)(z_i(t)-x_i(t))-V(t)(y_i(t)-x_i(t))]/M(t)=0,$$

and one of the differential equations must be eliminated, say the last one.

This leads to the following index-1 DAE model:

$$\frac{dM}{dt}=F(t)-L(t)-V(t), \tag{8.43a}$$

$$\frac{dx_i}{dt}=[F(t)(z_i-x_i)-V(t)(y_i-x_i)]/M(t), \quad i=1,\ldots,NC-1, \tag{8.43b}$$

$$y_i(t)=K_i(T,P)x_i(t), \quad i=1,\ldots,NC, \tag{8.43c}$$

$$L(t)=\psi^L(P,M), \tag{8.43d}$$

$$1=\sum_{i=1}^{NC}x_i, \tag{8.43e}$$

$$0=\sum_{i=1}^{NC}[F(t)(z_i(t)-x_i(t))-V(t)(y_i(t)-x_i(t))]/M(t). \tag{8.43f}$$

Note that x_{NC} is now an algebraic variable and there are no restrictions on specifying initial conditions for the differential variables. ■

8.4.2 Treatment of State Path Inequalities

Inequality constraints present two difficulties. First, as with other path inequalities, entry and exit points need to be determined for the time period over which the constraints are saturated. Second, if the control variable does not appear explicitly in these constraints, an additional formulation is required to apply the Euler–Lagrange equations. The optimality conditions for state path inequalities are given in [72, 307, 248].

For simplicity, we consider the optimal control problem (8.6) with only differential variables. This is still fairly general, as index-1 algebraic constraints lead to $y(t)$ as implicit functions of $z(t)$ and can therefore be eliminated implicitly. For the state variable inequalities, $g_I(z(t)) \le 0$, the inequality constraints are differentiated (with the differential equations substituted) until the controls $u(t)$ appear explicitly. The jth inequality $g_{I,j}(z(t)) \le 0$ is termed a qth *order inequality path constraint*, where the integer q represents the number of differentiations needed to recover an inequality that contains the control variable explicitly. Moreover, the differentiated constraint $g_{I,j}$ leads to the following array of equations:

$$N_j(z(t))=\left[g_{I,j}(z(t))^T, \frac{dg_{I,j}(z(t))}{dt}^T, \ldots, \frac{d^{q-1}g_{I,j}(z(t))}{dt^{q-1}}^T\right]^T=0,$$

$$\frac{d^q g_{I,j}(z(t))}{dt^q}=0, \quad t \in (t_{entry}, t_{exit}),$$

which must be satisfied when the path constraint is active. Since the control variable appears in the differentiated constraint, the Hamiltonian is now redefined as

$$H(t) = f(z,u,p)^T \lambda + g_E(z,u,p)^T \nu_E + \sum_j \frac{d^{q_j} g_{I,j}(z,u)}{dt^{q_j}} \nu_{I,j},$$

where we have distinguished the integer q_j for each path constraint. The Euler–Lagrange conditions can now be derived as in (8.23). In addition, state path constraints also require the additional corner conditions:

$$\lambda(t^+_{entry}) = \lambda(t^-_{entry}) - \frac{\partial N_j}{\partial z} \pi_j, \tag{8.44a}$$

$$\lambda(t^+_{exit}) = \lambda(t^-_{exit}) \tag{8.44b}$$

or

$$\lambda(t^+_{exit}) = \lambda(t^-_{exit}) - \frac{\partial N_j}{\partial z} \pi_j, \tag{8.45a}$$

$$\lambda(t^+_{entry}) = \lambda(t^-_{entry}). \tag{8.45b}$$

Here π_j is an additional multiplier on the corner conditions and t^+ and t^- represent points just before and just after the change in the active set. Note that choice of corner conditions emphasizes the nonuniqueness of the multipliers with path inequalities [72, 307]. In fact, there are a number of related conditions that can be applied to the treatment of path inequalities [188] and this confirms the ill-posedness of the modified Euler–Lagrange equations.

Characteristics of path inequalities will be considered again in Chapter 10. Because of the complexity in handling the above conditions along with finding entry and exit points, the treatment of these constraints is difficult for all but small problems. The following example illustrates the application of these conditions.

Example 8.8 (Car Problem Revisited). We now consider a slight extension to Example 8.5 by imposing a speed limit constraint, $g_I(z(t)) = z_2 - V \le 0$. As described above, the derivation of the Euler–Lagrange equations defines the adjoint system so that the objective and constraint functions are influenced directly by control variables. To recapture this influence, we differentiate $g_I(z)$ with respect to time in order to recover the control variable. For this example, we therefore have

$$\frac{dg_I}{dt} = \frac{\partial g_I}{\partial z}^T f(z,u) = \frac{dz_2}{dt} = u(t),$$

and we define a multiplier for the path constraint with the following Euler–Lagrange equations:

$$\begin{aligned} H &= \lambda^T f(z,u) - \nu_I (dg_I/dt) + \nu_U (u(t) - u_U) + \nu_L (u_L - u(t)) \\ &= \lambda_1 z_2(t) + \lambda_2 u(t) + 1 + \nu_I u(t) + \nu_U (u(t) - u_U) + \nu_L (u_L - u(t)) \end{aligned} \tag{8.46}$$

$$\frac{d\lambda_1}{dt} = 0, \tag{8.47}$$

$$\frac{d\lambda_2}{dt} = -\lambda_1, \tag{8.48}$$

$$\frac{\partial H}{\partial u} = \lambda_2 + \nu_I - \nu_L + \nu_U = 0, \tag{8.49}$$

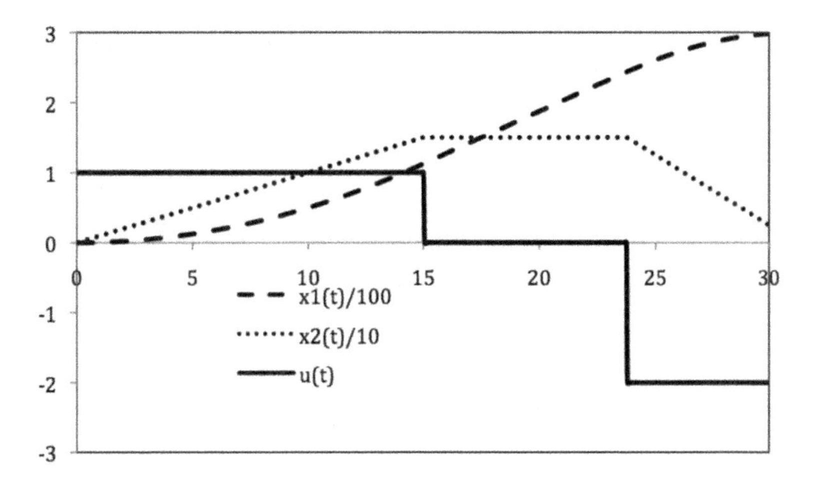

Figure 8.8. *State and control profiles for car problem with speed limit.*

$$0 \le (u(t) - u_L) \perp \nu_L \ge 0, \tag{8.50}$$
$$0 \le (u_U - u(t)) \perp \nu_U \ge 0, \tag{8.51}$$
$$0 \le (V - z_2(t)) \perp \nu_I \ge 0. \tag{8.52}$$

These conditions are more complex than in the previous examples. Moreover, because of the influence of the path constraint, an additional *corner condition* is required:

$$\lambda_2(t_1^+) = \lambda_2(t_1^-) - \pi,$$

where π is a constant and t_1 is the entry point of the path constraint. Solution of path-constrained problems through the above Euler–Lagrange equations is generally difficult, especially because the location of the active path constrained segments is not known a priori. Nevertheless, the following intuitive solution:

$$u(t) = \begin{cases} u_U, & \lambda_2 < 0, \quad t \in [0, t_1), \\ 0, & \lambda_2 = -\nu_I, \quad t \in (t_1, t_2), \\ u_L, & \lambda_2 > 0, \quad t \in (t_2, t_f], \end{cases} \tag{8.53}$$

can be easily checked with these Euler–Lagrange equations (see Exercise 5). The solution profiles for this system with $u_L = -2$, $u_U = 1$, $V = 15$, and $L = 300$ are shown in Figure 8.8. ■

8.5 Singular Control Problems

Singular optimal control problems are encountered often in process engineering. Applications include dynamic optimization of batch processes, batch reactor control [107, 368], and optimal mixing of catalysts [207]. These problems arise when the control appears only linearly in problem (8.6). In this case, the control does not appear explicitly in $\frac{\partial H}{\partial u}$.

Under these conditions, the control lies at its bounds if $\frac{\partial f}{\partial u}\lambda + \frac{\partial g_E}{\partial u}\nu_E$ is not identically zero. As shown in Example 8.5, this leads to "bang-bang" control. On the other hand, when the state equations are nonlinear in $z(t)$ and linear in $u(t)$, $\frac{\partial f}{\partial u}\lambda + \frac{\partial g_E}{\partial u}\nu_E$ may also be zero over a nonzero time period and this leads to a singular profile in $u(t)$. Singular arcs cannot be determined directly from the Euler–Lagrange equations. Instead, as with state path inequalities, repeated time differentiations of $\frac{\partial H}{\partial u}$ need to be performed to recover the optimal control.

For simplicity of exposition, we consider only a single control profile with simple bounds, $u(t) \in [u_L, u_U]$, neglect algebraic constraints and decisions p, and write the differential equations as

$$\frac{dz}{dt} = f_1(z(t)) + f_2(z(t))u(t).$$

As in Section 8.3, we define the Hamiltonian as

$$H(z,y,u,p) = (f_1(z) + f_2(z)u)^T\lambda + \nu_L^T(u_L - u(t)) + \nu_U^T(u(t) - u_U)$$

and

$$\frac{\partial H}{\partial u} = H_u(t) = f_2(z)\lambda - \nu_L + \nu_U = 0.$$

Repeated time derivatives of $H_u(t)$ are required to obtain an explicit function in $u(t)$. If such a function exists, then the condition for the singular arc over $t \in (t_{entry}, t_{exit})$ is given by

$$\frac{d^q H_u(t)}{dt^q} = \varphi(z,u) = 0, \tag{8.54}$$

where q is an *even integer* that represents the minimum number of times that H_u must be differentiated. The order of the singular arc is given by $q/2$ and, for a scalar control, a second order condition over the singular arc can be defined by the *generalized Legendre–Clebsch* condition [72].

- Necessary condition:

$$(-1)^{q/2}\frac{\partial}{\partial u}\left(\frac{d^q H_u(t)}{dt^q}\right) = \frac{\partial \varphi(z,u)}{\partial u} \geq 0 \quad \text{for } t \in (t_{entry}, t_{exit}).$$

- Sufficient condition:

$$(-1)^{q/2}\frac{\partial}{\partial u}\left(\frac{d^q H_u(t)}{dt^q}\right) = \frac{\partial \varphi(z,u)}{\partial u} > 0 \quad \text{for } t \in (t_{entry}, t_{exit}).$$

As with state path inequalities, entry and exit points must be found and the following stationarity conditions $\frac{d^l H_u(t)}{dt^l} = 0$, $l = 0, \ldots, q-1$, must also be satisfied for $t \in [t_{entry}, t_{exit}]$. On the other hand, there are no corner conditions, and both the Hamiltonian and the adjoint variables remain continuous over $[0, t_f]$.

To illustrate the application of the singular arc conditions, we consider a classical optimal control example in reaction engineering.

Example 8.9 (Singular Optimal Control: Catalyst Mixing Problem). Consider the catalyst mixing problem analyzed by Jackson [207]. The reactions $A \Longleftrightarrow B \rightarrow C$ take place in a tubular reactor at constant temperature. The first reaction is reversible and is catalyzed by Catalyst I, while the second irreversible reaction is catalyzed by Catalyst II. The goal of this problem is to determine the optimal mixture of catalysts along the length t of the reactor in order to maximize the amount of product C. Using $u(t) \in [0,1]$ as the fraction of Catalyst I, an intuitive solution would be to use Catalyst I at the beginning of the reactor with Catalyst II toward the end of the reactor, leading to a "bang-bang" profile. As with the solution in Example 8.5, the switching point would then be determined by the available length of the reactor, t_f. However, the bang-bang policy leads only to production of B in the first portion, and for large t_f this production is limited by equilibrium of the reversible reaction. As a result, production of C will be limited as well. Instead, the production of C can be enhanced through a mixture of catalysts over some internal location in the reactor, where the reversible reaction is driven forward by consumption, as well as production, of B. This motivates the singular arc solution.

The resulting optimal catalyst mixing problem can be stated as

$$\begin{aligned}
\max \quad & c(t_f) \\
\text{s.t.} \quad & \frac{da}{dt} = -u(k_1 a(t) - k_2 b(t)), \\
& \frac{db}{dt} = u(k_1 a(t) - k_2 b(t)) - (1-u)k_3 b(t), \\
& a_0 = a(t) + b(t) + c(t), \\
& a(0) = a_0, \quad b(0) = 0, \quad u(t) \in [0,1].
\end{aligned}$$

By eliminating the algebraic state $c(t)$ and its corresponding algebraic equation, an equivalent, but simpler, version of this problem is given by

$$\min \quad a(t_f) + b(t_f) - a_0 \tag{8.55a}$$

$$\text{s.t.} \quad \frac{da(t)}{dt} = -u(k_1 a(t) - k_2 b(t)), \tag{8.55b}$$

$$\frac{db(t)}{dt} = u(k_1 a(t) - k_2 b(t)) - (1-u)k_3 b(t), \tag{8.55c}$$

$$a(0) = a_0, \quad b(0) = 0, \quad u(t) \in [0,1]. \tag{8.55d}$$

The Hamiltonian can be written as

$$H(t) = (\lambda_2 - \lambda_1)(k_1 a(t) - k_2 b(t))u(t) - \lambda_2 k_3 b(t)(1 - u(t)) - \nu_0 u(t) + \nu_1(u(t) - 1). \tag{8.56}$$

The adjoint equations are given by

$$\frac{d\lambda_1}{dt} = -(\lambda_2 - \lambda_1)k_1 u(t), \tag{8.57}$$

$$\frac{d\lambda_2}{dt} = (\lambda_2 - \lambda_1)k_2 u(t) + (1 - u(t))\lambda_2 k_3, \tag{8.58}$$

$$\lambda_1(t_f) = 1, \quad \lambda_2(t_f) = 1, \tag{8.59}$$

and the stationarity conditions for the Hamiltonian are given by

$$\frac{\partial H}{\partial u} = J(t) - \nu_0 + \nu_1 = 0, \tag{8.60}$$

$$0 \leq \nu_0(t) \perp u(t) \geq 0, \tag{8.61}$$

$$0 \leq \nu_1(t) \perp (1 - u(t)) \geq 0, \tag{8.62}$$

where we define $J(t) = (\lambda_2 - \lambda_1)(k_1 a(t) - k_2 b(t)) + \lambda_2 k_3 b(t)$.

Note that $u(t) = 0$ implies that $\nu_0 = J \geq 0$. Also, $u(t) = 1$ implies $-\nu_1 = J \leq 0$, and $u \in (0,1)$ implies $J = 0$. Also, from (8.25) we know that the Hamiltonian $H(t)$ remains constant over the length of the reactor.

Based on the analysis in [207], we now deduce the form of the optimal control profile as follows:

- From the state equations we see that $b(t_f) \to 0$ only if $t_f \to \infty$ and $u(t_f) = 0$. Else, $b(t) > 0$ for $t > 0$. The final conditions for the adjoint variables, i.e., $\lambda_1(t_f) = \lambda_2(t_f) = 1$, allow the Hamiltonian to be written as

$$H(t_f) = -k_3 b(t_f)(1 - u(t_f)),$$

and since $J(t_f) = k_3 b(t_f) > 0$, we have $u(t_f) = 0$ and $H(t) = H(t_f) = -k_3 b(t_f) < 0$.

- Assume at $t = 0$ that $u(0) = 1$. For $u(t) = 1$, we note that $a_0 = a(t) + b(t)$ and

$$\frac{da}{dt} + \frac{db}{dt} = 0 = -k_3 b(t)(1 - u(t)).$$

Since

$$H(t) = J(t)u(t) - \lambda_2 k_3 b(t) < 0,$$

we have at $t = 0$,

$$-k_3 b(t_f) = H(0) = J(0)u(0) - \lambda_2 k_3 b(0) = J(0)u(0) < 0,$$

which is consistent with $u(0) = 1$. Alternately, if we had assumed that $u(0) \in [0,1)$, we would have $H(0) = 0$, which contradicts $H(0) = -k_3 b(t_f) < 0$.

- By continuity of the state and the adjoint variables, we know that $J(t)$ must switch sign at some point in $t \in (0, t_f)$. Moreover, to indicate the presence of a singular arc, we need to determine whether $J(t) = 0$ over a nonzero length.

Now if $J = 0$ over a nonzero length segment, we can apply the following time differentiations:

$$J = (\lambda_2 - \lambda_1)(k_1 a - k_2 b) + \lambda_2 k_3 b = 0, \tag{8.63a}$$

$$\begin{aligned}\frac{dJ}{dt} &= (\dot{\lambda}_2 - \dot{\lambda}_1)(k_1 a - k_2 b) + (\lambda_2 - \lambda_1)(k_1 \dot{a} - k_2 \dot{b}) + \dot{\lambda}_2 k_3 b + \lambda_2 k_3 \dot{b} \\ &= k_3(k_1 a \lambda_2 - k_2 b \lambda_1) = 0,\end{aligned} \tag{8.63b}$$

$$\begin{aligned}\frac{d^2 J}{dt^2} &= \lambda_2 k_3 (k_1 a - k_2 b) + u\,[\lambda_1(k_2 b(k_2 - k_3 - k_1) - 2k_1 k_2 a) \\ &\quad + \lambda_2(k_1 a(k_2 - k_3 - k_1) + 2k_1 k_2 b)] = 0,\end{aligned} \tag{8.63c}$$

where $\dot{\xi} = \frac{d\xi}{dt}$. Note that two time differentiations, $\frac{d^2 J}{dt^2} = 0$, are required to expose $u(t)$ and to yield an expression for $u(t)$ over this time segment. This leads to a singular arc of order $q/2 = 1$. Another approach used in [207] would be to reformulate the simpler expression in (8.63b) by defining a new variable $\bar{\lambda}$ as

$$\bar{\lambda} = \frac{\lambda_1}{\lambda_2} = \frac{k_1 a}{k_2 b}. \tag{8.64}$$

With $J = 0$, we can substitute $\bar{\lambda}$ into (8.63a) and obtain

$$J = \lambda_2(t) k_2(t) b(t)[k_3/k_2 - (1-\bar{\lambda})^2] = 0 \implies \bar{\lambda} = 1 \pm (k_3/k_2)^{1/2}, \tag{8.65}$$

so $\bar{\lambda}$ is a constant for the singular arc. Over this time period we can write

$$\dot{\lambda}_1 = \bar{\lambda}\dot{\lambda}_2 \implies (\lambda_1 - \lambda_2)k_1 u = \bar{\lambda}[(\lambda_2 - \lambda_1)k_2 u + (1-u)k_3\lambda_2]. \tag{8.66}$$

Dividing this equation by $(\lambda_2 k_2)$ and using (8.64) and (8.65) leads to the singular arc value of $u(t)$:

$$-(1-\bar{\lambda})u k_1/k_2 = \bar{\lambda}[(1-\bar{\lambda})u + (1-u)(1-\bar{\lambda})^2] \implies u = \frac{\bar{\lambda}(\bar{\lambda}-1)k_2}{k_1 + k_2\bar{\lambda}^2}. \tag{8.67}$$

Since $u(t) > 0$ for the singular arc, we require $\bar{\lambda} = 1 + (k_3/k_2)^{1/2}$ so that

$$u(t) = \frac{(k_3 k_2)^{1/2} + k_3}{k_1 + k_2 + k_3 + 2(k_3 k_2)^{1/2}} \tag{8.68}$$

whenever $J(t) = 0$. This leads to a candidate solution with

$$u(t) = \begin{cases} 1, & t \in [0, t_1), \text{ with increasing } t, \frac{k_1 a}{k_2 b} \text{ decreases to } \bar{\lambda}, \\ \frac{(k_3 k_2)^{1/2} + k_3}{k_1 + k_2 + k_3 + 2(k_3 k_2)^{1/2}}, & t \in (t_1, t_2), \ \bar{\lambda} = \frac{k_1 a}{k_2 b} = \frac{\lambda_1}{\lambda_2} \text{ is constant}, \\ 0, & t \in (t_2, t_f], \text{ with increasing } t, \frac{\lambda_1}{\lambda_2} \text{decreases from } \bar{\lambda} \text{ to } 1. \end{cases} \tag{8.69}$$

The switching points can now be determined by solving the state equations forward from $t = 0$ and the adjoint equations backward from $t = t_f$ as follows:

- From $t = 0$ we have $u(t) = 1$ and $\bar{\lambda} = \frac{k_1 a(t)}{k_2 b(t)}$. The state equations are then integrated forward until $\frac{k_1 a(t)}{k_2 b(t)} = 1 + (k_3/k_2)^{1/2}$, the value of $\bar{\lambda}$ in the singular arc. This defines the entry point, t_1, for the singular arc.

- For $t \in [t_2, t_f]$, we note that

$$\frac{d\bar{\lambda}}{dt} = -\frac{\dot{\lambda}_1\lambda_2 - \dot{\lambda}_2\lambda_1}{(\lambda_2)^2} = -\frac{\lambda_1\lambda_2 k_3}{(\lambda_2)^2} = -k_3\bar{\lambda}. \tag{8.70}$$

  Since $\lambda_1(t_f) = \lambda_2(t_f) = \bar{\lambda}(t_f) = 1$, equation (8.70) can be integrated backward until $\bar{\lambda}(t) = \exp((t_f - t)k_3) = 1 + (k_3/k_2)^{1/2}$. This defines the exit point t_2.

With this solution strategy, optimal control and state profiles are shown in Figure 8.9 for $a_0 = 1, k_1 = 1, k_2 = 10, k_3 = 1$, and $t_f = 4$. These reveal a "bang-singular-bang" control policy. ■

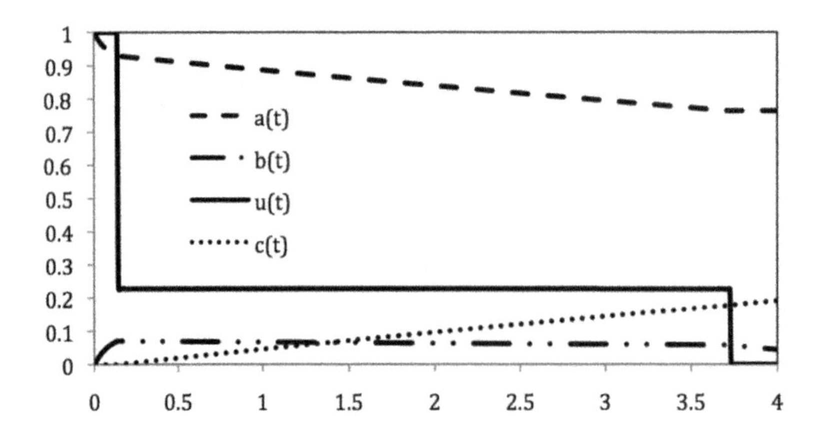

Figure 8.9. *State and control profiles for optimal catalyst mixing.*

8.6 Numerical Methods Based on NLP Solvers

The previous sections developed the necessary conditions for the optimization of dynamic process systems. These conditions follow from a natural progression of optimal control problems, starting from "unconstrained" problems to mixed path constraints and concluding with state path constraints and singular control. For this progression, the optimality conditions became increasingly harder to apply and the illustrative examples often rely on *analytic solutions* of the state and adjoint equations to facilitate the treatment of inequality constraints and singular arcs. Unfortunately, this is no longer possible for larger systems, and efficient numerical methods are required instead.

A wide variety of approaches has been developed to address the solution of (8.6). These strategies can be loosely classified as *Optimize then Discretize* and *Discretize then Optimize*. In the first case, the state equations and the Euler–Lagrange equations are discretized in time and solved numerically . In the second case, the state and control profiles are discretized and substituted into the state equations. The resulting algebraic problem is then solved with an NLP algorithm. There are a number of pros and cons to both approaches, and a roadmap of optimization strategies is sketched in Figure 8.10.

Optimize then Discretize (O-D)

The *indirect* or *variational approach* is based on the solution of the first order necessary conditions developed in this chapter. For problems without inequality constraints, the optimality conditions can be formulated as a set of DAEs with careful attention to the boundary conditions. Often the state variables have only initial conditions specified and the adjoint variables have final conditions. The resulting BVP can be addressed with different approaches, including single shooting, invariant embedding, multiple shooting, or discretization methods such as collocation on finite elements or finite differences. A review of these approaches can be found in [90]. In particular, two approaches have been widely used for optimal control.

In the first approach, *control vector iteration* (CVI) can be applied to state equations stated as initial value problems. CVI proceeds by solving the state variables forward in

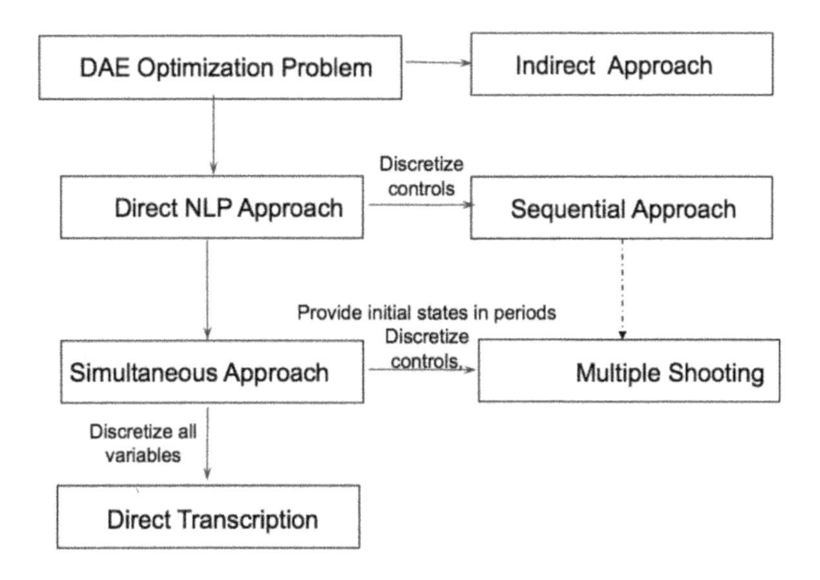

Figure 8.10. *Classification of DAE optimization strategies.*

time, and then solving the adjoint equations backward in time for some particular value of the control profile. Having the states and adjoints, the control variable is then updated by a steepest descent or Newton-based scheme to satisfy stationarity of the Hamiltonian. However, this approach requires repeated, nested solution of the state and adjoint equations and may be slow to converge. Moreover, inequalities are not straightforward to incorporate in this approach. Usually they are handled, with varying degrees of success, through projection methods (for control profile bounds) [279] or by applying a penalty function [72].

The second approach applies more modern methods of solving BVPs. Here the state and the Euler–Lagrange equations are solved simultaneously using multiple shooting or collocation-based methods for BVPs [16, 14]. On the other hand, the BVP approach does not ensure descent directions of the objective function, and therefore does not exploit globalization strategies developed for NLP algorithms. Moreover, path inequalities are harder to incorporate with this approach, as finding the correct switching structure along with suitable initial guesses for state and adjoint variables is often difficult. A promising approach for inequalities is to apply an interior point relaxation to the complementarity conditions in (8.23), and to incorporate these into the BVP solution strategy [407, 352]. Finally, as with CVI, singular control problems are difficult to solve with this approach.

Discretize then Optimize (D-O)

The difficulties associated with solving BVPs and handling inequality constraints motivate the formulation of constrained, finite-dimensional problems that can use the full machinery of large-scale NLP solvers. This approach has the advantage of directly finding good approximate solutions that are feasible to the state equations. On the other hand, the formulation requires an accurate level of discretization of the control (and possibly the state) profiles. This may be a daunting task particularly for path constrained and singular problems.

As seen in Figure 8.10, methods that apply NLP solvers can be separated into two groups: the *sequential* and the *simultaneous* strategies. In the *sequential methods*, only the control variables are discretized and the resulting NLP is solved with *control vector parameterization* (CVP) methods. In this formulation the control variables are represented as piecewise polynomials [398, 399, 28] and optimization is performed with respect to the polynomial coefficients. Given initial conditions and a set of control parameters, the DAE model is then solved in an inner loop, while the parameters representing the control variables are updated on the outside using an NLP solver. Gradients of the objective function with respect to the control coefficients and parameters are calculated either from direct sensitivity equations of the DAE system or by integration of adjoint sensitivity equations.

Sequential strategies are relatively easy to construct, as they link reliable and efficient codes for DAE and NLP solvers. On the other hand, they require repeated numerical integration of the DAE model and are not guaranteed to handle open loop unstable systems [16, 56]. Finally, path constraints can also be incorporated within this approach but are often handled approximately, within the limits of the control parameterization. These approaches will be developed in more detail in the next chapter.

Optimization with multiple shooting serves as a bridge between sequential and direct transcription approaches and was developed to handle unstable DAE systems. In this approach, the time domain is partitioned into smaller time elements and the DAE models are integrated separately in each element [60, 62, 250]. Control variables are treated in the same manner as in the sequential approach and sensitivities are obtained for both the control variables, as well as for the initial conditions of the states in each element. In addition, equality constraints are added to the nonlinear program in order to link the elements and ensure that the states remain continuous over time. Inequality constraints for states and controls can be imposed directly at the grid points, although constraints on state profiles may be violated between grid points. These methods will also be developed in the next chapter.

Direct transcription approaches deal with full discretization of the state and control profiles and the state equations. Typically the discretization is performed by using collocation on finite elements, a Runge–Kutta method. The resulting set of equations and constraints leads to a large nonlinear program that is addressed with large-scale NLP solvers. The approach is fully simultaneous and requires no nested calculations with DAE solvers. Moreover, both structure and sparsity of the KKT system can be exploited by modern NLP codes, such as those developed in Chapter 6. On the other hand, adaptive discretization of the state equations, found in many DAE solvers, is generally absent in direct transcription. Instead, the number of finite elements is fixed during the solution of the NLP and the mesh is either fixed or varies slightly during this solution. Nevertheless, an outer loop may be added to this approach to improve accuracy of the solution profiles, by refining the finite element mesh [40]. Finally, in some cases direct links can be made between the KKT conditions of the NLP and the discretized Euler–Lagrange equations. This offers the exciting possibility of relating *D-O* with *O-D* strategies. The direct transcription approach will be developed further in Chapter 10.

8.7 Summary and Conclusions

This chapter presents and analyzes a general formulation for optimization problems with DAEs. These problems can be found in a wide variety of process engineering applications, and their successful solution can lead to significant performance improvements in the chemical and energy industries. The solution of these dynamic optimization problems is explored through the machinery of optimal control theory. In particular, we deal with a variational approach of an adjoined system, and we develop adjoint and stationarity conditions as well as transversality conditions. The resulting Euler–Lagrange equations provide the necessary conditions for optimality. Coupled with the state equations, they yield a BVP, whose solution leads to optimal state and control profiles.

Treatment of path inequalities adds additional optimality conditions and poses some challenges as well. For these inequalities, the location of constrained arcs is a combinatorial process that requires calculation of constraint multipliers and checking complementarity. Moreover, because of the assumptions in developing the Euler–Lagrange equations, this problem becomes more difficult when path inequalities are not explicit functions of the control profile. In this case, state path constraints must be reformulated with appropriate corner conditions applied at the entry and exit points of the active constraint. Finally, singular control problems present an additional challenge. As with state path inequalities, the location of the singular arc requires reformulation of the stationary conditions for the Hamiltonian, checking the activity of the control variable constraints, and determining the entry and exit points of the singular arc.

The progression of dynamic optimization from unconstrained to control constrained, state path constrained, and singular problems is illustrated by small, real-world examples that show the application of the optimality conditions and demonstrate some of their challenges. On the other hand, larger problems require numerical solution strategies. As described in the previous section, efficient optimization algorithms are not straightforward to develop from the Euler–Lagrange and the state equations. Early algorithms based on this "indirect method" are neither fast nor reliable. On the other hand, more recent methods based on modern BVP solvers have led to promising performance results, even on large-scale problems [235]. Nevertheless, difficulties with indirect methods have motivated the development of direct NLP methods. While direct methods represent a departure from the optimality conditions in this chapter, the Euler–Lagrange equations still provide the concepts and insights for the successful formulation and solution of NLP strategies, and the interpretation of accurate solution profiles. These NLP strategies will be treated in the next two chapters.

8.8 Notes for Further Reading

The study of dynamic optimization problems originates from the rich field of *calculus of variations*. The first systematic theory in this field is attributed to John and James Bernoulli with the solution of the *brachistochrone* problem (find the shape of a wire trajectory between two points, on which a bead can descend in minimum time). The calculus of variations is treated in a number of classic texts including Bliss [59], Courant and Hilbert [104], Sagan [343], and Gelfand and Fomin [157].

The evolution of variational calculus to *optimal control* stems from Pontryagin's maximum principle discovered in the 1950s. Fundamental texts in this field include Pontryagin et al. [311], Lee and Markus [248], Fleming and Rishel [133], and Macki and Strauss [269]. Also, an excellent, concise set of course notes was developed by Evans [127]. The informal derivation of optimality conditions in this chapter stems from the classic text of Bryson and Ho [72] and the survey paper of Pesch [307]. In addition, optimal control has led to a broad set of applications including aeronautical [72, 325] and satellite applications

[40, 12], economics [359], and robotics [76, 370]. An interesting set of process control applications is presented in the text by Ray [329].

Indirect numerical algorithms that "optimize then discretize" are beyond the scope of this text. Nevertheless, the development of CVI algorithms is described in [72, 328]. Jones and Finch [212] review a number of advanced CVI methods including quasi-Newton methods and conjugate gradient methods [244]. More recent application of indirect methods and BVP solvers for optimal control can be found in [177]. An interesting synthesis of NLP and BVP strategies can be found in von Stryk and Bulirsch [400]. More recently, Weiser [406] extended BVP methods to include interior point relaxation to deal with the complementarity conditions for path inequalities.

The solution of high-index DAEs is a widely explored topic. A nice introduction is given in [16], and detailed treatment of this topics can be found in [237, 70]. Algorithms for reformulation of high-index DAEs to index-1 systems have been developed in [19, 277]. Related reformulation strategies based on sparsity structure can be found in [304, 234].

Finally, the Euler–Lagrange equations have been very useful in deriving optimum-seeking control strategies [233, 368]. Using the *necessary conditions of optimality* (NCO), optimal outputs can be tracked by determining inputs from conditions that require active constraints or maintain stationarity conditions. Based on measurements, active constraints are often straightforward to track by assigning them to appropriate inputs. On the other hand, sensitivity information is difficult to determine online. For singular problems, these can often be computed through feedback laws derived from Lie algebra [368], which still may be cumbersome to develop. Fortunately, even an approximation of these sensitivities is sufficient to develop reasonable control laws.

8.9 Exercises

1. Consider the dynamic flash systems in Example 8.7.

   (a) Verify that the DAE system given by (8.40)–(8.41) is index 1.

   (b) Verify that the DAE systems (8.42) and (8.43) are index 2 and index 1, respectively.

   (c) Discuss likely differences in the solutions of (8.43) and (8.40)–(8.41).

2. Show that the solution profiles for Example 8.3 satisfy the optimality conditions (8.16).

3. Show that the solution profiles for Example 8.4 satisfy the optimality conditions (8.23).

4. Consider the two tanks in Figure 8.11 with levels $h_1(t)$ and $h_2(t)$ and cross-sectional areas A_1 and A_2, respectively. The inlet flow rate to the first tank $F_0 = g(t)$ and $F_2(t)$ is given by $C_v h_2^{1/2}$, where C_v is a valve constant.

   (a) Write the DAE model so that $F_1(t)$ is adjusted to keep $h_1 = 2h_2$ over time. What is the index of the DAE system?

   (b) Reformulate the system in part (a) so that consistent initial conditions can be specified directly.

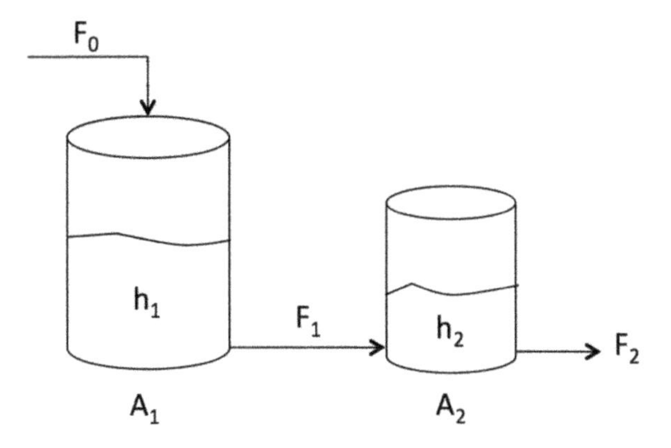

Figure 8.11. *DAE model for two tanks in problem* 4.

5. Verify that the solution profiles in Example 8.8 satisfy the Euler–Lagrange equations with state path inequalities.

6. For the singular control problem given by

$$\min \ \Phi(z(t)) \quad \text{s.t.} \quad \frac{dz(t)}{dt} = f_1(z) + f_2(z)u(t), \quad z(0) = z_0, \quad u(t) \in [u_L, u_U],$$

   show that $q \geq 2$ in (8.54).

7. Consider the mechanical system with time-varying forces, F and T. The system can be written as

$$\begin{aligned} m\frac{d^2x}{dt^2} &= -T(t)x(t), \\ m\frac{d^2y}{dt^2} &= F(t) - T(t)y(t), \\ x(t) = f(t), &\quad y(t) = g(t), \end{aligned}$$

   where m is the system mass, and x and y are the displacements in the horizontal and vertical directions.

   (a) Formulate this problem as a semiexplicit first order DAE system. What is the index of this system?

   (b) Reformulate this problem as an index-1 system using the index reduction approach presented in Section 8.4.1.

8. For the problem below, derive the two-point BVP and show the relationship of $u(t)$ to the state and adjoint variables:

$$\begin{aligned}
\min \quad & z_1(1)^2 + z_2(1)^2 \\
\text{s.t.} \quad & \frac{dz_1(t)}{dt} = -2z_2, \quad z_1(0) = 1, \\
& \frac{dz_2(t)}{dt} = z_1 u(t), \quad z_2(0) = 1.
\end{aligned}$$

9. Solve Example 8.9 for the case where the first reaction is irreversible ($k_2 = 0$). Show that this problem has a bang-bang solution.

Chapter 9

Dynamic Optimization Methods with Embedded DAE Solvers

This chapter develops dynamic optimization strategies that couple solvers for differential algebraic equations (DAEs) with NLP algorithms developed in the previous chapters. First, to provide a better understanding of DAE solvers, widely used Runge–Kutta and linear multistep methods are described, and a summary of their properties is given. A framework is then described for *sequential dynamic optimization* approaches that rely on DAE solvers to provide function information to the NLP solver. In addition, gradient information is provided by direct and adjoint sensitivity methods. Both methods are derived and described along with examples. Unfortunately, the sequential approach can fail on unstable and ill-conditioned systems. To deal with this case, we extend the sequential approach through the use of *multiple shooting* and concepts of boundary value solvers. Finally, a process case study is presented that illustrates both sequential and multiple shooting approaches.

9.1 Introduction

Chapter 8 provided an overview of the features of dynamic optimization problems and their optimality conditions. From that chapter, it is clear that the extension to larger dynamic optimization problems requires efficient and reliable numerical algorithms. This chapter addresses this question by developing a popular approach for the optimization of dynamic process systems. Powerful DAE solvers for large-scale initial value problems have led to widely used simulation environments for dynamic nonlinear processes. The ability to develop dynamic process simulation models naturally leads to their extension for optimization studies. Moreover, with the development of reliable and efficient NLP solvers, discussed in Chapter 6, an optimization capability can be implemented for dynamic systems along the lines of modular optimization modes discussed in Chapter 7. With robust simulation models, the implementation of NLP codes can be done in a reasonably straightforward way.

To develop sequential strategies that follow from the integration of DAE solvers and NLP codes, we consider the following DAE optimization problem:

$$\min \quad \varphi(\mathbf{p}) = \sum_{l=1}^{N_T} \Phi^l(z^l(t_l), y^l(t_l), p^l) \tag{9.1a}$$

$$\text{s.t.} \quad \frac{dz^l}{dt} = f^l(z^l(t), y^l(t), p^l), \quad z^l(t_{l-1}) = z_0^l, \tag{9.1b}$$

$$g^l(z^l(t), y^l(t), p^l) = 0, \quad t \in (t_{l-1}, t_l], \quad l = 1, \dots, N_T, \tag{9.1c}$$

$$p_L^l \le p^l \le p_U^l, \tag{9.1d}$$

$$y_L^l \le y^l(t_l) \le y_U^l, \tag{9.1e}$$

$$z_L^l \le z^l(t_l) \le z_U^l, \tag{9.1f}$$

$$h(p^1, \dots, p^{N_T}, z_0^1, z^1(t_1), z_0^2, z^2(t_2), \dots, z_0^{N_T}, z^{N_T}(t_{N_T})) = 0. \tag{9.1g}$$

This problem is related to problem (8.5) in Chapter 8. Again, we have differential variables $z(t)$ and algebraic variables $y(t)$ that appear in the DAE system (9.1b)–(9.1c) in semiexplicit form, and we assume that the invertibility of $g(\cdot, y(t), \cdot)$ permits an implicit elimination of the algebraic variables $y(t) = y[z(t), p]$. This allows us to consider the DAEs as equivalent ODEs. Moreover, while (9.1) still has N_T time periods, we no longer consider time-dependent bounds, or other path constraints on the state variables. Also, control profiles $u(t)$ are now represented as parameterized functions with coefficients that determine the optimal profile. Consequently, the decisions in (9.1) appear only in the time-independent vector p^l. Finally, algebraic constraints and terms in the objective function are applied only at the end of each period, t_l.

Problem (9.1) can be represented as the following nonlinear program:

$$\min \quad \varphi(\mathbf{p}) \tag{9.2a}$$

$$\text{s.t.} \quad c_E(\mathbf{p}) = 0, \tag{9.2b}$$

$$c_I(\mathbf{p}) \le 0, \tag{9.2c}$$

where $\mathbf{p} = [(p^1)^T, (p^2)^T, \dots, (p^{N_T})^T]^T \in \mathbb{R}^{n_p}$. The DAE system in (9.1b)–(9.1c) is solved in an inner loop, and all the constraint and objective functions are now implicit functions of $\mathbf{p}$.

Figure 9.1 provides a sketch of the sequential dynamic optimization strategy for problem (9.1). At a given iteration of the optimization cycle, decision variables p^l are specified by the NLP solver. With these values of p^l we treat the DAE system as an initial value problem and integrate (9.1b)–(9.1c) forward in time for periods $l = 1, \dots, N_T$. This integration provides the state profiles that determine the objective and constraint functions. The next component evaluates the gradients of the objective and constraint functions with respect to p^l. These are usually provided through the solution of DAE sensitivity equations. The function and gradient information is then passed to the NLP solver so that the decision variables can be updated. The cycle then continues with the NLP solver driving the convergence of problem (9.2).

While any of the NLP solvers from Chapter 6 can be applied to this strategy, it is important to note that problem (9.2) is relatively small and the constraint gradients are dense. For this approach, there are few opportunities for the NLP solver to exploit sparsity or structure of the dynamic model or the KKT system. Instead, the dominant calculation cost lies in the solution of the DAE system and the sensitivity equations. Consequently, SQP methods, with codes such as *NPSOL, NLPQL,* or *fmincon*, are generally well suited for this task, as they require relatively few iterations to converge.

This chapter provides the necessary background concepts that describe sequential optimization strategies. As discussed in Chapter 7, accurate function and gradient values

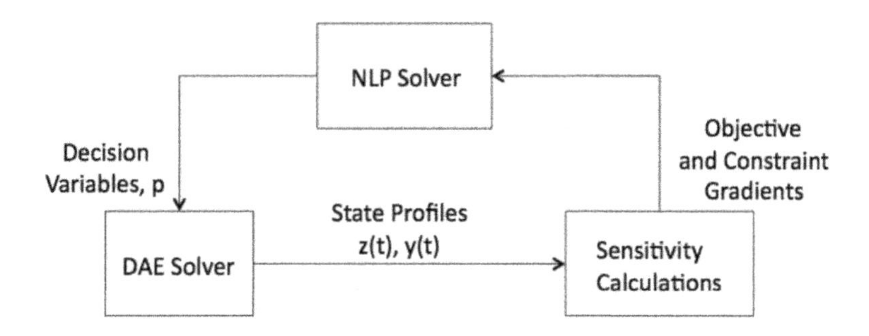

Figure 9.1. *Sketch of sequential dynamic optimization strategy.*

are required in order for the NLP code to perform reliably. This is enabled by state-of-the-art DAE solvers as well as the formulation and efficient solution of sensitivity equations. A successful sequential algorithm requires careful attention to the problem formulation, as failure of the DAE solver or the sensitivity component is fatal for the optimization loop. For instance, the sequential approach may fail for unstable systems, as the forward integration of the DAE may lead to unbounded state profiles. For these systems, failure can be avoided through the use of boundary value methods.

The next section provides a brief survey of DAE solvers and shows how they have been developed to solve challenging, large-scale (index-1) DAE systems. Section 9.3 describes the calculation of sensitivity information from the DAE system. Both *direct* and *adjoint* sensitivity equations are derived and relative advantages of both are discussed. The application of sensitivity calculations is also demonstrated on formulations of (9.1) that approximate the multiperiod optimal control problem (8.5) from the previous chapter. Next, Section 9.4 extends the sequential approach to multiple shooting formulations which deal with unstable dynamic process systems. In this section we outline methods for BVPs that allow the solution of dynamic systems that cause initial value DAE solvers to fail. Section 9.5 then presents a detailed case study that illustrates both sequential and multiple shooting optimization strategies.

9.2 DAE Solvers for Initial Value Problems

Numerical methods for the solution (or integration) of initial value problems have a history that extends back over a century and covers a very broad literature. This section provides a descriptive presentation that summarizes concepts and properties related to the most widely used DAE solvers. It also provides the background to understand their influence on the resulting optimization strategies. For simplicity of presentation, the methods will be developed for ODEs with initial values of the form

$$\frac{dz}{dt} = f(z, p), \quad z(0) = z_0. \tag{9.3}$$

As shown in Section 9.2.3, these methods can easily be extended to index-1 DAEs. Because of the invertibility of the algebraic system, algebraic variables can be represented as implicit functions of differential variables. We also note that semiexplicit high-index DAEs can be reformulated to index 1 as demonstrated in Chapter 8.

ODE solvers for initial value problems proceed in increments of time $t_{i+1} = t_i + h_i$, $h_i > 0$, $i = 0, \ldots, N$, by approximating the ODE solution, $z(t_i)$, by z_i at each time point. Discretization of the ODE model leads to difference equations of the form

$$z_{i+1} = \phi(z_{i-j+1}, f_{i-j+1}, h_i;\ j = j_0, \ldots, n_q) \tag{9.4}$$

with the approximated state z_{i+1} based on previous values, z_{i-j} and $f_{i-j} = f(z_{i-j}, p)$. With $j_0 = 1$, we have an explicit solution strategy, and z_{i+1} can be determined directly. Setting $j_0 = 0$ leads to an implicit strategy which requires an iterative solution for z_{i+1}.

Initial value solvers can be classified into two types. *Single-step methods* take the form (9.4) with $n_q = 1$; these are typically Runge–Kutta methods with a (possibly variable) time step h_i. *Multistep methods* require $n_q > 1$ in (9.4) and therefore contain some history of the state profile. Here, the difference equations are derived so that all of the time steps h_i in (9.4) are the same size for $i = 1, \ldots, n_q$. Both types of methods need to satisfy the following properties:

- The difference equation must be consistent. Rewriting (9.4) as

$$\frac{z_{i+1} - z_i}{h_i} - \bar{\phi}(z_{i-j+1}, f_{i-j+1}, h_i) = 0 \tag{9.5}$$

  and substituting the true solution $z(t)$ into (9.5) at time points t_i, leads to the *local truncation error*:

$$d_i(z) = (z(t_{i+1}) - z(t_i))/h_i - \bar{\phi}(z(t_{i-j+1}), f(z(t_{i-j+1}), p), h_i) = O(h_i^q) \tag{9.6}$$

  as $h_i \to 0$. This condition corresponds to *consistency of order* q.

- The approximate solution trajectory must also satisfy the *zero-stability* property

$$\|z_i - \bar{z}_i\| \le K\{\|z_0 - \bar{z}_0\| + \max_{i \in \{1,N\}} \|d_i(z) - d_i(\bar{z})\|\}, \quad i \in \{1, N\}$$

  for some constant $K > 0$, for all $h_i \in [0, h_0]$ and for two neighboring solutions z_i and $\bar{z}_i$. This property implies bounded invertibility of the difference formula (9.4) at all t_i and time steps $0 \le h_i \le h_0$.

- Consistency and zero-stability imply a *convergent method of order* q.

These conditions govern the choice of time steps h_i that are needed to obtain accurate and stable solutions to (9.3). Certainly h_i must be chosen so that the local error is small relative to the solution profile. In addition, the stability condition can also limit the choice of time step, and the stability condition can be violated even when the local truncation error is small.

If a small time step is dictated by the stability condition and not the accuracy criterion, then we characterize the DAEs as a *stiff system*. For example, the scalar ODE

$$\frac{dz}{dt} = -10^6(z - \cos(t))$$

has a very fast transient, after which the solution essentially becomes $z(t) = \cos(t)$. However, while a reasonably large time step ensures an accurate approximation to $\cos(t)$, all explicit methods (and many implicit ones) are stable only if h_i remains small, i.e., $h_i = O(10^{-6})$. Stiff systems usually arise in dynamic processes with multiple time scales, where the fastest component dies out quickly but the step restriction still remains to prevent an unstable integration.

Because the zero-stability property does not directly determine appropriate time steps for h_i, more specific stability properties have been established with scalar homogeneous and nonhomogeneous test ODEs

$$\frac{d\zeta}{dt} = \lambda \zeta, \tag{9.7a}$$

$$\frac{d\zeta}{dt} = \lambda(\zeta - \gamma(t)), \tag{9.7b}$$

respectively, where λ is a complex constant, and $\gamma(t)$ is a forcing function. It can be argued that these test equations provide a reasonable approximation to (9.3) after linearization about a solution trajectory, and decomposition into orthogonal components. For these test equations we define the following properties:

- For a specified value of $h_i = h$, the difference formula (9.4) is *absolutely stable* if for all $i = 1, 2, \ldots$, it satisfies $|\zeta_{i+1}| \leq |\zeta_i|$ for (9.7a).
- If the difference formula (9.4) is *absolutely stable* for all $\mathrm{Re}(h\lambda) \leq 0$, then it is *A-stable*. This is an especially desirable property for stiff systems, as h is no longer limited by a stability condition.
- A slightly stronger property holds if we replace λ in (9.7a) by $\lambda(t)$, and define analogous regions of absolute stability, $|\zeta_{i+1}| \leq |\zeta_i|$. The system is characterized as *AN-stable*, or equivalently *algebraically stable*, if absolute stability holds for all $\mathrm{Re}(h\lambda(t)) < 0$.
- The difference formula (9.4) has *stiff decay* if $|\zeta_i - \gamma(t_i)| \to 0$ as $\mathrm{Re}(h\lambda) \to -\infty$ for the test equation (9.7b).

9.2.1 Runge–Kutta Methods

Runge–Kutta methods are single-step methods that take the form

$$z_{i+1} = z_i + \sum_{k=1}^{n_s} b_k f(t_i + c_k h_i, \hat{z}_k), \tag{9.8a}$$

$$\hat{z}_k = z_i + \sum_{j=1}^{n_{rk}} a_{kj} f(t_i + c_j h_i, \hat{z}_j), \quad k = 1, \ldots, n_s, \tag{9.8b}$$

where n_s is the number of stages, and intermediate stage variables are given by $\hat{z}_k$ for the profile z_i. For the summation in (9.8b), explicit Runge–Kutta methods are characterized by $n_{rk} < k$, semi-implicit methods have $n_{rk} = k$, and fully implicit methods have $n_{rk} = n_s$.

Note that all of the evaluations occur within the time step with $c_k \in [0,1]$, and the method is characterized by the matrix A and vectors b and c, commonly represented by the Butcher block:

$$\begin{array}{c|c} c & A \\ \hline & b^T \end{array} \tag{9.9}$$

The coefficients satisfy the relations $\sum_{k=1}^{n_s} b_k = 1$ and $\sum_{j=1}^{n_s} a_{kj} = c_k$, and they are derived from Taylor series expansions of (9.8a)–(9.8b). These single-step methods have a number of interesting features and advantages:

- Because Runge–Kutta methods are single step, adjustment, of the time step h_i is easy to implement and independent of previous profile information.
- The ability to control the time step directly allows accurate location of nonsmooth events in the state profiles. Because of the single-step nature, accurate solutions require that the state profile be smooth only within a step, and continuous across steps.
- Explicit methods have bounded, well-defined stability regions that increase in size with order q. These are determined by applying (9.8) to the test equation (9.7a) and determining where absolute stability holds for a grid of $h\lambda$ values in the complex plane.
- *A-stable implicit Runge–Kutta methods can be found for any order q.* Implicit methods with A nonsingular, and b^T equal to the last row of A, also satisfy the stiff decay property. Consequently, these implicit methods are well suited for stiff problems.
- Calculation of the local truncation error (9.6) is generally not straightforward for Runge–Kutta methods. The local error is often estimated by embedding a high-order Runge–Kutta method with a lower order method, using the same c and A values for both methods, and different b values. This allows two methods to be executed in parallel, and values of z_{i+1} to be compared, without additional calculation. The popular RKF45 formulae (see [16]) are based on such an embedded method.
- For an explicit Runge–Kutta method, increasing the order q of the approximation error tends to be expensive, as $n_s \geq q$.
- Most semi-implicit methods are of order $n_s + 1$, and fully implicit methods have orders up to $2n_s$. Of course, implicit methods tend to be more expensive to apply than explicit ones.

9.2.2 Linear Multistep Methods

As illustrated in Figure 9.2, linear multistep methods are based on the following form:

$$z_{i+1} = \sum_{j=0}^{n_s} \alpha_j z_{i-j+1} + \sum_{j=0}^{n_s} \beta_j h f_{i-j+1}. \tag{9.10}$$

These methods are explicit if $\alpha_0 \neq 0$ or $\beta_0 \neq 0$. Popular linear multistep methods usually take two forms. First, the *Adams* form has $\alpha_1 = 1$, $\alpha_0 = 0$, $\alpha_j = 0, j = 1,\ldots,n_s$,

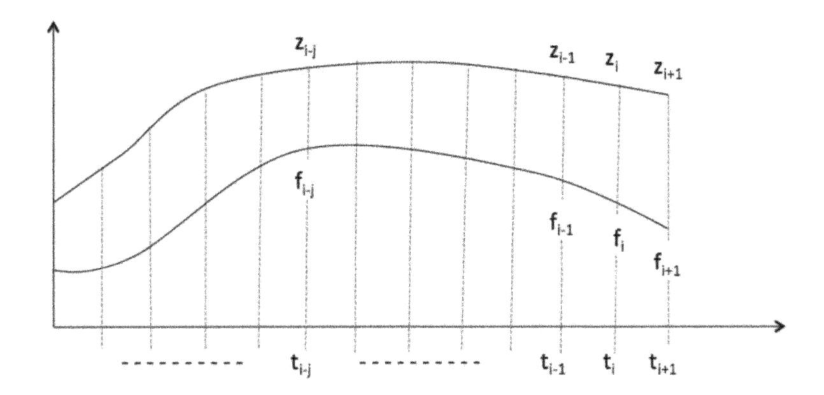

Figure 9.2. *Sketch of linear multistep strategy.*

with nonzero values of β_j. On the other hand, the *Gear* or *backward difference formula* (BDF) is implicit, and has $\alpha_0 = 0, \beta_j = 0, j = 1,\dots,n_s$, with nonzero values for β_0 and $\alpha_j, j = 1,\dots,n_s$. Derivation of these methods is based on approximating the profile history for both z_i and $f(z_i, p)$ with an interpolating polynomial, which is then used to carry the integration forward. In particular, with z_{i-j} and f_{i-j} calculated from n_s previous steps, the coefficients α_j and β_j are derived by integrating, from t_i to t_{i+1}, an interpolating polynomial for $f(z(t), p)$ that is represented by the Newton divided difference formula [16]:

$$\begin{aligned} f(z(t), p) \approx f_{n_s}(t) = f[t_i] + f[t_i, t_i](t - t_i) + \cdots \\ + f[t_i, t_{i-1}, t_{i-n_s}](t - t_i)(t - t_{i-1}) \cdots (t - t_{i-n_s+1}), \end{aligned} \tag{9.11}$$

where $h = t_i - t_{i-1}$, $f[t_i] = f(z_i, p)$, and

$$f[t_i, \dots, t_{i-j}] = \frac{f[t_i, \dots, t_{i-j-1}] - f[t_{i-1}, \dots, t_{i-j}]}{t_i - t_{i-1}}. \tag{9.12}$$

For this interpolating polynomial, the interpolation error is given by

$$f(z(t), p) - f_{n_s}(t) = f[t_i, \dots, t_{i-n_s}, t] \prod_{j=0}^{n_s} (t - t_{i-j}) = O(h^{n_s+1}), \tag{9.13}$$

and if $f(z(t), p)$ has bounded derivatives, then for small h,

$$(n_s + 1)! f[t_i, \dots, t_{i-n_s}, t] \approx \frac{d^{n_s+1} f}{dt^{n_s+1}}.$$

A key advantage of linear multistep methods is that higher order methods can be applied simply by considering a longer profile history, and this requires no additional calculation. On the other hand, multistep methods are not self-starting, and single-step methods must be used to build up the n_s-step history for the polynomial representation (9.10). Moreover, nonsmooth events due to discontinuous control profiles, or a change in period, also require the profile history to be restarted. If these events occur frequently, repeated construction of profile histories may lead to an inefficient low-order method. Finally, the derivation is based on an interpolating polynomial with a constant time step h. Consequently, adjusting h to satisfy an error tolerance affects the sequence of n_s steps and requires a new polynomial interpolation and reconstruction of the profile history.

The explicit and implicit Adams methods can be characterized by the following properties:

- Derived from interpolation with the Newton divided difference formula, the explicit *Adams–Bashforth* method

$$z_{i+1} = z_i + \sum_{j=1}^{n_s} \beta_j h f_{i-j+1} \tag{9.14}$$

has order $q = n_s$, while the implicit *Adams–Moulton* method

$$z_{i+1} = z_i + \sum_{j=0}^{n_s} \beta_j h f_{i-j+1} \tag{9.15}$$

has order $q = n_s + 1$. Moreover, the local truncation term can be estimated with the aid of the residual of the polynomial interpolant (9.13).

- Combining both implicit and explicit formulae leads to a predictor-corrector method, where the explicit formula provides a starting guess for z_{i+1} used in the implicit formula. For h suitably small, successive substitution of z_{i+1} into $f(z, p)$ and updating z_{i+1} with (9.15) leads to a cheap fixed-point convergence strategy.

- Regions of absolute stability tend to be smaller than for single-step methods. Moreover, these regions *shrink in size* with increasing order. Hence, smaller steps are required to satisfy stability requirements. This arises because the associated difference equation derived from (9.10) has multiple roots; all of these roots must satisfy the absolute stability criterion.

- There are no explicit linear multistep methods that are A-stable. There are no implicit linear multistep methods above order 2 that are A-stable.

- Because of the stability limitations on h (and also because predictor-corrector methods converge only for small h), Adams methods should not be used for stiff problems.

Stiff problems are handled by BDF methods. While these methods are not A-stable above order 2, they do have *stiff decay* up to order 6. Moreover, even for order 6, the regions of absolute stability cover most of the left half plane in the phase space of $h\lambda$. The only excluded regions are near the imaginary axis with $|\text{Im}(h\lambda)| > 1$ for $\text{Re}(h\lambda) > -5$. These regions correspond to oscillatory solutions that normally require small steps to satisfy accuracy requirements anyway. Consequently, large time steps h can then be selected once fast transients have died out. The BDF method has the following features:

- The BDF is given by

$$z_{i+1} - \sum_{j=1}^{n_s} \alpha_j z_{i-j+1} - \beta_0 h f_{i+1} = \phi(z_{i+1}) = 0 \tag{9.16}$$

with order $q = n_s$. Note that f_{i+1} is an implicit function of z_{i+1}.

- Because large steps are taken, z_{i+1} is determined from (9.16) using the Newton iteration,

$$z_{i+1}^{m+1} = z_{i+1}^{m} - J(z_{i+1}^{m})^{-1}\phi(z_{i+1}^{m}),$$

where m is the iteration counter. For large systems, calculation of the Newton step can be the most time-consuming part of the integration.

- The Jacobian is given by $J(z_{i+1}) = I - h\beta_0 \frac{\partial f}{\partial z}^T$ and $\frac{\partial f}{\partial z}$ is generally required from the dynamic model. Note that $J(z)$ is nonsingular for h sufficiently small. To save computational expense, a factorized form of $J(z)$ is used over multiple Newton iterations and time steps, t_{i+j}. It is refactorized only if the Newton iteration fails to converge quickly.

9.2.3 Extension of Methods to DAEs

Both single-step and multistep methods can be extended in a straightforward manner to systems of semiexplicit, index-1 DAEs. Since we require invertibility of the algebraic equations, the simultaneous solution of $z(t)$ and $y(t)$ is often more efficient when combined with implicit ODE formulae.[5]

For semiexplicit DAEs of the form (9.1b)–(9.1c), single-step methods can be extended to the form

$$z_{i+1} = z_i + \sum_{k=1}^{n_s} b_k f(t_i + c_k h_i, \hat{z}_k, \hat{y}_k), \tag{9.17a}$$

$$\hat{z}_k = z_i + \sum_{j=1}^{n_{rk}} a_{kj} f(t_i + c_j h_i, \hat{z}_j, \hat{y}_j), \quad k = 1, \ldots, n_s, \tag{9.17b}$$

$$0 = g(\hat{z}_k, \hat{y}_k), \tag{9.17c}$$

where $\hat{z}_k$ and $\hat{y}_k$ are intermediate stage variables and $n_{rk} \geq k$ is defined from (9.7b). If the DAE is index 1, the algebraic equations can be eliminated with the algebraic variables determined implicitly. Consequently, the single-step formula generally has the same order and stability properties as with ODEs. This is especially true for methods with stiff decay. A more general discussion of order reduction and stability limitations of (9.17) for Runge–Kutta methods and high-index DAEs can be found in [70].

Since predictor-corrector Adams methods require explicit formulae, we do not consider them for DAEs and focus instead on the BDF method, which is the most popular method for DAE systems. The extension of the BDF method to semiexplicit index-1 DAEs is given by

$$z_{i+1} = \sum_{j=1}^{n_s} \alpha_j z_{i-j+1} + \beta_0 h f(z_{i+1}, y_{i+1}), \tag{9.18a}$$

$$g(z_{i+1}, y_{i+1}) = 0. \tag{9.18b}$$

[5] There seem to be no explicit DAE solvers, although half-explicit Runge–Kutta methods have been proposed [16].

Again, as the last equation leads to implicit solution of the algebraic equations, the extended BDF formula has the same order and stability properties as with ODEs. More discussion of order reduction and stability limitations of (9.17) for linear multistep methods and high-index DAEs can be found in [70].

Also, the BDF equations (9.18) are also solved with Newton's method using the Jacobian given by

$$J(z_{i+1}, y_{i+1}) = \begin{bmatrix} I - h\beta_0 \frac{\partial f}{\partial z}^T & h\beta_0 \frac{\partial f}{\partial y}^T \\ \frac{\partial g}{\partial z}^T & \frac{\partial g}{\partial y}^T \end{bmatrix}. \tag{9.19}$$

For semiexplicit index-1 systems, $\frac{\partial g}{\partial y}$ is always nonsingular and therefore $J(z, y)$ remains nonsingular for sufficiently small values of h. Since calculation of the Newton step is an expensive part of the DAE solution, a factorized form of $J(z, y)$ is often carried over multiple Newton iterations and time steps, t_{i+j}. It is refactorized only if the Newton iteration experiences convergence difficulties.

9.3 Sensitivity Strategies for Dynamic Optimization

The third component in Figure 9.1 deals with efficient and accurate gradient calculations for the NLP solver. For this task, derivatives need to be calculated implicitly from the DAE model, often in tandem with the solution of the DAEs. This section develops the approaches for sensitivity with respect to decisions $\mathbf{p}$ in (9.2). Since optimal control problems are already parameterized, this approach is easily generalized to problem (9.1).

Calculation of gradients with DAE models can be done in three ways: perturbation, direct sensitivity, and adjoint sensitivity. Perturbation is the easiest to implement but may lead to unacceptable errors in the gradients. For (9.2) solved by an NLP solver at a given iteration, we can apply a forward difference perturbation to the vector $\mathbf{p}$ and define

$$\mathbf{p}_j = \mathbf{p} + \delta e_j, \quad j = 1, n_p,$$

where the jth element of vector e_j is 1 and the other elements are zero, and δ is a small perturbation size. Solving (9.1b)–(9.1c) with values of $\mathbf{p}_j$ leads to objective and constraint values:

$$\Psi(\mathbf{p}_j) = [\varphi(\mathbf{p}_j),\, c_E(\mathbf{p}_j)^T,\, c_I(\mathbf{p}_j)^T], \tag{9.20}$$

and the approximate gradient

$$\frac{1}{\delta}(\Psi(\mathbf{p}_j) - \Psi(\mathbf{p})) = \nabla_{\mathbf{p}_j} \Psi(\mathbf{p}) + O(\delta) + O(\delta^{-1}).$$

As noted in Chapter 7, gradients evaluated by perturbation are often plagued by truncation errors and roundoff errors. Truncation errors result from neglect of higher order Taylor series terms that vanish as $\delta \to 0$. On the other hand, roundoff errors that result from internal calculation loops (i.e., convergence noise from the DAE solver) are independent of δ and lead to large gradient errors as $\delta \to 0$. While δ can be selected carefully and roundoff errors can be controlled, the gradient error cannot be eliminated completely. As seen in Chapter 7, this can lead to inefficient and unreliable performance by the NLP solver. For the rest of this section, we therefore focus on two sensitivity approaches that provide gradients with the same level of accuracy as the state profiles.

9.3.1 Direct Sensitivity Calculations

Consider the semiexplicit, index-1 DAE system given in (9.1b)–(9.1c) and only one time period in (9.1) written over $t \in [0, t_f]$. Assuming that the functions $f(\cdot)$ and $g(\cdot)$ are sufficiently smooth in all of their arguments, we formally differentiate the DAE system with respect to the decisions p as follows:

$$\frac{d}{dp}\left\{ \begin{array}{c} \frac{dz}{dt} = f(z(t), y(t), p), \quad z(0) = z_0(p) \\ g(z(t), y(t), p) = 0 \end{array} \right\}. \tag{9.21}$$

Since $z(t)$ is sufficiently smooth in t and p, we can exchange the order of differentiation, define the sensitivity matrices $S(t) = \frac{dz}{dp}^T$ and $R(t) = \frac{dy}{dp}^T$, and obtain the following sensitivity equations:

$$\frac{dS}{dt} = \frac{\partial f}{\partial z}^T S(t) + \frac{\partial f}{\partial y}^T R(t) + \frac{\partial f}{\partial p}^T, \quad S(0) = \frac{\partial z_0}{\partial p}^T, \tag{9.22a}$$

$$0 = \frac{\partial g}{\partial z}^T S(t) + \frac{\partial g}{\partial y}^T R(t) + \frac{\partial g}{\partial p}^T. \tag{9.22b}$$

Note that for $n_z + n_y$ states and n_p decision variables, we have $(n_z + n_y) \times n_p$ sensitivity equations. These equations are linear time variant (LTV), index-1 DAEs and require the state profile solution in their right-hand sides. Also, decision variables that arise in the initial conditions also influence the initial conditions of the sensitivity equations. Once the sensitivities are determined, we calculate the gradients for the objective and constraint functions in (9.2) by using (9.20) and the chain rule to yield

$$\nabla_p \Psi = S(t_f)^T \frac{\partial \Psi}{\partial z} + R(t_f)^T \frac{\partial \Psi}{\partial y} + \frac{\partial \Psi}{\partial p}$$

The formulation of the sensitivity equations is illustrated with the following example.

Example 9.1 Consider the following DAEs:

$$\begin{aligned} \frac{dz_1}{dt} &= (z_1)^2 + (z_2)^2 - 3y(t), \quad z_1(0) = 5, \\ \frac{dz_2}{dt} &= z_1 z_2 + z_1(y(t) + p_2), \quad z_2(0) = p_1, \\ 0 &= z_1 y(t) + p_3 z_2. \end{aligned}$$

Defining $\frac{dS_{ij}}{dt} = \dot{S}_{ij}$ for z_i and p_j, the corresponding sensitivity equations (9.22) are given by

$$\begin{bmatrix} \dot{S}_{11} & \dot{S}_{12} & \dot{S}_{13} \\ \dot{S}_{21} & \dot{S}_{22} & \dot{S}_{23} \end{bmatrix} = \begin{bmatrix} 2z_1 & 2z_2 \\ z_2 + y + p_2 & z_1 \end{bmatrix} S(t) + \begin{bmatrix} -3 \\ z_1 \end{bmatrix} R(t) + \begin{bmatrix} 0 & 0 & 0 \\ 0 & z_1 & 0 \end{bmatrix},$$

$$\begin{bmatrix} S_{11}(0) & S_{12}(0) & S_{13}(0) \\ S_{21}(0) & S_{22}(0) & S_{23}(0) \end{bmatrix} = \begin{bmatrix} 0 & 0 & 0 \\ 1 & 0 & 0 \end{bmatrix},$$

$$z_1 [\, R_1 \quad R_2 \quad R_3 \,] + [y(t) \;\; p_3] S(t) + [0 \;\; 0 \;\; z_2] = 0.$$

With $n_z = 2$, $n_y = 1$, and $n_p = 3$, the original DAE system has two differential equations and one algebraic equation. The $n_p(n_z + n_y)$ sensitivity equations consist of six differential equations, with initial conditions, and three algebraic equations. ■

To avoid the need to store the state profiles, sensitivity equations are usually solved simultaneously with the state equations. Because of their desirable stability properties, BDF methods are often adapted to the solution of the combined state-sensitivity system. This implicit method applies Newton's method to determine the states at the next time step. With the addition of the sensitivity equations, the same Jacobian

$$J(z,y) = \begin{bmatrix} I - h\beta_0 \frac{\partial f}{\partial z}^T & h\beta_0 \frac{\partial f}{\partial y}^T \\ \frac{\partial g}{\partial z}^T & \frac{\partial g}{\partial y}^T \end{bmatrix} \tag{9.23}$$

can be used to solve them as well, due to the structure of (9.22). This leads to a number of ways to automate the sensitivity calculation and make it more efficient. Moreover, a number of large-scale DAE solvers takes advantage of sparsity and special structures (e.g., bandedness) of the Jacobian.

Sensitivity methods can be classified into the *staggered direct* [85], *simultaneous corrector* [272], and *staggered corrector* options [128]. At each time step, the simultaneous corrector method solves the entire dynamic system of state (9.1b)–(9.1c) and sensitivity equations (9.22). Instead, the staggered methods first solve the DAEs for the state variables at each time step. Here the accuracy of the state and the sensitivity equations must be checked separately, and a shorter step may be required if the sensitivity error is not acceptable. After the Newton method has converged for the state variables, sensitivity equations are solved at the current step. In solving the sensitivity equations, the staggered direct method updates the factorization of the Jacobian (9.23) at every step, while the staggered corrector refactorizes this Jacobian matrix only when necessary. Since this is often the most expensive step in the sensitivity algorithm, considerable savings can be made. More details on these three options can be found in [258]. The partial derivatives in the sensitivity equations can be determined either by a finite difference approximation or through automatic differentiation. Finally, several automatic differentiation tools [58, 306, 173] are available to provide exact partial derivatives. When implemented with BDF solvers, they lead to faster and more accurate sensitivities and fewer Newton solver failures.

9.3.2 Adjoint Sensitivity Calculations

While the direct sensitivity approach has been implemented in a number of dynamic optimization applications, it may be inefficient when n_p is large. A complementary approach that handles many decision variables can be derived based on the variational approach considered in Chapter 8.

For the current set of decisions p and state variables $z(t)$ and $y(t)$, we again derive sensitivities based on perturbations about these profiles. However, here we consider the sensitivity of the objective and constraint functions separately. As defined in (9.20), we denote ψ as an element of the n_Ψ-vector $\Psi(\mathbf{p})$, which represents either an objective or constraint function at the end of the time period.

We adjoin the DAEs and define *adjoint variables* as follows:

$$\psi(z(t),y(t),p) \equiv \psi(z(t_f),p) + \int_0^{t_f} \left[\lambda(t)^T \left(f(z(t),y(t),p) - \frac{dz}{dt}(t) \right) \right. \\ \left. + \nu^T g(z(t),y(t),p) \right] dt. \tag{9.24}$$

Note that the differential and algebraic adjoints, $\lambda(t)$ and $\nu(t)$, respectively, serve as multipliers to influence $\psi(z(t_f),p)$. Applying integration by parts to $\int \lambda(t)^T \frac{dz}{dt}(t)dt$ and substituting into (9.24) leads to the equivalent expression

$$\psi(z(t),y(t),p) \equiv \psi(z(t_f),p) - z(t_f)^T \lambda(t_f) + z(0)^T \lambda(0) \\ + \int_0^{t_f} \left[\lambda(t)^T f(z(t),y(t)p) + z(t)^T \frac{d\lambda}{dt}(t) + \nu^T g(z(t),y(t),p) \right] dt. \tag{9.25}$$

As in Chapter 8, we apply perturbations $\delta z(t)$, $\delta y(t)$, dp about the current point. Applying these perturbations to (9.25) leads to

$$d\psi = \left[\frac{\partial \psi}{\partial z} - \lambda(t_f) \right]^T \delta z(t_f) + \lambda(0)^T \delta z(0) + \frac{\partial \psi}{\partial p}^T dp \\ + \int_0^{t_f} \left[\frac{\partial f}{\partial z}\lambda + \frac{d\lambda}{dt} + \frac{\partial g}{\partial z}\nu \right]^T \delta z(t) + \left[\frac{\partial f}{\partial y}\lambda + \frac{\partial g}{\partial y}\nu \right]^T \delta y(t) \\ + \left[\frac{\partial f}{\partial p}\lambda + \frac{\partial g}{\partial p}\nu \right]^T dp\, dt. \tag{9.26}$$

We now define the adjoint variables so that only the perturbations dp have a direct influence on $d\psi$:

1. From perturbation of the final state, $\delta z(t_f)$, we have

$$\lambda(t_f) = \frac{\partial \psi}{\partial z}. \tag{9.27}$$

   This leads to a boundary or *transversality* condition for $\lambda(t)$.

2. From perturbation of the differential state, $\delta z(t)$, we have

$$\frac{d\lambda}{dt} = -\frac{\partial f}{\partial z}\lambda - \frac{\partial g}{\partial z}\nu \tag{9.28}$$

   which yields a differential equation for $\lambda(t)$.

3. From perturbation of the algebraic state, $\delta y(t)$, we have

$$\frac{\partial f}{\partial y}\lambda + \frac{\partial g}{\partial y}\nu = 0, \tag{9.29}$$

   which leads to an algebraic equation with the algebraic variable $\nu(t)$.

4. We assume that the perturbation of the initial state, $\delta z(0)$, is governed by the variable p in the optimization problem, $z(0) = z_0(p)$, and we define

$$\lambda(0)^T \delta z(0) = \lambda(0)^T \frac{\partial z_0}{\partial p}^T dp$$

and group this term with dp.

By eliminating the state perturbation terms and suitably defining the adjoint variables above, (9.26) is now simplified as follows:

$$d\psi = \left\{ \left[\frac{\partial \psi}{\partial p} + \frac{\partial z_0}{\partial p} \lambda(0) \right]^T + \int_0^{t_f} \left[\frac{\partial f}{\partial p} \lambda + \frac{\partial g}{\partial p} \nu \right]^T dt \right\} dp. \tag{9.30}$$

Calculation of adjoint sensitivities requires the solution of a DAE system with final values specified for the differential variables. Note that if the state equations are semiexplicit, index 1, then the adjoint system is also semiexplicit and index 1. Once the state and adjoint variables are obtained, the integrals allow the direct calculation of the gradients $\nabla_p \Psi$.

We now revisit Example 9.1 and contrast the adjoint approach with direct sensitivity.

Example 9.2 For the DAEs

$$\begin{aligned}
\frac{dz_1}{dt} &= (z_1)^2 + (z_2)^2 - 3y, \quad z_1(0) = 5, \\
\frac{dz_2}{dt} &= z_1 z_2 + z_1(y + p_2), \quad z_2(0) = p_1, \\
0 &= z_1 y + p_3 z_2,
\end{aligned}$$

we write

$$\begin{aligned}
\lambda(t)^T f(z, y, p) + \nu(t)^T g(z, y, p) = \lambda_1((z_1)^2 + (z_2)^2 - 3y(t)) \\
+ \lambda_2(z_1 z_2 + z_1(y(t) + p_2)) + \nu(z_1 y(t) + p_3 z_2).
\end{aligned}$$

Using the above adjoint equations, we have from (9.28) and (9.27),

$$\frac{d\lambda_1}{dt} = -(2z_1\lambda_1 + (z_2 + y + p_2)\lambda_2 + y\nu), \quad \lambda_1(t_f) = \frac{\partial \psi(z(t_f))}{\partial z_1}, \tag{9.31}$$

$$\frac{d\lambda_2}{dt} = -(2z_2\lambda_1 + z_1\lambda_2 + p_3\nu), \quad \lambda_2(t_f) = \frac{\partial \psi(z(t_f))}{\partial z_2}. \tag{9.32}$$

From (9.29) we have the algebraic equation

$$0 = -3\lambda_1 + z_1\lambda_2 + z_1\nu. \tag{9.33}$$

Solving this system of DAEs leads to the adjoint profiles $\lambda(t)$ and $\nu(t)$. The gradients are then obtained from (9.30) as follows:

$$\begin{aligned}
\nabla_{p_1} \psi &= \frac{\partial \psi}{\partial p_1} + \lambda_2(0), \\
\nabla_{p_2} \psi &= \frac{\partial \psi}{\partial p_2} + \int_0^{t_f} z_1 \lambda_2 \, dt, \\
\nabla_{p_3} \psi &= \frac{\partial \psi}{\partial p_3} + \int_0^{t_f} z_2 \nu \, dt.
\end{aligned}$$

The original DAE system has two differential equations and one algebraic equation with $n_z = 2$, $n_y = 1$, and $n_p = 3$. In addition, we have n_Ψ objective and constraint functions at final time. Therefore, the adjoint sensitivities require the solution of $n_\Psi(n_z + n_y)$ DAEs and $n_\Psi(n_p)$ integrals (which are generally less expensive to solve). From this example, we see that if $n_p > n_\Psi$, then the adjoint approach is more efficient than solving the direct sensitivity equations. ∎

On the other hand, the adjoint sensitivity approach is more difficult to implement than direct sensitivity; hence, there are fewer adjoint sensitivity codes [259, 197] than direct sensitivity codes and the adjoint approach is less widely applied, except for systems with many decision variables. Solution of the adjoint equations requires storage of state profiles, and retrieval is needed for the backward integration. To avoid the storage requirement, especially for large systems, an adjoint approach is usually implemented with a *checkpointing* scheme. Here the state variables are stored at only a few *checkpoints* in time. Starting from the checkpoint closest to t_f, the state is reconstructed by integrating forward from this checkpoint, while the adjoint variable is calculated by integrating backward up to this checkpoint. Once the adjoint is calculated at this point, we back up to an earlier checkpoint and the state and adjoint calculation repeats until $t = 0$. The checkpointing scheme offers a trade-off between repeated adjoint calculation and state variable storage. Moreover, strategies have been developed for the optimal distribution of checkpoints [172] that lead to very efficient adjoint sensitivity strategies.

9.3.3 Evolution to Optimal Control Problems

With the development of accurate sensitivity strategies for the DAE system, we now apply the strategy shown in Figure 9.1 for time-independent decision variables. To illustrate this approach, we consider a small parameter estimation example for a batch reactor.

Example 9.3 The parameter estimation problem is formulated as follows:

$$\min \quad \varphi(p) = \sum_{l=1}^{N_D} (x(t_l) - \bar{x}_l)^T W (x(t_l) - \bar{x}_l) \tag{9.34a}$$

$$\text{s.t.} \quad \frac{dz(t)}{dt} = f(z(t), y(t), p), \quad z(0) = z_0, \tag{9.34b}$$

$$0 = g(z(t), y(t), p), \tag{9.34c}$$

$$p^L \leq p \leq p^U, \tag{9.34d}$$

where $\bar{x}_l$ are N_D sets of experimental measurements obtained at time t_l, and $x(t_l)^T = [z(t_l)^T \; y(t_l)^T]$ are the corresponding calculated differential and algebraic states. W is the weighting matrix for the least squares function; it is usually set to the inverse of the covariance matrix of the measurements. Parameter estimation problems are typically solved using the scheme in Figure 9.1 and are described in a large number of studies (e.g., [295, 85, 23, 49]). This formulation is a special case of the multiperiod problem (9.1) where the data sets define the periods, the state variables remain continuous over all periods, and the decisions p are the same in all periods.

NLP algorithms for parameter estimation frequently apply the Gauss–Newton assumption. Here we consider a Newton-type method with optional bounds where the calculation of the (reduced) Hessian is simplified by assuming that the residual terms $x(t_l) - \bar{x}_l \approx$

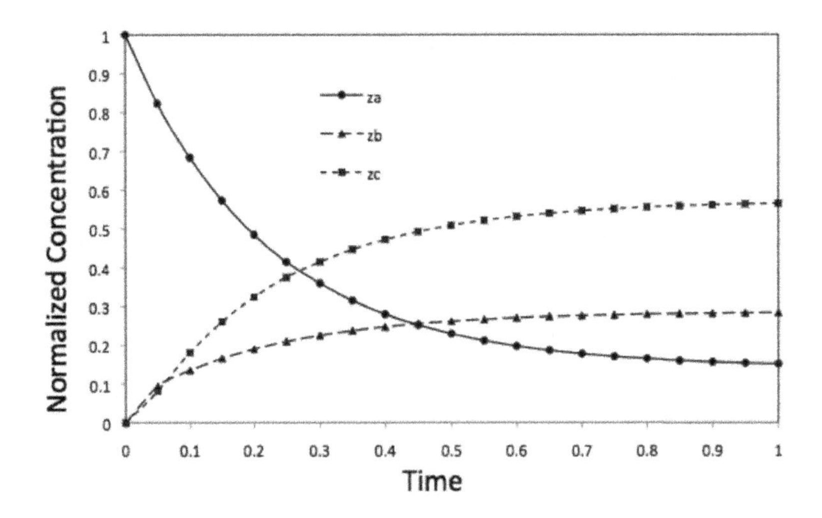

Figure 9.3. *Batch reactor data in Example* 9.3.

0 at the solution, and can be neglected. This simplification leads to the Hessian approximation

$$\frac{d^2\varphi(p)}{dp^2} \approx \sum_{l=1}^{N_D} \frac{dx(t_l)}{dp} W \frac{dx(t_l)}{dp}^T = \sum_{l=1}^{N_D} [S(t_l)^T \, R(t_l)^T] W \begin{bmatrix} S(t_l) \\ R(t_l) \end{bmatrix}. \tag{9.35}$$

where $S(t_l) = \frac{dz(t_l)}{dp}^T$ and $R(t_l) = \frac{dy(t_l)}{dp}^T$.

With the assumption of small residuals, this simplification leads to a quadratically convergent method while calculating only first derivatives. Should this assumption be violated, then the Gauss–Newton method has only a linear rate of convergence.

To illustrate, we consider the batch reactor data in Figure 9.3 for the first order reversible reactions:

$$A \Longleftrightarrow B \Longleftrightarrow C$$

in liquid phase. The DAE model for the reactor system is given by

$$\frac{dz_A}{dt} = -p_1 z_A + p_2 z_B, \quad z_A(0) = 1, \tag{9.36a}$$

$$\frac{dz_B}{dt} = p_1 z_A - (p_2 + p_3) z_B + p_4 y_C, \quad z_B(0) = 0, \tag{9.36b}$$

$$z_A + z_B + y_C = 1. \tag{9.36c}$$

The 12 sensitivity equations (9.22) are written as

$$\frac{dS_{A1}}{dt} = -S_{A1} p_1 - z_A + p_2 S_{B1}, \quad S_{A1}(0) = 0, \tag{9.37a}$$

$$\frac{dS_{B1}}{dt} = S_{A1} p_1 + z_A - (p_2 + p_3) S_{B1} + p_4 R_{C1}, \quad S_{B1}(0) = 0, \tag{9.37b}$$

$$S_{A1} + S_{B1} + R_{C1} = 0, \tag{9.37c}$$

$$\frac{dS_{A2}}{dt} = -S_{A2}p_1 + p_2 S_{B2} + z_B, \quad S_{A2}(0) = 0, \tag{9.37d}$$

$$\frac{dS_{B2}}{dt} = S_{A2}p_1 - (p_2 + p_3)S_{B2} - z_B + p_4 R_{C2}, \quad S_{B2}(0) = 0, \tag{9.37e}$$

$$S_{A2} + S_{B2} + R_{C2} = 0, \tag{9.37f}$$

$$\frac{dS_{A3}}{dt} = -S_{A3}p_1 + p_2 S_{B3}, \quad S_{A3}(0) = 0, \tag{9.37g}$$

$$\frac{dS_{B3}}{dt} = S_{A3}p_1 - (p_2 + p_3)S_{B3} - z_B + p_4 R_{C3}, \quad S_{B3}(0) = 0, \tag{9.37h}$$

$$S_{A3} + S_{B3} + R_{C3} = 0, \tag{9.37i}$$

$$\frac{dS_{A4}}{dt} = -S_{A4}p_1 + p_2 S_{B4}, \quad S_{A4}(0) = 0, \tag{9.37j}$$

$$\frac{dS_{B4}}{dt} = S_{A4}p_1 - (p_2 + p_3)S_{B4} + p_4 R_{C3} + y_C, \quad S_{B4}(0) = 0, \tag{9.37k}$$

$$S_{A4} + S_{B4} + R_{C4} = 0, \tag{9.37l}$$

and the weighting matrix in the objective function is set to $W = I$.

The parameter estimation problem is solved using the scheme in Figure 9.1. A trust region SQP method is applied, similar to the algorithm described in Section 6.2.3, with the QP subproblem given by (6.38) and the Hessian given by (9.35). This approach is implemented in the GREG parameter estimation code [372, 384]. Starting from $p^0 = [10, 10, 30, 30]^T$ the algorithm converges after 8 iterations and 20 DAE evaluations to the optimal parameters given by

$$(p^*)^T = [3.997, 1.998, 40.538, 20.264]$$

with $\varphi^* = 4.125 \times 10^{-5}$. At the solution the reduced Hessian has eigenvalues that range from 10^{-2} to 10^3; such ill-conditioning is frequently encountered in parameter estimation problems. ■

To extend the sequential approach to optimal control problems, we rely on the multiperiod problem (9.1) and represent the control profile as piecewise polynomials in each period. In addition, the length of each period may be variable. The decisions are still represented by p^l and the state variables remain continuous across the periods.

For the gradient calculations, both the direct and adjoint sensitivity equations are easily modified to reflect parameters that are active only in a given period.

- For the direct sensitivity approach, we define the sensitivity matrices $S_{p^l}(t) = \frac{dz(t)}{dp^l}^T$ and $R_{p^l}(t) = \frac{dy(t)}{dp^l}^T$ and modify the sensitivity equations (9.22) as follows. For $t \in [t_{l-1}, t_l]$,

$$\frac{dS^l}{dt} = \frac{\partial f}{\partial z}^T S^l(t) + \frac{\partial f}{\partial y}^T R^l(t) + \frac{\partial f}{\partial p^l}^T, \quad S^l(0) = 0, \tag{9.38}$$

$$0 = \frac{\partial g}{\partial z}^T S^l(t) + \frac{\partial g}{\partial y}^T R^l(t) + \frac{\partial g}{\partial p^l}^T, \quad t \in [t_{l-1}, t_l]; \tag{9.39}$$

and for $t \notin [t_{l-1}, t_l]$,

$$\frac{dS^l}{dt} = \frac{\partial f}{\partial z}^T S^l(t) + \frac{\partial f}{\partial y}^T R^l(t), \quad S^l(0) = 0, \quad t \notin [t_{l-1}, t_l], \tag{9.40}$$

$$0 = \frac{\partial g}{\partial z}^T S^l(t) + \frac{\partial g}{\partial y}^T R^l(t), \quad t \notin [t_{l-1}, t_l]. \tag{9.41}$$

- For the adjoint sensitivity approach, the adjoint equations (9.28), (9.27), and (9.29) remain unchanged. The only change occurs in (9.30), which is now rewritten for the objective function φ as:

$$\frac{d\varphi}{dp^l} = \frac{\partial \varphi}{\partial p^l} + \int_{t_{l-1}}^{t_l} \left[\frac{\partial f}{\partial p^l}\lambda + \frac{\partial g}{\partial p^l}\nu \right] dt. \tag{9.42}$$

Finally, path inequality constraints are usually approximated by enforcing them only at t_l. Because the periods can be made as short as needed, this approximation is often a practical approach to maintain near feasibility of these constraints.

To demonstrate the (approximate) solution of optimal control problems with the sequential approach, we revisit Example 8.4 from Chapter 8.

Example 9.4 For the nonisothermal batch reactor with first order parallel reactions $A \to B, A \to C$, we maximize the final amount of product B subject to an upper temperature bound. The optimal control problem over $t \in [0,1]$ can be stated as

$$\begin{aligned} \min \quad & -b(1) \\ \text{s.t.} \quad & \frac{da}{dt} = -a(t)(u(t) + k u(t)^\beta), \\ & \frac{db}{dt} = a(t)u(t), \\ & a(0) = 1, \quad b(0) = 0, \quad u(t) \in [0, U]. \end{aligned}$$

Here we represent the controls as piecewise constants over the time periods $l = 1, \ldots, N_T$, with each period having a variable duration. We therefore define $p^l = [u_l, h_l]$ and redefine time with the mapping $t = t_{l-1} + h_l(\tau - (l-1))$, so that t_l is replaced by l and $t \in [0,1]$ is replaced by $\tau \in [0, N_T]$. This allows us to rewrite the optimal control problem as a multiperiod problem of the form (9.1):

$$\min \quad -b(N_T) \tag{9.43a}$$

$$\text{s.t.} \quad \frac{da}{d\tau} = -h_l a(\tau)(u_l + k u_l^\beta), \tag{9.43b}$$

$$\frac{db}{d\tau} = h_l a(\tau) u_l, \tag{9.43c}$$

$$u_l \in [0, U], \quad \tau \in (l-1, l], \quad l = 1, \ldots, N_T, \tag{9.43d}$$

$$a(0) = 1, \quad b(0) = 0, \quad \sum_{l=1}^{N_T} h_l = 1, \; h_l \geq 0, \tag{9.43e}$$

and $a(\tau)$ and $b(\tau)$ are continuous across period boundaries. The direct sensitivity equations for (9.43) are given for $\tau \in (l-1, l]$:

$$\frac{dS_{a,h_l}}{d\tau} = -a(\tau)(u_l + ku_l^\beta) - h_l S_{a,h_l}(\tau)(u_l + ku_l^\beta), \tag{9.44a}$$

$$\frac{dS_{a,u_l}}{d\tau} = -h_l a(\tau)(1 + \beta k u_l^{\beta-1}) - h_l S_{a,u_l}(\tau)(u_l + ku_l^\beta), \tag{9.44b}$$

$$\frac{dS_{b,h_l}}{d\tau} = a(\tau)u_l + h_l S_{a,h_l}(\tau)u_l, \tag{9.44c}$$

$$\frac{dS_{b,u_l}}{d\tau} = h_l a(\tau) + h_l S_{a,u_l}(\tau)u_l, \tag{9.44d}$$

and for $\tau \in (l'-1, l')$, $l' \neq l$:

$$\frac{dS_{a,h_l}}{d\tau} = -h_{l'} S_{a,h_l}(\tau)(u_{l'} + ku_{l'}^\beta), \tag{9.45a}$$

$$\frac{dS_{a,u_l}}{d\tau} = -h_{l'} S_{a,u_l}(\tau)(u_{l'} + ku_{l'}^\beta), \tag{9.45b}$$

$$\frac{dS_{b,h_l}}{d\tau} = h_{l'} S_{a,h_l}(\tau)u_{l'}, \tag{9.45c}$$

$$\frac{dS_{b,u_l}}{d\tau} = h_{l'} S_{a,u_l}(\tau)u_{l'}. \tag{9.45d}$$

Also for $\tau = 0$,

$$S_{a,h_l}(0) = 0, \quad S_{b,h_l}(0) = 0, \quad S_{a,u_l}(0) = 0, \quad S_{b,u_l}(0) = 0,$$

and $S(\tau)$ is also continuous across period boundaries. As a result, we now have $4N_T$ sensitivity equations.

We also derive the related adjoint sensitivity system. Starting from

$$\lambda^T f(z,p) = -\lambda_a(h_l a(\tau)(u_l + ku_l^\beta)) + \lambda_b h_l a(\tau)u_l,$$

the adjoint sensitivity equations for (9.43) are given by

$$\frac{d\lambda_a}{d\tau} = \lambda_a h_l(u_l + ku_l^\beta) - \lambda_b h_l u_l, \quad \lambda_a(N_T) = 0, \tag{9.46a}$$

$$\frac{d\lambda_b}{d\tau} = 0, \quad \lambda_b(N_T) = -1 \quad \text{for } \tau \in (l-1, l], \quad l = 1, \dots, N_T, \tag{9.46b}$$

and $\lambda_a(\tau)$ and $\lambda_b(\tau)$ are continuous across period boundaries. The corresponding integrals for the derivatives of the objective function ($\varphi = -b(N_T)$), with respect to decision variables, are given by

$$\frac{\partial \varphi}{\partial h_l} = \int_{l-1}^{l} -a(\tau)(u + ku_l^\beta)\lambda_a + a(\tau)u_l\lambda_b d\tau, \tag{9.47a}$$

$$\frac{\partial \varphi}{\partial u_l} = \int_{l-1}^{l} -\lambda_a(h_l a(\tau)(1 + \beta k u_l^{\beta-1})) + \lambda_b h_l a(\tau) d\tau, \quad l = 1, \dots, N_T. \tag{9.47b}$$

Note that only two adjoint equations and $2N_T$ integrals need to be evaluated.

Table 9.1. *NLP solution of problem* (9.43) *with SNOPT.*

N_T	$-\varphi^*$	# NLP variables	# Iterations	# Function evaluations
5	0.57177	10	25	52
10	0.57305	20	47	93
20	0.57343	40	81	163
50	0.57352	100	131	227
100	0.57354	200	227	378

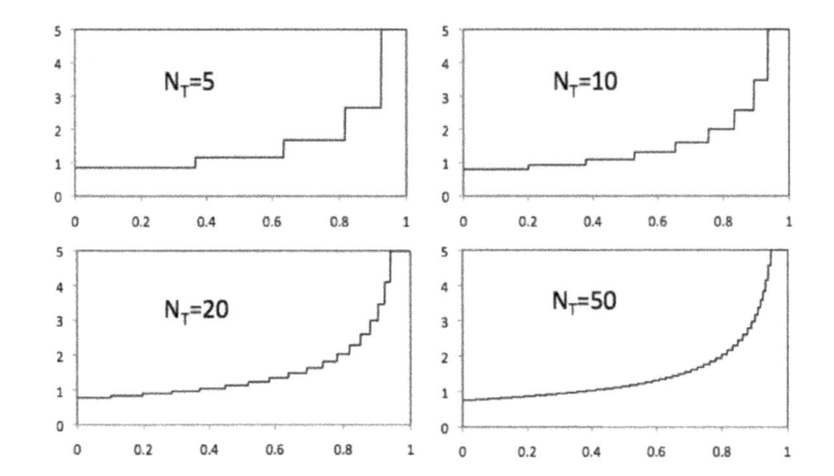

Figure 9.4. *Optimal temperature profiles for Example* 9.4 *with increasing numbers of elements* N_T.

Both the direct and adjoint sensitivity equations can be applied using the approach of Figure 9.1. On the other hand, because this example is linear in the states, the state equations can be solved *analytically* once p^l are fixed. This allows us to solve this problem under *ideal* conditions, without any errors from the DAE or the sensitivity equation solvers.

For this example we set $U = 5$, $k = 0.5$, and $\beta = 2$ and add the bounds, $0 \leq h_l \leq 2/N_T$, to the NLP (9.43). With starting values of $h_l^0 = 1/N_T$ and $u_l^0 = 1$, we apply the SNOPT NLP solver to (9.43) and apply the approach in Figure 9.1. The results are shown in Table 9.1 with control profiles in Figure 9.4 for increasing values of N_T. Comparing these profiles to the solution in Example 8.4, we see that even for the coarse profile with $N_T = 5$, the same trend can be observed as in Figure 8.5. The temperature starts at a low temperature and gradually increases, with a final acceleration to the upper bound. As N_T is increased, the control profile is refined with smaller steps that maintain this shape. Finally, with $N_T = 100$, the temperature profile (not shown) cannot be distinguished from the solution in Figure 8.5. On the other hand, more accurate solutions with larger values of N_T come at an additional cost. Each of the function evaluations in Table 9.1 requires the solution of the DAE system, and each SQP iteration requires the solution of the sensitivity equations as well. From this table, increasing N_T leads to an increase in the number of SQP iterations and function values. In particular, without exact Hessians, the quasi-Newton update may require a large number of iterations, particularly on problems with many decision variables. Note also that the work per iteration increases with N_T as well. This is especially true if direct sensitivity calculations are implemented. ∎

9.4 Multiple Shooting

The approach in Figure 9.1 is not well suited to unstable and poorly conditioned dynamic systems. These can lead to unbounded state profiles, failure of the DAE solver, or convergence difficulties in the NLP solver. Under such conditions, the dynamic optimization must be restarted with the (often frustrated) hope that the DAE solver will not fail again. This problem can be avoided by incorporating more state variable information into the NLP formulation. In particular, the general form of the multiperiod problem (9.1) allows a *multiple shooting* formulation that successfully avoids problems with instability.

Multiple shooting methods for dynamic optimization are derived from their counterparts in the solution of BVPs. As noted in Theorem 8.2, BVPs have more limited existence properties than initial value problems (IVPs). Moreover, they also have different stability properties. This has interesting implications for the *single shooting approach*, where boundary conditions in (8.4) are matched through iterative application of IVP solvers and adjustment of the unknown initial conditions. Because of unstable dynamic modes in the IVP, such an approach may fail for a broad set of BVPs. We will also see that the analogous sequential approach in Figure 9.1 can suffer the same consequences for dynamic optimization problems.

In the multiple shooting approach, the time domain is partitioned into smaller time periods or *elements* and the DAE models are integrated separately in each element [60, 62, 250]. Control variables are treated in the same manner as in the sequential approach. Moreover, gradient information is obtained through sensitivity equations for both the control variables as well as the initial conditions of the states in each element. To provide continuity of the states across elements, equality constraints are added to the nonlinear program. Inequality constraints for states and controls are then imposed directly at the grid points t_l. As with the sequential approach, this may be adequate for inequalities on control profiles approximated by piecewise constant or linear functions. However, path constraints on state profiles may be violated between grid points, especially if $t_{l+1} - t_l$ is not small. For optimal control problems, we consider the multiple shooting formulation with the following formulation:

$$\begin{aligned}
\min \quad & \varphi(\mathbf{p}) && (9.48a)\\
\text{s.t.} \quad & z^{l-1}(t_{l-1}) - z_0^l = 0, \quad l = 2,\dots,N_T, && (9.48b)\\
& z^{N_T}(t_{N_T}) - z_f = 0, \quad z^1(0) = z_0^1, && (9.48c)\\
& p_L^l \le p^l \le p_U^l, && (9.48d)\\
& y_L^l \le y^l(t_l) \le y_U^l, && (9.48e)\\
& z_L^l \le z^l(t_l) \le z_U^l, \quad l = 1,\dots,N_T, && (9.48f)
\end{aligned}$$

with the DAE system

$$\frac{dz^l}{dt}(t) = f^l(z^l(t), y^l(t), p^l), \quad z^l(t_{l-1}) = z_0^l,$$
$$g^l(z^l(t), y^l(t), p^l) = 0, \quad t \in (t_{l-1}, t_l], \quad l = 1,\dots,N_T,$$

embedded in an inner loop for each element l. We note that sensitivities are required for z_0^l as well as p^l. On the other hand, direct and adjoint sensitivity equations need only be solved over the elements where their variables are present.

At first glance, multiple shooting seems to offer few advantages over the sequential approach. A larger NLP is formed, more sensitivity information is required, and the number of SQP iterations may not be reduced. However, for ill-conditioned and unstable systems, the advantages become clear, as the sequential approach may be incapable of considering such systems. Unstable systems arise frequently in reactive systems and control problems. In fact, even though stable solutions may exist at the optimum, the initial value solver may fail to provide bounded state profiles over the sequence of NLP iterates. This is shown in the next example.

Example 9.5 Consider the simple dynamic optimization problem discussed in [61, 89] and given by

$$\min \quad (z_1(1.0))^2 \tag{9.49a}$$

$$\text{s.t.} \quad \frac{dz_1}{dt} = z_2, \quad z_1(0) = 0, \tag{9.49b}$$

$$\frac{dz_2}{dt} = \tau^2 z_1 - (\pi^2 + \tau^2)\sin(\pi t), \quad z_2(0) = p. \tag{9.49c}$$

Because this problem has only a single decision variable and can be solved analytically, it can be verified that the optimum value of $p = \pi$ and that $z_1(t) = \sin(\pi t)$ and $z_2(t) = \pi\cos(\pi t)$ at the solution. However, this problem has an unstable mode proportional to $\exp(\tau t)$, and for large values of τ, say 50, an initial value solver will fail even if the initial condition p is initialized within 10^{-10} of π.

On the other hand, if we apply the multiple shooting approach using problem (9.48), we find that for sufficiently small elements, say $t_l - t_{l-1} \leq 1/\tau$, the multiple shooting approach easily converges to the analytic solution. Moreover, since the variables z_0^l and p appear linearly in the constraints (9.48b), the objective is actually quadratic in these variables and the resulting nonlinear program can be solved directly as a quadratic program. Using a value of $\tau = 50$, Figure 9.5 compares the analytic solution with multiple shooting with the unbounded solution that would result from a sequential approach based on initial value solvers. ∎

Motivated by Example 9.5 we now investigate why the multiple shooting approach is successful on unstable problems. Unstable state profiles typically encountered with multiple shooting are sketched in Figure 9.6. Here we see that unbounded solutions can be prevented in the multiple shooting formulation (9.48), because it introduces two important features not present in the sequential approach.

First, by choosing the length of the period or element to be sufficiently small, the multiple shooting approach limits the escape associated with an unstable dynamic mode. Instead, the imposition of bounds on the state at the end of the elements forces the state to remain in a bounded region. In contrast, with the sequential approach, an unstable state profile can become unbounded before t_f is reached.

Second, a subtle, but more important, feature is the prevention of ill-conditioning in the NLP problem. Even when the state profile remains bounded, unstable dynamic modes can lead to serious ill-conditioning in the Jacobian matrices in (9.48) or the Hessian matrix

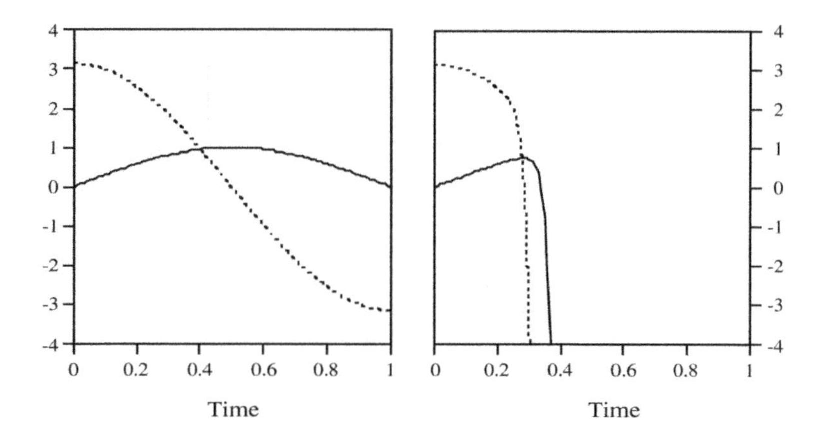

Figure 9.5. *Comparison of multiple shooting solution (left) with initial value profiles (right) for Example* 9.5 *with* $\tau = 50$. *The solid line shows* $z_1(t)$ *while the dotted line shows* $z_2(t)$.

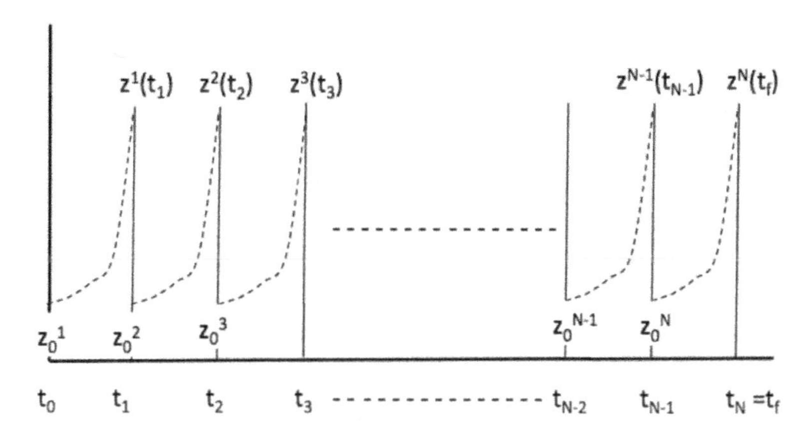

Figure 9.6. *Multiple shooting formulation for unstable systems.*

in (9.2). By imposing endpoint constraints within (9.48), or simply by imposing sufficiently tight bounds on all of the states at the end of the element, this formulation has the property of "pinning down the unstable modes." This property is known as *dichotomy* and it is crucial to ensure the stability of BVP solvers.

To develop the concepts of this important property, we first describe how dichotomy is related to stability of BVPs and how this influences the multiple shooting formulation.

9.4.1 Dichotomy of Boundary Value Problems

In general, dichotomy is difficult to show for nonlinear BVPs. However, in analogy to Theorem 8.2, a related property can be shown for linearized BVPs about a fixed trajectory.

For this, we start with the linear IVP:

$$\frac{dz}{dt} = A(t)z + q(t), \quad z(0) = c, \quad t \geq 0, \tag{9.50}$$

with the solution

$$z(t) = Z(t)\left[c + \int_0^t Z^{-1}(s)q(s)ds\right], \tag{9.51}$$

where $Z(t)$ is the fundamental solution to

$$\frac{dZ}{dt} = A(t)Z(t), \quad Z(0) = I. \tag{9.52}$$

We write the corresponding BVP as

$$\frac{dz}{dt} = A(t)z + q(t), \quad B_0 z(0) + B_1 z(1) = b, \quad t \in [0,1], \tag{9.53}$$

and we find the corresponding IVP solution by substituting (9.51) into the boundary condition to obtain

$$b = [B_0 + B_1 Z(1)]c + B_1 Z(1)\int_0^1 Z^{-1}(s)q(s)ds \tag{9.54}$$

$$= Qc + B_1 Z(1)\int_0^1 Z^{-1}(s)q(s)ds \tag{9.55}$$

$$\Longrightarrow c = Q^{-1}\left[b - B_1 Z(1)\int_0^1 Z^{-1}(s)q(s)ds\right], \tag{9.56}$$

where $Q = B_0 + B_1 Z(1)$. From Theorem 8.1 we know that the IVP has a unique solution, and from Theorem 8.2 a unique solution exists for the BVP if and only if Q is nonsingular.

To assess the well-posedness of (9.53), we need to ensure that the solution remains bounded under perturbations of the problem data (i.e., b and $q(t)$). For this we define $\Phi(t) = Z(t)Q^{-1}$ as the fundamental solution to

$$\frac{d\Phi}{dt} = A(t)\Phi, \; B_0\Phi(0) + B_1\Phi(1) = I, \quad t \in [0,1], \tag{9.57}$$

and rewrite (9.51) as

$$\begin{aligned}
z(t) &= Z(t)\left[c + \int_0^t Z^{-1}(s)q(s)ds\right] \\
&= Z(t)Q^{-1}\left[b - B_1 Z(1)\int_0^1 Z^{-1}(s)q(s)ds\right] + Z(t)\int_0^t Z^{-1}(s)q(s)ds \\
&= \Phi(t)b - \Phi(t)B_1\Phi(1)\int_0^1 \Phi^{-1}(s)q(s)ds + \Phi(t)\int_0^t \Phi^{-1}(s)q(s)ds \\
&= \Phi(t)b + \Phi(t)\left[(I - B_1\Phi(1))\int_0^t \Phi^{-1}(s)q(s)ds - B_1\Phi(1)\int_t^1 \Phi^{-1}(s)q(s)ds\right] \\
&= \Phi(t)b + \Phi(t)\left[B_0\Phi(0)\int_0^t \Phi^{-1}(s)q(s)ds - B_1\Phi(1)\int_t^1 \Phi^{-1}(s)q(s)ds\right],
\end{aligned}$$

where we use $Z(1)Z(s)^{-1} = \Phi(1)\Phi(s)^{-1}$ and $B_0\Phi(0) + B_1\Phi(1) = I$. This allows us to write the solution to (9.53) as

$$z(t) = \Phi(t)b + \int_0^1 G(t,s)q(s)ds, \tag{9.58}$$

where we define the Green's function as:

$$G(t,s) = \begin{cases} G_0(t,s) = \Phi(t)P\Phi^{-1}(s), & s \leq t, \\ G_1(t,s) = -\Phi(t)(I-P)\Phi^{-1}(s), & s > t, \end{cases} \tag{9.59}$$

and the matrices $P = B_0\Phi(0)$ and $I - P = B_1\Phi(1)$ correspond to the initial and final conditions that are "pinned down." The dichotomy property of pinning down the unstable modes is analogous to the zero-stability property presented in Section 9.2. In this case, we invert (9.53) and obtain (9.58) and the Green's function. Moreover, we define the BVP as stable and *well-conditioned* if there exists a constant κ of moderate size (relative to the problem data $A(t), q(t)$, and b) such that the following inequalities hold:

$$\begin{aligned} \|z(t)\| &\leq \kappa \left[\|b\| + \int_0^1 \|q(s)\|ds \right], \\ \|\Phi(t)\| &\leq \kappa, \ t \geq 0, \\ \|G_0(t,s)\| &\leq \kappa, \ s \leq t, \\ \|G_1(t,s)\| &\leq \kappa, \ s > t. \end{aligned} \tag{9.60}$$

Under these conditions, the BVP has dichotomy.

To illustrate, we consider the following linear time invariant, homogeneous BVP with $\tau > 0$:

$$\begin{bmatrix} \frac{dz_1}{dt} \\ \frac{dz_2}{dt} \end{bmatrix} = \begin{bmatrix} \tau & 0 \\ 0 & -\tau \end{bmatrix} \begin{bmatrix} z_1 \\ z_2 \end{bmatrix}, \quad \begin{bmatrix} z_1(1) \\ z_2(0) \end{bmatrix} = \begin{bmatrix} 1 \\ 1 \end{bmatrix}.$$

The fundamental solution of the IVP is given by

$$Z(t) = \begin{bmatrix} e^{\tau t} & 0 \\ 0 & e^{-\tau t} \end{bmatrix}$$

with the following information from the BVP:

$$B_0 = \begin{bmatrix} 0 & 1 \\ 0 & 0 \end{bmatrix}; \quad B_1 = \begin{bmatrix} 0 & 0 \\ 1 & 0 \end{bmatrix};$$

$$Q = \begin{bmatrix} 0 & 1 \\ e^{\tau} & 0 \end{bmatrix}; \quad \Phi(t) = \begin{bmatrix} 0 & e^{\tau(t-1)} \\ e^{-\tau t} & 0 \end{bmatrix};$$

$$\Phi^{-1}(s) = \begin{bmatrix} 0 & e^{\tau s} \\ e^{\tau(1-s)} & 0 \end{bmatrix}; \quad \Phi(0) = \begin{bmatrix} 0 & e^{-\tau} \\ 1 & 0 \end{bmatrix}; \quad \Phi(1) = \begin{bmatrix} 0 & 1 \\ e^{-\tau} & 0 \end{bmatrix}.$$

From the dynamic models and the boundary conditions, we see that the unstable dynamic mode $z_1(t)$ is pinned down by its boundary condition. The corresponding Green's functions are given by

$$G_0(t,s) = \begin{bmatrix} 0 & 0 \\ 0 & e^{\tau(s-t)} \end{bmatrix} \text{ for all } t \geq s; \;\; G_1(t,s) = -\begin{bmatrix} e^{\tau(t-s)} & 0 \\ 0 & 0 \end{bmatrix} \text{ for all } t < s,$$

and from (9.60), $\kappa \geq 1$, the BVP is stable.

However, if the boundary conditions are changed to

$$\begin{bmatrix} z_1(0) \\ z_2(1) \end{bmatrix} = \begin{bmatrix} 1 \\ 1 \end{bmatrix},$$

then the unstable mode is no longer pinned down and the Green's functions become

$$G_0(t,s) = \begin{bmatrix} e^{\tau(t-s)} & 0 \\ 0 & 0 \end{bmatrix} \text{ for all } t \geq s; \;\; G_1(t,s) = -\begin{bmatrix} 0 & 0 \\ 0 & e^{\tau(s-t)} \end{bmatrix} \text{ for all } t < s.$$

From (9.60) we require $\kappa \geq e^{\tau}$. Since this constant may be much larger than τ, the BVP is considered unstable.

The dichotomy property applies not only to the BVP but also to the discretization used for its numerical solution. As shown in [16], the inverse of the Jacobian matrix associated with the matching constraints (9.48b) in the multiple shooting formulation consists of the quantities $G(t^{l-1}, t^l)$ and $\Phi(t^l)$. Therefore, the boundedness of these quantities has a direct influence on the *discretized* BVP that arises in the multiple shooting formulation. Moreover, de Hoog and Mattheij [109] investigated the relationship between the discretized BVP and the fundamental BVP solution and showed that the conditioning constants for the Jacobian are closely related to κ in (9.60). Therefore to promote a stable and well-conditioned solution, the appropriate dichotomous boundary condition is required in (9.48). For this NLP, the required boundary condition can be added explicitly through a final time constraint that could include additional decision variables, or through a sufficiently tight bound on the dynamic modes at final time. In either case, a well-conditioned discretized system results. This property is illustrated by the case study in the next section.

9.5 Dynamic Optimization Case Study

The sequential and multiple shooting methods for dynamic optimization have been applied to a wide variety of real-world process systems. In this section we demonstrate both of these approaches on an industrially motivated example for nonlinear model predictive control. Here we consider the reactor system in the dynamic process model proposed and developed at Eastman Chemical Company and described in [118]. The following gas-phase reactions take place in the nonisothermal reactor shown in Figure 9.7:

$$\begin{aligned} A + C + D &\rightarrow G, \\ A + C + E &\rightarrow H, \\ A + E &\rightarrow F, \\ 3D &\rightarrow 2F. \end{aligned}$$

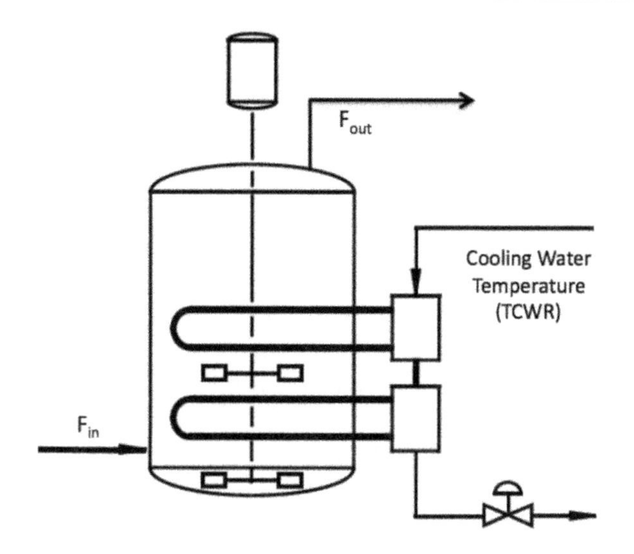

Figure 9.7. *Diagram of the Tennessee Eastman reactor.*

These reactions lead to two liquid products, G and H, and an unwanted liquid by-product, F. The reactor model is given by the following ODEs:

$$\frac{dN_{A,r}}{dt} = y_{A,in}F_{in} - y_{A,out}F_{out} - R_1 - R_2 - R_3, \tag{9.61a}$$

$$\frac{dN_{B,r}}{dt} = y_{B,in}F_{in} - y_{B,out}F_{out}, \tag{9.61b}$$

$$\frac{dN_{C,r}}{dt} = y_{C,in}F_{in} - y_{C,out}F_{out} - R_1 - R_2, \tag{9.61c}$$

$$\frac{dN_{D,r}}{dt} = y_{D,in}F_{in} - y_{D,out}F_{out} - R_1 - \frac{3}{2}R_4, \tag{9.61d}$$

$$\frac{dN_{E,r}}{dt} = y_{E,in}F_{in} - y_{E,out}F_{out} - R_2 - R_3, \tag{9.61e}$$

$$\frac{dN_{F,r}}{dt} = y_{F,in}F_{in} - y_{F,out}F_{out} + R_3 + R_4, \tag{9.61f}$$

$$\frac{dN_{G,r}}{dt} = y_{G,in}F_{in} - y_{G,out}F_{out} + R_1, \tag{9.61g}$$

$$\frac{dN_{H,r}}{dt} = y_{H,in}F_{in} - y_{H,out}F_{out} + R_2, \tag{9.61h}$$

$$\left(\sum_{i=A}^{H} N_{i,r}C_{p,i}\right)\frac{dT_r}{dt} = y_{i,in}C_{p,vap,i}F_{in}(T_{in} - T_r) \\ - Q_{CW}(T_r, T_{CW}, F_{CW}) - \Delta H_{Rj}(T_r)R_j, \tag{9.61i}$$

$$(\rho_{CW}V_{CW}C_{p,CW})\frac{dT_{CW}}{dt} = F_{CW}C_{p,CW}(T_{CW,in} - T_{CW}) \\ + Q_{CW}(T_r, T_{CW}, F_{CW}). \tag{9.61j}$$

The differential equations represent component mass balances in the reactor as well as energy balances for the reactor vessel and heat exchanger. The additional variables are defined by the following equations:

$$
\begin{aligned}
x_{i,r} &= N_{i,r} / \left(\sum_{i=D}^{H} N_{i,r} \right), \quad i = D, E, \ldots, H, \\
y_{i,out} &= P_{i,r} / P_r, \quad i = A, B, \ldots, H, \\
P_r &= \sum_{i=A}^{H} P_{i,r}, \\
P_{i,r} &= N_{i,r} R T_r / V_{Vr}, \quad i = A, B, C, \\
V_{Lr} &= \sum_{i=D}^{H} N_{i,r} / \rho_i, \\
V_{Vr} &= V_r - V_{Lr}, \\
P_{i,r} &= \gamma_{ir} x_{ir} P_i^{sat}, \quad i = D, E, F, G, H, \\
R_1 &= \alpha_1 V_{Vr} \exp[\mu_1 - \nu_1 / T_r] \; P_{A,r}^{1.15} \; P_{C,r}^{0.370} \; P_{D,r}, \\
R_2 &= \alpha_2 V_{Vr} \exp[\mu_2 - \nu_2 / T_r] \; P_{A,r}^{1.15} \; P_{C,r}^{0.370} \; P_{E,r}, \\
R_3 &= \alpha_3 V_{Vr} \exp[\mu_3 - \nu_3 / T_r] \; P_{A,r} P_{E,r}, \\
R_4 &= \alpha_4 V_{Vr} \exp[\mu_4 - \nu_4 / T_r] \; P_{A,r} P_{D,r}, \\
F_{out} &= \beta \sqrt{P_r - P_s}.
\end{aligned}
$$

Here F_{in} and F_{out} are the inlet and outlet flow rates. In addition, P_r and P_s are the reactor and system pressures, and $N_i, y_i, x_i, P_i^{sat}, P_{ir}, \rho_i$ are reactor holdups, vapor and liquid mole fractions, vapor pressures, partial pressures, and liquid densities of component i. R_j are the reaction rates, V_{Lr}, V_{Vr}, V_{CW} are the liquid and vapor reactor volumes and the heat exchanger volume, respectively. In addition, $\gamma_i, \alpha_j, \mu_j, \nu_j$ are constants, $\Delta H_{R,j}, C_{p,i}$ are heats of reaction and heat capacities, β is the valve constant, and Q_{CW} is a function that models the heat removed by the heat exchanger.

This system has three output variables $y(t)$ (reactor pressure P_r, level V_L, and temperature T_r), and the control (or manipulated) variables u_l are the reactor feed rate F_{in}, the agitator speed β, and the cooling water flow rate F_{CW}. For the purpose of this study, we assume that complete and accurate state information is available over all time. More details on the reactor model can be found in [118, 345, 211].

The NMPC controller needs to maintain conditions of reactor operation that correspond to equal production of products G and H. However, at this operating point the reactor is open-loop unstable, and without closed-loop stabilizing control, the state profiles become unbounded.

The dynamic model (9.61) can be represented as the following ODE system along with output equations:

$$\frac{dz(t)}{dt} = f(z(t), u), \quad y(t) = g(z(t), u).$$

The control problem is described over a prediction horizon with N_T time periods of 180 seconds each. The objective function is to keep the reactor operating at the steady state point

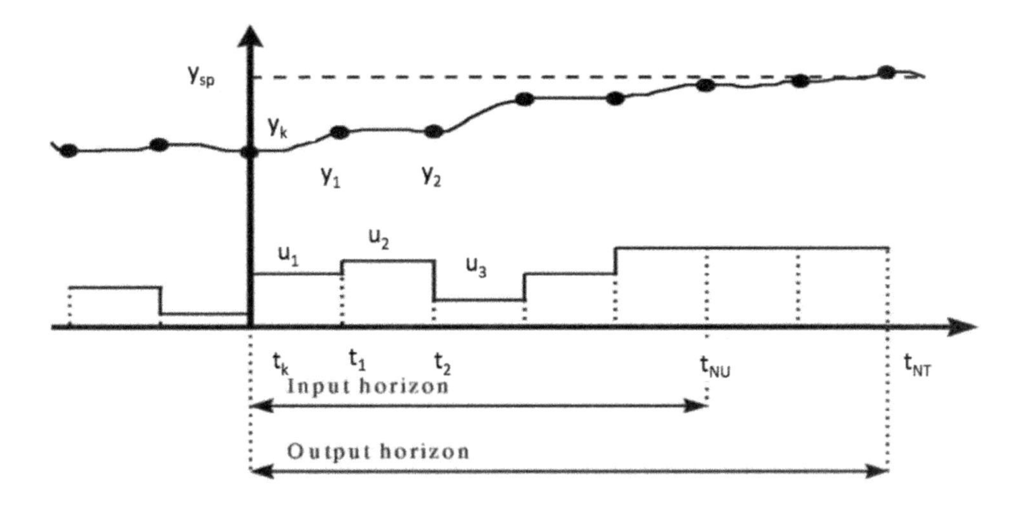

Figure 9.8. *NMPC optimization over a predictive horizon at time k.*

without violating the operating limits and constraints on the process variables. This problem takes the form of (9.2) and is written over the predictive horizon shown in Figure 9.8 as

$$\min \quad \varphi(\mathbf{p}) = \sum_{l=1}^{N_T} (y(t_l) - y_{sp})^T Q_y (y(t_l) - y_{sp}) + (u_l - u_{l-1})^T Q_u (u_l - u_{l-1}) \tag{9.62a}$$

$$\text{s.t.} \quad \frac{dz(t)}{dt} = f(z(t), u_l), \quad z(0) = z_k, \tag{9.62b}$$

$$y(t) = g(z(t), u_l), \tag{9.62c}$$

$$u_{L,l} \leq u_l \leq u_{U,l}, \quad l = 1, \ldots, N_T, \tag{9.62d}$$

where weighting matrices for the controller are $Q_y = I$ and $Q_u = 10^{-4} \times I$. By adjusting the bounds $u_{L,l}$ and $u_{U,l}$, the control variables can be varied over the input horizon with N_U periods, while the process is simulated for several more time periods (the output horizon with N_T periods) with constant controls. The output horizon must be long enough so that the process will arrive at the steady state at the end of the simulation. After the NMPC problem (9.62) is solved at time k with initial state z_k, the control variables in the first time period u_1 are injected into the process. At time $k+1$, the dynamic model is updated with new measurements from the process, and problem (9.62) is set up and solved again with z_{k+1} as the initial condition.

Problem (9.62) was solved with both sequential and multiple shooting approaches. Moreover, with multiple shooting we consider the option of adding constraints at final time to force the model to steady state values. This constraint guarantees the dichotomy property for the dynamic system. The NMPC problem was solved with the NEWCON package, which applied DASSL and DDASAC to solve the DAEs and the direct sensitivity equations, respectively. The SQP solver incorporates the QPSOL package to solve the QP subproblem, and the Gauss–Newton approximation (9.35) was used for the Hessian.

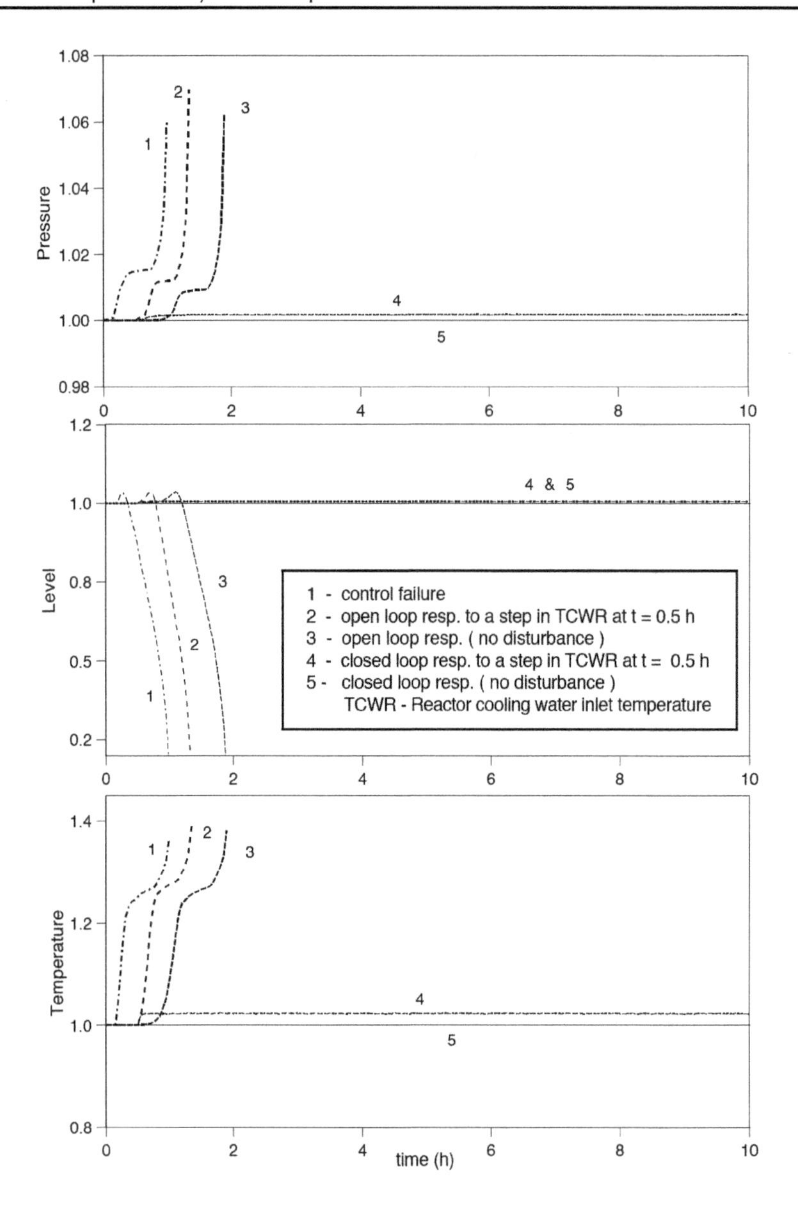

Figure 9.9. *Normalized output variables.*

For this problem, open-loop instability of the reactor has a strong influence on the convergence properties of these algorithms. In particular, for predictive horizons of length $N_T = 10$, and $N_U = 5$, the condition number of the Hessian is on the order of 10^8. With an output horizon at $N_T = 20$, the condition number grows to approximately 10^{12}.

Figure 9.9 shows the open-loop and closed-loop responses of the normalized output variables. Here the G/H mass ratio is also shown in Figure 9.10. These graphs describe the

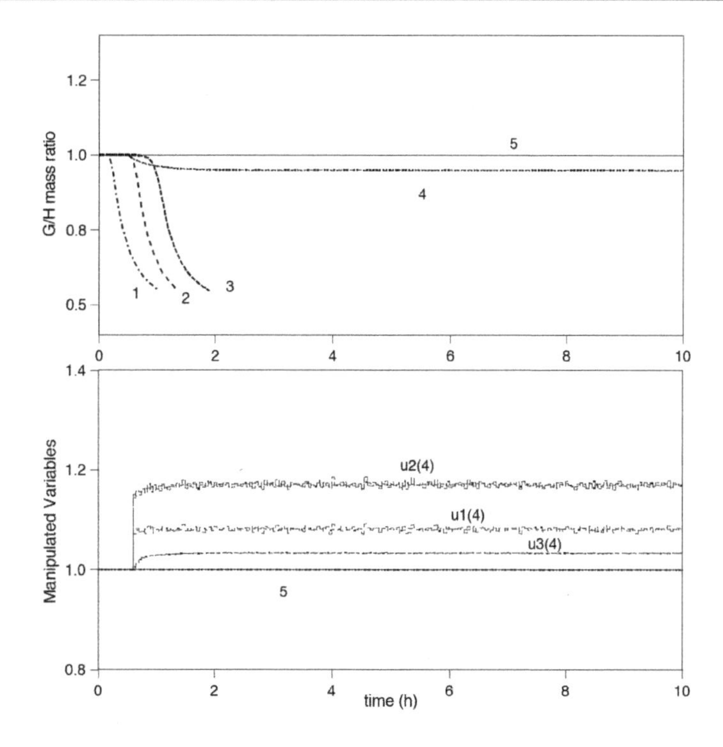

Figure 9.10. *Normalized G/H mass ratio and control variables.*

following two different situations:

- Responses are obtained without any disturbances. This case deals with regulation at an unstable setpoint and corresponds to curve 5 in Figures 9.9 and 9.10.

- A response is obtained for a $+5°C$ disturbance in the cooling water inlet temperature $T_{CW,in}$, introduced after 10 time periods (0.5 h). This corresponds to curve 4 in Figures 9.9 and 9.10.

As seen from this figure, the process is clearly open-loop unstable (as indicated by curves 2 and 3, with and without temperature disturbances, respectively). Curves 4 and 5 show the successful closed-loop results, with and without disturbances. A slight offset is observed in curve 4 because the unmeasured disturbance is not included in the model (9.62). Using the sequential approach, the algorithm is able to solve the sequence of control problems with $N_T = 10$ and $N_U = 5$ (curves 4 and 5), with an average of 13 SQP iterations required for each horizon problem. On the other hand, for a longer output horizon $N_T = 20$ the ODE solver fails at $t = 0.05$ h and the sequential approach terminates. The trajectory of the output predictions for this case is represented in Figure 9.9 by curve 1. This case illustrates the limitations induced by large roundoff errors from unstable modes.

On the other hand, the multiple shooting formulation was successful in solving these problems. Curves 4 and 5 were obtained for horizons of $N_U = 5$, $N_T = 10$ regardless of whether the dichotomous endpoint constraints were used or not. The multiple shooting formulation was twice as fast as the sequential approach and required an average of only 6 SQP iterations to solve problem (9.62). For the larger output horizon with $N_T = 20$, the disturbance case (curve 4) was also obtained successfully with and without endpoint constraints. However, without endpoint constraints, ill-conditioning often led to error messages from the QP solver. Finally, for the case of no disturbances (curve 5), only the endpoint formulation was able to solve this problem. Without endpoint constraints, the QP solver fails at time $t = 7$ h and the controller terminates. This failure is due to ill-conditioning of the constraint gradients when endpoint constraints are absent. This difficulty also increases with longer predictive horizons. On the other hand, by using multiple shooting with terminal output constraints for $N_T = 30$ and $N_U = 5$, the flat curves 4 and 5 can still be obtained for simulation times of 20 hours (i.e., a sequence of 400 horizon problems (9.62)).

9.6 Summary and Conclusions

This chapter provides an overview of strategies for the solution of dynamic optimization problems with embedded DAE solvers. As seen in Figure 9.1, this approach requires three components: the NLP solver, the DAE solver, and sensitivity calculations to provide gradients. For the first component, gradient-based NLP solvers from Chapter 6 can be applied directly and the preference is for SQP-type methods; these generally require fewer function and gradient evaluations. For the second component, Section 9.2 presents an overview of DAE solvers and briefly describes Runge–Kutta and linear multistep methods, along with their convergence properties. Finally, accurate sensitivity calculations are required for the optimization and Section 9.3 presents both direct and adjoint sensitivity approaches. These approaches complement each other. Direct sensitivity is easier to apply and is efficient with few decisions and many constraints. On the other hand, the adjoint approach efficiently handles problems with many decision variables, but is best suited for few constraints and requires storage of the state profiles. Based on these components, the sequential optimization strategy is reasonably straightforward to implement and the approach has been applied within a number of commercial simulation platforms. However, in the presence of instability, the sequential strategy may fail and an alternative approach is needed. Section 9.4 therefore presents the multiple shooting formulation based on concepts from the solution of BVPs. This approach relies on the same components in Figure 9.1, but by solving the DAEs separately within each period, it can restrict the escape of unstable modes and avoid ill-conditioning through the dichotomy property. Nevertheless, the multiple shooting formulation may be more expensive to apply.

DAE sensitivity and optimization approaches are illustrated through a number of examples. In particular, Section 9.5 presents a case-study comparison between the sequential and multiple shooting approaches for an unstable reactor system. The embedded DAE optimization methods in this chapter have a number of advantages, particularly for large-scale simulation models that require only a few degrees of freedom for optimization. Nevertheless, the computational cost of the sequential and multiple shooting methods is strongly dependent on the efficiency of the DAE and sensitivity solvers, and their exploitation of the structure of the dynamic system. A number of initial value DAE solvers are well suited for large-scale dynamic systems, as they exploit the structure and sparsity of the Jacobian matrix (9.19). This topic is beyond the scope of this chapter.

On the other hand, difficulties with embedded optimization methods include the following issues.

- Both sequential and multiple shooting approaches require the repeated solution of DAE systems. The DAE solution and sensitivity calculations represent the dominant costs of the optimization. In particular, the direct sensitivity approach can become an overwhelming component of the computation if the problem has many variables in the nonlinear program.
- Both approaches are heavily dependent on the reliability of the DAE solver. Failure in the DAE solver or the corresponding sensitivity calculation will cause the optimization strategy to fail. This issue was explored for unstable DAEs, but failure may also be due to nonsmoothness, loose tolerances, and other features of DAE systems and solvers.
- Control profiles and path constraints must be approximated in the sequential and multiple shooting approaches. While this is often adequate for many practical problems, a close approximation requires many more decision variables and more expensive solutions.
- By embedding the DAE solver and sensitivity into the optimization strategy, the constraint gradients in problem (9.1) are no longer sparse, but contain dense blocks. The linear algebra associated with these blocks may be expensive for problems with many decision or state variables, especially in the multiple shooting formulation.

Overcoming these issues will be addressed in the next chapter, where large-scale, sparse NLP formulations are developed that directly incorporate the discretized DAE system.

9.7 Notes for Further Reading

The summary on DAE solvers for both initial value and boundary value problems is based on the excellent text by Ascher and Petzold [16]. In addition to the references cited in this chapter, there is a wealth of literature on this topic. Monographs and textbooks, including [360, 182, 183, 77] for ODE solvers, [14] for BVPs, and [70, 237, 183] for DAEs, are great places to start.

In addition, there are a number of excellent ODE, BVP, and DAE codes, including DASSL [308], CVODES [197], COLDAE [15], and PASVA [253]. Moreover, suites of software packages include ODEPACK (www.netlib.org) and SUNDIALS [196]. All of these are FORTRAN or C codes that are open source and can be downloaded through the Internet. Similarly, a number of direct sensitivity codes are available including ODESSA [252], DDASAC [85], and DASPK [258]. There are also a few adjoint sensitivity codes, including CVODES [197] and DASPK/Adjoint [259], that also apply checkpointing (see [172]).

The sequential optimization approach has its roots in the mid 1960s with the development of control vector parameterization (CVP) [72]. Gradient-based approaches based on adjoint methods were developed in the 1970s and can be found in [189, 164, 349]. With the development of direct sensitivity, these approaches were updated and developed in the 1980s and can be found in [397, 299, 258]. This approach is now incorporated into a number of dynamic process simulation packages including ASPEN Custom Modeler [2], gPROMS [27, 3], and JACOBIAN [4]. Moreover, the sequential approach for process optimization continues to be developed by a number of groups including Tolsma and Barton [386]; Feehery, Tolsma, and Barton [128]; Franke and coworkers [150, 149]; Mangold et al. [274]; Schlegel et al. [355]; Prata et al. [317]; Kadam and Marquardt [214, 215]; Santos et al. [346, 345]; Romanenko and Santos [337]; and Oliveira and Biegler [299].

Similarly, the multiple shooting method has seen significant recent development through the work of Leineweber et al. [250, 251]; Bock [61]; and Diehl, Bock, and Schlöder [115]. In particular, the MUSCOD code implements large-scale SQP methods with advanced DAE solvers. Building on the concepts in Section 9.4, sophisticated advances have been implemented in MUSCOD that accelerate the direct sensitivity step and promote reliable solutions, even for unstable and chaotic systems. More information on the MUSCOD algorithm can be found in [251].

Finally, as shown in Chapters 5 and 6, the performance of the NLP solver in Figure 9.1 can be greatly enhanced if second derivatives can be made available from the DAE solver. Recent work (see [186, 301]) has led to efficient strategies to calculate Hessian vector products from the sensitivity equations. Here second order adjoint equations are applied to the direct sensitivity equations (or vice versa) for search directions supplied by the NLP solver. This approach leads to an accurate Hessian vector product at the cost of an additional adjoint sensitivity step.

9.8 Exercises

1. Consider the reactor optimization problem given by

$$\begin{aligned} \min \quad & L - 500 \int_0^L (T(t) - T_S) dt \\ \text{s.t.} \quad & \frac{dq}{dt} = 0.3(1 - q(t)) \exp(20(1 - 1/T(t))), \quad q(0) = 0, \\ & \frac{dT}{dt} = -1.5(T(t) - T_S) + 2/3 \frac{dq}{dt}, \quad T(0) = 1, \end{aligned}$$

where $q(t)$ and $T(t)$ are the normalized reactor conversion and temperature, respectively, and the decision variables are $T_S \in [0.5, 1]$ and $L \in [0.5, 1.25]$.

   (a) Derive the direct sensitivity equations for the DAEs in this problem.

   (b) Using MATLAB or a similar package, apply the sequential approach to find the optimum values for the decision variables.

   (c) How would you reformulate the problem so that the path constraint $T(t) \leq 1.45$ can be enforced?

2. Consider the optimal control problem in Example 9.4.

   (a) For the sequential optimization formulation, derive the analytical solution for substitution in (9.2) and solve the problem in GAMS or AMPL for various values of N_T.

   (b) For the multiple shooting optimization formulation, derive the analytical solution for substitution in (9.48) and solve the problem in GAMS or AMPL for various values of N_T.

3. Consider the optimal control problem in Example 8.3.

   (a) Derive the direct sensitivity and adjoint sensitivity equations required for the sequential formulation (9.2).

   (b) Derive the direct sensitivity and adjoint sensitivity equations required for the multiple shooting formulation (9.48).

   (c) For the sequential optimization formulation, derive the analytical solution for substitution in (9.2) and solve the problem in GAMS or AMPL for various values of N_T.

   (d) For the multiple shooting optimization formulation, derive the analytical solution for substitution in (9.48) and solve the problem in GAMS or AMPL for various values of N_T.

4. Develop the adjoint sensitivity equations for Example 9.3.

5. Consider the system of differential equations in Example 9.5.

$$\frac{dz_1}{dt} = z_2,$$
$$\frac{dz_2}{dt} = 1600 z_1 - (\pi^2 + 1600)\sin(\pi t).$$

   (a) Show that the analytic solution of these differential equations are the same for the initial conditions $z_1(0) = 0, z_2(0) = \pi$ and the boundary conditions $z_1(0) = z_1(1) = 0$.

   (b) Find the analytic solution for the initial and boundary value problems. Comment on the dichotomy of each system.

   (c) Consider the optimal control problem:

$$\begin{aligned} \min \;\; & (z_1(1.0))^2 \\ \text{s.t.} \;\; & \frac{dz_1}{dt} = z_2, \; z_1(0) = 0, \\ & \frac{dz_2}{dt} = 1600 z_1 - (p^2 + 1600)\sin(pt), \quad z_2(0) = \pi. \end{aligned}$$

   Formulate this problem with the multiple shooting approach and solve the resulting nonlinear programming problem.

6. Consider the following reactor optimization problem:

$$\begin{aligned} \max \;\; & c_2(1.0) \\ \text{s.t.} \;\; & \frac{dc_1}{dt} = -k_1(T)c_1^2, \quad c_1(0) = 1, \\ & \frac{dc_2}{dt} = k_1(T)c_1^2 - k_2(T)c_2, \quad c_2(0) = 0, \end{aligned}$$

where $k_1 = 4000\,\exp(-2500/T)$, $k_2 = 62000\,\exp(-5000/T)$, and $T \in [298, 398]$. Discretize the temperature profile as piecewise constants over N_T periods and perform the following:

(a) Derive the direct sensitivity equations for the DAEs in this problem.

(b) Cast this example in the form of (9.2) and solve using the sequential strategy with MATLAB or a similar package.

(c) Cast this example in the form of (9.48) and solve using the multiple shooting approach with MATLAB or a similar package.

7. Consider the following optimal control problem:

$$\begin{aligned} \max \ & z_3(1.0) \\ \text{s.t.} \ & \frac{dz_1}{dt} = z_2, \quad z_1(0) = 0, \\ & \frac{dz_2}{dt} = -z_2 + u(t), \quad z_2(0) = -1, \\ & \frac{dz_3}{dt} = z_1^2 + z_2^2 + 0.005\,u(t)^2, \quad z_3(0) = 0. \end{aligned}$$

Discretize the control profile as piecewise constants over N_T periods and perform the following:

(a) Derive the adjoint sensitivity equations for this problem.

(b) Cast this example in the form of (9.2) and solve using the sequential strategy with MATLAB or a similar package.

(c) Cast this example in the form of (9.48) and solve using the multiple shooting approach with MATLAB or a similar package.

Chapter 10

Simultaneous Methods for Dynamic Optimization

Following on embedded methods for dynamic optimization, this chapter considers "all-at-once" or *direct transcription* methods that allow a simultaneous approach for this optimization problem. In particular, we consider formulations based on orthogonal collocation methods. These methods can also be represented as a special class of implicit Runge–Kutta (IRK) methods, and concepts and properties of IRK methods apply directly. Large-scale optimization formulations are then presented with the aim of maintaining accurate state and control profiles and locating potential break points. This approach is applied to consider a number of difficult problem classes including unstable systems, path constraints, and singular controls. Moreover, a number of real-world examples are featured that demonstrate the characteristics and advantages of this approach. These include batch crystallization processes, grade transitions in polymer processes, and large-scale parameter estimation for complex industrial reactors.

10.1 Introduction

This chapter evolves from sequential and multiple shooting approaches for dynamic optimization by considering a large nonlinear programming (NLP) formulation *without* an embedded DAE solver. Instead, we consider the multiperiod dynamic optimization problem (8.5) where the periods themselves are represented by finite elements in time, with piecewise polynomial representations of the state and controls in each element. This approach leads to a discretization that is equivalent to the Runge–Kutta methods described in the previous chapter. Such an approach leads to a fully open formulation, represented in Figure 7.1, and has a number of pros and cons. The large-scale NLP formulation allows a great deal of sparsity and structure, along with flexible decomposition strategies to solve this problem efficiently. Moreover, convergence difficulties in the embedded DAE solver are avoided, and sensitivity calculations from the solver are replaced by direct gradient and Hessian evaluations within the NLP formulation. On the other hand, efficient large-scale NLP solvers are required for efficient solutions, and accurate state and control profiles require careful formulation of the nonlinear program.

To address these problems, we rely on efficient large-scale NLP algorithms developed in Chapter 6. In particular, methods that accept exact second derivatives have fast convergence properties. Moreover, exploitation of the structure of the KKT matrix leads to efficient large-scale algorithms. Nevertheless, additional concerns include choosing an accurate and stable discretization, selecting the number and the length of finite elements, and dealing with unstable dynamic modes. Finally, the fundamental relation of the NLP formulation and the original dynamic optimization problem needs to be analyzed. All of these issues will be explored in this chapter, and properties of the resulting *simultaneous collocation* formulation will be demonstrated on a number of real-world process examples.

The next section describes the collocation approach and motivates its properties, particularly high-order approximations and the relation to IRK methods. Section 10.3 incorporates the discretized DAEs into an NLP formulation and discusses the addition of finite elements and the incorporation of discontinuous control profiles. In addition, the calculation of error bounds is discussed and the extension to variable finite elements is developed. Moreover, we also analyze the treatment of unstable modes. Section 10.4 then explores the relation of the NLP formulation with the dynamic optimization problem. A key issue to establishing this relationship is regularity of the optimal control problem that translates into nonsingularity of the KKT matrix. Subsections 10.4.4 and 10.4.5 deal with open questions where these regularity conditions are violated. This section also deals with problems with inequality constraints on state profiles. As seen in Chapter 8, these problems are difficult to handle, but with the simultaneous collocation method, they may be treated in a more straightforward way. We also deal with singular control problems, discuss related convergence difficulties, both with indirect and direct approaches, and present heuristic regularization approaches to overcome these difficulties.

10.2 Derivation of Collocation Methods

We first consider the differential-algebraic system (8.2) given by

$$\frac{dz}{dt} = f(z(t), y(t), u(t), p), \quad z(0) = z_0, \tag{10.1a}$$

$$g(z(t), y(t), u(t), p) = 0. \tag{10.1b}$$

The simultaneous approach requires a discretization of the state $z(t), y(t)$ and control profiles $u(t)$. A number of suitable approaches were explored in Section 9.2, but unlike discretizations for IVPs, we require the following features to yield an efficient NLP formulation:

- Since the nonlinear program requires an iterative solution of the KKT conditions, there is no computational advantage to an explicit ODE discretization.

- Because the NLP formulation needs to deal with discontinuities in control profiles, a single-step method is preferred, as it is self-starting and does not rely on smooth profiles that extend over previous time steps. As shown in the next section, the collocation formulation requires smooth profiles only *within the finite element*.

- The high-order implicit discretization provides accurate profiles with relatively few finite elements. As a result, the number of finite elements need not be excessively large, particularly for problems with many states and controls.

With these features in mind, we consider a piecewise polynomial representation of the state and control profiles.

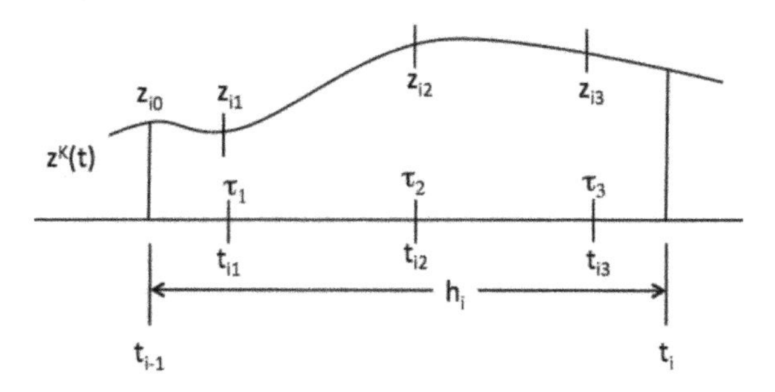

Figure 10.1. *Polynomial approximation for state profile across a finite element.*

10.2.1 Polynomial Representation for ODE Solutions

We consider the following ODE:

$$\frac{dz}{dt} = f(z(t),t), \quad z(0) = z_0, \tag{10.2}$$

to develop the *collocation method*, where we solve this differential equation at selected points in time. For the state variable, we consider a polynomial approximation of order $K+1$ (i.e., degree $\leq K$) over a single finite element, as shown in Figure 10.1. This polynomial, denoted by $z^K(t)$, can be represented in a number of equivalent ways, including the power series representation

$$z^K(t) = \alpha_0 + \alpha_1 t + \alpha_2 t^2 + \cdots + \alpha_K t^K, \tag{10.3}$$

the Newton divided difference approximation (9.11) in Chapter 9, or B-splines [14, 40]. To develop the NLP formulation, we prefer representations based on Lagrange interpolation polynomials, because the polynomial coefficients have the same variable bounds as the profiles themselves. Here we choose $K+1$ interpolation points in element i and represent the state in a given element i as

$$\left.\begin{aligned} t &= t_{i-1} + h_i \tau, \\ z^K(t) &= \sum_{j=0}^{K} \ell_j(\tau) z_{ij}, \end{aligned}\right\} t \in [t_{i-1}, t_i], \quad \tau \in [0,1], \tag{10.4}$$

$$\text{where } \ell_j(\tau) = \prod_{k=0,\neq j}^{K} \frac{(\tau - \tau_k)}{(\tau_j - \tau_k)},$$

$\tau_0 = 0$, $\tau_j < \tau_{j+1}$, $j = 0, \ldots, K-1$, and h_i is the length of element i. This polynomial representation has the desirable property that $z^K(t_{ij}) = z_{ij}$, where $t_{ij} = t_{i-1} + \tau_j h_i$.

Equivalently, the *time derivative* of the state in element i can be represented as a Lagrange polynomial with K interpolation points, i.e.,

$$\frac{dz^K(t)}{d\tau} = \sum_{j=1}^{K} \bar{\ell}_j(\tau) \dot{z}_{ij},$$

where $\dot{z}_{ij}$ represents $\frac{dz^K(t_{ij})}{d\tau}$ and $\bar{\ell}_j(\tau) = \prod_{k=1,\neq j}^{K} \frac{(\tau-\tau_k)}{(\tau_j-\tau_k)}$. For element i this leads to the *Runge–Kutta* basis representation for the differential state:

$$z^K(t) = z_{i-1} + h_i \sum_{j=1}^{K} \Omega_j(\tau)\dot{z}_{ij} \tag{10.5}$$

where z_{i-1} is a coefficient that represents the differential state at the beginning of element i and $\Omega_j(\tau)$ is a polynomial of order K, satisfying

$$\Omega_j(\tau) = \int_0^{\tau} \bar{\ell}_j(\tau')d\tau', \quad t \in [t_{i-1}, t_i], \quad \tau \in [0,1].$$

To determine the polynomial coefficients that approximate the solution of the DAE, we substitute the polynomial into (10.2) and enforce the resulting algebraic equations at the interpolation points τ_k. This leads to the following *collocation equations*:

$$\frac{dz^K(t_{ik})}{dt} = f(z^K(t_{ik}), t_{ik}), \quad k = 1, \ldots, K, \tag{10.6}$$

with $z^K(t_{i-1})$ determined separately. For the polynomial representations (10.4) and (10.5), it is convenient to normalize time over the element, write the state profile as a function of τ, and apply $\frac{dz^K}{d\tau} = h_i \frac{dz^K}{dt}$. For the Lagrange polynomial (10.4), the collocation equations become

$$\sum_{j=0}^{K} z_{ij} \frac{d\ell_j(\tau_k)}{d\tau} = h_i f(z_{ik}, t_{ik}), \quad k = 1, \ldots, K, \tag{10.7}$$

while the collocation equations for the Runge–Kutta basis are given by

$$\dot{z}_{ik} = f(z_{ik}, t_{ik}), \tag{10.8a}$$

$$z_{ik} = z_{i-1} + h_i \sum_{j=1}^{K} \Omega_j(\tau_k)\dot{z}_{ij}, \quad k = 1, \ldots, K, \tag{10.8b}$$

with z_{i-1} determined from the previous element $i-1$ or from the initial condition on the ODE.

10.2.2 Collocation with Orthogonal Polynomials

Once the interpolation points τ_k are determined, the collocation equations (10.7) and (10.8) are merely algebraic equations that can be incorporated directly within an NLP formulation. What remains is the determination of τ_k that allows the most accurate approximation of the state variables for (10.2). The solution of (10.2) can also be represented by the implicit integral

$$z(t_i) = z(t_{i-1}) + \int_{t_{i-1}}^{t_i} f(z(t), t)dt, \tag{10.9}$$

and numerical solution of this integral is given by the quadrature formula

$$z(t_i) = z(t_{i-1}) + \sum_{j=1}^{K} \omega_j h_i f(z(t_{ij}), t_{ij}), \quad t_{ij} = t_{i-1} + h_i \tau_j. \tag{10.10}$$

The optimal choice of interpolation points τ_j and quadrature weights ω_j leads to $2K$ degrees of freedom with the result that (10.10) will provide the *exact* solution to (10.9) as long as $f(z(t),t)$ is a polynomial in t of order $2K$ (degree $\leq 2K-1$). The optimal choice of interpolation points is given by the following theorem.

Theorem 10.1 (Accuracy of Gaussian Quadrature). The quadrature formula (10.10) provides the exact solution to the integral (10.9) if $f(z(t),t)$ is a polynomial in t of order $2K$ and τ_j are the roots of a Kth degree polynomial, $P_K(\tau)$ with the property

$$\int_0^1 P_j(\tau)P_{j'}(\tau)d\tau = 0, \quad j=0,\ldots,K-1,\ j'=1,\ldots,K, \text{ for indices } j \neq j'. \qquad (10.11)$$

Proof: Without loss of generality, we consider only scalar profiles in (10.9) and (10.10) and define z and f directly as functions of τ with the domain of integration $\tau \in [0,1]$. We expand the integrand as a polynomial and write

$$\begin{aligned} f(\tau) &= \sum_{j=1}^{K} \ell_j(\tau) f(\tau_j) + \frac{d^K f(\bar{\tau})}{d\tau^K} \prod_{j=1}^{K} (\tau - \tau_j)/K! \\ &= \sum_{j=1}^{K} \ell_j(\tau) f(\tau_j) + q_{K-1}(\tau) \prod_{j=1}^{K} (\tau - \tau_j) + Q(\tau), \end{aligned}$$

where $\bar{\tau} \in [0,1]$, $\prod_{j=1}^{K}(\tau - \tau_j)$ is of degree K, $Q(\tau)$ is the residual polynomial of order $2K+1$, and $q_{K-1}(\tau)$ is a polynomial of degree $\leq K-1$, which can be represented as $q_{K-1}(\tau) = \sum_{j=0}^{K-1} \alpha_j P_j(\tau)$, with coefficients α_j.

Now if $f(\tau)$ is a polynomial of order $2K$ we note that $Q(\tau)=0$. Choosing τ_j so that $\prod_{j=1}^{K}(\tau - \tau_j) = \kappa P_K(\tau)$, for some $\kappa > 0$, then leads to

$$\begin{aligned} \int_0^1 f(\tau)d\tau &= \int_0^1 \sum_{j=1}^{K} \ell_j(\tau) f(\tau_j) + q_{K-1}(\tau) \prod_{j=1}^{K} (\tau - \tau_j) d\tau \\ &= \int_0^1 \sum_{j=1}^{K} \ell_j(\tau) f(\tau_j) d\tau, \end{aligned}$$

where the last integral follows from (10.11) with $j' = K$, and $j = 0,\ldots,K-1$. The desired result then follows by setting $\omega_j = \int_0^1 \ell_j(\tau)d\tau$ and noting that $t_{ij} = t_{i-1} + h_i\tau_j$. □

This result justifies the choice of collocation points τ_j as the roots of $P_K(\tau)$, the shifted Gauss–Legendre polynomial[6] with the orthogonality property (10.11). This polynomial belongs to the more general class of Gauss–Jacobi polynomials that satisfy

$$\int_0^1 (1-\tau)^{\alpha}\tau^{\beta} P_j(\tau)P_{j'}(\tau)d\tau = 0, \quad j \neq j'. \qquad (10.12)$$

[6]Gauss–Legendre polynomials are normally defined with $\tau \in [-1,1]$ as the domain of integration in (10.11). In this chapter, we will only consider $\tau \in [0,1]$.

Table 10.1. *Shifted Gauss–Legendre and Radau roots as collocation points.*

Degree K	Legendre Roots	Radau Roots
1	0.500000	1.000000
2	0.211325	0.333333
	0.788675	1.000000
3	0.112702	0.155051
	0.500000	0.644949
	0.887298	1.000000
4	0.069432	0.088588
	0.330009	0.409467
	0.669991	0.787659
	0.930568	1.000000
5	0.046910	0.057104
	0.230765	0.276843
	0.500000	0.583590
	0.769235	0.860240
	0.953090	1.000000

Using these polynomials, Theorem 10.1 can be suitably modified to allow an exact quadrature for $f(\tau)$ with degree $\leq 2K-1-\alpha-\beta$. Gauss–Jacobi polynomials of degree K can be written as

$$P_K^{(\alpha,\beta)} = \sum_{j=0}^{K} (\tau-1)^j \gamma_j \tag{10.13}$$

with

$$\gamma_j = \frac{(\alpha+K)!(\alpha+\beta+K+j)!}{(\alpha+j)!(\alpha+\beta+K)!(K-j)!j!}, \quad j=0,\dots,K.$$

With this result, we note from (10.8) that

$$z^K(t_i) = z^K(t_{i-1}) + h_i \sum_{j=1}^{K} \Omega_j(1) f(z_{ij}),$$

with $\omega_j = \Omega_j(1)$, corresponds directly to the quadrature formula (10.10) of appropriate accuracy. Because Gauss–Jacobi polynomials lead to high-order approximations, we choose their roots as the collocation points in (10.7) or (10.8). This leads to the following cases:

- $\alpha=0, \beta=0$ leads to *Gauss–Legendre* collocation with truncation error (9.6) that is $O(h^{2K})$;
- $\alpha=1, \beta=0$ leads to *Gauss–Radau* collocation with truncation error (9.6) that is $O(h^{2K-1})$;

- $\alpha = 1, \beta = 1$ leads to *Gauss–Lobatto* collocation with truncation error (9.6) that is $O(h^{2K-2})$.

We will focus on Gauss–Legendre and Radau collocation because of their compatibility with the NLP formulations and desirable stability properties seen below. Table 10.1 lists values of τ_j for Gauss–Legendre and Radau collocation for various values of K. These values allow the construction of the collocation equations (10.7) and (10.8).

So far, the collocation equations (10.7) and (10.8) are written over a single element with $z^K(t_{i-1})$ specified. If only a single element is used over the entire time domain (and K is large), then $z^K(t_0) = z_0$. For multiple elements, with $N > 1$, we enforce continuity of the state profiles across element boundaries. With Lagrange interpolation profiles, this is written as

$$z_{i+1,0} = \sum_{j=0}^{K} \ell_j(1) z_{ij}, \quad i = 1, \dots, N-1, \tag{10.14a}$$

$$z_f = \sum_{j=0}^{K} \ell_j(1) z_{Nj}, \quad z_{1,0} = z_0, \tag{10.14b}$$

and with the Runge–Kutta basis, continuity conditions are given by

$$z_i = z_{i-1} + h_i \sum_{j=1}^{K} \Omega_j(1) \dot{z}_{ij}, \quad i = 1, \dots, N-1, \tag{10.15a}$$

$$z_f = z_{N-1} + h_N \sum_{j=1}^{K} \Omega_j(1) \dot{z}_{Nj}. \tag{10.15b}$$

Finally, we note that by using the Runge–Kutta basis, equations (10.8) and (10.15) show that the collocation approach is an IRK method in the form of equations (9.8). This can be seen directly by noting the following equivalences: $n_s = K$, $c_k = \tau_k$, $a_{kj} = \Omega_j(\tau_k)$, and $b_k = \Omega_k(1)$ of appropriate accuracy. Because collocation methods are IRK methods, they enjoy the following properties described in Section 9.2.

- Collocation methods are A-stable, and both Gauss–Legendre and Radau collocation are *AN-stable*, or equivalently *algebraically stable*. As a result, there is no stability limitation on h_i for stiff problems.

- Radau collocation has stiff decay. Consequently, large time steps h_i are allowed for stiff systems that capture steady state components and slow time scales.

- Both Gauss–Legendre and Radau collocations are among the highest order methods. The truncation error (9.6) is $O(h^{2K})$ for Gauss–Legendre and $O(h^{2K-1})$ for Radau. This high-order error applies to z_i, but not to the intermediate points, z_{ij}.

To illustrate the use of collocation methods, we consider the following small IVP.

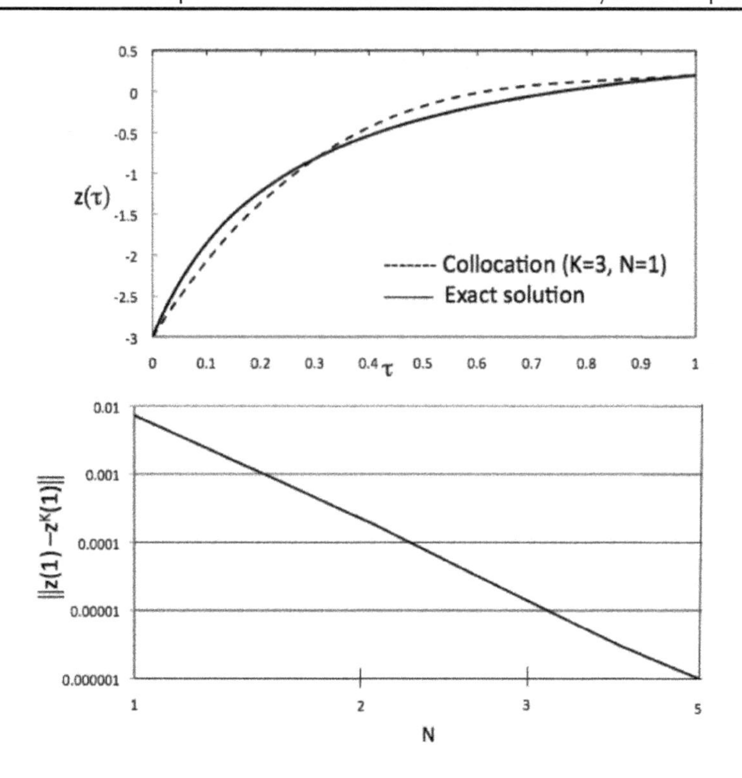

Figure 10.2. *Comparison of Radau collocation solution ($K = 3$) with exact solution for $\frac{dz}{dt} = z^2 - 2z + 1$, $z(0) = -3$.*

Example 10.2 (Demonstration of Orthogonal Collocation). Consider a single differential equation:

$$\frac{dz}{dt} = z^2 - 2z + 1, \quad z(0) = -3, \tag{10.16}$$

with $t \in [0, 1]$. This equation has an analytic solution given by $z(t) = (4t - 3)/(4t + 1)$. Using Lagrange interpolation and applying the collocation and continuity equations (10.7) and (10.14), respectively, with $K = 3$ collocation points, N elements, and $h = 1/N$ leads to

$$\sum_{j=0}^{3} z_{ij} \frac{d\ell_j(\tau_k)}{d\tau} = h(z_{ik}^2 - 2z_{ik} + 1), \quad k = 1, \ldots, 3, \; i = 1, \ldots, N,$$

and

$$z_{i+1,0} = \sum_{j=0}^{3} \ell_j(1) z_{ij}, \quad i = 1, \ldots, N - 1,$$

$$z_f = \sum_{j=0}^{K} \ell_j(1) z_{Nj}, \quad z_{1,0} = -3.$$

Using Radau collocation, we have $\tau_0 = 0$, $\tau_1 = 0.155051$, $\tau_2 = 0.644949$, and $\tau_3 = 1$. For $N = 1$ and $z_0 = -3$, the collocation equations are given by

$$\sum_{j=0}^{3} z_j \frac{d\ell_j(\tau_k)}{d\tau} = (z_k^2 - 2z_k + 1), \quad k = 1,\ldots,3,$$

which can be written out as

$$\begin{aligned}
&z_0(-30\tau_k^2 + 36\tau_k - 9) + z_1(46.7423\tau_k^2 - 51.2592\tau_k + 10.0488) \\
&+ z_2(-26.7423\tau_k^2 + 20.5925\tau_k - 1.38214) + z_3\left(10\tau_k^2 - \frac{16}{3}\tau_k + \frac{1}{3}\right) \\
&= (z_k^2 - 2z_k + 1), \quad k = 1,\ldots,3.
\end{aligned}$$

Solving these three equations gives $z_1 = -1.65701$, $z_2 = 0.032053$, $z_3 = 0.207272$ with $z_0 = -3$, along with the polynomial approximation presented in Figure 10.2 for $N = 1$. Successively increasing $N = 1/h$ and solving the corresponding collocation equations shows rapid convergence to the true solution. As seen in Figure 10.2, the error $\|z(1) - z^K(1)\|$ is less than 10^{-6} for $N = 5$ and converges with $O(h^5)$, which is consistent with the expected order $2K - 1$. ■

10.3 NLP Formulations and Solution

With the collocation discretization of the state equations, we now consider the multiperiod dynamic optimization problem given in (8.5). With either representation of the differential state profiles (10.7) and (10.8), the control variables and algebraic states can also be represented by Lagrange interpolation profiles

$$u(t) = \sum_{j=1}^{K} \bar{\ell}_j(\tau) u_{ij}, \quad y(t) = \sum_{j=1}^{K} \bar{\ell}_j(\tau) y_{ij},$$
$$\text{where } \bar{\ell}_j(\tau) = \prod_{k=1,\neq j}^{K} \frac{(\tau - \tau_k)}{(\tau_j - \tau_k)}, \tag{10.17}$$

and the collocation equations for the DAEs can be written as

$$\sum_{j=0}^{K} \dot{\ell}_j(\tau_k) z_{ij} - h_i f(z_{ik}, y_{ik}, u_{ik}, p) = 0, \quad i \in \{1,\ldots,N\},\ k \in \{1,\ldots,K\}, \tag{10.18a}$$

$$g(z_{ik}, y_{ik}, u_{ik}, p) = 0, \quad i \in \{1,\ldots,N\},\ k \in \{1,\ldots,K\}, \tag{10.18b}$$

where $\dot{\ell}(\tau) = \frac{d\ell_j(\tau)}{d\tau}$. The NLP formulation corresponding to the single-period version ($N_T = 1$) of (8.5) can be written as

$$\min \Phi(z_f) \tag{10.19a}$$

$$\text{s.t.} \sum_{j=0}^{K} \dot{\ell}_j(\tau_k) z_{ij} - h_i f(z_{ik}, y_{ik}, u_{ik}, p) = 0, \tag{10.19b}$$

$$g(z_{ik}, y_{ik}, u_{ik}, p) = 0, \tag{10.19c}$$
$$z_L \le z_{ik} \le z_U, \quad u_L \le u_{ik} \le u_U, \tag{10.19d}$$
$$y_L \le y_{ik} \le y_U, \quad p_L \le p \le p_U, \tag{10.19e}$$
$$k \in \{1, \dots, K\}, \quad i \in \{1, \dots, N\},$$
$$z_{i+1,0} = \sum_{j=0}^{K} \ell_j(1) z_{ij}, \quad i = 1, \dots, N-1, \tag{10.19f}$$
$$z_f = \sum_{j=0}^{K} \ell_j(1) z_{Nj}, \quad z_{1,0} = z(t_0), \tag{10.19g}$$
$$h_E(z_f) = 0. \tag{10.19h}$$

Problem (10.19) allows for a number of formulation options, especially when extended to multiperiod problems. The simplest case, where the dynamic system is described by a single finite element ($N = 1$), leads to the class of *pseudospectral methods* [327, 342]. Such methods can be very accurate for dynamic optimization problems that have smooth profile solutions. On the other hand, if the solution profiles have changes in active sets or the control profiles are discontinuous over time, then these solutions are not sufficiently accurate to capture the solution of the state equations, and multiple elements need to be introduced.

For the multielement formulation, it is important to note from (10.17) that the algebraic and control profiles are *not* specified at $\tau = 0$. Moreover, continuity is not enforced at the element boundary for these profiles. For algebraic profiles defined by index-1 constraints (10.18b), continuity of these profiles can be determined directly from the continuous differential variables at t_i. On the other hand, control profiles are allowed to be discontinuous at element boundaries and this allows us to capture interesting solutions of optimal control problems accurately, including those observed in Examples 8.5, 8.8, and 8.9. Finally, for process control applications, the control profile is often represented as a piecewise constant profile, where the breakpoints are defined by finite element boundaries. This profile description is straightforward to incorporate within (10.19). Moreover, as seen next, accurate control profiles can be determined with the aid of variable finite elements.

10.3.1 Treatment of Finite Elements

An important concern in the direct transcription approach is the appropriate selection of time steps h_i. Clearly, if h_i is fixed in the NLP, then the resulting problem is less nonlinear and often easier to solve. In fact, for a linear DAE system, fixing h_i leads to a linearly constrained nonlinear program. On the other hand, treating h_i as variables in (10.19), where they can "move" during the solution of the nonlinear program, provides a number of advantages. Provided that values of h_i remain suitably small, variable finite elements can locate breakpoints in the optimal control profile as well as profile segments with active bounds. Moreover, as Runge–Kutta methods require state profiles to remain smooth *only within a finite element*, only a few guidelines are needed to extend problem (10.19) to deal with variable finite elements.

- The number of finite elements N must be chosen sufficiently large. This is often estimated by trial and error through comparison with numerically integrated state profiles evaluated at different values of the decision variables p and $u(t)$.

- Moving finite element formulations are described extensively in [14, 341] for the solution of BVPs. Moreover, assuming that the state profiles are smooth over the domain of interest, the *global error* from the polynomial approximation, $e(t) = z(t) - z^K(t)$, can be estimated from

$$\max_{t\in[0,t_f]} \|e(t)\| \le C_1 \max_{i\in[1,\ldots,N]} (\|T_i(t)\|) + O(h_i^{K+2}), \tag{10.20}$$

  where C_1 is a computable constant and $T_i(t)$ can be computed from the polynomial solution. Choices for $T_i(t)$ are reviewed in Russell and Christensen [341]. In particular, the error estimate given by

$$T_i(t) = \frac{d^{K+1}z(t)}{dt^{K+1}} h^{K+1}$$

  is widely used. This quantity can be estimated from discretization of $z^K(t)$ over elements that are neighbors of element i.

- Alternately, one can obtain an error estimate from

$$T_i(t) = \frac{dz^K(t)}{d\tau} - h_i f(z^K(t), y(z^K(t)), u_i(t), p).$$

  This residual-based estimate can be calculated directly from the discretized DAEs. Here, $T(t_{ik}) = 0$ at collocation points, but choosing a noncollocation point $t_{i,nc} = t_{i-1} + h_i \tau_{nc}$, $\tau_{nc} \in [0,1]$, leads to $\|e_i(t)\| \le \bar{C}\|T_i(t_{i,nc})\|$ with the constant $\bar{C}$ given by

$$\bar{C} = \left| \frac{1}{A} \int_0^{\tau_{nc}} \prod_{j=1}^{K} (s - \tau_j) ds \right|, \quad A = \prod_{j=1}^{K} (\tau_{nc} - \tau_j).$$

  With this estimate, N sufficiently large, and a user-specified error tolerance ϵ, appropriate values of h_i can be determined by adding the constraints

$$\sum_{i=1}^{N} h_i = t_f, \quad h_i \ge 0, \tag{10.21a}$$

$$\bar{C}\|T_i(t_{i,nc})\| \le \epsilon \tag{10.21b}$$

  to (10.19). This extended formulation has been developed in [396] and demonstrated on a number of challenging reactor optimization problems. In particular, this approach allows variable finite elements to track and adapt to steep profiles encountered over the course of an optimization problem.

- On the other hand, the variable finite element formulation with (10.21) does lead to more constrained and difficult NLPs, with careful initializations needed to obtain good solutions. A simpler approach is to choose a sufficiently large number of elements and a nominal set of time steps $\bar{h}_i$ by trial and error, and to first solve (10.19) with these fixed values. Using this solution to initialize the variable element problem, we then relax h_i, add the following constraints to (10.19),

$$\sum_{i=1}^{N} h_i = t_f, \quad h_i \in [(1-\gamma)\bar{h}_i, (1+\gamma)\bar{h}_i], \tag{10.22}$$

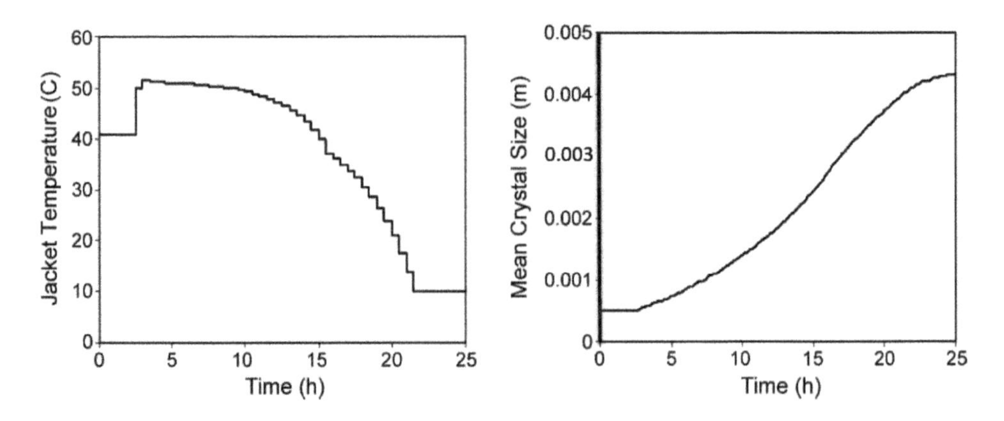

Figure 10.3. *Crystallizer growth problem with optimal cooling temperature and crystal size profiles.*

with a constant $\gamma \in [0, 1/2)$, and solve the resulting NLP again. This approach allows the time steps to remain suitably small and provides sufficient freedom to locate breakpoints in the control profiles.

- One drawback to this NLP formulation is that addition of h_i as variables may lead to additional zero curvature in the reduced Hessian for time steps that have little effect on the solution profiles. This issue will be addressed in Section 10.4. Finally, optimal control solutions can be further improved by monitoring the solution and adding additional time steps if several h_i have optimal values at their upper bounds [48, 378].

To illustrate the problem formulation in (10.19) along with variable elements, we consider the following crystallization example.

Example 10.3 (Optimal Cooling Profile for Crystallization). In this example, we consider the dynamic optimization of a crystallizer described with a simple population balance model. Here we seek to find the optimum temperature profile to cool the crystallizer and grow the largest crystals. Conventional crystallization kinetics are characterized in terms of two dominant phenomena: nucleation (the creation of new particles) and crystal growth of existing particles. Both competing phenomena consume desired solute material during the crystallization process. To obtain larger (and fewer) crystals, nucleation needs to be avoided, and the goal of the optimization is to find operating strategies that promote growth of existing crystals.

The dynamic optimization problem for the crystallizer consists of a DAE model, a lower bound on the jacket temperature as a function of the solute concentration, and an objective to maximize the crystal length. This objective also corresponds to minimizing the crystal surface area in order to obtain higher purity of the crystals. The crystallizer operates over 25 hours and the control variable is the cooling profile in the crystallizer jacket. The dynamic optimization problem can be stated as

$$\max \quad L_s(t_f) \tag{10.23a}$$

$$\text{s.t.} \quad \frac{dL_s}{dt} = K_g L_s^{0.5} \Delta T^{\eta_1}, \; L_s(0) = 0.0005 \text{ m}, \tag{10.23b}$$

$$\frac{dN_c}{dt} = B_n \Delta T^{\eta_2}, \; N_c(0) = 0, \tag{10.23c}$$

$$\frac{dL}{dt} = N_c \frac{dL_s}{dt} + L_0 \frac{dN_c}{dt}, \; L(0) = 0, \tag{10.23d}$$

$$\frac{dA_c}{dt} = 2\alpha N_c \frac{dL_s}{dt} + L_0^2 \frac{dN_c}{dt}, \; A_c(0) = 0, \tag{10.23e}$$

$$\frac{dV_c}{dt} = 3\beta A_c \frac{dL_s}{dt} + L_0^3 \frac{dN_c}{dt}, \; V_c(0) = 0, \tag{10.23f}$$

$$\frac{dM_c}{dt} = 3(W_{s0}/L_{s0}^3) L_s^2 \frac{dL_s}{dt} + \rho V_s \frac{dV_c}{dt}, \; M_c(0) = 2.0 \text{ kg}, \tag{10.23g}$$

$$\frac{dC_c}{dt} = -\frac{1}{V_s} \frac{dM_c}{dt}, \; C_c(0) = 5.4 \text{ kg/l}, \tag{10.23h}$$

$$\frac{dT_c}{dt} = \left(K_c \frac{dM_c}{dt} - K_e (T_c - T_j) \right) / (W C_p), \; T_c(0) = 75^\circ \text{ C} \tag{10.23i}$$

$$T_a \le T_j \in [10, 100^\circ \text{ C}] \tag{10.23j}$$

where L_s is the mean crystal size, N_c is the number of nuclei per liter (l) of solvent, L is the total length of the crystals per liter of solvent, A_c the total surface area of the crystals per liter of solvent, V_c is the total volume of the crystals per liter of solvent, C_c is the solute concentration, M_c is the total mass of the crystals, and T_c is the crystallizer temperature. Additional constants include $V_s = 300l$, the volume of the solvent, $W = 2025\,\text{kg}$, the total mass in the crystallizer, $\Delta T = \max(0, T_{equ} - T_c)$, the degree of supercooling,[7] $T_{equ} = \sum_{i=1}^{4} a_i (\bar{C})^{i-1}$, the equilibrium temperature, $\bar{C} = 100 C_c / (1.35 + C_c)$, the weight fraction, $L_{s0} = 5 \times 10^{-4} m$, the initial crystal size, $L_0 = 5 \times 10^{-5} m$, the nucleate crystal size, $W_{s0} = 2\,\text{kg}$, the weight of seed crystals, $\rho = 1.58\,\text{kg}/l$, the specific gravity of the crystals, and $\alpha = 0.2$, $\beta = 1.2$, the shape factors for area and volume of the crystals, respectively. The control variable is the jacket temperature, T_j, which has a lower bound, $T_a = \sum_{i=1}^{4} b_i (\bar{C})^{i-1}$. The remaining parameter values are $K_g = 0.00418$, $B_n = 385$, $C_p = 0.4$, $K_c = 35$, $K_e = 377$, $\eta_1 = 1.1$, and $\eta_2 = 5.72$. Finally the polynomial coefficients for T_{equ} and T_a are:

$$a = [-66.4309, 2.8604, -0.022579, 6.7117 \times 10^{-5}],$$
$$b = [16.08852, -2.708263, 0.0670694, -3.5685 \times 10^{-4}].$$

Applying the formulations (10.19) and (10.22) to this model, with a piecewise constant control profile, three-point Radau collocation, and 50 variable finite elements, leads to an NLP with 1900 variables and 1750 equality constraints. The optimal solution was obtained in 12.5 CPUs (1.6 MHz, Pentium 4 PC running Windows XP) using 105 iterations of a reduced-space interior point solver (IPOPT v2.4). The optimal profiles of the mean crystal size and the jacket temperature are given in Figure 10.3. Over the 25-hour operation, the mean crystal size increases by over eight times, from 0.5 mm to 4.4 mm. Also, note

[7] The max function is replaced by a smoothed approximation.

that in order to maximize the crystal size, the jacket *cooling* temperature must first *increase* to reduce the number of nucleating particles. Additional information on this optimization study can be found in [242]. ■

10.3.2 Treatment of Unstable Dynamic Systems

The discretization of the DAEs in the simultaneous collocation approach can be viewed as a direct extension of the multiple shooting approach discussed in Section 9.4. As a result, it inherits the dichotomy property, which is essential for the treatment of unstable systems. As noted in Section 9.4.1, the NLP formulation (10.19) obtains stable, optimal solutions because

- state variable bounds imposed over a finite element limit the escape associated with unstable (i.e., increasing) dynamic modes,
- the increasing modes are pinned down by boundary conditions. As a result, the Green's function associated with the corresponding BVP for the optimal control problem is well-conditioned [109]. Moreover, de Hoog and Mattheij [109] noted that stability bounds for the collocation equations approach κ in (9.60) as the time steps approach zero.

To develop the second concept for the NLP formulation (10.19), we note that the dichotomy property can be enforced by maintaining a stable pivot sequence for the associated KKT matrix. To see this, consider the Jacobian of the collocation equations given by

$$A^T = \left[\begin{array}{cccccc|ccc} I & & & & & & 0 & & \\ T^1 & C^1 & & & & & U^1 & & \\ \bar{D}^1 & D^1 & -I & & & & & & \\ & & T^2 & C^2 & & & & U^2 & \\ & & \bar{D}^2 & D^2 & -I & & & & \\ & & & & T^3 & C^3 & & & U^3 \\ & & & & & \ddots \quad \ddots & & & \ddots \end{array}\right] = [A_z | A_u], \tag{10.24}$$

where T^i is the Jacobian of z_{i0}, C^i is the Jacobian of the state variables z_{ik}, y_{ik}, and U^i is the Jacobian of the control variables u_{ik} for (10.19b)–(10.19c) in element i. Similarly, $\bar{D}^i$ and D^i are the Jacobians with respect to z_{i0} and z_{ik} in (10.19f). The partition of A^T into A_z and A_u can be interpreted as the selection of dependent (basic) and decision (superbasic) variables in the reduced-space optimization strategy.

With initial conditions specified in the first row of A^T, u_{ik} fixed, and A_z square and nonsingular, solving for the state variables leads to an ill-conditioned system in the presence of increasing (i.e., unstable) dynamic modes. This essentially mimics the solution strategy of the sequential optimization approach in Figure 9.1. On the other hand, if the columns of U^i span the range of the rows of $-I$ that correspond to unstable modes, then by repartitioning A_z and A_u, the control variable could be shifted from A_u to A_z, and the corresponding columns of $-I$ (in an element $j > i$) could be shifted from A_z to A_u. Moving these unstable states (columns of $-I$) to A_u has the effect of providing boundary conditions that pin down the unstable modes. In addition, moving control variables into A_z

Table 10.2. *HIPS kinetic mechanism.*

Initiation reactions		
Thermal	$3M_S \xrightarrow{k_{i0}} 2R_S^1$	
Chemical	$I \xrightarrow{k_d} 2R$	$R + M_S \xrightarrow{k_{i1}} R_S^1$
	$R + B_0 \xrightarrow{k_{i2}} B_R$	$B_R + M_S \xrightarrow{k_{i3}} B_R^1$
Propagation reactions	$R_S^j + M_S \xrightarrow{k_p} R_S^{j+1}$	$B_{RS}^j + M_S \xrightarrow{k_p} B_{RS}^{j+1}$
Termination reactions		
Homopolymer	$R_S^j + R_S^m \xrightarrow{k_t} P_M^j$	
Grafting	$R_S^j + B_R \xrightarrow{k_t} B_P^j$	$R_S^j + B_{RS}^m \xrightarrow{k_t} B_P^{j+m}$
Crosslinking	$B_R + B_R \xrightarrow{k_t} B_{EB}$	$B_{RS}^j + B_R \xrightarrow{k_t} B_{PB}^j$
	$B_{RS}^j + B_{RS}^m \xrightarrow{k_t} B_{PB}^{j+m}$	
Transfer reactions		
Monomer	$R_S^j + M_S \xrightarrow{k_{fs}} P^j + R_S^1$	$B_{RS}^j + M_S \xrightarrow{k_{fs}} B_P^j + R_S^1$
Grafting sites	$R_S^j + B_0 \xrightarrow{k_{fb}} P^j + B_R$	$B_{RS}^j + B_0 \xrightarrow{k_{fb}} B_P^j + B_R$

leads to corresponding free states starting from element $i+1$. Moreover, this repartitioning strategy can be left to the KKT matrix solver alone, as long as a reliable pivoting strategy is applied. Because of the relation between dichotomous BVPs and well-conditioned discretized systems [109], the pivoting strategy itself should then lead to the identification of a stable repartition for A_z.

To demonstrate this approach we consider a challenging dynamic optimization problem that deals with an unstable polymerization reactor.

Example 10.4 (Grade Transition for High-Impact Polystyrene). We consider the dynamic optimization of a polymerization reactor that operates at unstable conditions. Polymerization reactors typically manufacture a variety of products or grades, and an effective grade transition policy must feature minimum transition time in order to minimize waste product and utility consumption. The minimum transition policy can be determined from the discretized dynamic optimization problem given by (10.19) with the objective given by

$$\min \int_0^\theta \|z(t) - \hat{z}\|^2 + \|u(t) - \hat{u}\|^2 dt, \tag{10.25}$$

where $\hat{z}$ and $\hat{u}$ are the states and inputs for the desired operating point of the new product grade, and θ is the transition horizon length. We consider the grade transition in [141] that deals with free-radical bulk polymerization of styrene/polybutadiene, using a monofunctional initiator (I) to form high-impact polystyrene (HIPS). The polymerization occurs in a nonisothermal stirred tank reactor assuming perfect mixing, constant reactor volume and physical properties, and quasi steady state and the long chain assumptions for the polymerization reactions. The reaction mechanism involves the initiation, propagation, transfer, and termination reactions shown in Table 10.2. Polybutadiene is also added in

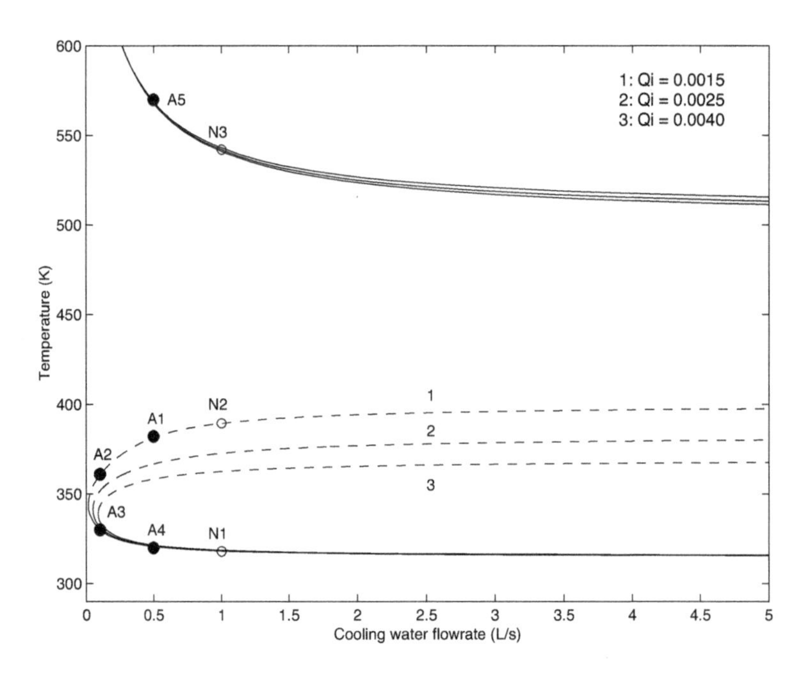

Figure 10.4. *Multiple steady states for HIPS reactor. The dashed lines indicate unstable branches for operation.*

order to guarantee desired mechanical properties by promoting grafting reactions. In Table 10.2, the superscript on each species refers to the length of the polymer chain, B_0 is the polybutadiene unit, B_{EB} is the cross-linked polybutadiene, B_P is the grafted polymer, B_{PB} is the cross-linked polymer, B_R is an activated polybutadiene unit, B_{RS} is grafted radical with a styrene end group, I is the initiator, M_S is styrene monomer, P is the homopolymer, and R is the free radical. The dynamic model can be found in [393, 141] along with rate constants and reactor data. This model includes a mass balance for the initiator, monomer, butadiene, and radical species. Also, included are zeroth-moment models for the polymer products and an energy balance over the stirred tank reactor. Finally, the manipulated variables are cooling water flow rate F_j and initiator volumetric flow rate F_i.

For the grade transition problem, Figure 10.4 displays the multiple steady states for this reactor in the space of one of the input parameters: cooling water flow rate. Under nominal operating conditions ($F_j = 1\, l/s$ and $F_i = 0.0015\, l/s$), the reactor exhibits three steady state values (denoted by N1, N2, and N3). The lower and upper steady state branches are stable but represent undesired operating conditions. On the intermediate unstable branch, the monomer conversion rises to around 30%, which is given by the nominal, but unstable, operating point N2. Note that a sequential approach cannot be applied to transitions on the intermediate branches, as numerical errors in the integration may lead to drift of the transition to the stable outer branches.

In [141], 14 grade transitions were considered from N2 to either stable or unstable steady states, using either cooling water and/or initiator flow as control variables. NLP for-

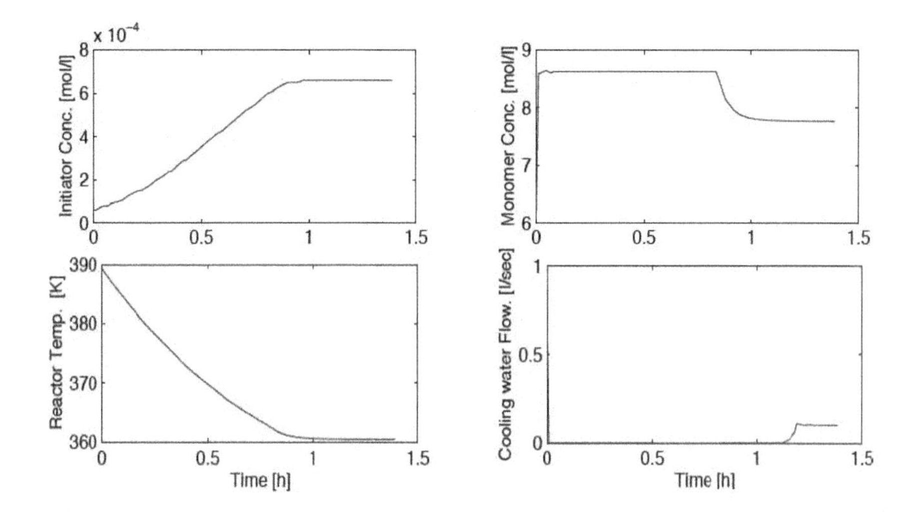

Figure 10.5. *Optimal state and control profiles for HIPS grade transition:* N2 → A2.

mulations of the form (10.19) and (10.25) were solved with the polymerization model using the IPOPT code and Radau collocation with $K = 3$ and finite elements up to $N = 40$. To demonstrate this approach, we consider the transition from N2 to A2 in Figure 10.4. For this transition, cooling water flow rate was the manipulated variable, and the resulting nonlinear program has 496 variables, 316 equations, and 20 finite elements. The optimization problem required 223 iterations and 5.01 CPUs (1.6 MHz Pentium 4 PC running Linux) to solve.

At the desired state A2, the reactor operates close to a turning point. To achieve this state, the optimal cooling water flow rate suddenly drops from $1\,l/s$ to its lower bound and remains there for most of the transition time. It then rises to its final desired value at the end. From Figure 10.5 we see that the states reach their desired end values in less than 1.3 h. From the expected behavior of nonisothermal reactors with multiple steady states, we note that at the *unstable point* N2, the loss of cooling leads to a *decrease in temperature.* With this decrease in reactor temperature, the propagation rate step is reduced, less initiator is consumed, and monomer and initiator concentration increase. When the flow rate increases at the end, the monomer concentration decreases as well, and conversion increases to a value higher than that at the initial point. ∎

10.3.3 Large-Scale NLP Solution Strategies

With the simultaneous collocation approach, an important consideration is the efficient solution of large nonlinear programs in the form of (10.19). From the development of NLP strategies in Chapter 6 and the hierarchy of optimization models in Figure 7.2, it is clear that a large-scale NLP strategy is needed that exploits problem structure and sparsity. In particular, we consider a Newton-based algorithm that requires few iterations, inexpensive solution of the Newton step, and low complexity in handling inequality constraints. These advantages can be realized by a full-space NLP formulation with exact first and second derivatives and a barrier method to handle inequality constraints. These methods are described in Section 6.3 and include the popular IPOPT, KNITRO, and LOQO codes.

Table 10.3. *Computational complexity/NLP iteration (with $n_w = n_z + n_y$ state variables, n_u control variables, N time steps, $2 \le \alpha \le 3$, $1 \le \beta \le 2$).*

	Sequential	Multiple Shooting	Simultaneous Collocation
(i) Integration	$n_w^\beta N$	$n_w^\beta N$	—
(ii) Sensitivity	$n_w n_u N^2$	$n_w(n_w + n_u)N$	$(n_w + n_u)N$
(iii) Hessian Evaluation	$n_w n_u^2 N^3$	$n_w(n_w + n_u)^2 N$	$(n_w + n_u)N$
(iv) Decomposition	—	$n_z^3 N$	—
(v) Factorization/Backsolve	$(n_u N)^\alpha$	$(n_u N)^\alpha$	$((n_w + n_u)N)^\beta$

With the aid of Newton-based barrier solvers, we note that while simultaneous collocation approaches lead to larger NLP formulations, the effort to solve them still remains quite reasonable. The computational complexity of this approach is explored through a comparison with dynamic optimization strategies developed in Chapter 9. Table 10.3 lists the complexity of the major algorithmic steps for dynamic optimization of (9.1) using the sequential, multiple shooting, and simultaneous collocation strategies. While a detailed comparison of these methods is often problem dependent, the table allows a brief overview of the computational effort for each method as well as a discussion of distinguishing features in each algorithm.

Step (i) requires sequential and multiple shooting methods to invoke an embedded DAE solver that *integrates* forward in time. The integration is performed with a Newton solver at each time step, and often with a sparse matrix routine embedded within the Newton solver. Sparse factorization of the Newton step occurs at a cost that scales slightly more than linearly with problem size (i.e., with exponent $\beta \in [1,2]$). For the simultaneous approach, this step is replaced by the optimization step (v). In Step (ii) both multiple shooting and sequential approaches obtain reduced gradients through direct *sensitivity calculations* of the DAE system. While this calculation is often implemented efficiently, the cost scales linearly with the number of inputs times the size of the DAE system, since previous factorizations can be reused. With the sequential approach, the number of inputs is $n_u N$; with multiple shooting, sensitivity is calculated separately in each time step and the number of inputs is $n_w + n_u$. For the simultaneous approach the gradient calculation (through automatic differentiation) usually scales linearly with the problem size. Step (iii) deals with the calculation of second derivatives, which is rarely used for the embedded optimization approaches. For both multiple shooting and sequential approaches the cost of reduced Hessians scales roughly with the number of inputs times the *sensitivity* cost. Instead, quasi-Newton methods are employed, but often at the cost of additional iterations of the NLP solver, as seen in Example 9.4. On the other hand, the calculation cost of the sparse Hessian for the simultaneous collocation approach usually scales linearly with the problem size.

In addition, multiple shooting executes a *decomposition* (step (iv)) which requires projection of the Hessian to dimension $n_u N$, through the factorization of *dense* matrices. With the collocation approach, the Hessian remains sparse and its calculation (through automatic differentiation) also scales with the problem size. Step (v) deals with the optimization step determination; sequential and multiple shooting methods require the solution of a quadratic program (QP) with $n_u N$ variables, and dense constraint and reduced Hessian

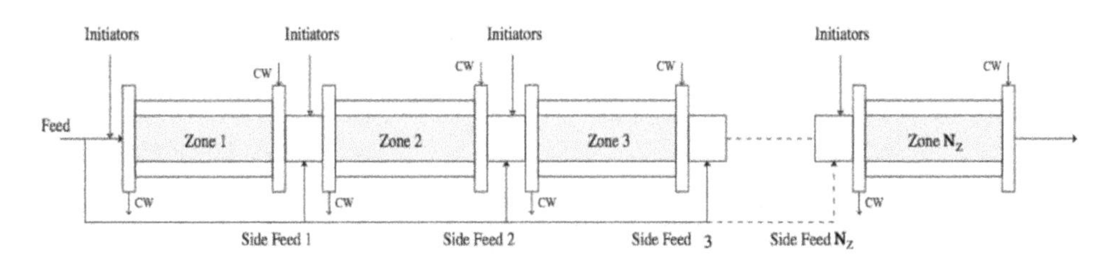

Figure 10.6. *Schematic representation of a typical high-pressure LDPE tubular reactor.*

matrices. These require dense factorizations and matrix updates that scale with the exponent $\alpha \in [2,3]$. The QP also chooses an active constraint set, a combinatorial step which is often accelerated by a *warm start.* On the other hand, with a barrier approach applied to simultaneous collocation, the active set is determined from the solution of nonlinear (KKT) equations through a Newton method. The corresponding Newton step is obtained through sparse factorization and a backsolve of the KKT system in step (v).

Table 10.3 shows that as the number of inputs $n_u N$ increases, one sees a particular advantage to the simultaneous collocation approach incorporated with a barrier NLP solver. Moreover, the specific structure of the KKT matrix can be exploited with the simultaneous collocation approach as demonstrated in the following case study.

10.3.4 Parameter Estimation for Low-Density Polyethylene Reactors

To demonstrate the solution of large-scale NLP formulations for dynamic optimization, we consider a challenging parameter estimation problem for a low-density polyethylene (LDPE) reactor. LDPE is one of the highest volume manufactured polymers and is widely used in films, sheets, and packaging. LDPE polymer grades are produced in gas-phase, multizone tubular reactors at high temperature (130–300 °C) and pressure (1500–3000 atm). Here, ethylene is polymerized through a free-radical mechanism [229] in the presence of complex mixtures of peroxide initiators. A typical tubular reactor can be described as a jacketed, multizone device with a predefined sequence of reaction and cooling zones. Different configurations of monomer and initiator mixtures enter in feed and multiple sidestreams, and they are selected to maximize reactor productivity and obtain the desired polymer properties. The total reactor length ranges between 0.5 to 2 km, while its internal diameter does not exceed 70–80 mm. A schematic representation of a typical tubular reactor is presented in Figure 10.6.

DAE models for LDPE tubular reactors comprise detailed polymerization kinetic mechanisms and rigorous methods for the prediction of the reacting mixture as well as thermodynamic and transport properties at extreme conditions. For this case study, a first-principles model is applied that describes the gas-phase free-radical homopolymerization of ethylene in the presence of several different initiators and chain-transfer agents at supercritical conditions. Details of this model are presented in [418], and the reaction mechanism for the polymerization is presented in Table 10.4. Here, $I_i, i = 1,\dots,N_I$, $R\cdot$, M, and $S_i, i = 1,\dots,N_S$, denote the initiators, radicals, monomer, and chain-transfer agent (CTA) molecules, respectively. The symbol η_i represents the efficiency of initiator i, P_r represents "live" polymer chains, and M_r are "dead" polymer chains with r monomer units. The corresponding reaction rates for the monomers, initiators, CTAs, and "live" and "dead" polymer

Table 10.4. *Ethylene homopolymerization kinetic mechanism.*

Initiator(s) decomposition	**Incorporation of CTAs**
$I_i \xrightarrow{\eta_i k_{d_i}} 2R \quad i = 1, N_I$	$P_r + S_i \xrightarrow{k_{spi}} P_{r+1} \; i = 1, N_s$
Chain initiation	**Termination by combination**
$R^{\cdot} + M \xrightarrow{k_{I_1}} P_1$	$P_r + P_x \xrightarrow{k_{tc}} M_{r+x}$
Chain propagation	**Termination by disproportionation**
$P_r + M \xrightarrow{k_p} P_{r+1}$	$P_r + P_x \xrightarrow{k_{td}} M_r + M_x$
Chain transfer to monomer	**Backbiting**
$P_r + M \xrightarrow{k_{fm}} P_1 + M_r$	$P_r \xrightarrow{k_b} P_r'$
Chain transfer to polymer	**β-scission of sec- and tert-radicals**
$P_r + M_x \xrightarrow{k_{fp}} P_x + M_r$	$P_r \xrightarrow{k_\beta} M_r^{=} + P_1$
Chain transfer to CTAs	
$P_r + S_i \xrightarrow{k_{si}} P_1 + M_r \; i = 1, N_s$	

chains can be obtained by combining the reaction rates of the elementary reactions describing their production and consumption.

To simplify the description of the polymer molecular weight distribution and related properties, we apply the method of moments. This method represents average polymer molecular weights in terms of the leading moments of the polymer chain-length distributions. Additional simplifying assumptions are described in [418, 419] to develop *steady state* differential molar and energy balances that describe the evolution of the reacting mixture along each reactor zone. The detailed design equations are reported in [418].

For high-pressure LDPE reactors, the determination of the kinetic rate constants in Table 10.4 remains a key challenge. These rate constants are both pressure and temperature dependent, using a general Arrhenius form with parameters determined from reactor data. In addition to estimating kinetic parameters, adjustable parameters must also be determined that account for uncertainty in reactor conditions. These are mainly due to fouling of the inner reactor wall from continuous polymer build-up and the decomposition of reaction initiators in each reaction zone. Because both phenomena are difficult to predict by a mechanistic model, heat-transfer coefficients (HTCs) that account for fouling are estimated to match the plant reactor temperature profile in each zone. Similarly, efficiency factors η_i are associated with the decomposition of each initiator i. Because initiator efficiencies vary widely along the reactor, they are estimated for each reaction zone in order to match the plant reactor temperature profile. This approach was demonstrated in [418] and a typical model fit to the data is presented in Figure 10.7.

The combined reactor model with multiple zones can be formulated as the following multiperiod index-1 DAE system:

$$
\begin{aligned}
\mathbf{F}_{s,l}\left[\frac{dz_{s,l}(t)}{dt}, z_{s,l}(t), y_{s,l}(t), p_{s,l}, \Pi\right] &= 0, \\
\mathbf{G}_{s,l}\left[z_{s,l}(t), y_{s,l}(t), p_{s,l}, \Pi\right] &= 0, \\
z_{s,l}(0) &= \phi(z_{s,l-1}(t_{L_{s,l-1}}), u_{s,l}), \\
s = 1, \ldots, NS, \quad & l = 1, \ldots, NZ_s.
\end{aligned} \tag{10.26}
$$

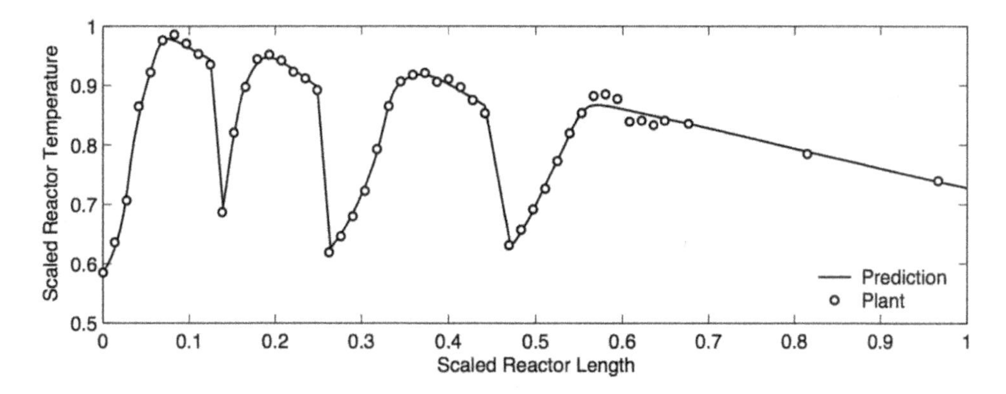

Figure 10.7. *Temperature profile and model fit for LDPE reactor.*

The subscript l refers to the reactor zone defined for an operating scenario s for which a data set is provided. Also, NZ_s is the total number of zones in scenario s; this allows parameter estimation with data from different reactor configurations. In addition, note that the DAE models are coupled across zones through material and energy balances $\phi(\cdot)$. This coupling expression contains input variables $u_{s,l}$ that include flow rates and temperatures for the monomer, initiator, and cooling water sidestreams along the reactor. Also, $t_{L_{s,l}}$ denotes the total length of zone l in scenario s. The decision variables $p_{s,l}$ represent local parameters corresponding to the HTCs and initiator efficiencies for each zone l and scenario s, and the variables Π correspond to the kinetic rate constants. As developed in [418], the reactor model contains around 130 ODEs and 500 algebraic equations for each scenario s, and the total number of DAEs in (10.26) increases linearly with the number of scenarios. In addition to the large number of equations, the reactor model is highly nonlinear and stiff.

The objective of this dynamic optimization problem is to estimate the (global) kinetic parameters, Π, as well as (local) HTCs and initiator efficiencies, $p_{s,l}$, that match the plant reactor operating conditions and polymer properties. For this we consider a multiscenario errors-in-variables-measured (EVM) parameter estimation problem of the form

$$
\begin{aligned}
\min_{\Pi,\, p_{s,l},\, u_{s,l}} \quad & \sum_{s=1}^{NS} \sum_{l=1}^{NZ_s} \sum_{i=1}^{NM_{s,l}} \left(y_{s,l}(t_i) - y^M_{s,l,i} \right)^T \mathbf{V}_{\mathbf{y}}^{-1} \left(y_{s,l}(t_i) - y^M_{s,l,i} \right) \\
& + \sum_{s=1}^{NS} \sum_{l=1}^{NZ_s} \left(u_{s,l} - u^M_{s,l} \right)^T \mathbf{V}_{\mathbf{u}}^{-1} \left(u_{s,l} - u^M_{s,l} \right) \\
\text{s.t.} \quad & \mathbf{F}_{s,l}\left[\frac{dz_{s,l}(t)}{dt}, z_{s,l}(t), y_{s,l}(t), u_{s,l}, p_{s,l}, \Pi \right] = 0, \\
& \mathbf{G}_{s,l}\left[z_{s,l}(t), y_{s,l}(t), u_{s,l}, p_{s,l}, \Pi \right] = 0, \\
& \mathbf{H}_{s,l}\left[z_{s,l}(t), y_{s,l}(t), p_{s,l}, \Pi \right] \le 0, \\
& z_{s,l}(0) = \phi(z_{s,l-1}(t_{L_{s,l-1}}), u_{s,l}), \\
& s = 1, \dots, NS, \quad l = 1, \dots, NZ_s,
\end{aligned}
\tag{10.27}
$$

where the input variables u include flow rates and temperatures for the monomer, initiator, and cooling water streams in each zone. The output variables y contain the reactor temperature profile along each zone, jacket outlet temperatures at each zone, and molecular weight distribution and product quality (melt index and density) at the reactor outlet. All of these are matched to the corresponding plant measurements (with covariances $\mathbf{V_u}$ and $\mathbf{V_y}$) for each operating scenario s.

Solving (10.27) was effected through the simultaneous collocation approach. For each scenario (or data set), 16 finite elements were used for the reaction zones, 2 finite elements were used for the cooling zones, and 3 Radau collocation points were used in each element. This leads to around 12,000 constraints, 32 local parameters p_s, and 35 input variables u_s for each scenario, as well as 25 global variables. The resulting nonlinear program was solved with IPOPT (version 3.2) using a general-purpose linear solver for the KKT system. Using this solution, it was shown in [418] that the 95% confidence regions for the estimated parameters can be reduced substantially as the number of data sets N_s is increased. On the other hand, as N_s increases, the general-purpose linear solver becomes the key bottleneck to efficient solution of the nonlinear program.

Instead, the KKT matrix for (10.27) can be reordered to expose an efficient arrowhead structure of the form

$$\begin{bmatrix} K_1 & & & & & A_1 \\ & K_2 & & & & A_2 \\ & & K_3 & & & A_3 \\ & & & \ddots & & \vdots \\ & & & & K_{NS} & A_{NS} \\ A_1^T & A_2^T & A_3^T & \cdots & A_{NS}^T & 0 \end{bmatrix} \begin{bmatrix} \Delta v_1 \\ \Delta v_2 \\ \Delta v_3 \\ \vdots \\ \Delta v_{NS} \\ \Delta \Pi \end{bmatrix} = \begin{bmatrix} r_1 \\ r_2 \\ r_3 \\ \vdots \\ r_{NS} \\ r_\Pi \end{bmatrix}, \qquad (10.28)$$

where v_s are the primal and dual variables in each scenario s, r_s corresponds to the corresponding KKT conditions in each scenario, and r_Π corresponds to the KKT condition with respect to Π. The system (10.28) contains diagonal blocks K_s that represent the KKT matrix for each scenario, and the matrices A_s represent the equations that link the global variables to each scenario. Moreover, by replacing Π by local variables in each scenario and introducing additional linear equations that link these variables to Π, the matrices A_s have a simple, sparse structure that can be automated in a general-purpose way [419]. With the structure in (10.28), one can apply a Schur complement decomposition strategy that avoids the full factorization of the KKT matrix on a single processor. Instead, when distributed over multiple processors, one avoids memory bottlenecks and can handle a large number of scenarios (i.e., data sets) in the parameter estimation problem. Moreover, to obtain exact first and second derivative information, each scenario was implemented as a separate AMPL model that indicates internally the set of variables corresponding to the global parameters Π. This allows the construction of a linking variable vector that is passed to the IPOPT solver.

Figure 10.8 presents computational results for the solution of multiscenario nonlinear programs with up to 32 data sets. The results were obtained in a Beowulf-type cluster using standard Intel Pentium IV Xeon 2.4 GHz, 2 GB RAM processors running Linux. These results are also compared with the serial solution of the multiscenario problems on a single processor. As seen in the figure, the serial solution of the multiscenario nonlinear programs

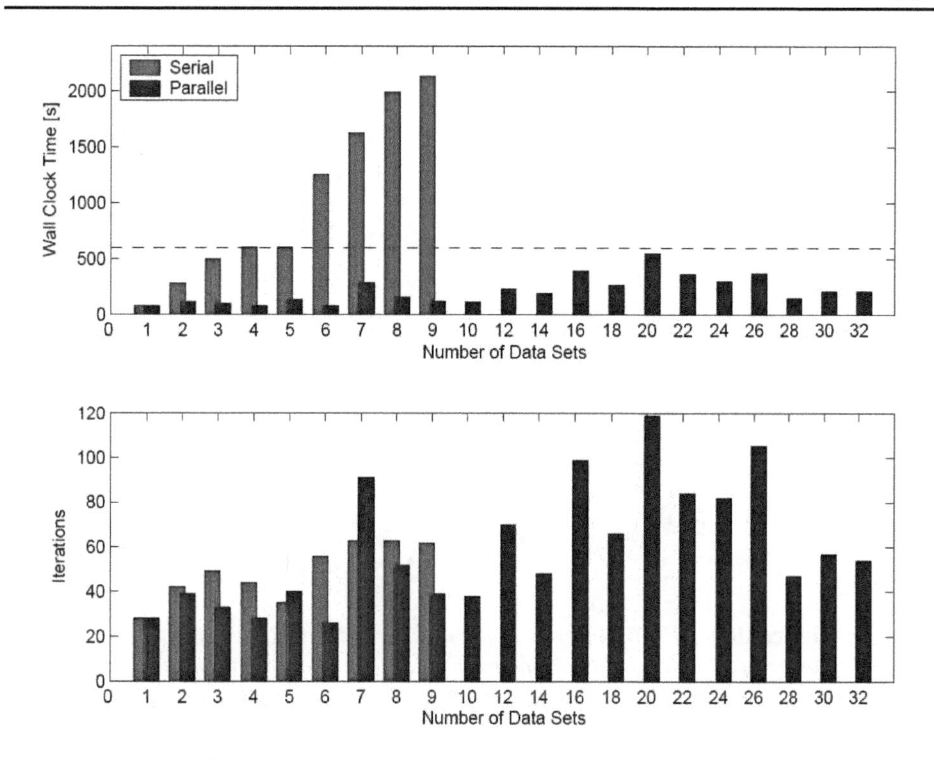

Figure 10.8. *Total CPU time and iterations for the solution of multiscenario nonlinear programs with serial and parallel implementations.*

has memory available for only 9 data sets, while the parallel implementation overcomes this memory bottleneck and solves problems with over 32 data sets. In all of the analyzed cases, sufficient second order KKT conditions are satisfied and, consequently, the full parameter set can be estimated uniquely. The largest problem solved contains around 4,100 differential and 16,000 algebraic equations, over 400,000 variables, and 2,100 degrees of freedom. The solution time in the serial implementation increases significantly with the addition of more data sets; the 9 data set problem requires over 30 wall clock CPU minutes. In contrast, the parallel solution takes consistently less than 10 wall clock minutes regardless of the number of data sets. On the other hand, it is important to emphasize that this behavior is problem (and data) dependent. In fact, the solution of the 32 data set problem requires fewer iterations than with 20 data sets. This behavior is mainly attributed to the nonlinearity of the constraints and the influence of the initialization with different N_S.

To provide more insight into the performance of the algorithm, Figure 10.9 presents computational results for the time required per iteration and per factorization of the KKT matrix. This leads to a more consistent measure of the scalability of the proposed strategy. For the parallel approach, a near-perfect scaling is reflected in the time per factorization. Also, the time per iteration can be consistently kept below 5 wall clock CPU seconds, while the factorization in the serial approach can take as much as 35 wall clock seconds before running out of memory. Additional information on this case study can be found in [419].

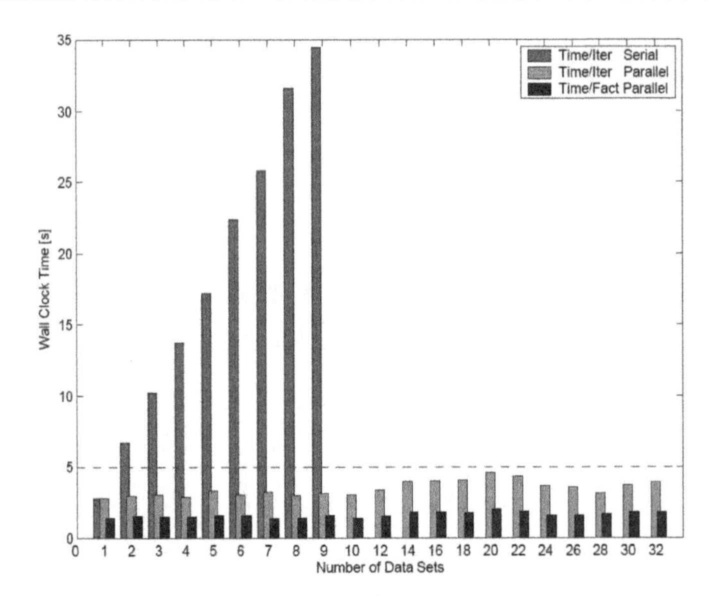

Figure 10.9. *CPU time per iteration and per factorization of the KKT matrix during the solution of multiscenario nonlinear programs with serial and parallel implementations.*

10.4 Convergence Properties of Simultaneous Approach

In Chapter 8, the Euler–Lagrange equations were formulated in an infinite-dimensional space, as optimality conditions for the solution of the optimal control problem (8.6). From these conditions, indirect methods were proposed that led to an "optimize then discretize" approach, where the DAEs representing the Euler–Lagrange equations were discretized to finite dimension and solved to a desired level of accuracy.

In contrast, direct approaches based on a "discretize then optimize" strategy were developed in Chapter 9 and also in this chapter. For the direct approaches in Chapter 9, we have focused on the solution of nonlinear programs with a finite number of decision variables, either through the variables p or through discretization of the control profiles. On the other hand, simultaneous collocation methods have the ability to discretize both the control and state profiles at the same level of accuracy. Consequently, it is important to consider the characteristics of the simultaneous approach and the conditions under which it can converge to the solution of (8.6). In this section, we analyze these properties for Gauss–Legendre and Radau collocation with simultaneous NLP formulations. We then consider some open questions when the nonlinear program is no longer well-posed, and we offer some guidelines and directions for future work.

For this analysis we consider a simpler optimal control problem of the form

$$\min\ \Phi(z(t_f)) \tag{10.29a}$$

$$\text{s.t. } \frac{dz}{dt} = f(z(t),u(t)), \quad z(t_0) = z_0, \tag{10.29b}$$

$$h_E(z(t_f)) = 0. \tag{10.29c}$$

Note that in (10.29) we have removed the algebraic variables $y(t)$ as these are implicit functions of $z(t)$ and $u(t)$ from the (index-1) algebraic equations. To focus on convergence of the control profiles, we also removed decision variables p of finite dimension. Finally, we have removed the inequality constraints, as they present additional difficulties for the optimality conditions in Chapter 8. These will be discussed later in this section.

As derived in Chapter 8, the Euler–Lagrange equations are given by

$$\frac{dz}{dt} = f(z,u), \quad z(t_0) = z_0, \tag{10.30a}$$

$$h_E(z(t_f)) = 0, \tag{10.30b}$$

$$\frac{d\lambda}{dt} = -\frac{\partial f(z,u)}{\partial z}\lambda(t), \tag{10.30c}$$

$$\lambda(t_f) = \frac{\partial \Phi(z(t_f))}{\partial z} + \frac{\partial h_E(z(t_f))}{\partial z}\eta_E, \tag{10.30d}$$

$$\frac{\partial f(z,u)}{\partial u}\lambda = 0. \tag{10.30e}$$

These conditions form a BVP, and to obtain a well-posed solution, we require the following assumption, which will be examined later in this section.

Assumption I: The BVP (10.30a)–(10.30e) is dichotomous with an index-1 algebraic equation (10.30e) that allows unique solution of the algebraic variable $u(t)$.

One can either solve the BVP (10.30a)–(10.30e) with some discretization method or instead discretize (10.29) over time and solve the related nonlinear program, as is done with the direct transcription approach discussed in this chapter. This section shows the relationship between both methods and also addresses related issues such as convergence rates (as a function of time step h) and estimation of adjoint profiles from KKT multipliers.

The collocation discretization of problem (10.29) is given by

$$\min \Phi(z_f) \tag{10.31a}$$

$$\text{s.t.} \sum_{j=0}^{K} \dot{\ell}_j(\tau_k) z_{ij} - h_i f(z_{ik}, u_{ik}) = 0, \tag{10.31b}$$

$$k \in \{1,\dots,K\}, \quad i \in \{1,\dots,N\},$$

$$z_{i+1,0} = \sum_{j=0}^{K} \ell_j(1) z_{ij}, \quad i = 1,\dots,N-1, \tag{10.31c}$$

$$z_f = \sum_{j=0}^{K} \ell_j(1) z_{Nj}, \quad z_{1,0} = z(t_0), \tag{10.31d}$$

$$h_E(z_f) = 0 \tag{10.31e}$$

with the corresponding Lagrange function given by

$$\mathcal{L} = \varphi(z_f) + \eta_E^T h_E(z_f) + \sum_{i=1}^{N}\sum_{j=1}^{K}\left\{\bar{\lambda}_{ij}^T\left[h_i f(z_{ij}, u_{ij}) - \sum_{k=0}^{K}\dot{\ell}_k(\tau_j) z_{ik}\right]\right\}$$
$$+\sum_{i=1}^{N-1}\bar{\nu}_i^T\left(z_{i+1,0} - \sum_{j=0}^{K}\ell_j(1) z_{ij}\right) + \bar{\nu}_0^T(z_{1,0} - z(t_0)) + \bar{\nu}_N^T\left(z_f - \sum_{j=0}^{K}\ell_j(1) z_{N,j}\right)$$

For the solution of (10.31) we also require the following assumption.

Assumption II: For all suitable small values of h_i, the nonlinear program (10.31) has a solution where the sufficient second order KKT conditions and LICQ are satisfied. This guarantees a unique solution for the primal and dual variables.

We redefine the NLP multipliers for (10.31b) $\bar{\lambda}_{ij}$, as $\omega_j \lambda_{ij} = \bar{\lambda}_{ij}$ and $\omega_j > 0$ is the quadrature weight. With this transformation we can write the KKT conditions as

$$\nabla_{z_f}\mathcal{L} = \nabla_z \Phi(z_f) + \nabla_z h_E(z_f)\eta_E + \bar{\nu}_N = 0, \tag{10.32a}$$

$$\nabla_{z_{ij}}\mathcal{L} = \omega_j h_i \nabla_z f(z_{ij}, u_{ij})\lambda_{ij} - \sum_{k=1}^{K}\omega_k \lambda_{ik}\dot{\ell}_j(\tau_k) - \bar{\nu}_i \ell_j(1) = 0, \tag{10.32b}$$

$$\nabla_{u_{ij}}\mathcal{L} = \omega_j h_i \nabla_u f(z_{ij} u_{ij})\lambda_{ij} = 0, \tag{10.32c}$$

$$\nabla_{z_{i,0}}\mathcal{L} = \bar{\nu}_{i-1} - \bar{\nu}_i \ell_0(1) - \sum_{k=1}^{K}\omega_k \lambda_{ik}\dot{\ell}_0(\tau_k) = 0. \tag{10.32d}$$

The KKT conditions are similar to the discretization applied to (10.30a)–(10.30e) but with some important differences. In particular, (10.32c) is a straightforward collocation discretization of (10.30e). On the other hand, multipliers, $\bar{\nu}_i$, are added that correspond to continuity conditions on the state variables, and these also appear in (10.32a) and (10.32b). To reconcile this difference with (10.30d) and (10.30c), we first consider collocation at Gauss–Legendre roots and then take advantage of properties of Legendre polynomials. These will redefine the corresponding terms in (10.32b) and (10.32d).

10.4.1 Optimality with Gauss–Legendre Collocation

Consider the Gauss–Legendre quadrature for

$$\sum_{k=1}^{K}\omega_k \lambda_{ik}\dot{\ell}_j(\tau_k), \quad j = 0, \ldots, K, \; \tau_k \in (0,1). \tag{10.33}$$

From Theorem 10.1, this quadrature formula is exact for polynomial integrands up to degree $2K-1$. Since $\dot{\ell}_j(\tau)$ is a polynomial of degree $K-1$, we may represent λ_{ij} as coefficients of an interpolating polynomial evaluated at Gauss–Legendre roots. Because we can choose a polynomial, $\lambda_i(\tau)$, up to degree K (order $K+1$), we are free to choose an additional interpolation point for this polynomial.

Taking advantage of the order property for Gauss quadrature, we integrate (10.33) by parts to obtain the following relations:

$$\begin{aligned}\sum_{k=1}^{K} \omega_k \lambda_{ik} \dot{\ell}_j(\tau_k) &= \int_0^1 \lambda_i(\tau)\dot{\ell}_j(\tau)d\tau \\ &= \lambda_i(1)\ell_j(1) - \lambda_i(0)\ell_j(0) - \int_0^1 \dot{\lambda}_i(\tau)\ell_j(\tau)d\tau \\ &= \lambda_i(1)\ell_j(1) - \lambda_i(0)\ell_j(0) - \sum_{k=1}^{K} \omega_k \dot{\lambda}_i(\tau_k)\ell_j(\tau_k). \end{aligned} \tag{10.34}$$

Since $\ell_j(\tau_k) = 1$ for $j = k$ and zero otherwise, these relations simplify to

$$\sum_{k=1}^{K} \omega_k \lambda_{ik} \dot{\ell}_j(\tau_k) = \lambda_i(1)\ell_j(1) - \omega_j \dot{\lambda}_i(\tau_j), \quad j = 1,\ldots,K, \tag{10.35}$$

$$\sum_{k=1}^{K} \omega_k \lambda_{ik} \dot{\ell}_0(\tau_k) = \lambda_i(1)\ell_0(1) - \lambda_i(0). \tag{10.36}$$

Substituting these relations into (10.32b) and (10.32d) leads to

$$\nabla_{z_{ij}} \mathcal{L} = \omega_j [h_i \nabla_z f(z_{ij}, u_{ij})\lambda_{ij} + \dot{\lambda}_i(\tau_j)] - (\bar{\nu}_i + \lambda_i(1))\ell_j(1) = 0, \tag{10.37}$$

$$\nabla_{z_{i,0}} \mathcal{L} = \lambda_i(0) + \bar{\nu}_{i-1} - (\bar{\nu}_i + \lambda_i(1))\ell_0(1) = 0. \tag{10.38}$$

Note that the NLP multipliers provide information only at the collocation points for λ_{ij} and $\bar{\nu}_i$. Moreover, from Assumption II, the primal and dual variables are unique. Finally, because we are free to choose an additional coefficient for the $(K+1)$th order polynomial $\lambda_i(\tau)$, we may impose $(\bar{\nu}_i + \lambda_i(1)) = 0$. From (10.38), this leads to continuity of the λ_i profiles, i.e., $\lambda_i(0) = -\bar{\nu}_{i-1} = \lambda_{i-1}(1)$. With these simplifications, the KKT conditions (10.32) now become

$$\nabla_{z_f} \mathcal{L} = \nabla_{z_f} \Phi + \nabla_{z_f} h_E \eta_E - \lambda_N(1) = 0, \tag{10.39a}$$

$$\nabla_{z_{ij}} \mathcal{L} = \omega_j [\dot{\lambda}_i(\tau_j) + h_i \nabla_z f(z_{ij}, u_{ij})\lambda_{ij}] = 0, \tag{10.39b}$$

$$\nabla_{u_{ij}} \mathcal{L} = \omega_j h_i \nabla_u f(z_{ij}, u_{ij})\lambda_{ij} = 0, \tag{10.39c}$$

$$\nabla_{z_{i,0}} \mathcal{L} = \lambda_{i-1}(1) - \lambda_i(0) = 0. \tag{10.39d}$$

Note that these transformed KKT conditions are *equivalent to discrete approximations of the Euler–Lagrange equations* (10.30c)–(10.30e). Therefore, (10.31b)–(10.31e) and (10.39a)–(10.39c) can be solved directly as a discretized BVP, and convergence rates for (10.30a)–(10.30e) are directly applicable to the simultaneous NLP formulation (without inequality constraints). In particular, Reddien [331] shows that these calculated state, adjoint, and control profiles converge at the rate of $O(h^{K+1})$ with superconvergence (i.e., $O(h^{2K})$) at the element endpoints.

10.4.2 Optimality with Radau Collocation

In Chapter 9 we saw that Radau collocation has stronger stability properties than Gauss–Legendre collocation. While its approximation error is one order less, it is stiffly stable and also A-stable for index-2 DAEs. Moreover, on direct transcription problems Radau collocation leads to much smoother and less oscillatory profiles [34]. As a result, it is well suited for the simultaneous collocation approach.

On the other hand, the equivalence between the KKT conditions and the discretized Euler–Lagrange equations relies on the Gauss quadrature property (10.34) that holds for Gauss–Legendre collocation. Unfortunately, this property does not hold for Radau collocation, because of the choice of $\tau_K = 1$.

Instead, we consider a more detailed analysis of the KKT conditions for Radau collocation (10.32), as provided in [217]. This is summarized as follows:

- We first consider the order of accuracy for the KKT system and examine how the continuous-time solution $(z^*(t), u^*(t), \lambda^*(t))$ approximates (10.31b)–(10.31e) and (10.32). This approximation error then becomes the residual error term $\mathbf{r}$. We define the discretized true solution for $i = 1,\dots,N, j = 1,\dots,K$ as

$$\left.\begin{array}{l} z^*_{ij} = z^*(t_{i-1} + h_i\tau_j) \\ \lambda^*_{ij} = \lambda^*(t_{i-1} + h_i\tau_j) \\ u^*_{ij} = u^*(t_{i-1} + h_i\tau_j) \end{array}\right\} = w^*. \tag{10.40}$$

  Examining these equations shows that the residual error at any collocation point is $O(h^K)$ for Radau collocation and $O(h^{K+1})$ for Gauss–Legendre collocation.

- Writing (10.19b)–(10.19h), (10.32a)–(10.32d) in vector form as $\phi(\cdot) = 0$ leads to

$$\phi(w^*) - \phi(w) = \mathbf{r}, \tag{10.41}$$

  where w is a vector that represents all the primal and dual variables in $\phi(\cdot) = 0$ and w^* is a vector that represents the solution of the optimal control problem discretized at the collocation points. We also apply the mean value theorem:

$$\phi(w^*) - \phi(w) = \mathbf{J}(w^* - w), \tag{10.42}$$

  where $\mathbf{J} = \int_0^1 \nabla\phi^T(\zeta w^* + (1-\zeta)w)d\zeta$. If $\mathbf{J}^{-1}$ exists and is invertible, then

$$\Delta w = w^* - w = \mathbf{J}^{-1}\mathbf{r}. \tag{10.43}$$

  The mean value theorem allows a symmetric $\mathbf{J}$ matrix to be constructed which has the same symbolic form and structure as the KKT matrix of (10.19a)–(10.19c), (10.19f), (10.19g), (10.19h).

10.4.3 Convergence Orders for NLP Solutions

We now re-examine Assumption I in light of (10.43) and perform a closer analysis of the Jacobian $\mathbf{J}$ given by

$$\mathbf{J} = \begin{pmatrix} \mathcal{H} & \mathcal{A} \\ \mathcal{A}^T & 0 \end{pmatrix}, \tag{10.44}$$

where $\mathcal{H}$ is the Hessian of the Lagrange function and $\mathcal{A}^T$ is the Jacobian of the equality constraints of the NLP.

- For the analysis we assume equally spaced time steps ($h_1 = \cdots = h_N = h$) and that $\|w - w^*\|_\infty$ remains bounded for sufficiently small h. (This assumption is based on the fact that the KKT matrices at w^* and w also satisfy these two assumptions with w^* and w "close" to each other [217].) From Theorem 5.4, $\mathbf{J}$ is guaranteed to be invertible if the constraint gradients $\mathcal{A}$ are linearly independent and the reduced Hessian $\mathcal{H}_r$ (i.e., $\mathcal{H}$ projected into the null space of $\mathcal{A}^T$) is positive definite.

- The similarity of $\mathbf{J}$ to the KKT matrix allows us to perform a reduced-space decomposition similar to the one developed in Section 5.3.2. By partitioning the constraint Jacobian between state and control variables, $\mathcal{A}^T = (\mathcal{A}_B \mid \mathcal{A}_S)$, we define

$$\mathcal{Y} = \begin{pmatrix} \mathbf{I}_1 \\ 0 \end{pmatrix}, \qquad \mathcal{Z} = \begin{pmatrix} -\mathcal{A}_B{}^{-1}\mathcal{A}_S \\ \mathbf{I}_2 \end{pmatrix}, \tag{10.45}$$

where $\mathbf{I}_1$ and $\mathbf{I}_2$ are identity matrices of appropriate dimensions. These matrices satisfy

$$\mathcal{A}^T\mathcal{Y} = \mathcal{A}_B, \quad \mathcal{A}^T\mathcal{Z} = 0, \quad (\mathcal{Y} \mid \mathcal{Z}) \text{ is nonsingular,} \tag{10.46}$$

and we partition Δw into subvectors for the state, control, and adjoint variables with

$$[\Delta w_z^T \ \Delta w_u^T] = (\mathcal{Y} \mid \mathcal{Z}) \begin{pmatrix} \mathbf{d}_Y \\ \mathbf{d}_Z \end{pmatrix}. \tag{10.47}$$

Also, by partitioning the residual vector into $\mathbf{r}^T = \left[\mathbf{r}_z^T, 0^T, \mathbf{r}_\lambda^T\right]$, the modified linear system then becomes

$$\begin{pmatrix} \mathcal{Y}^T\mathcal{H}\mathcal{Y} & \mathcal{Y}^T\mathcal{H}\mathcal{Z} & \mathcal{A}_B^T \\ \mathcal{Z}^T\mathcal{H}\mathcal{Y} & \mathcal{Z}^T\mathcal{H}\mathcal{Z} & 0 \\ \mathcal{A}_B & 0 & 0 \end{pmatrix} \begin{pmatrix} \mathbf{d}_Y \\ \mathbf{d}_Z \\ \Delta w_\lambda \end{pmatrix} = \begin{pmatrix} \mathbf{r}_z \\ -\mathcal{A}_S^T\mathcal{A}_B^{-T}\mathbf{r}_z \\ \mathbf{r}_\lambda \end{pmatrix}. \tag{10.48}$$

- The orders of the vectors and matrices in (10.48) increase with $\frac{1}{h}$ and this must be considered when analyzing the normed quantities with respect to the time step, h. Also, the analysis in [217] follows an assumption in Hager [178] that

$$\|\mathcal{H}_r^{-1}\| = \|(\mathcal{Z}^T\mathcal{H}\mathcal{Z})^{-1}\|_\infty = O\left(\frac{1}{h}\right). \tag{10.49}$$

With this property, (10.48) can be solved to provide order results for Δw. For Radau collocation, $\|\mathbf{d}_Y\|_\infty = O(h^{K+1})$, $\|\mathbf{d}_Z\|_\infty = O(h^K)$, and $\|\Delta w_\lambda\|_\infty = O(h^K)$ for sufficiently small h.

- From this analysis, the states, the multipliers, and the controls with Radau collocation converge to the true solution at the rate of $O(h^K)$, and the Lagrange multipliers scaled by ω_j provide estimates of the true adjoints. Following the same analysis, convergence to the true solution with Gauss–Legendre collocation occurs at the rate of $O(h^{K+1})$.

The above results were originally developed for unconstrained optimal control problems and were extended in [219] to account for the final-time equality constraints (10.29c). This extension requires that the dynamic system be controllable. These results provide the framework for obtaining results for the optimal control problem (10.29) with the direct transcription approach, even for large systems.

On the other hand, there are two classes of optimal control problems where Assumptions I and II are not satisfied. First, singular control problems do not satisfy Assumption I and, in particular, the assumption that $\|\mathcal{H}_r^{-1}\|_\infty = O(\frac{1}{h})$. In the next subsection, we will examine the behavior of direct NLP methods for this problem class and consider some open questions for their solution. Second, state inequality constrained problems do not satisfy Assumption II as h_i becomes small, because constraint gradients do not satisfy the LICQ as the time step $h_i \to 0$. To conclude this section, we will examine the behavior of direct NLP methods and consider open questions for this class as well.

10.4.4 Singular Optimal Controls

As seen in Chapter 8, singular optimal control problems arise when the control appears only linearly in the differential equations and in the performance index. These problems are encountered often in process engineering, including dynamic optimization of batch processes, batch reactor control [107, 368], and optimal mixing of catalysts [207]. As seen in Chapter 8, the optimal control profile cannot be determined directly from stationarity of the Hamiltonian. Because the Euler–Lagrange equations for (10.29) are no longer index 1, we find that the DAEs become ill-conditioned. Instead, if the control is not at the bounds, then repeated time differentiations of (10.30e) must be performed to recover the control, as demonstrated in Section 8.5. Moreover, because the singular arc is often present for only part of the time domain, identifying the entrance and exit points of the arc is often difficult.

On the other hand, direct methods (whether sequential or simultaneous) are often applied successfully to singular control problems, *provided that the control profile is represented by a low-dimensional parameterization.* Difficulties arise as a more accurate control profile is sought and a finer discretization is applied. The simultaneous collocation method applied to singular problems can also reflect this ill-conditioned behavior. In particular, applying direct methods to singular problems shows that (10.49) is no longer satisfied and that the reduced Hessian becomes very ill-conditioned as the time step goes to zero, $h \to 0$. Tests on a number of singular examples (see [222]) reveal, that $\|\mathcal{H}_r^{-1}\|_\infty = O(h^{-j})$, where j is greater than 2 and has been observed as high as 5 (with the value of j less for Radau collocation than for Gauss collocation). To illustrate this behavior, we consider the following test example [8].

Example 10.5 (Singular Control with Direct Transcription). Consider the singular optimal control problem given by

$$\min_u z_3\left(\frac{\pi}{2}\right) \tag{10.50a}$$

$$\text{s.t.} \frac{dz_1}{dt} = z_2, \quad z_1(0) = 0, \tag{10.50b}$$

$$\frac{dz_2}{dt} = u, \quad z_2(0) = 1, \tag{10.50c}$$

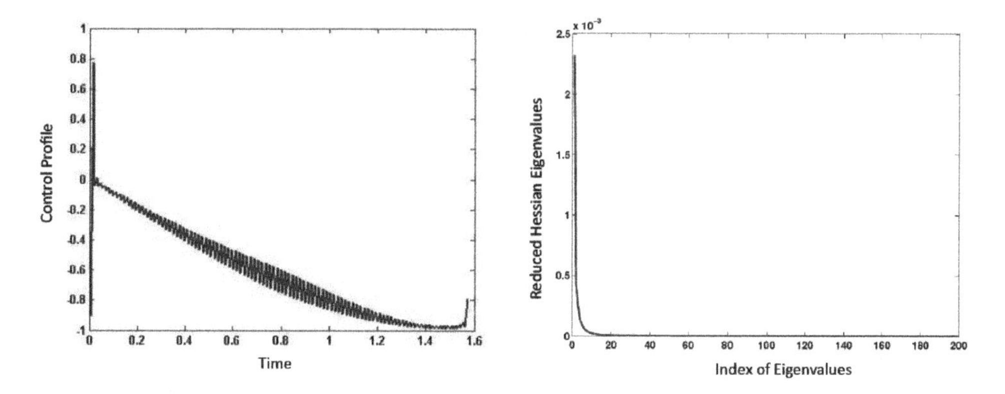

Figure 10.10. *Control profile obtained for $N = 100$, $K = 2$ for problem* (10.50).

$$\frac{dz_3}{dt} = \frac{1}{2}z_2^2 - \frac{1}{2}z_1^2, \quad z_3(0) = 0, \tag{10.50d}$$

$$-1 \leq u \leq 1. \tag{10.50e}$$

The analytic solution of this problem shows a control profile that is singular over the entire time domain, $t \in [0, \pi/2]$.

Figure 10.10 presents the results of solving problem (10.50) using the simultaneous approach with $K = 2$ and $N = 100$. We see that for large N, the control profile is highly oscillatory with spikes at initial and final time. Moreover, note that the spectrum of eigenvalues for the reduced Hessian $\mathcal{H}_r$ quickly drops to zero, and for $N = 100$ its condition number is 1.5×10^9. Similar behaviors have been observed with other discretizations as well. ∎

To address singular problems, Jacobson, Gershwin, and Lele [208] proposed to solve the singular control problem by adding the regularization $\epsilon \int_0^{t_f} u(t)^T S u(t) dt$ to the objective function (with S positive definite, $\epsilon > 0$). The resulting problem is nonsingular and is solved for a monotonically decreasing sequence $\{\epsilon_k\}$. Similar ideas are given in [414, 357]; the latter source also proves consistency of this approach. However, as ϵ is decreased, problems tend to become increasingly ill-conditioned, and the quality of the optimal control deteriorates. In addition, regularizations have been performed through coarse discretizations of the control profile and application of an NLP-based strategy [355]. Finally, several methods are based on analytical calculation of the singular optimal controls (see [368] for a summary). However, many of these methods rely on tools from Lie algebra and involve cumbersome calculations, especially for large problems.

All of these approaches aim at a *regularization* not only to improve the spectrum of eigenvalues for the reduced Hessian and to guarantee a unique solution, but also to recover the true solution of the singular problem as well. A recent approach developed in Kameswaran [222] relies on properties of the Hamiltonian ((8.25) and (8.54)) and applies an indirect approach that explicitly recovers the states and adjoint profiles. With these quantities, an accurate approximation of the Hamiltonian and $\partial H/\partial u$, and their stationarity with respect to time, can be enforced within an extended NLP formulation. While this approach is more complex than a nonlinear program derived for direct transcription, it has been very effective in solving challenging singular control problems of orders one and two.

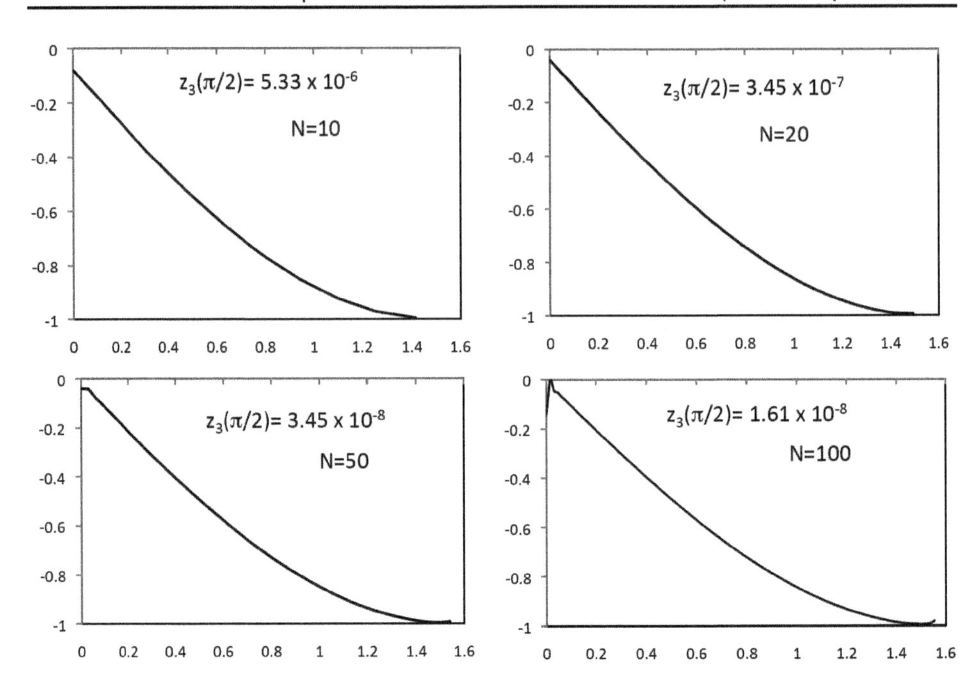

Figure 10.11. *Singular control profiles obtained for piecewise constant controls in* (10.50).

Consequently, accurate determination of control profiles for these singular problems still remains an open question and must be handled carefully within the context of direct NLP-based methods. For the simultaneous collocation approach, a simple heuristic strategy is to apply a control discretization that is coarser than the state profile but still can be refined by decreasing the time step. This has the effect of retarding the rapid decay in the spectrum of eigenvalues in the reduced Hessian so that reasonable solutions can still be obtained. For instance, problem (10.19) can be modified so that $u(t)$ is represented by $u_i = u_{ij}$, $j = 1, \ldots, K$ is piecewise constant in an element. Applying this modification to Example 10.5 leads to the control profiles shown in Figure 10.11. From the shapes of the profile and the value of the objective function as N is increased, an accurate approximation of the solution profiles may be obtained before the onset of ill-conditioning.

Finally, the heuristic approach with coarse control profiles in (10.19) can be extended to variable finite elements as well. By using relatively few elements and a high-order collocation formula, singular arcs can be located, and breakpoints can be determined for control profiles that have bang-bang and singular characteristics. For instance, the catalyst mixing profiles in Figure 8.9 match the analytical solution in Example 8.9 and were determined using this approach with up to 100 elements and 3 collocation points.

10.4.5 High-Index Inequality Path Constraints

Optimality conditions for high-index inequality path constraints were discussed in Section 8.4. As with singular control problems, these optimality conditions deal with high-index algebraic constraints, which require repeated differentiation to expose the control profile. Moreover, junction conditions must be imposed to define the boundary conditions for the path constraint multiplier. As seen in Example 8.8, these conditions are difficult to enforce within an optimization algorithm, even for a small, linear problem. Moreover, different choices of corner conditions demonstrate the nonuniqueness of the multipliers for path inequalities [72, 307] and ill-posedness in their determination.

In the direct simultaneous approach, path constraints can be imposed easily as bounds on the variable coefficients. A typical selection of bounds includes all mesh points and collocation points. For time steps that are sufficiently small and profiles that are sufficiently smooth within the element, this approach can ensure the feasibility of path constraints in practice. Nevertheless, open questions remain on the influence of these high-index inequalities on the solution of the NLP problem (10.19).

While much work still needs to be done to analyze the behavior of path constraints within collocation formulations, we observe that these constraints lead to violations of constraint qualifications in the nonlinear program along with undesirable behavior of the KKT multipliers. To demonstrate this observation, we analyze the problem presented in the following example from an NLP perspective.

Example 10.6 (Betts–Campbell Problem [45]). Consider a PDE constrained optimization problem given by

$$\min \int_0^5 \int_0^\pi z^2(x,t)dxdt + q\int_0^5 (u_0^2(t)+u_\pi^2(t))dt \tag{10.51a}$$

$$\text{s.t.} \quad \frac{\partial z}{\partial t} = \frac{\partial^2 z}{\partial x^2}; \quad 0 \le x \le \pi;\ 0 \le t \le 5; \tag{10.51b}$$

$$z(x,0) = 0; \quad z(0,t) = u_0(t); \tag{10.51c}$$

$$z(\pi,t) = u_\pi(t); \tag{10.51d}$$

$$\sin(x)\sin\left(\frac{\pi t}{5}\right) - 0.7 \le z(x,t) \tag{10.51e}$$

with q fixed at 10^{-3}. This problem represents a heat conduction problem where the boundary controls are used to minimize the field temperature subject to a minimum temperature profile constraint.

Problem (10.51) is discretized spatially using a central difference scheme and equidistant spacing with $n+1$ points. A trapezoidal scheme is used to discretize the double integral in the objective function, and this converts it into a single integral. The spatial mesh size is then defined as $\delta = \frac{\pi}{n}$. With suitable time scaling, we obtain the following optimal control problem:

$$\min \frac{1}{2}\int_0^{5\delta^{-2}} 2\delta^3 \sum_{k=1}^{n-1} z_k^2(\tau)d\tau + \frac{\delta^3 + 2q\delta^2}{2}\int_0^{5\delta^{-2}} (u_0^2(\tau)+u_\pi^2(\tau))d\tau \tag{10.52a}$$

$$\text{s.t.} \quad \frac{dz_1}{d\tau} = z_2 - 2z_1 + u_0; \quad z_1(0) = 0; \tag{10.52b}$$

$$\frac{dz_2}{d\tau} = z_3 - 2z_2 + z_1; \quad z_2(0) = 0; \tag{10.52c}$$

$$\vdots$$

$$\frac{dz_{n-1}}{d\tau} = u_\pi - 2z_{n-1} + z_{n-2}; \quad z_{n-1}(0) = 0; \tag{10.52d}$$

$$u_0(\tau) \geq \sin(0)\sin\left(\frac{\pi\delta^2\tau}{5}\right) - 0.7 = -0.7; \tag{10.52e}$$

$$z_k(\tau) \geq \sin(k\delta)\sin\left(\frac{\pi\delta^2\tau}{5}\right) - 0.7; \quad k = 1,\dots,n-1; \tag{10.52f}$$

$$u_\pi(\tau) \geq \sin(\pi)\sin\left(\frac{\pi\delta^2\tau}{5}\right) - 0.7 = -0.7. \tag{10.52g}$$

Note that the control profiles u_0 and u_π are applied only at the boundaries and directly influence z_1 and z_{n-1}. Moreover, computational experience [45, 220] confirms that only the inequality constraints in the spatial center (at $n/2$) become active at some point in time. Using this observation, and the symmetry of the resulting DAE optimization problem, the active path constraint requires $n/2$ time differentiations to expose the boundary control. With finer spatial discretization, the index of this constraint can be made arbitrarily high. Here, cases are considered with $n = 10$ and $n = 20$.

Nevertheless, the simultaneous approach can provide accurate results irrespective of the temporal discretization scheme. For any temporal discretization (including explicit Euler with N equally spaced steps), problem (10.52) is transformed into the following quadratic program:

$$\begin{aligned} \min\ & \frac{1}{2}\mathbf{u}^T\mathbf{H}\mathbf{u} + \frac{1}{2}\mathbf{z}^T\mathbf{Q}\mathbf{z} \\ \text{s.t.}\ & \mathbf{A}\mathbf{z} + \mathbf{B}\mathbf{u} = 0, \\ & \mathbf{s} = \mathbf{C}\mathbf{z} + \mathbf{d}, \\ & \mathbf{s} \geq 0, \end{aligned} \tag{10.53}$$

where, **z**, **u**, **s** are now written as vectors of discretized values of $z(\tau), u_0$, and the slack in the inequality constraint, respectively. The matrices **H** and **Q** are positive definite for all temporal and spatial mesh sizes, **A** is invertible, and **C** is full row rank [217]. With these assumptions the above QP can be reduced to the following smaller QP:

$$\begin{aligned} \min_{\mathbf{u}}\ & \frac{1}{2}\mathbf{u}^T\left(\mathbf{H} + \mathbf{B}^T\mathbf{A}^{-T}\mathbf{Q}\mathbf{A}^{-1}\mathbf{B}\right)\mathbf{u} \\ \text{s.t.}\ & \mathbf{C}\mathbf{A}^{-1}\mathbf{B}\mathbf{u} - \mathbf{d} \leq 0. \end{aligned} \tag{10.54}$$

As the above QP is strictly convex, it has a unique optimal solution $\mathbf{u}^*$. On the other hand, with increasing temporal discretization, the fraction of dependent rows (for a given tolerance level) in matrix $\mathbf{CA}^{-1}\mathbf{B}$ also increases, as seen from the singular values plotted in Figure 10.12. Hence, the LICQ fails to hold for a given tolerance level and calculation of the QP multipliers is ill-conditioned. In fact, the profile of the path constraint multipliers exhibits inconsistent behavior and appears to go unbounded as N is increased. Neverthe-

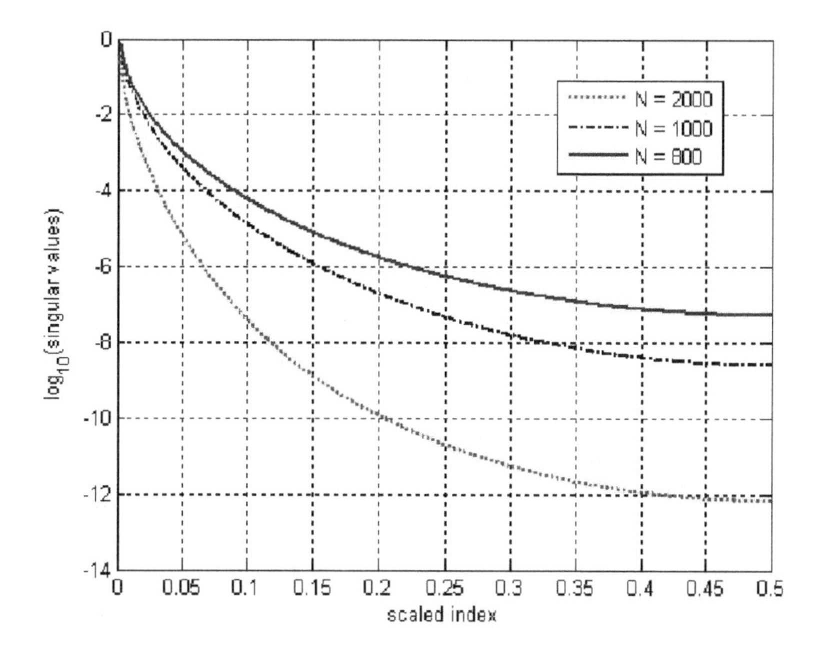

Figure 10.12. *Singular values for Jacobian of active constraints in the QP* (10.54) *with* $n = 20$. *Note that the fraction of constraints with singular values below a given tolerance increases with* N.

less, as observed from Figure 10.13, the optimal control profile is well defined and appears to converge for increasing values of N. ■

From Figures 10.12 and 10.13 it is clear that the failure of LICQ triggers the ill-conditioned behavior of the multipliers. This has a strong impact on the estimation of the adjoint profiles even though the control profile is well defined. The reason for this behavior can be deduced from Theorem 4.9. For the linearly constrained problem (10.54) neither unique multipliers nor constraint qualifications are necessary to solve the QP, as the optimum is defined by the gradient of the objective function and the cone of the active constraints.

On the other hand, the indirect approach (using a multipoint BVP with the path constraints reformulated to index 1) is more difficult to apply for this problem. This method is applied in [220, 45] but fails to converge. Moreover, it is noted in [220] that the constrained arc is unstable. This leads to an inability to obtain the states, the adjoints, and the controls with the indirect approach.

From this illustrative example, it is also clear that the simultaneous approach can produce meaningful solutions for inequality path constrained optimal control problems. On the other hand, open questions still remain including (a) convergence and convergence rates of the state and control profiles as $h \to 0$, (b) location of entrance and exit points for the active constraints (e.g., through variable finite elements), and (c) the consequence of failure of LICQ on nonlinearly constrained problems. Additional discussion on this problem can be found in [40].

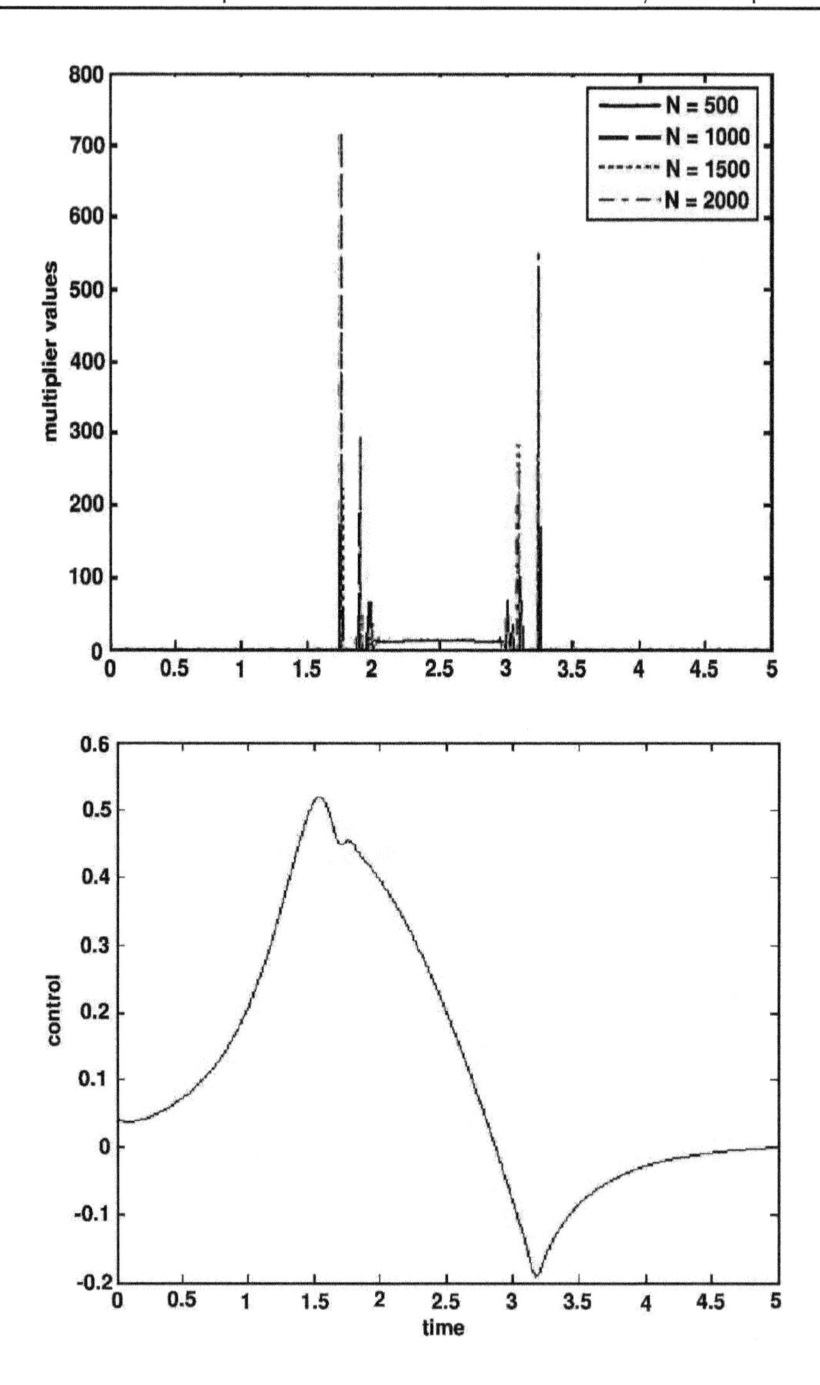

Figure 10.13. *Active constraint multipliers and control variable profiles for the QP* (10.54) *with $n = 20$. Note inconsistency of the multipliers with increasing N while the control profile remains essentially unchanged.*

10.5 Summary and Conclusions

This chapter considers simultaneous collocation formulations for dynamic optimization problems. These direct transcription formulations are based on full discretizations of the DAEs and do not rely on embedded DAE solvers. Instead, the entire problem is addressed at the level of the NLP solver. Because of this open formulation, exact first and second derivatives can be made available through the optimization modeling system, and both structure and sparsity can be exploited. On the other hand, an efficient large-scale NLP solver is needed to take full advantage of this approach. As shown in Section 10.3, Newton-based barrier methods are well suited for this task as they have low computational complexity.

The simultaneous approach is based on an IRK scheme. Using collocation on finite elements, the discretization is A-stable, high order, and can handle nonsmooth events at element boundaries. Moreover, because the approach can be viewed as an extended boundary value solver, it also works well with open loop unstable systems. Finally, the NLP formulation can be extended to deal with variable finite elements in order to find optimal breakpoint locations. All of these features were demonstrated on large-scale process examples.

Lastly, the ability of the simultaneous approach to handle many degrees of freedom allows the representation of the control profile at the same level of accuracy as the state profiles. With both Gauss–Legendre and Radau collocation, convergence rates were shown on optimal control problems without inequalities. For these systems a direct link can be made between indirect and direct simultaneous methods, and the equivalence of these methods can be shown. Moreover, as demonstrated on two examples, the NLP formulation allows for the easy incorporation of inequalities on the state and control profiles and for some options in handling singular control problems. Nevertheless, for these two problem classes, a number of open questions still exist, particularly with respect to convergence and reliable behavior for NLP formulations.

10.6 Notes for Further Reading

As direct transcription relies on efficient, large-scale NLP solvers, its development has been somewhat slower and less widespread than the methods in Chapters 8 and 9. Nevertheless, the method dates back to the 1970s [388, 290, 84, 55, 296, 61] and was prompted by the realization that these methods were especially suitable for unstable systems and systems with path constraints. Moreover, there is a growing literature for this dynamic optimization approach including two excellent monographs [40, 318] that describe related direct transcription approaches with Runge–Kutta discretizations. In addition, related *quasi-sequential* approaches have been developed that solve the collocation equations in an inner loop and therefore apply a nested optimization strategy [201, 257].

The development of the simultaneous collocation approach is taken from a number of papers [106, 396, 264, 89, 377, 401, 378, 217]. More detail can also be found in the following Ph.D. theses [394, 92, 222]. In addition, the properties of the collocation method and related boundary value approaches can be found in [341, 14, 13, 16]. Convergence results for simultaneous methods have been developed for a class of discretization schemes: Gauss collocation [331, 106], explicit Runge–Kutta methods satisfying a symmetry condition [178], Radau collocation [217], and general IRK methods [179, 65].

Moreover, the simultaneous approach has been applied widely in aeronautical and astronautical applications. A cursory literature search reveals several hundred publications that apply simultaneous approaches in this area. Specific applications include collision avoidance for multiple aircraft [42, 325] and underwater vehicles [365], trajectories for satellites and earth orbiters [351, 103], and the design of multiple paths and orbits for multibody dynamics [68], including interplanetary travel [43]. An overview of these applications is given in [40]. Moreover, the SOCS (sparse optimal control software) package [40], a commercial software package developed and marketed by Boeing Corporation, has been widely used for these and other engineering applications. Also, GESOP, a graphical environment that incorporates SOCS as well as multiple shooting methods, can be found in [158].

In process engineering, applications of the simultaneous approach include the design and optimal operation of batch processes. These include optimization of operating policies for fermentors [107] and bioreactors [334], flux balance models for metabolic systems [232, 271, 321], batch distillation columns [264, 298], membrane separators [125], polymerization reactors [141, 209], crystallization [242], freeze-drying processes [67], and integrated multiunit batch processes [46]. Other off-line applications include parameter estimation of reactive systems [384, 126], design of periodic separation processes including pressure swing adsorption [291] and simulated moving beds [230, 225], optimal grade transitions in polymer processes [91, 141], reactor network synthesis [21, 302], and economic performance analysis of batch systems [263].

Online applications include dynamic data reconciliation algorithms for batch processes [7, 260], state estimation and process identification [375], optimal startup policies for distillation columns [320], optimal feed policies for direct methanol fuel cells [412], and a number of algorithms and case studies for nonlinear model predictive control (NMPC) [364, 211]. Moreover, commercial applications of NMPC include several applications at ExxonMobil and Chevron Phillips, which use the NLC package and NOVA solver both from Honeywell, Inc. [333, 417]. Other software implementations of the simultaneous approach include *DynoPC* [240], a Windows-based platform; the MATLAB-based *OptimalControlCentre* [211] and *dynopt* [95] packages, and the pseudospectral packages PROPT [342] and GPOPS [327].

10.7 Exercises

1. Apply two-point Gauss–Legendre collocation to Example 10.2 and find the solution for up to $N = 10$ finite elements. Calculate the global error and determine the order of the method.

2. Using the information reported in [242], resolve the crystallizer optimization problem described in Example 10.3 by applying two-point Radau collocation to the state and control profiles. How does the solution compare with Example 10.3?

3. Consider the unstable dynamic optimization problem given in Example 9.5. Formulate this problem using two-point collocation and compare the solutions with Radau and Gauss–Legendre collocation. How many finite elements are needed in each case?

4. Consider the parallel batch reactor system in Example 9.4. Formulate this problem using two-point collocation and variable finite elements. Compare the solutions with Radau and Gauss–Legendre collocation. How many finite elements are needed in each case?

5. Consider the batch reactor system in Example 8.3. Formulate this problem using three point Radau collocation and variable finite elements.

   (a) Solve this problem without the control bound and compare with the solution in Example 8.3.

   (b) Solve the problem with $u(t) \leq 2$. Compare the state profiles with a profile determined by a DAE solver. How many finite elements are needed to achieve a solution with less than 10^{-4} error in the states?

6. Write the KKT conditions for (10.19). Using Gauss–Legendre collocation, apply the quadrature formulation and extend the derivation in Section 10.4 to deal with algebraic variables and decision variables p.

   (a) Without considering inequality constraints, compare the transformed KKT system to the optimality conditions developed in Chapter 8.

   (b) Using the KKT multipliers, write a discrete version of the Hamiltonian function. If the finite elements are variables in (10.19) and the constraints (10.22) are added, what additional conditions arise on the Hamiltonian from the KKT system?

   (c) Explain why changes in active constraints and discontinuities in control profiles must be confined to the boundaries of the finite elements. Discuss how this can be enforced in the KKT conditions.

7. Consider the singular catalyst mixing problem in Example 8.9.

   (a) Apply three-point Gauss collocation and solve with piecewise constant controls for increasing values of N.

   (b) Apply three-point Radau collocation and solve with piecewise constant controls for increasing values of N. Compare this solution with Gauss collocation.

   (c) Apply three-point Radau collocation and solve with control coefficients at each collocation point for increasing values of N. Compare this solution to those with piecewise controls.

Chapter 11

Process Optimization with Complementarity Constraints

Steady state and dynamic process models frequently deal with switches and other nonsmooth decisions that can be represented through complementarity constraints. If these formulations can be applied at the NLP level, they provide an "all-at-once" strategy that can be addressed with state-of-the-art NLP solvers. On the other hand, these mathematical programs with complementarity constraints (MPCCs) have not been widely studied in process engineering and must be handled with care. These formulations are nonconvex and violate constraint qualifications, including LICQ and MFCQ, which are required for well-posed, nonlinear programs. This chapter deals with properties of MPCCs including concepts of stationarity and linear independence that are essential for well-defined NLP formulations. NLP-based solution strategies for MPCCs are then reviewed along with examples of complementarity drawn from steady state chemical engineering applications. In addition, we extend these MPCC formulations for the optimization of a class of hybrid dynamic models, where the differential states remain continuous over time. These involve differential inclusions of the Filippov type, and a formulation is developed that preserves the piecewise smoothness properties of the dynamic system. Results on several process examples drawn from distillation optimization, process control, and hybrid systems are used to illustrate MPCC formulations and demonstrate the proposed optimization approach.

11.1 Introduction

Optimization problems were introduced in Chapter 1 with discrete and continuous decision variables. This mixed integer nonlinear programming formulation (1.1) leads to the most general description of process optimization problems. To solve these problems, NLP formulations are usually solved at a lower level for fixed values of the discrete variables. Based on information from the NLP solution, a search of the discrete variables is then conducted at a higher level.

While the scope of this book is restricted to nonlinear programs with smooth objective and constraint functions, the ability to deal with (some) discrete decisions within an "all-at-once" NLP formulation can have an advantage over nested MINLP strategies. This motivates the modeling of these decisions with complementarity constraints. Complementarity is a relationship between variables where either one (or both) must be at its bound. In principle, these relationships can be embedded within an NLP formulation, but these complements introduce an inherent nonconvexity as well as linear dependence of constraints, which make the nonlinear program harder to solve.

Complementarity problems arise in a large number of applications in engineering and economic systems. They include contact and friction problems in computational mechanics, equilibrium relations in economics, and a wide variety of discrete events in process systems. Many of these models embed themselves within optimization problems [130, 267]. However, these mathematical programming formulations and related solution strategies are not yet fully developed in process engineering.

In process optimization, complementarity models allow a number of options when dealing with phase changes, flow reversal, safety valve operation, and other discrete events. These events are often embedded within modular process simulators (as described in Chapter 7) that model flowsheets through conditional (IF-THEN) statements embedded in the software. In fact, depending on the value of computed variables, different sets of equations may even be used for the simulation. However, with embedded conditional statements, the modular approach has significant limitations in dealing with process optimization. Instead, equation-based flowsheet models are much better suited to an optimization framework and have the flexibility to exploit complex specifications and recycle streams. Moreover, equation-based process optimization, also described in Chapter 7, must deal with conditional statements in a different way. In this context, complementarity models offer an efficient alternative.

Complementarities also arise naturally in the solution of multilevel optimization problems. Such problems are members of a more general problem class called mathematical programs with equilibrium constraints (MPECs) [267]. For bilevel optimization problems of the form

$$\min_{x} \quad f(x,y) \tag{11.1a}$$

$$\text{s.t.} \quad (x,y) \in Z, \tag{11.1b}$$

$$y = \arg\min_{\hat{y}}\{\theta(x,\hat{y}) : \hat{y} \in C(x)\}, \tag{11.1c}$$

we define Z and $C(x)$ as feasible regions for the upper and lower level problems, respectively, and $f(x,y)$ and $\theta(x,y)$ as objective functions for the upper and lower level problems, respectively.

Bilevel optimization problems can be reformulated as mathematical programs with complementarity constraints (MPCCs) by writing the optimality conditions of the inner optimization problem as constraints on the outer problem. An MPCC takes the following general form:

$$\min \quad f(x,y,z) \tag{11.2a}$$

$$\text{s.t.} \quad h(x,y,z) = 0, \tag{11.2b}$$

$$g(x,y,z) \leq 0, \tag{11.2c}$$

$$0 \leq x \perp y \geq 0, \tag{11.2d}$$

where $\perp$ is the complementarity operator enforcing at least one of the bounds to be active. (Note that we have classified the complementing variables as x and y, both in $\mathbb{R}^{n_c}$ with the remaining variables, $z \in \mathbb{R}^{n-2n_c}$.) The complementarity constraint (11.2d) implies the following:

$$x_{(i)} = 0 \quad \text{OR} \quad y_{(i)} = 0, \quad i = 1,\dots,n_c,$$
$$x \geq 0,\; y \geq 0,$$

for each vector element i. Here the OR operator is inclusive, as both variables may be zero. Alternately, the complementarity constraint may be written in several equivalent ways:

$$x^T y = 0,\; x \geq 0,\; y \geq 0, \tag{11.3}$$
$$x_{(i)} y_{(i)} = 0, \quad i = 1,\dots,n_c, \qquad x \geq 0,\; y \geq 0, \tag{11.4}$$
$$x_{(i)} y_{(i)} \leq 0, \quad i = 1,\dots,n_c, \qquad x \geq 0,\; y \geq 0. \tag{11.5}$$

These alternate forms are particularly useful when applying existing NLP solution strategies to solve MPCCs. In addition, every MPCC problem can be rewritten as the equivalent MPEC problem:

$$\min_{x,y,z} \quad f(x,y,z) \tag{11.6a}$$
$$\text{s.t.} \quad h(x,y,z) = 0, \tag{11.6b}$$
$$g(x,y,z) \leq 0, \quad x \geq 0, \tag{11.6c}$$
$$y = \arg\left\{ \min_{\hat{y}} x^T \hat{y} \quad \text{s.t. } \hat{y} \geq 0 \right\}. \tag{11.6d}$$

Section 11.2 summarizes MPCC properties and develops related NLP formulations that lead to standard NLP problems that satisfy constraint qualifications. Also included is a numerical case study comparison of these formulations with standard NLP solvers. Section 11.3 then considers the modeling of discrete decisions with MPCC formulations and describes several important chemical engineering applications of MPCCs. To demonstrate both the models and solution strategies, we consider a distillation optimization case study in Section 11.4. Section 11.5 extends complementarity concepts to the optimization of a class of hybrid dynamic systems, with continuous states and discontinuous right-hand sides. Known as Filippov systems [132], these are considered with the *complementarity class of hybrid systems* described and analyzed in [193], where the differential states remain continuous over all time. For this class, we adopt the simultaneous collocation approach from Chapter 10 along with variable finite elements to track the location of these nonsmooth features. Two case studies illustrate this approach in Section 11.6 and demonstrate the advantages of the MPCC formulation.

11.2 MPCC Properties and Formulations

MPCCs of the form (11.2) are *singular* optimization problems that cannot be solved directly with most NLP solvers. Difficulties associated with the solution of MPCCs can be seen from the following conditions:

- If we rewrite $0 \le x \perp y \ge 0$ equivalently as $x \ge 0,\ y \ge 0$, and $x^T y \le 0$, then the KKT conditions for (11.2) are given by

$$\nabla_x f(x,y,z) + \nabla_x g(x,y,z)u + \nabla_x h(x,y,z)v + (\delta y - \alpha) = 0,$$
$$\nabla_y f(x,y,z) + \nabla_y g(x,y,z)u + \nabla_y h(x,y,z)v + (\delta x - \beta) = 0,$$
$$\nabla_z f(x,y,z) + \nabla_z g(x,y,z)u + \nabla_z h(x,y,z)v = 0,$$
$$0 \ge g(x,y,z) \perp u \ge 0,\ h(x,y,z) = 0, \quad 0 \le \delta \perp x^T y \le 0,$$
$$0 \le x \perp \alpha \ge 0, \quad 0 \le y \perp \beta \ge 0.$$

  From Theorem 4.14 we know that if constraint qualifications (CQs) hold, the KKT conditions are necessary conditions for optimality. However, these CQs are violated for MPCCs and it is easy to formulate MPCC examples where the multipliers α, β, and δ are unbounded or do not exist.

- At feasible points that satisfy

$$h(x,y,z) = 0, \quad g(x,y,z) \le 0, \quad 0 \le x \perp y \ge 0,$$

  we have for all $x_{(i)} = 0$ that $x_{(i)} y_{(i)} = 0$ as well; a similar condition holds for $y_{(i)} = 0$. As a result, the LICQ is violated, and failure of the LICQ implies that the multipliers, if they exist, are nonunique.

- The weaker MFCQ condition from Definition 4.16 is also violated. MFCQ requires linearly independent gradients for the equality constraints and a feasible direction into the interior of the cone of inequality constraint gradients. This constraint qualification is a necessary and sufficient condition for boundedness of the multipliers. On the other hand, $0 \le x \perp y \ge 0$ can be written equivalently as $x \ge 0,\ y \ge 0$, and $x^T y \le 0$. Therefore, at a feasible point $(\bar{x},\bar{y},\bar{z})$, there is no feasible search direction that satisfies $\bar{x}(\Delta y) + \bar{y}(\Delta x) < 0$. Consequently, MFCQ cannot hold either, and multipliers of the MPCC (11.2) will be nonunique and unbounded.

Because these constraint qualifications do not hold, it is not surprising that an MPCC may be difficult to solve. In order to classify an MPCC solution, we introduce the concept of *B-stationarity*.[8] A point $w^* = [x^{*,T}, y^{*,T}, z^{*,T}]^T$ is a *B-stationary point* if it is feasible to the MPCC, and $d = 0$ is a solution to the following linear program with equilibrium constraints (LPEC) [267, 326, 350]:

$$\min_{d} \quad \nabla f(w^*)^T d \tag{11.7a}$$
$$\text{s.t.} \quad g(w^*) + \nabla g(w^*)^T d \le 0, \tag{11.7b}$$
$$h(w^*) + \nabla h(w^*)^T d = 0, \tag{11.7c}$$
$$0 \le x^* + d_x \perp y^* + d_y \ge 0. \tag{11.7d}$$

The LPEC therefore verifies that locally there is no feasible direction that improves the objective function. On the other hand, verification of this condition may require the solution

[8] There are weaker stationarity conditions that identify *weak*, *A*-, *C*-, and *M-stationary* points. However, these conditions are not sufficient to identify local optima, as they allow negative multipliers and have feasible descent directions [256].

of 2^m linear programs where m is the cardinality of the biactive set $I_X \cap I_Y$, where

$$I_X = \left\{ i : x^{*,(i)} = 0 \right\},$$
$$I_Y = \left\{ i : y^{*,(i)} = 0 \right\}.$$

Strong stationarity is a more useful, but less general, definition of stationarity for the MPCC problem. A point is strongly stationary if it is feasible for the MPCC and $d = 0$ solves the following relaxed problem:

$$\min_{d} \quad \nabla f(w^*)^T d \tag{11.8a}$$
$$\text{s.t.} \quad g(w^*) + \nabla g(w^*)^T d \leq 0, \tag{11.8b}$$
$$h(w^*) + \nabla h(w^*)^T d = 0, \tag{11.8c}$$
$$d_{x(i)} = 0, \quad i \in I_X \setminus I_Y, \tag{11.8d}$$
$$d_{y(i)} = 0, \quad i \in I_Y \setminus I_X, \tag{11.8e}$$
$$d_{x(i)} \geq 0, \quad i \in I_X \cap I_Y, \tag{11.8f}$$
$$d_{y(i)} \geq 0, \quad i \in I_X \cap I_Y. \tag{11.8g}$$

Strong stationarity can also be related to stationarity of an NLP relaxation of (11.2), abbreviated RNLP and defined as

$$\min_{x,y,z} \quad f(x,y,z) \tag{11.9a}$$
$$\text{s.t.} \quad h(x,y,z) = 0, \tag{11.9b}$$
$$g(x,y,z) \leq 0, \quad x \geq 0, \tag{11.9c}$$
$$x_{(i)} = 0, \quad i \in I_X \setminus I_Y, \tag{11.9d}$$
$$y_{(i)} = 0, \quad i \in I_Y \setminus I_X, \tag{11.9e}$$
$$x_{(i)} \geq 0, \quad i \in I_X \cap I_Y, \tag{11.9f}$$
$$y_{(i)} \geq 0, \quad i \in I_X \cap I_Y. \tag{11.9g}$$

This property implies that there is no feasible descent direction at the solution of either (11.8) or (11.9). This condition is equivalent to B-stationarity if the biactive set $I_X \cap I_Y$ is empty, or if the MPEC-LICQ property holds [10]. MPEC-LICQ requires that the following set of vectors be linearly independent:

$$\left\{ \nabla g_i(w^*) | i \in I_g \right\} \cup \left\{ \nabla h(w^*) \right\} \cup \left\{ \nabla x_{(i)} | i \in I_X \right\} \cup \left\{ \nabla y_{(i)} | i \in I_Y \right\}, \tag{11.10}$$

where

$$I_g = \left\{ i : g_i(w^*) = 0 \right\}.$$

MPEC-LICQ is equivalent to LICQ for (11.9) and implies that the multipliers of either (11.8) or (11.9) are bounded and unique. Satisfaction of MPEC-LICQ leads to the following result.

Theorem 11.1 [10, 255, 350] If w^* is a solution to the MPCC (11.2) and MPEC-LICQ holds at w^*, then w^* is strongly stationary.

Strong stationarity is the key assumption that allows the solution of MPCCs through NLP formulations. With this property, a well-posed set of multipliers for (11.9) verifies optimality of (11.2), and a suitable reformulation of MPCC can lead to a number of equivalent, well-posed nonlinear programs. As a result, MPCCs can be addressed directly through these reformulations.

Finally, a related constraint qualification is MPEC-MFCQ, i.e., MFCQ applied to (11.9). As noted above, this property requires that all the equality constraints in (11.9) have linearly independent gradients and there exists a nonzero vector $d \in \mathbb{R}^n$ such that

$$d_{x(i)} = 0, \quad i \in I_X \setminus I_Y, \tag{11.11a}$$
$$d_{y(i)} = 0, \quad i \in I_Y \setminus I_X, \tag{11.11b}$$
$$\nabla h(w^*)^T d = 0, \tag{11.11c}$$
$$\nabla g_i(w^*)^T d < 0, \quad i \in I_g, \tag{11.11d}$$
$$d_{x(i)} > 0,\ d_{y(i)} > 0, \quad i \in I_X \cap I_Y. \tag{11.11e}$$

MPEC-MFCQ also implies that the multipliers of (11.9) will be bounded.

Similarly, second order conditions can be defined for MPCCs that extend analogous conditions developed for constrained nonlinear programs in Chapter 4. In Theorem 4.18, sufficient second order conditions were developed with the use of constrained (or allowable) search directions. For MPCCs, the allowable search directions d are defined in [326, 350] for the following cones:

- $d \in \bar{S}$, where d is tangential to equality constraints and inequality constraints with positive multipliers, and it forms an acute angle to active constraints with zero multipliers.
- $d \in S^*$, where $d \in \bar{S}$ and also tangential to at least one of the branches for $i \in I_X \cup I_Y$.
- $d \in \bar{T}$, where d is tangential to equality constraints and inequality constraints with nonzero multipliers.
- $d \in T^*$, where $d \in \bar{T}$ and also tangential to at least one of the branches for $i \in I_X \cup I_Y$.

We define the MPCC Lagrange function given by

$$\mathcal{L}_C = f(w) + g(w)^T u + h(w)v - \alpha^T x - \beta^T y$$

and with w^* as a strongly stationary point and multipliers $u^*, v^*, \alpha^*, \beta^*$ that satisfy the KKT conditions for (11.9). The following second order sufficient conditions (SOSC) and strong second order sufficient conditions (SSOSC), defined in [326], require

$$d^T \nabla_{ww} \mathcal{L}_C(w^*, u^*, v^*, \alpha^*, \beta^*) d \geq \sigma$$

if there is a $\sigma > 0$ for the following allowable directions:

- RNLP-SOSC for $d \in \bar{S}$,
- MPEC-SOSC for $d \in S^*$,
- RNLP-SSOSC for $d \in \bar{T}$,
- MPEC-SSOSC for $d \in T^*$.

MPEC-SOSC, the weakest of these conditions, is sufficient to define w^* as a strict local minimizer of (11.2).

With these conditions, we now explore the reformulation of MPCCs to well-posed nonlinear programs.

11.2.1 Solution Strategies

If no reformulation of an MPCC is made and the complementarity is represented by (11.3), (11.4), or (11.5), the constraint multipliers will be nonunique and unbounded, with dependent constraints at every feasible point. Some active set NLP strategies may be able to overcome this difficulty if they are developed to handle degenerate NLP formulations [75]. In particular, encouraging results have been reported with the Filter-SQP algorithm [138]. Moreover, elastic mode SQP algorithms have finite termination and global convergence properties when solving MPCCs [10]. On the other hand, the performance of an active set algorithm is still affected by combinatorial complexity as the problem size increases.

The following MPCC reformulations have been analyzed in [10, 204, 255, 324, 326] and allow standard NLP tools to be applied:

$$\begin{aligned} \text{Reg}(\epsilon): \quad & \min \quad f(w) && (11.12a)\\ & \text{s.t.} \quad h(w) = 0, && (11.12b)\\ & \qquad g(w) \leq 0, && (11.12c)\\ & \qquad x, y \geq 0, && (11.12d)\\ & \qquad x_{(i)} y_{(i)} \leq \epsilon, \quad i = 1, \ldots, n_c; && (11.12e) \end{aligned}$$

$$\begin{aligned} \text{RegComp}(\epsilon): \quad & \min \quad f(w) && (11.13a)\\ & \text{s.t.} \quad h(w) = 0, && (11.13b)\\ & \qquad g(w) \leq 0, && (11.13c)\\ & \qquad x, y \geq 0, && (11.13d)\\ & \qquad x^T y \leq \epsilon; && (11.13e) \end{aligned}$$

$$\begin{aligned} \text{RegEq}(\epsilon): \quad & \min \quad f(w) && (11.14a)\\ & \text{s.t.} \quad h(w) = 0, && (11.14b)\\ & \qquad g(w) \leq 0, && (11.14c)\\ & \qquad x, y \geq 0, && (11.14d)\\ & \qquad x_{(i)} y_{(i)} = \epsilon, \quad i = 1, \ldots, n_c; && (11.14e) \end{aligned}$$

$$\begin{aligned} \text{PF}(\rho): \quad & \min \quad f(w) + \rho x^T y && (11.15a)\\ & \text{s.t.} \quad h(w) = 0, && (11.15b)\\ & \qquad g(w) \leq 0, && (11.15c)\\ & \qquad x, y \geq 0. && (11.15d) \end{aligned}$$

For the first three, *regularized* formulations, the complementarity conditions are relaxed and the MPCC is reformulated in Reg(ϵ), RegComp(ϵ), or RegEq(ϵ) with a positive relaxation parameter ϵ. The solution of the MPCC, w^*, can be obtained by solving a series of relaxed solutions, $w(\epsilon)$, as ϵ approaches zero.

The convergence properties of these NLP formulations can be summarized by the following theorems, developed in [326].

Theorem 11.2 Suppose that w^* is a strongly stationary solution to (11.2) at which MPEC-MFCQ and MPEC-SOSC are satisfied. Then there exist $r_0 > 0, \bar{\epsilon}$, and $\epsilon \in (0, \bar{\epsilon}]$ so that $w(\epsilon)$, the global solution of Reg(ϵ) with constraint $\|w(\epsilon) - w^*\| \leq r_0$ that lies closest to w^*, satisfies $\|w(\epsilon) - w^*\| = O(\epsilon^{1/2})$. If the stronger conditions MPEC-LICQ and RNLP-SOSC hold, then under similar conditions we have $\|w(\epsilon) - w^*\| = O(\epsilon)$.

Uniqueness properties have also been shown for Reg(ϵ). Also, a property similar to Theorem 11.2 holds for the RegComp(ϵ) formulation, where the individual complementarity constraints are replaced by the single constraint $x^T y \leq \epsilon$. However, local uniqueness of the solutions to RegComp(ϵ) cannot be guaranteed.

For the RegEq(ϵ) formulation the following convergence property holds.

Theorem 11.3 Suppose that w^* is a strongly stationary solution to (11.2) at which MPEC-LICQ and MPEC-SOSC are satisfied. Then there exist $r_0 > 0$, $\bar{\epsilon}$, and $\epsilon \in (0, \bar{\epsilon}]$ so that $w(\epsilon)$, the global solution of RegEq(ϵ) with the constraint $\|w(\epsilon) - w^*\| \leq r_0$ that lies closest to w^*, satisfies $\|w(\epsilon) - w^*\| = O(\epsilon^{1/4})$.

Note that even with slightly stronger assumptions than in Theorem 11.2, the RegEq(ϵ) formulation will exhibit slower convergence. Despite this property, the RegEq formulation has proved to be popular because simpler equations replace the complementarity conditions. In particular, the related nonlinear complementarity problem (NCP) functions and smoothing functions have been widely used to solve MPCCs [94, 167, 241, 369]. A popular NCP function is the Fischer–Burmeister function

$$\phi(x, y) = x + y - \sqrt{x^2 + y^2} = 0. \tag{11.16}$$

As this function has an unbounded derivative at $x = y = 0$, it is usually smoothed to the form

$$\phi(x, y) = x + y - \sqrt{x^2 + y^2 + \epsilon} = 0 \tag{11.17}$$

for some small $\epsilon > 0$. The solution to the original problem is then recovered by solving a series of problems as ϵ approaches zero. An equivalence can be made between this method and RegEq(ϵ). Accordingly, the problem will converge at the same slow convergence rate.

In contrast to the regularized formulations, we also consider the exact ℓ_1 penalization shown in PF(ρ). Here the complementarity can be moved from the constraints to the objective function, and the resulting problem is solved for a particular value of ρ. If $\rho \geq \rho_c$, where ρ_c is the critical value of the penalty parameter, then the complementarity constraints will be satisfied at the solution. Similarly, Anitescu, Tseng, and Wright [10] considered a related "elastic mode" formulation, where artificial variables are introduced to relax the constraints in PF(ρ) and an additional ℓ_∞ constraint penalty term is added. In both cases the resulting NLP formulation has the following properties.

Theorem 11.4 [326] If w^* is a strongly stationary point for the MPCC (11.2), then for all ρ sufficiently large, w^* is a stationary point for PF(ρ). Also, if MPEC-LICQ, MPEC-MFCQ, or MPEC-SOSC hold for (11.2), then the corresponding LICQ, MFCQ, or SOSC properties hold for PF(ρ).

Theorem 11.5 [10] If w^* is a solution to PF(ρ) and w^* is feasible for the MPCC (11.2), then w^* is a strongly stationary solution to (11.2). Moreover, if LICQ or SOSC hold for PF(ρ), then the corresponding MPEC-LICQ or MPEC-SOSC properties hold for (11.2).

Theorem 11.4 indicates that the solution strategy is attracted to strongly stationary points for sufficiently large values of ρ. However, it is not known beforehand how large ρ must be. If the initial value of ρ is too small, a series of problems with increasing ρ values may need to be solved. Also, Theorem 11.5 indicates that a local solution of PF(ρ) is a solution to the MPCC if it is also feasible to the MPCC. This condition is essential as there is no guarantee that PF(ρ) will not get stuck at a local solution that does not satisfy the complementarity, as observed by [204]. Nevertheless, with a finite value of $\rho > \rho_c$ this ℓ_1 penalization ensures that strongly stationary points of the MPCC are local minimizers to PF(ρ).

The PF(ρ) formulation has a number of advantages. Provided that the penalty parameter is large enough, the MPCC may then be solved as a single problem, instead of a sequence of problems. Moreover, PF(ρ) allows any NLP solver to be used to solve a complementarity problem, without modification of the algorithm. On the other hand, using the PF(ρ) formulation the penalty parameter can also be changed during the course of the optimization, for both active set and interior point optimization algorithms. This modification was introduced and demonstrated in [10, 255].

Finally, there is a strong interaction between the MPCC reformulation and the applied NLP solver. In particular, as noted in [255], if a barrier NLP method (like LOQO, KNITRO, or IPOPT) is applied to PF(ρ), then there is a correspondence between RegComp(ϵ) and intermediate PF(ρ) solutions with the barrier parameter $\mu_l \rightarrow 0$. This relation can be seen by comparing the first order KKT conditions of RegComp(ϵ) and intermediate PF(ρ) subproblems, and identifying a corresponding set of parameters ρ, μ, and ϵ.

11.2.2 Comparison of MPCC Formulations

The formulations of the previous section provide a straightforward way to represent the MPCC through NLP, as long as it has a strongly stationary solution. An important component of this task is the efficiency and reliability of the NLP solver. In this section we compare and evaluate each of these formulations through a library test set provided with the GAMS modeling environment [71]. This can be accomplished through the NLPEC metasolver [71] which reformulates the complementarity constraints to a user-specified NLP reformulation, including (11.12)–(11.15). NLPEC then calls a user-specified NLP solver to solve the reformulated model. The final results from the NLP solver are then translated back into the original MPCC model. NLPEC verifies that the complementarities are satisfied in the resulting solution.

For this comparison we consider the MPECLib collection of MPCC test problems maintained by Dirkse [284]. This test set consists of 92 problems including small-scale models from the literature and several large industrial models. The performance of the NLP reformulations PF(ρ) with $\rho = 10$, Reg(ϵ) with $\epsilon = 10^{-8}$, and NCP formulation $x^T y = 0$, $x, y \geq 0$ were compared with CONOPT (version 3.14) and IPOPT (version 3.2.3) used to solve the resulting NLP problems. Also included is the IPOPT-C solver, which applies the Reg(ϵ) formulation within IPOPT [324] and coordinates the adjustment of the relaxation parameter ϵ with the barrier parameter μ.

The results of this comparison are presented as Dolan–Morè plots in Figures 11.1 and 11.2. All results were obtained on an Intel Pentium 4, 1.8 GHz CPU with 992 MB of RAM. The plots portray both robustness and relative performance of a set of solvers on a given problem set. Each problem is assigned its minimum solution time among the algorithms compared. The figure then plots the fraction of test problems solved by a particular algorithm within *Time Factor* of this minimum CPU time.

Figure 11.1 shows the performance plot of different reformulations and solvers tested on all 92 problems of the test set. Since the problems are inherently nonconvex, not all of the different reformulation and solver combinations converged to the same local solutions. While this makes a direct comparison difficult, this figure is useful to demonstrate the robustness of the methods (as *Time Factor* becomes large). With one exception, all of the reformulation and solver combinations are able to solve at least 84% of the test problems. The most reliable solvers were IPOPT-Reg(ϵ) (99%), CONOPT-Penalty (98%), and IPOPT-Penalty (94%). In contrast, IPOPT-Mult (IPOPT with the complementarity written as (11.3) and no further reformulation) was able to solve only 57% of all of the problems. This is not unexpected, as IPOPT is an interior point algorithm and the original MPCC problem has no interior at the solution.

Figure 11.2 shows the same reformulations and solvers for only 22 of the 92 test problems. For these 22 problems, all reformulation-solver combinations gave the same solutions, when they were successful. This allows for a more accurate comparison of the solvers' performance. Table 11.1 displays these 22 test problems and their respective objective function values. From Figure 11.2 we note that CONOPT-Reg(ϵ) performed the fastest on 73% of the problems, but it turns out to solve only 76% of them. Instead, CONOPT-Penalty and IPOPT-Penalty (PF(ρ)) are the next best in performance and they also prove to be the most robust, solving over 90% of the problems. These turn out to be the best all-around methods.

Because the test problems are not large, CONOPT provides the best performance on this test set. Also, the CONOPT formulations appear to take advantage of the CONOPT active set strategy in the detection and removal of dependent constraints. As a result, CONOPT-Mult (with no reformulation) performs well. On the other hand, the IPOPT formulations (IPOPT-Penalty, IPOPT-Reg, IPOPT-C) follow similar trends within Figure 11.2, with the penalty formulation as the most robust. This can be explained by the similarities among these methods, as analyzed in [255]. Finally, IPOPT-Mult (with no reformulation) is the worst performer as it cannot remove dependent constraints, and therefore suffers most from the inherent degeneracies in MPCCs.

From this brief set of results, we see the advantages of the PF(ρ) strategy, particularly since it is easier to address with general NLP solvers such as CONOPT [119] and IPOPT [206]. These solvers have competing advantages; CONOPT quickly detects active sets and handles dependent constraints efficiently, while IPOPT has low computational complexity in handling large-scale problems with many inequality constraints.

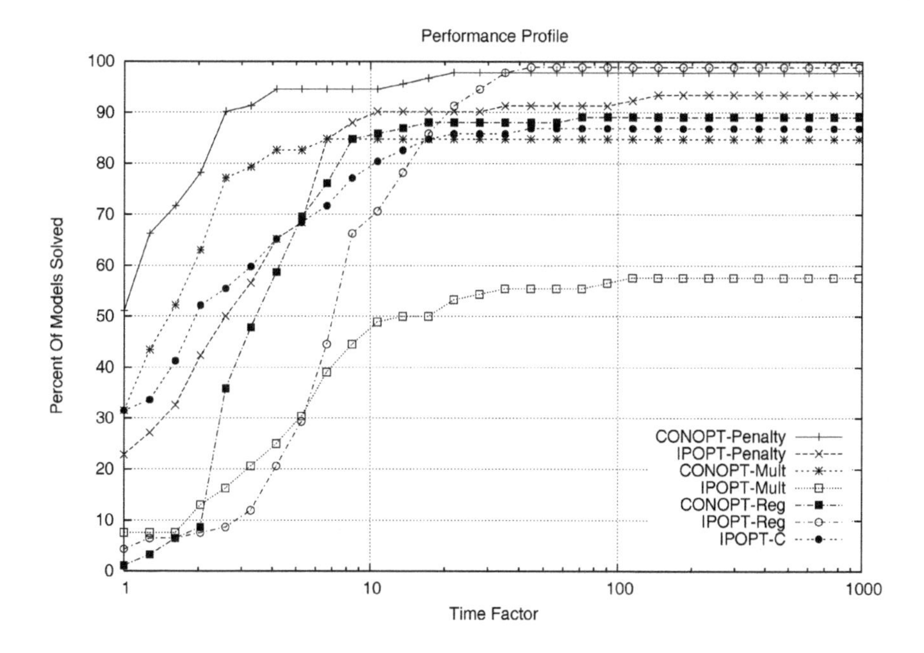

Figure 11.1. *Performance profiles of MPCC reformulations for 92 MPECLib test problems.*

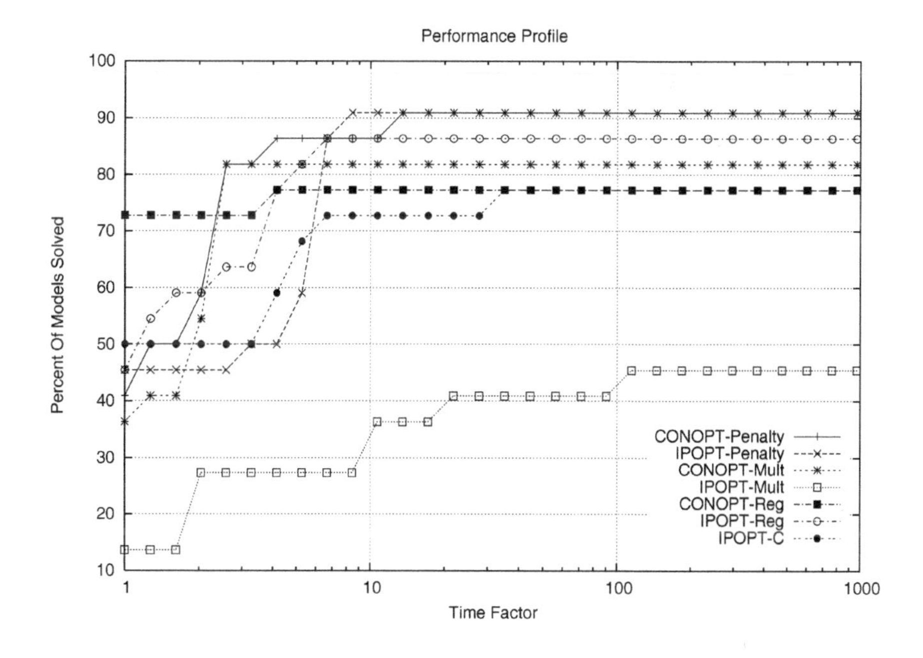

Figure 11.2. *Performance profiles for 22 selected test problems.*

Table 11.1. *Objective function values of the 22 MPCCLib test problems that converged to the same solutions.*

Problem Name	Objective Function Value	Constraints	Variables	Complementarities
bard2	-6600	10	13	8
bard3	-12.67872	6	7	4
bartruss3_0	3.54545×10^{-7}	29	36	26
bartruss3_1	3.54545×10^{-7}	29	36	11
bartruss3_2	10166.57	29	36	6
bartruss3_3	3×10^{-7}	27	34	26
bartruss3_4	3×10^{-7}	27	34	11
bartruss3_5	10166.57	27	34	6
desilva	-1	5	7	4
ex9_1_4m	-61	5	6	4
findb10s	2.02139×10^{-7}	203	198	176
fjq1	3.207699	7	8	6
gauvin	20	3	4	2
kehoe1	3.6345595	11	11	5
outrata31	2.882722	5	6	4
outrata33	2.888119	5	6	4
outrata34	5.7892185	5	6	4
qvi	7.67061×10^{-19}	3	5	3
three	2.58284×10^{-20}	4	3	1
tinque_dhs2	N/A	4834	4805	3552
tinque_sws3	12467.56	5699	5671	4480
tollmpec	-20.82589	2377	2380	2376

11.3 Complementary Models for Process Optimization

Complementarity applications in chemical engineering are widespread for modeling discrete decisions. Their use in modeling nonsmooth controllers and tiered splitters is well known and a key component of equation-based models for real-time optimization [151]. For process optimization, both for steady state and dynamic systems, complementarity models also allow a number of options when dealing with phase changes, flow reversal, safety valve operation, and other discrete events, as discussed in [33, 324].

Moreover, with a better understanding of NLP-based solution strategies for MPCCs, it is essential to consider the formulation of well-posed complementarity models. In this section, we consider a systematic MPCC modeling strategy and apply it to a number of process examples.

To develop these concepts, we again caution that complementarity formulations are nonconvex problems that may lead to multiple local solutions. Consequently, with the application of standard NLP solvers, the user will need to be satisfied with local solutions, unless additional problem-specific information can be applied. Despite the nonconvexity in all MPCC formulations, it is clear that poorly posed MPCC formulations also need to be avoided.

For instance, the feasible region for

$$0 \le x \perp (1-x) \ge 0 \tag{11.18}$$

consists of only two isolated points, and the disjoint feasible region associated with this example often leads to convergence failures, unless the model is initialized close to a solution with either $x = 0$ or $x = 1$. Moreover, without careful consideration, it is not difficult to create MPCCs with similar difficulties.

Because the MPCC can be derived from an associated MPEC, we consider the formulation in (11.1). To avoid disjoint regions, we require that both $\theta(x, y)$ and $C(x)$ be convex in y for all $(x, y) \in Z$. This leads to a well-defined solution for the inner problem. We use this observation to develop the following guidelines for the examples in this study.

- Start with a convex inner level problem such as

$$\min_y \varphi(x)y \quad \text{s.t.} \quad y_a \le y \le y_b, \tag{11.19}$$

  where $\varphi(x)$ is a switching function for the discrete decision. Often a linear program in y is a good choice.

- When possible, formulate (11.19) so that the upper level constraints in (11.1), i.e., $(x, y) \in Z$, do not interfere with the selection of any value of $y \in C(x)$.

- The resulting MPEC is then converted to an MPCC. For instance, applying the KKT conditions to (11.19), we obtain the complementarities

$$\varphi(x) - s_a + s_b = 0, \tag{11.20a}$$
$$0 \le (y - y_a) \perp s_a \ge 0, \tag{11.20b}$$
$$0 \le (y_b - y) \perp s_b \ge 0. \tag{11.20c}$$

- Simplify the relations in (11.20) through variable elimination and application of the complementarity conditions.

- Finally, incorporate the resulting complementarity conditions within the NLP reformulations described in Section 11.2.

Note that in the absence of upper level constraints in y, the solution of (11.20) allows y to take any value in $[y_a, y_b]$ when $\varphi(x) = 0$. Thus, it is necessary to avoid disjoint feasible regions for the resulting MPCC, and this is frequently a problem-specific modeling task. For instance, complementarity formulations should not be used to model logical disjunctions such as *exclusive or* (EXOR) operators as in (11.18) because they lead to disjoint regions for y. On the other hand, logical disjunctions such as an *inclusive or* operator can be modeled successfully with MPCCs.

In the remainder of this section, we apply these guidelines to develop complementarity models that arise in process applications.

Commonly Used Functions

The following nonsmooth functions can be represented by complementarity formulations.

- The absolute value operator $z = |f(x)|$ can be rewritten as

$$z = f(x)y, \tag{11.21}$$

$$y = \arg\left\{\min_{\hat{y}} -f(x)\hat{y} \quad \text{s.t.} \ -1 \le \hat{y} \le 1\right\}. \tag{11.22}$$

The corresponding optimality conditions are

$$f(x) = s_b - s_a, \tag{11.23a}$$
$$0 \le s_b \perp (1-y) \ge 0, \tag{11.23b}$$
$$0 \le s_a \perp (y+1) \ge 0. \tag{11.23c}$$

By substituting (11.23a) into (11.21) and by manipulating (11.23b), (11.23c) to eliminate y, we obtain the simplified model

$$z = s_b + s_a, \tag{11.24a}$$
$$f(x) = s_b - s_a, \tag{11.24b}$$
$$0 \le s_b \perp s_a \ge 0. \tag{11.24c}$$

- The max operator $z = \max\{f(x), z_a\}$ can be rewritten as

$$z = f(x) + (z_a - f(x))y, \tag{11.25}$$

$$y = \arg\left\{\min_{\hat{y}} (f(x) - z_a)\hat{y} \quad \text{s.t.} \ 0 \le \hat{y} \le 1\right\}. \tag{11.26}$$

The corresponding optimality conditions are

$$z = f(x) + (z_a - f(x))y, \tag{11.27a}$$
$$f(x) - z_a = s_a - s_b, \tag{11.27b}$$
$$0 \le s_b \perp (1-y) \ge 0, \tag{11.27c}$$
$$0 \le s_a \perp y \ge 0. \tag{11.27d}$$

These relations can be simplified by using (11.27b), (11.27c), (11.27d) and substituting into (11.27a) to eliminate y, leading to

$$z = f(x) + s_b, \tag{11.28a}$$
$$f(x) - z_a = s_a - s_b, \tag{11.28b}$$
$$0 \le s_b \perp s_a \ge 0. \tag{11.28c}$$

- The min operator $z = \min(f(x), z_b)$ can be treated in a similar way by defining the problem

$$z = f(x) + (z_b - f(x))y, \tag{11.29}$$

$$y = \arg\left\{\min_{\hat{y}} (z_b - f(x))\hat{y} \quad \text{s.t.} \ 0 \le \hat{y} \le 1\right\}. \tag{11.30}$$

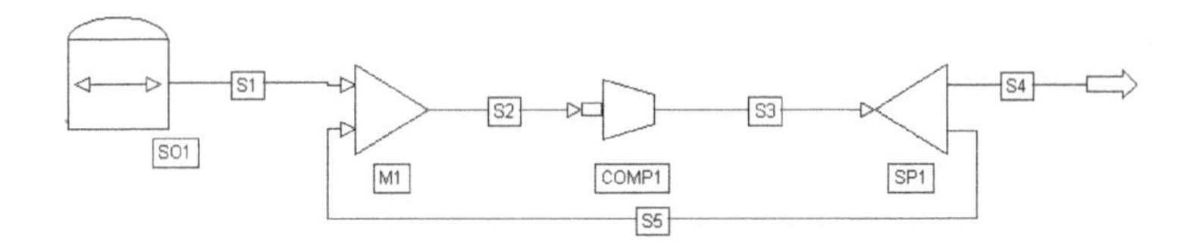

Figure 11.3. *Flowsheet example of compressor kick-back.*

Applying the optimality conditions and simplifying leads to the following complementarity system:

$$z = f(x) - s_a, \tag{11.31a}$$
$$z_b - f(x) = s_b - s_a, \tag{11.31b}$$
$$0 \le s_b \perp s_a \ge 0. \tag{11.31c}$$

- The signum or sign operation $z = \text{sgn}(x)$ can be represented by a complementarity formulation:

$$\min \quad -y \cdot x \tag{11.32a}$$
$$\text{s.t.} \quad -1 \le y \le 1, \tag{11.32b}$$

and the corresponding complementarity relations can be written as

$$x = s_b - s_a, \tag{11.33a}$$
$$0 \le s_a \perp (y+1) \ge 0, \tag{11.33b}$$
$$0 \le s_b \perp (1-y) \ge 0. \tag{11.33c}$$

Further simplification of these conditions is left to Exercise 11.32(a).

Flow Reversal

Flow reversal occurs in fuel headers and pipeline distribution networks. This is problematic in simulation models as physical properties and other equations implicitly assume the flow rate to be positive. These situations may be modeled with the absolute value operator, using complementarities as in (11.24). The magnitude of the flow rate can then be used in the sign sensitive equations.

Relief Valves, Check Valves, Compressor Kickback

Relief valves, check valves, and compressor kick-back operations are all related phenomena that can be modeled using complementarities in a manner similar to the functions used above.

- Check valves prevent fluid flows in reverse directions that may arise from changes in differential pressure. Assuming that the directional flow is a monotonic function

$f(\Delta p)$ of the differential pressure, we model the flow F through the check valve as

$$F = \max\{0, f(\Delta p)\}$$

and rewrite the max operator as the complementarity system given in (11.28).

- Relief valves allow flow only when the pressure, p, or pressure difference, Δp, is larger than a predetermined value. Once open, the flow F is some function of the pressure, $f(p)$. This can be modeled as $F = f(p)y$ with

$$\min_y \quad (p_{\max} - p)y \text{ s.t. } \quad 0 \le y \le 1.$$

  The inner minimization problem sets $y = 1$ if $p > p_{\max}$, $y = 0$ if $p < p_{\max}$, and $y \in [0,1]$ if $p = p_{\max}$. This behavior determines if the flow rate should be zero, when the valve is closed, or determined by the expression $f(p)$, when the valve is open. The related complementarity conditions are

$$\begin{aligned} F &= f(p)y, \\ (p_{\max} - p) &= s_0 - s_1, \\ 0 \le y &\perp s_0 \ge 0, \\ 0 \le (1-y) &\perp s_1 \ge 0. \end{aligned}$$

- As shown in Figure 11.3, compressor kick-back, where recirculation is required when the compressor feed drops below a critical value, can be modeled in a similar manner. Following the stream notation in Figure 11.3, we can write

$$F_{S2} = \max(F_{S1}, f_{crit}), \quad F_{S5} = f_{crit} - F_{S1}$$

  and rewrite the max operator using the complementarities in (11.28).

Piecewise Functions

Piecewise smooth functions are often encountered in physical property models, tiered pricing, and table lookups. This composite function can be represented by a scalar ξ along with the following inner minimization problem and associated equation [323]:

$$\min_y \quad \sum_{i=1}^{N} (\xi - a_i)(\xi - a_{i-1}) y_{(i)} \tag{11.34a}$$

$$\text{s.t.} \quad \sum_{i=1}^{N} y_{(i)} = 1, \; y_{(i)} \ge 0, \tag{11.34b}$$

$$z = \sum_{i=1}^{N} f_i(\xi) y_{(i)}, \tag{11.35}$$

where N is the number of piecewise segments, $f_i(\xi)$ is the function over the interval $\xi \in [a_{i-1}, a_i]$, and z represents the value of the piecewise function. This LP sets $y_{(i)} = 1$

and $y_{j \neq i} = 0$ when $\xi \in (a_{i-1}, a_i)$. The associated equation will then set $z = f_i(\xi)$, which is the function value on the interval. The NLP (11.34) can be rewritten as the following complementarity system:

$$\sum_{i=1}^{N} y_{(i)} = 1, \tag{11.36a}$$

$$(\xi - a_i)(\xi - a_{i-1}) - \gamma - s_i = 0, \tag{11.36b}$$

$$0 \leq y_{(i)} \perp s_i \geq 0. \tag{11.36c}$$

If the function $z(\xi)$ is piecewise smooth but not continuous, then $y_{(i)}$ can "cheat" at $\xi = a_i$ or $\xi = a_{i-1}$ by taking fractional values. For instance, cost per unit may increase stepwise over different ranges, and jumps in costs that occur at a_i may be replaced by arbitrary intermediate values. A way around this problem is to define $z(\xi)$ as a continuous, but nonsmooth, function (e.g., unit cost times quantity), so that fractional values of $y_{(i)}$ will still represent $z(a_i)$ accurately.

PI Controller Saturation

PI (proportional plus integral) controller saturation has been studied using complementarity formulations [416]. The PI control law takes the following form:

$$u(t) = K_c\left(e(t) + \frac{1}{\tau_I}\int_0^t e(t')dt'\right), \tag{11.37}$$

where $u(t)$ is the control law output, $e(t)$ is the error in the measured variable, K_c is the controller gain, and τ_I is the integral time constant. The controller output $v(t)$ is typically subject to upper and lower bounds, v_{up} and v_{lo}, i.e., $v(t) = \max(v_{lo}, \min(v_{up}, u(t)))$. The following inner minimization relaxes the controller output to take into account the saturation effects:

$$\begin{aligned} \min_{y_{up}, y_{lo}} \; & (v_{up} - u)y_{up} + (u - v_{lo})y_{lo} \\ \text{s.t.} \quad & 0 \leq y_{lo},\ y_{up} \leq 1 \end{aligned} \tag{11.38}$$

with $v(t) = u(t) + (v_{up} - u(t))y_{up} + (v_{lo} - u(t))y_{lo}$. Suppressing the dependence on t and applying the KKT conditions to (11.38) leads to

$$\begin{aligned} v &= u + (v_{up} - u)y_{up} + (v_{lo} - u)y_{lo}, \\ (v_{up} - u) &- s_{0,up} + s_{1,up} = 0, \\ (u - v_{lo}) &- s_{0,lo} + s_{1,lo} = 0, \\ 0 \leq s_{0,up} \perp y_{up} \geq 0, &\quad 0 \leq s_{1,up} \perp (1 - y_{up}) \geq 0, \\ 0 \leq s_{0,lo} \perp y_{lo} \geq 0, &\quad 0 \leq s_{1,lo} \perp (1 - y_{lo}) \geq 0. \end{aligned} \tag{11.39}$$

Eliminating the variables y_{up} and y_{lo} and applying the complementarity conditions leads to the following simplification:

$$\begin{aligned} v &= u + s_{1,lo} - s_{1,up}, \\ (v_{up} - u) &- s_{0,up} + s_{1,up} = 0, \\ (u - v_{lo}) &- s_{0,lo} + s_{1,lo} = 0, \\ 0 \leq s_{1,up} \perp s_{0,up} \geq 0, &\quad 0 \leq s_{0,lo} \perp s_{1,lo} \geq 0. \end{aligned} \tag{11.40}$$

Phase Changes

As described in Example 7.1 in Chapter 7, flash separators operate at vapor-liquid equilibrium by concentrating high-boiling components in the liquid stream and low-boiling components in the vapor stream. The process model (7.3) that describes this system is derived from minimization of the Gibbs free energy of the system [344]. Flash separations operate at vapor-liquid equilibrium (i.e., boiling mixtures) between the so-called dew and bubble point conditions. Outside of these ranges, one of the phases disappears and equilibrium no longer holds between the phases. To model these phase changes, Gibbs minimization at a fixed temperature T and pressure P is written as a constrained optimization problem of the form

$$\min_{l_i, v_i} \quad G(T,P,l_i,v_i) = \sum_{i=1}^{NC} l_i\, \bar{G}_i^L + \sum_{i=1}^{NC} v_i\, \bar{G}_i^V$$

$$\text{s.t.} \quad \sum_{i=1}^{NC} l_i \geq 0, \quad \sum_{i=1}^{NC} v_i \geq 0,$$

$$l_i + v_i = m_i^T > 0, \quad i = 1,\dots,NC,$$

where NC refers to the number of chemical components with index i, $G(T,P,l_i,v_i)$ is the total Gibbs free energy,

$$\bar{G}_i^L = \bar{G}_i^{ig}(T,P) + RT\ \ln(f_i^L)),$$
$$\bar{G}_i^V = \bar{G}_i^{ig}(T,P) + RT\ \ln(f_i^V)),$$

$\bar{G}_i^{ig}$ is the ideal gas free energy per mole for component i, f_i^L and f_i^V are the mixture liquid and vapor fugacities for component i, l_i and v_i are the moles of component i in the liquid and vapor phase, R is the gas constant, and m_i^T are the total moles of component i. The first order KKT conditions for this problem are given by

$$\bar{G}_i^{ig}(T,P) + RT\ \ln(f_i^L) + \left\{ \sum_{j=1}^{NC} \left(l_j \frac{\partial \bar{G}_j^L}{\partial l_i} + v_j \frac{\partial \bar{G}_j^V}{\partial l_i} \right) \right\} - \alpha_L - \gamma_i = 0, \tag{11.41a}$$

$$\bar{G}_i^{ig}(T,P) + RT\ \ln(f_i^V) + \left\{ \sum_{j=1}^{NC} \left(l_j \frac{\partial \bar{G}_j^L}{\partial v_i} + v_j \frac{\partial \bar{G}_j^V}{\partial v_i} \right) \right\} - \alpha_V - \gamma_i = 0, \tag{11.41b}$$

$$0 \leq \alpha_L \perp \sum_{i=1}^{NC} l_i \geq 0, \tag{11.41c}$$

$$0 \leq \alpha_V \perp \sum_{i=1}^{NC} v_i \geq 0, \tag{11.41d}$$

$$l_i + v_i = m_i^T, \quad i = 1,\dots,NC. \tag{11.41e}$$

The bracketed terms in (11.41a) and (11.41b) are equal to zero from the Gibbs–Duhem equation [344], and subtracting (11.41a) from (11.41b) leads to

$$RT\ \ln(f_i^V / f_i^L) - \alpha_V + \alpha_L = 0.$$

Moreover, defining $f_i^V = \phi_i^V(T,P,y_i)y_i$, $f_i^L = \phi_i^L(T,P,x_i)x_i$, and $K_i = \phi_i^L/\phi_i^V$, where ϕ_i^L and ϕ_i^V are fugacity coefficients, leads to

$$y_i = \exp\left(\frac{\alpha_V - \alpha_L}{RT}\right) K_i x_i.$$

From (11.41c) and (11.41d) we can deduce $0 \le \alpha_L \perp \alpha_V \ge 0$ and by defining $\beta = \exp(\frac{\alpha_V - \alpha_L}{RT})$, we have

$$\alpha_V > 0 \Longrightarrow \beta > 1, \quad \alpha_L > 0 \Longrightarrow \beta < 1$$

leading to the following complementarity system:

$$y_j = \beta K_j x_j, \tag{11.42a}$$
$$-s_L \le \beta - 1 \le s_V, \tag{11.42b}$$
$$0 \le L \perp s_L \ge 0, \tag{11.42c}$$
$$0 \le V \perp s_V \ge 0. \tag{11.42d}$$

In this manner, phase existence can be determined within the context of an MPCC. If a slack variable (s_L or s_V) is positive, either the corresponding liquid or vapor phase is absent and $\beta \ne 1$ relaxes the phase equilibrium condition, as required in (11.41). As seen in the next section, (11.42) can also be applied to the optimization of distillation columns, and, in Exercise 5, these conditions can also be extended as necessary conditions for multiphase equilibrium.

11.4 Distillation Case Studies

Phase behavior of a vapor liquid system is governed by the minimization of Gibbs free energy and can be modeled as complementarity constraints using (11.42). Moreover, these conditions allow phase changes not only in flash separators, but also distillation systems. The complementarity system (11.42) has been applied to the optimization of both steady state and dynamic tray columns [241, 324, 320]. In these studies, regularization methods were applied for distillation optimization, with the reflux ratio, feed tray location, and tray number as decision variables. This section considers steady state distillation optimization in two parts. First, complementarity models are developed that demonstrate phase changes on distillation trays during the optimization process. Second, we provide a performance comparison of MPCC formulations for distillation optimization.

11.4.1 Multicomponent Column Optimization with Phase Changes

We consider a distillation column for the separation of a feed containing a five-component hydrocarbon mixture: $j \in NC = \{C_3H_8$ (light key), $i-C_4H_{10}$, $n-C_4H_{10}$, $n-C_5H_{12}$, $i-C_5H_{12}$ (heavy key)$\}$. We assume that ideal vapor-liquid equilibrium holds and that there is no mass transfer resistance, with physical property data obtained from [140]. The distillation column model is similar to the ones presented in Sections 1.4.2 and 7.4.1, with the equilibrium equations modified to complementarity conditions (11.42) as shown below.

Overall Mass Balance

$$\begin{aligned}
L_1 + V_1 - L_2 &= 0, \\
L_i + V_i - L_{i+1} - V_{i-1} &= 0, \quad i = 2, \dots, N+1,\ i \notin \mathcal{S}, \\
L_i + V_i - L_{i+1} - V_{i-1} - F_i &= 0, \quad i \in \mathcal{S}, \\
R + D - V_{N+1} &= 0.
\end{aligned}$$

Energy Balance

$$\begin{aligned}
L_1 H_{L,1} + V_1 H_{V,1} - L_2 H_{L,2} - Q_R &= 0, \\
L_i H_{L,i} + V_i H_{V,i} - L_{i+1} H_{L,i+1} - V_{i-1} H_{V,i-1} &= 0, \quad i = 2, \dots, N+1, \\
L_i H_{L,i} + V_i H_{V,i} - L_{i+1} H_{L,i+1} - V_{i-1} H_{V,i-1} - F_i H_F &= 0, \quad i \in \mathcal{S}, \\
V_{N+1} H_{V,N+1} - (R + D) H_{L,D} - Q_C &= 0.
\end{aligned}$$

Component Mass Balance

$$\begin{aligned}
L_1 x_{1,j} + V_1 y_{1,j} - L_2 x_{2,j} &= 0, \quad j \in NC, \\
L_i x_{i,j} + V_i y_{i,j} - L_{i+1} x_{i+1,j} - V_{i-1,j} y_{i-1,j} &= 0, \quad j \in NC,\ i = 2, \dots, N+1,\ i \notin \mathcal{S}, \\
L_i x_{i,j} + V_i y_{i,j} - L_{i+1} x_{i+1,j} - V_{i-1} y_{i-1,j} - F,i x_{F,j} &= 0, \quad j \in NC,\ i \in \mathcal{S}, \\
(R + D) x_{D,j} - V_{N+1} y_{N+1,j} &= 0, \quad j \in NC.
\end{aligned}$$

Complementarities

$$\begin{aligned}
0 \le L_i \perp s_{L,i} &\ge 0, \quad i = 1, \dots, N+1, \\
0 \le V_i \perp s_{V,i} &\ge 0, \quad i = 1, \dots, N+1, \\
\beta_i - 1 + s_{L,i} - s_{V,i} &= 0, \quad i = 1, \dots, N+1.
\end{aligned}$$

Relaxed Phase Equilibrium

$$\begin{aligned}
y_{i,j} - \frac{\beta_i}{P} \exp\left(A_j + \frac{B_j}{C_j + T_i} \right) x_{i,j} &= 0, \quad j \in NC,\ i = 1, \dots, N+1, \\
\sum_{j \in NC} y_{i,j} - \sum_{j \in NC} x_{i,j} &= 0, \quad i = 1, \dots, N+1.
\end{aligned}$$

Bounds

$$\begin{aligned}
D, R, Q_R, Q_C &\ge 0, \\
1 \ge y_{i,j}, x_{i,j} &\ge 0, \quad j \in NC,\ i = 1, \dots, N+1,
\end{aligned}$$

where i are trays numbered starting from reboiler ($= 1$), j are components in the feed, P is the column pressure, $\mathcal{S}$ is the set for feed tray location, D is the distillate flow rate, R is the reflux flow rate, N is the number of trays in the column, F_i is feed flow rate, L_i/V_i is flow rate of liquid/vapor leaving tray i, T_i is temperature of tray i, H_F is feed enthalpy, $H_{L/V,i}$

is enthalpy of liquid/vapor leaving tray i, x_F, x_D are feed and distillate composition, $x/y_{i,j}$ is mole fraction of j in liquid/vapor leaving tray i, β_i is the relaxation parameter, $s_{L/V,i}$ are slack variables, A_j, B_j, C_j are Antoine coefficients, D is distillate flow rate, and $Q_{R/C}$ is the heat load on reboiler/condenser.

The feed is a saturated liquid of mole fractions [0.05,0.15,0.25,0.2,0.35] with the components in the order given above. The column has $N = 20$ trays, and feed enters on tray 12. The column is operated at a pressure of 725 kPa and we neglect pressure drop across the column. This problem has 2 degrees of freedom. The objective is to minimize the reboiler heat duty, which accounts for most of the energy costs. We are interested in observing whether the proposed algorithm can identify (possibly) dry and vaporless trays at the optimal solution. For this purpose we study three cases that differ in the recovery of key components:

(1) $x_{bottom,lk} \leq 0.01 x_{top,lk}$ and $x_{top,hk} \leq 0.01 x_{bottom,hk}$, where $lk = C_3H_8$ is the light key and $hk = iC_5H_{12}$ is the heavy key,

(2) $x_{bottom,hk} \geq 0.86$ and no top specification is given.

(3) $x_{bottom,hk} \geq 0.35$. This loose specification corresponds to no need for separation.

Since reboiler duty increases if we achieve more than the specified recovery, we expect these constraints to be active at the optimal solution. Figure 11.4 shows the liquid and vapor flow rates from all of the trays for these three cases. In case (1), we require high recovery in the top as well as the bottom, and the column operates with a high-reboiler load with liquid and vapor present on all trays. In case (2), we require high recovery of the heavy key only. Since we have no specification for the top products, there is vapor on all trays, but the trays above the feed tray run dry to attain minimum reboiler heat duty. Case (3) requires no more separation than is already present in the feed. The column is rewarded for not operating and has no vapor on its trays. Instead, the feed runs down the column as a liquid and leaves without change in composition.

All three cases were solved with IPOPT-C, a modification of IPOPT based on Reg(ϵ) with adaptive adjustment of ϵ (see [319]). The size of the problem in the three cases and the performance results are provided in Table 11.2. The number of constraints includes the number of complementarity constraints as well. All problems have been solved to a tolerance of less than 10^{-6} in the KKT error and required less that 2 CPU seconds on a Pentium III, 667 MHz CPU running Linux).

11.4.2 Tray Optimization

Tray optimization has been traditionally addressed using an MINLP approach. Instead, we consider a formulation that uses a differentiable distribution function (DDF) to direct all feed, reflux, and intermediate product streams to the column trays. As shown in Figure 11.5, streams for the feed and the reflux are fed to all trays as dictated by two discretized Gaussian distribution functions. Note that the gray area in Figure 11.5 consists only of vapor traffic, and consequently, each tray model must include complementarities (11.42) that allow for disappearance of the liquid phase. We modify the feed tray set $\mathcal{S} = \{2, \ldots, N+1\}$ and choose two additional continuous optimization variables N_f, the feed location, and N_t, the total number of trays, with $N_t \geq N_f$. We also specify feed and reflux flow rates for $i \in \mathcal{S}$

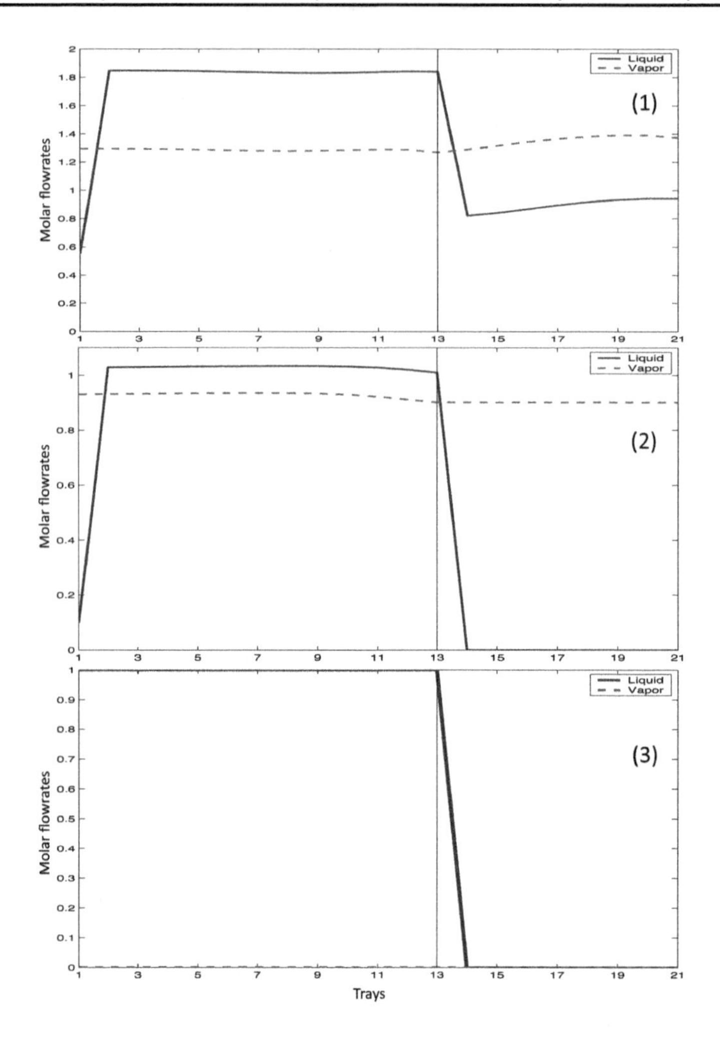

Figure 11.4. *Minimum energy distillation with liquid (solid lines) and vapor (dashed lines) molar flow rates for* (1) *top and bottom specifications,* (2) *bottom specification,* (3) *no specifications.*

based on the value of the DDF at that tray, given by

$$F_i = \frac{\exp\left(\frac{-(i-N_f)^2}{\sigma_f}\right)}{\sum\limits_{j\in\mathcal{S}} \exp\left(\frac{-(j-N_f)^2}{\sigma_f}\right)}, \quad i \in \mathcal{S}, \tag{11.43}$$

$$R_i = \frac{\exp\left(\frac{-(i-N_t)^2}{\sigma_t}\right)}{\sum\limits_{j\in\mathcal{S}} \exp\left(\frac{-(j-N_t)^2}{\sigma_t}\right)}, \quad i \in \mathcal{S}, \tag{11.44}$$

Table 11.2. *Results of distillation with energy minimization.*

Case	# Variables	# Constraints	# Complementarity Constraints	CPU Time (s)
1	480	436	42	0.5
2	478	434	42	1
3	495	455	42	2

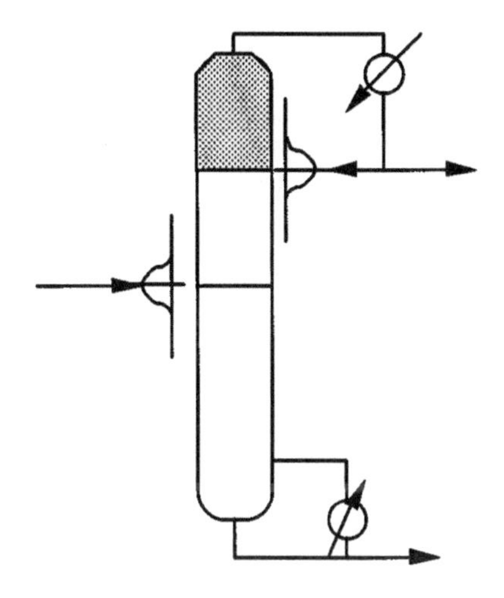

Figure 11.5. *Diagram of distillation column showing feed and reflux flows distributed according to the DDF. The gray section of the column is above the primary reflux location and has negligible liquid flows.*

where $\sigma_f, \sigma_t = 0.5$ are parameters in the distribution. Finally, we modify the overall mass, energy, and component balances in Section 11.4.1 by allowing feed and reflux flow rates on all trays $i \in \mathcal{S}$.

This modification enables the placement of feeds, sidestreams, and number of trays in the column to be continuous variables in the DDF. On the other hand, this approach leads to the upper trays with no liquid flows, and this requires the complementarity constraints (11.42). The resulting model is used to determine the optimal number of trays, reflux ratio, and feed tray location for a benzene/toluene separation. The distillation model uses ideal thermodynamics as in Section 11.4.1. Three MPCC formulations, modeled in GAMS and solved with CONOPT, were considered for the following cases:

- Penalty formulation, PF(ρ) with $\rho = 1000$.
- Relaxed formulation, Reg(ϵ). This strategy was solved in two NLP stages with $\epsilon = 10^{-6}$ followed by $\epsilon = 10^{-12}$.
- NCP formulation using the Fischer–Burmeister function (11.17). This approach is equivalent to RegEq(ϵ) and was solved in three NLP stages with $\epsilon = 10^{-4}$ followed by $\epsilon = 10^{-8}$ and $\epsilon = 10^{-12}$.

Table 11.3. *Distillation optimization: Comparison of MPCC formulations. All computations were performed on a Pentium 4 with 1.8 GHz and 992 MB RAM. (*CONOPT terminated at feasible point with negligible improvement of the objective function, but without satisfying KKT tolerance.)*

Cases	1	2	3
iter-PF(ρ)	356	395	354
iter-Reg(ϵ)	538	348	370
iter-NCP	179	119	254
CPUs-PF(ρ)	7.188	7.969	4.875
CPUs-Reg(ϵ)	14.594	7.969	7.125
CPUs-NCP	7.906	3.922	9.188
obj-PF(ρ)	9.4723	6.8103	3.0053
obj-Reg(ϵ)	9.5202*	6.8103*	2.9048*
obj-NCP	11.5904*	9.7288*	2.9332*
reflux-PF(ρ)	1.619	4.969	1.485
reflux-Reg(ϵ)	1.811	4.969	1.431
reflux-NCP	1.683	2.881	1.359
N_t-PF(ρ)	6.524	5.528	8.844
N_t-Reg(ϵ)	6.645	5.528	9.794
N_t-NCP	9.437	9.302	9.721

The binary column has a maximum of 25 trays, its feed is 100 mol/s of a 70%/30% mixture of benzene/toluene, and distillate flow is specified to be 50% of the feed. The reflux ratio is allowed to vary between 1 and 20, the feed tray location varies between 2 and 20, and the total tray number varies between 3 and 25. The objective function for the benzene-toluene separation minimizes:

$$objective = wt \cdot D \cdot x_{D,Toluene} + wr \cdot r + wn \cdot N_t, \tag{11.45}$$

where N_t is the number of trays, $r = R/D$ is the reflux ratio, D is the distillate flow, $x_{D,Toluene}$ is the toluene mole fraction, and wt, wr, and wn are the weighting parameters for each term; these weights allow the optimization to trade off product purity, energy cost, and capital cost. The column optimizations were initialized with 21 trays, a feed tray location at the seventh tray, and a reflux ratio at 2.2. Temperature and mole fraction profiles were initialized with linear interpolations based on the top and bottom product properties. The resulting GAMS models consist of 353 equations and 359 variables for the Reg(ϵ) and NCP formulations, and 305 equations and 361 variables for the PF(ρ) formulation. The following cases were considered:

- *Case* 1 *($wt = 1, wr = 1, wn = 1$)*: This represents the base case with equal weights for toluene in distillate, reflux ratio, and tray count. As seen in the results in Table 11.3, the optimal solution has an objective function value of 9.4723 with intermediate values of r and N_t. This solution is found quickly by the PF(ρ) formulation. On the other hand, the Reg(ϵ) formulation terminates close to this solution, while the NCP formulation terminates early with poor progress.

- *Case* 2 *($wt = 1, wr = 0.1, wn = 1$)*: In this case, less emphasis is given to energy cost. As seen in the results in Table 11.3, the optimal solution now has a lower objective function value of 6.8103 along with a higher value of r and lower value of N_t. This is found quickly by both PF(ρ) and Reg(ϵ), although only the former satisfies the convergence tolerance. On the other hand, the NCP formulation again terminates early with poor progress.

- *Case* 3 *($wt = 1, wr = 1, wn = 0.1$)*: In contrast to Case 2, less emphasis is now given to capital cost. As seen in the results in Table 11.3, the optimal solution now has an objective function value of 2.9048 with lower values of r and higher values of N_t. This is found quickly by the Reg(ϵ) formulation. Although Reg(ϵ) does not satisfy the convergence tolerance, the optimum could also be verified by PF(ρ). On the other hand, PF(ρ) quickly converges to a slightly different solution, which it identifies as a local optimum, while the NCP formulation requires more time to terminate with poor progress.

Along with the numerical study in Section 11.2.2, these three cases demonstrate that optimization of detailed distillation column models can be performed efficiently with MPCC formulations. In particular, it can be seen that the penalty formulation (PF(ρ)) represents a significant improvement over the NCP formulation both in terms of iterations and CPU seconds; PF(ρ) offers advantages over the Reg(ϵ) formulation as well.

11.5 Optimization of Hybrid Dynamic Systems

We now consider dynamic systems modeled with the simultaneous collocation formulation developed in Chapter 10. In that chapter we considered dynamic optimization problems of the form (10.19) with smooth state and control profiles within each period. Here we extend this formulation to consider switches in the dynamic system, governed by conditional inequalities. This class of *hybrid dynamic systems* has much in common with optimal control problems, but additional conditions are required to handle the switching behavior.

To illustrate a hybrid dynamic system modeled with complementarity conditions, we consider a switched system with ODE models and apply a change of notation from the previous sections. This system has discontinuous right-hand sides and is of the form considered by Filippov [132]:

$$\frac{dz}{dt} = \begin{cases} f_-(z(t), u(t)), & \sigma(z(t)) < 0, \\ \nu(t) f_-(z(t), u(t)) + (1 - \nu(t)) f_+(z(t), u(t)), \ \nu(t) \in [0, 1], & \sigma(z(t)) = 0, \\ f_+(z(t), u(t)), & \sigma(z(t)) > 0. \end{cases} \tag{11.46}$$

In this representation, $t, z(t), u(t), \nu(t)$, and $\sigma(z(t))$ are time, differential state variables, control variables, switching profiles, and guard (or switching) function, respectively. The scalar switching function, $\sigma(z(t))$, determines transitions to different state models, represented by a scalar switching profile $\nu(t)$ set to zero or one. At the transition point, where $\sigma(t) = 0$, a convex combination of the two models is allowed with $\nu(t) \in [0, 1]$. Note that if $\sigma(z(t)) = 0$ over a nonzero period of time, $\nu(t)$ can be determined from smoothness properties of $\sigma(z(t))$; i.e.,

$$\frac{d\sigma}{dt} = \nabla_z \sigma(z(t))^T [\nu(t) f_-(z(t), u(t)) + (1 - \nu(t)) f_+(z(t), u(t))] = 0. \tag{11.47}$$

Also, for this problem class, we assume that the differential states $z(t)$ remain continuous over time. Existence and uniqueness properties of (11.46) have been analyzed in [132, 249]. We can express (11.46) equivalently through the addition of slack variable profiles and complementarity constraints as

$$\sigma(z(t)) = s^p(t) - s^n(t), \tag{11.48a}$$

$$\frac{dz}{dt} = \nu(t) f_-(z,u) + (1-\nu(t)) f_+(z,u), \quad \nu(t) \in [0,1], \tag{11.48b}$$

$$0 \le s^p(t) \perp \nu(t) \ge 0, \tag{11.48c}$$

$$0 \le s^n(t) \perp (1-\nu(t)) \ge 0. \tag{11.48d}$$

Using the simultaneous collocation approach from Chapter 10, the DAE is converted into an NLP by approximating state and control profiles by a family of polynomials on finite elements, defined by $t_0 < t_1 < \cdots < t_N$. In addition, the differential equation is discretized using K-point Radau collocation on finite elements. As in Chapter 10, one can apply either Lagrange interpolation polynomials (10.4) or the Runge–Kutta representation (10.5) for the differential state profiles, with continuity enforced across element boundaries. The switching $\nu(t)$ and control profiles $u(t)$ are approximated using Lagrange interpolation polynomials (10.17).

A straightforward substitution of the polynomial representations allows us to write the collocation equations (10.6) for (11.48) as

$$\begin{aligned} \frac{dz^K}{dt}(t_{ik}) &= \nu_{ik} f_-(z_{ik},u_{ik}) + (1-\nu_{ik}) f_+(z_{ik},u_{ik}), \quad \nu_{ik} \in [0,1], \\ \sigma(z_{ik}) &= s^p_{ik} - s^n_{ik}, \\ 0 &\le s^p_{ik} \perp \nu_{ik} \ge 0, \\ 0 &\le s^n_{ik} \perp (1-\nu_{ik}) \ge 0, \\ i &= 1,\dots,N, \quad k = 1,\dots,K. \end{aligned}$$

However, this formulation is not sufficient to enforce smoothness within an element for $z(t), \nu(t)$, and $u(t)$. For this, we allow a variable length $h_i \in [h_L, h_U]$ for each finite element (determined by the NLP) and allow sign changes in $\sigma(z(t))$ only at t_i, the boundary of the finite element. Also, a positive lower bound on h_i is required to ensure that it does not go to zero, and an upper bound on h_i is chosen to limit the approximation error associated with the finite element. Permitting sign changes in $\sigma(t)$ only at the boundary of a finite element is enforced by choosing $\nu(t)$ to complement the L_1 norm, $\int_{t_{i-1}}^{t_i} |s^p(t)|\,dt$, and $1-\nu(t)$ to complement $\int_{t_{i-1}}^{t_i} |s^n(t)|\,dt$. With this modification, the discretized formulation becomes

$$\frac{dz^K}{dt}(t_{ik}) = \nu_{ik} f_-(z_{ik},u_{ik}) + (1-\nu_{ik}) f_+(z_{ik},u_{ik}), \quad \nu_{ik} \in [0,1], \tag{11.49a}$$

$$\sigma(z_{ik}) = s^p_{ik} - s^n_{ik}, \quad 0 \le \sum_{k'=0}^{K} s^p_{ik'} \perp \nu_{ik} \ge 0, \tag{11.49b}$$

$$0 \le \sum_{k'=0}^{K} s^n_{ik'} \perp (1-\nu_{ik}) \ge 0, \quad i = 1,\dots,N, k = 1,\dots,K, \tag{11.49c}$$

where we define $s^p_{i0} = s^p_{i-1,K}$ and $s^n_{i0} = s^n_{i-1,K}$ for $i = 2,\dots,N$. Note that the complementarities are now formulated so that only one branch of the complement is allowed over the element i, i.e.,

- $\nu(t) = 0$ only if $\sigma(z(t)) \geq 0$ for the entire finite element,
- $\nu(t) = 1$ only if $\sigma(z(t)) \leq 0$ for the entire finite element,
- $\nu(t) \in (0,1)$ only if $\sigma(z(t)) = 0$ over the entire finite element.

Moreover, for the last condition, we have an index-2 path constraint $\sigma(z(t)) = 0$ over the finite element. In our direct transcription approach, the Radau collocation scheme allows us to handle the high-index constraint directly as

$$\sigma(z_{ik}) = 0, \quad i = 1,\dots,N,\ k = 1,\dots,K,$$

and to obtain $\nu(t)$ implicitly through the solution of (11.49). As noted in Chapter 10 and in [16], Radau collocation is stable and accurate for index-2 systems; the error in the differential variables is $O(h^{2K-1})$ and the error in the algebraic variables is reduced only to $O(h^K)$.

We now generalize this formulation to multiple guard functions and switches that define N_T periods of positive length, indexed by $l = 1,\dots,N_T$, along with hybrid index-1 DAE models given by

$$\left.\begin{array}{l} F\left(\frac{dz}{dt}, z(t), y(t), u(t), \nu(t), p\right) = 0 \\ g(z(t), y(t), u(t), \nu(t), p) \leq 0 \end{array}\right\} t \in [t_{l-1}, t_l],\ l = 1,\dots,N_T,$$

with algebraic variables $y(t)$, time-independent optimization variables p, and inequality constraints g that may include the vector of switching functions. For this problem class, we further assume that periods can be of variable time, that transitions occur only at period boundaries, and that differential states remain continuous over all time. This leads to a set of switching profiles, indexed by $m \in \mathcal{M}$, and represented through a relation between guard functions, $\sigma_m(z(t))$, as follows:

$$\left.\begin{array}{l} F\left(\frac{dz}{dt}, z(t), y(t), u(t), \nu(t), p\right) = 0 \\ g(z(t), y(t), u(t), \nu(t), p) \leq 0 \\ \sigma_m(z(t)) < 0 \implies \nu_m(t) = 1 \\ \sigma_m(z(t)) > 0 \implies \nu_m(t) = 0 \\ \sigma_m(z(t)) = 0 \implies \nu_m(t) \in [0,1] \end{array}\right\} t \in (t_{l-1}, t_l],\ m \in \mathcal{M}, \tag{11.50a}$$

$$z(t_0) = z_0, \quad z(t^+_{N_T}) = z_f, \tag{11.50b}$$

$$z(t^-_l) = z(t^+_l), \quad l = 1,\dots,N_T - 1. \tag{11.50c}$$

In a manner similar to reformulating (11.48) to form the complementarity system (11.49), we apply Radau collocation on finite elements. Because one or more finite elements are allowed to make up a period l, we readopt the finite element index i for periods as well and apply the complementarity conditions within each finite element. This leads to the following formulation:

$$F\left(\frac{dz^K}{dt}(t_{ik}), z_{ik}, y_{ik}, u_{ik}, v_{ik}, p\right) = 0, \tag{11.51a}$$

$$g(z_{ik}, y_{ik}, u_{ik}, v_{ik}, p) \leq 0, \tag{11.51b}$$

$$z(t_i^-) = z(t_i^+), \tag{11.51c}$$

$$\sigma_m(z_{ik}) = s^p_{m,ik} - s^n_{m,ik}, \tag{11.51d}$$

$$0 \leq \left(\sum_{k'=0}^{K} s^p_{m,ik'}\right) \perp v_{m,ik} \geq 0, \tag{11.51e}$$

$$0 \leq \left(\sum_{k'=0}^{K} s^n_{m,ik'}\right) \perp (1 - v_{m,ik}) \geq 0, \tag{11.51f}$$

$$i = 1, \ldots, N, \; k = 1, \ldots, K, \; m \in \mathcal{M}, \tag{11.51g}$$

$$z(t_0) = z_0, \quad z(t_N^+) = z_f. \tag{11.51h}$$

Equations (11.51) constitute the constraints for the discretized hybrid dynamic optimization problem represented by the MPCC (11.2). Moreover, because a higher order IRK discretization is used within smooth finite elements, we are able to enforce accurate solutions to the hybrid dynamic system, with a relatively small number of finite elements.

11.6 Dynamic MPCC Case Studies

We now consider two case studies that illustrate the complementarity formulation for hybrid dynamic systems. The first example considers frequent discrete decisions in a dynamic system through the reformulation of the signum function. This case demonstrates the advantage of the MPCC reformulation strategy to obtain accurate optimal solutions. The second case study considers a hybrid dynamic system in a cascading tank process with checkvalves. Through the proposed MPCC formulation, optimal profiles can be determined that accommodate valve switching to provide desired setpoints.

11.6.1 Reformulation of a Differential Inclusion

The first example problem minimizes an arbitrary objective function

$$\phi = (z_{end} - 5/3)^2 + \int_{t_0}^{t_{end}} z^2(t)dt$$

subject to a differential inclusion $\frac{dz(t)}{dt} \in \mathrm{sgn}(z(t)) + 2$ with initial condition $z_0 = -2$, where the derivative can be defined by a complementarity system [371]. Because the differential inclusion has an analytical solution, there are no degrees of freedom in this problem. Hence, a unique value of the objective function can be found in a straightforward manner.

Nevertheless, the differential inclusion can be reformulated as the following optimal control problem:

$$\min \quad \phi = (z_{end} - 5/3)^2 + \int_{t_0}^{t_{end}} z^2 \cdot dt \tag{11.52a}$$

$$\text{s.t.} \quad \frac{dz(t)}{dt} = u(t) + 2, z(0) = z_0, \tag{11.52b}$$

$$z = s^+ - s^-, \tag{11.52c}$$

$$0 \leq 1 - u(t) \perp s^+(t) \geq 0, \tag{11.52d}$$

$$0 \leq u(t) + 1 \perp s^-(t) \geq 0. \tag{11.52e}$$

We present this case study in two parts. First, we demonstrate the NLP reformulation on an example with increasingly many complementarity conditions. In the second part, we apply the dynamic reformulation (11.51) to determine accurate solutions for this hybrid system. A related discussion of this problem can also be found in [40].

Optimization with Fixed Finite Elements

Applying the implicit Euler method with a fixed step size, the discretized problem becomes

$$\min \quad \phi = (z_{end} - 5/3)^2 + \sum_{i=1}^{N} h\,(z_i)^2 \tag{11.53a}$$

$$\text{s.t.} \quad \dot{z}_i = u_i + 2, \tag{11.53b}$$

$$z_i = z_{i-1} + h \cdot \dot{z}_i, \tag{11.53c}$$

$$z_i = s_i^+ - s_i^-, \tag{11.53d}$$

$$0 \leq 1 - u_i \perp s_i^+ \geq 0, \tag{11.53e}$$

$$0 \leq u_i + 1 \perp s_i^- \geq 0, \quad i = 1, \ldots, N. \tag{11.53f}$$

The resulting discretized problem can be scaled up to be arbitrarily large by increasing the number of finite elements N, with a discrete decision in each element. This system was modeled in GAMS and solved via automatic reformulation using the NLPEC package. The discretized MPCC was solved using the penalty reformulation (PF(ρ) with CONOPT using $\rho = 1000$). The results are shown in Table 11.4 for complementarities representing up to 8000 discrete decisions. It can be seen that the number of iterations increases almost linearly with the number of finite elements, and the computation time per iteration increases linearly as well. Consequently, the total computational time for CONOPT for PF(ρ) grows approximately quadratically with problem size.

IPOPT-C was also used to solve this problem and the results are shown in Table 11.4. It can be seen that the number of iterations increases only weakly as a function of problem size. In addition, the computational time per iteration increases linearly with the number of finite elements, as the largest computational cost of IPOPT is due to the linear solver. As a result, the total computational time increase is only slightly superlinear with problem size. Moreover, it is also notable that while CONOPT is faster for the smaller problems, IPOPT-C becomes faster for the larger problems.

Table 11.4. *Solution times (Pentium 4, 1.8 GHz, 992 MB RAM) for different solution strategies.*

N	Objective Function	CONOPT/PF CPU s.	CONOPT/PF Iterations	IPOPT-C CPU s.	IPOPT-C Iterations
10	1.4738	0.047	15	0.110	17
100	1.7536	0.234	78	1.250	41
1000	1.7864	9.453	680	28.406	78
2000	1.7888	35.359	1340	14.062	25
3000	1.7894	112.094	2020	106.188	84
4000	1.7892	211.969	2679	84.875	56
5000	1.7895	340.922	3342	199.391	87
6000	1.7898	468.891	3998	320.140	115
7000	1.7896	646.953	4655	457.984	141
8000	1.7898	836.891	5310	364.937	98

Optimization with Variable Finite Elements

Applying Radau collocation with the Runge–Kutta polynomial representation (10.5), $t_0 = 0, t_f = 2$, and with variable element lengths restricted to $1.6/N \leq h_i \leq 2.4/N$, the discretized problem for (11.52) can be written in the form of (11.51) as the following MPCC:

$$\min \quad \phi = \left(z_f - 5/3\right)^2 + \sum_{i=1}^{N}\sum_{k=1}^{K} \omega_k h_i \cdot (z_{ik})^2 \tag{11.54a}$$

$$\text{s.t.} \quad \dot{z}_{ik} = 2 - \nu_{ik}, \quad i = 1,\dots,N, k = 1,\dots,K, \tag{11.54b}$$

$$z_i = z_{i-1} + h_i \sum_{k=1}^{K} \Omega_K(1)\dot{z}_{i,k}, \quad i = 1,\dots,N, \tag{11.54c}$$

$$z_{ik} = z_{i-1} + \sum_{k'=1}^{K} \Omega_{k'}(\tau_k)\dot{z}_{i,k'}, \quad i = 1,\dots,N, k = 1,\dots,K, \tag{11.54d}$$

$$z_{ik} = s_{ik}^p - s_{ik}^n, \quad i = 1,\dots,N, k = 1,\dots,K, \tag{11.54e}$$

$$0 \leq 1 - \nu_{ik} \perp \sum_{k'=0}^{K} s_{ik'}^p \geq 0, \quad i = 1,\dots,N, \tag{11.54f}$$

$$0 \leq \nu_{ik} + 1 \perp \sum_{k'=0}^{K} s_{ik'}^n \geq 0, \quad i = 1,\dots,N, k = 1,\dots,K, \tag{11.54g}$$

$$1.6/N \leq h_i \leq 2.4/N, \quad i = 1,\dots,N, \tag{11.54h}$$

$$\sum_{i=1}^{N} h_i = 2. \tag{11.54i}$$

For this formulation, choosing $K = 1$ leads to an implicit Euler method with first order accuracy. Since the differential equation is piecewise constant, implicit Euler integrates the

Table 11.5. *Solution times (Pentium 4, 1.8 GHz, 992 MB RAM) MPCC formulation with variable finite elements.*

	MPCC Formulation		
N	Objective	Iters.	CPU s.
10	1.5364	25	0.063
100	1.7766	97	0.766
1000	1.7889	698	23.266
2000	1.7895	1345	77.188
3000	1.7897	2009	166.781
4000	1.7898	2705	343.016

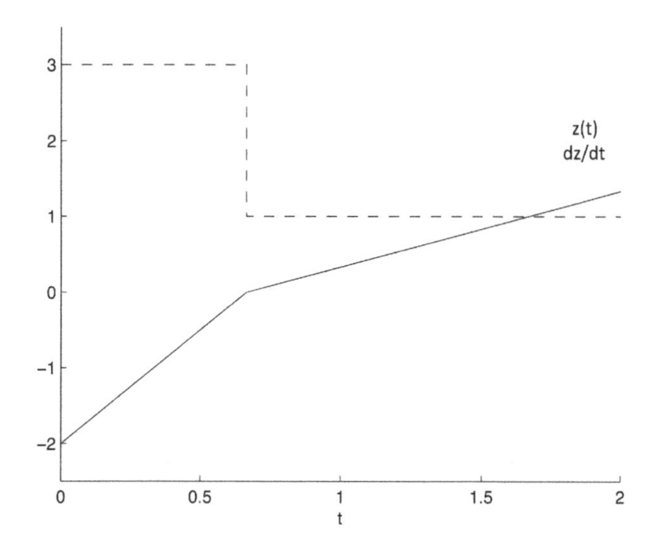

Figure 11.6. *Solution profiles of hybrid dynamic system.*

differential inclusion exactly and leads to exact identification of the switching locations. On the other hand, the integrand in the objective function is not integrated exactly.

The discretized MPCC was solved using the NLP penalty reformulation with CONOPT using $\rho = 1000$. The results are shown in Table 11.5. As in the case with fixed elements, the total computational time for CONOPT grows approximately quadratically with problem size.

The analytic solution for the hybrid dynamic system is plotted in Figure 11.6. Starting from $z(0) = -2$, $z(t)$ is piecewise linear, and the influence of $sgn(z)$ can be seen clearly from the plot of dz/dt. Moreover, from the analytic solution, it can be shown that $z(t)$ and the objective function, ϕ, are both differentiable in $z(0)$. On the other hand, a discretized problem with h_i fixed does not locate the transition point accurately and this leads to inaccurate profiles for $z(t)$ and $v(t)$. As discussed in [371], the application of fixed finite elements also leads to a nonsmooth dependence of the solution on $z(0)$. In Figure 11.7 the plot for $N = 100$ fixed elements shows a sawtooth behavior of ϕ versus $z(0)$. In con-

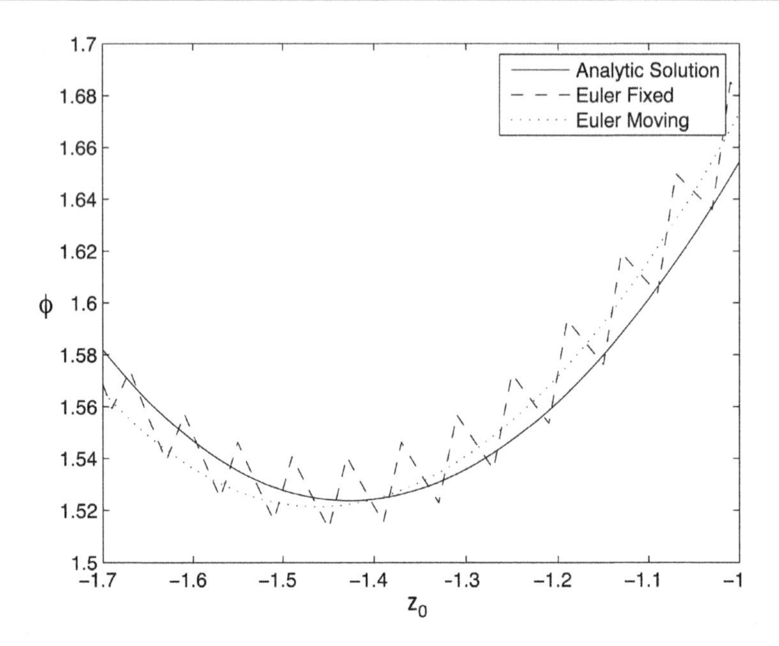

Figure 11.7. *Sensitivity of objective function with respect to z_0.*

trast with variable finite elements the objective function varies smoothly with $z(0)$. This occurs because the NLP solver can now locate the switching points accurately, and the complementarity formulation requires the differential state $z(t)$ to remain smooth within an element.

Moreover, the Euler discretization captures the piecewise linear $z(t)$ profile exactly and varies smoothly with $z(0)$. On the other hand, because ϕ still has local errors of $O(h_i^2)$, Figure 11.7 shows that the plot of the *analytically determined* objective function still differs from the Euler discretization, despite the accurate determination of $z(t)$. When the problem is resolved with $K = 3$, Radau collocation has fifth order accuracy and the numeric solution matches the analytical solution within machine precision. Nevertheless, for both $K = 1$ and $K = 3$ smoothness is still maintained in Figure 11.7 because the discontinuity is located exactly, and the integral is evaluated over the piecewise smooth portions, not across the discontinuity.

11.6.2 Cascading Tank Problem

The second case study computes an optimal control trajectory of valve positions for a series of cascading tanks. The problem is made arbitrarily large by changing various model parameters to investigate the growth of solution times. The system is illustrated in Figure 11.8 and was originally presented in [381].

The inlet to the first tank is located at the top of the tank and the inlet for each subsequent tank is located at an intermediate height H_i, where $i \in \{1,2,\ldots,n\}$ is the tank index, n is the total number of tanks, and L_i is the level in tank i. The outlet for all tanks is located at the bottom of the tank. The initial feed inlet to the first tank and all outlets are

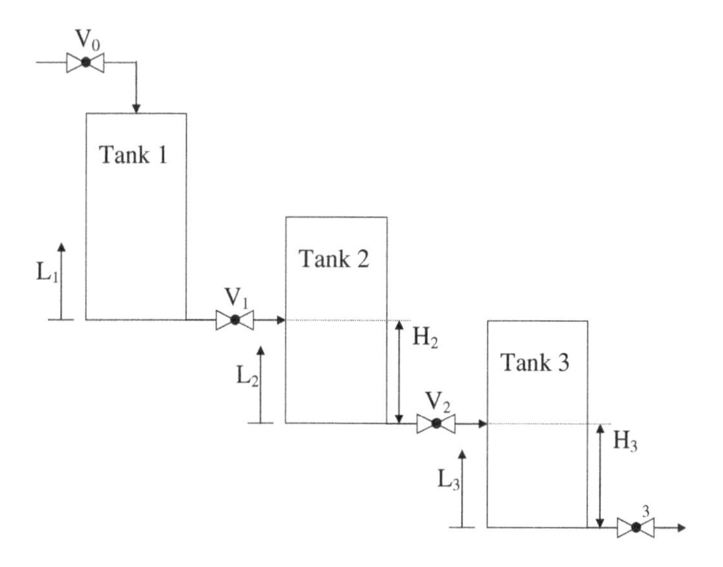

Figure 11.8. *Diagram of base case system.*

equipped with valves V_i to control the flow rates:

$$\frac{dL_i(t)}{dt} = \frac{1}{A_i}(F_{i-1}(t) - F_i(t)), \quad i = 1,\ldots,n_c, \tag{11.55}$$

$$F_0(t) = w_0(t)k_0, \tag{11.56}$$

$$F_i(t) = w_i(t)k_i(t)\sqrt{\Delta L_i(t)}, \quad i = 1,\ldots,n_c, \tag{11.57}$$

$$\Delta L_i(t) = \begin{cases} L_i(t) - L_{i+1}(t) - H_{i+1}, & L_{i+1}(t) > H_{i+1}, \\ L_i(t), & L_{i+1}(t) < H_{i+1}, \end{cases} \quad i \in \{1,2,\ldots,n-1\}, \tag{11.58}$$

$$\Delta L_n(t) = L_n(t). \tag{11.59}$$

Equation (11.55) is a material balance equation for each tank, while (11.56) and (11.57) describe the flow rates entering the first tank and leaving each tank, respectively. The flow rate out of tank i is governed by the liquid height above the valve. When the level in the next tank L_{i+1} is above its inlet level H_{i+1}, it influences the dynamics of L_i. This leads to the piecewise function (11.58) describing the difference in tank levels $\Delta L_i(t)$. Additionally, reverse flow (flow from tank i to tank $i-1$) is prohibited, so we include $\Delta L_i(t) \geq 0$.

We also select the following quadratic penalty as the objective function:

$$\phi = \sum_i \int_0^T (L_i(t) - 0.75)^2 dt. \tag{11.60}$$

Implicit Euler ($K = 1$) was used to discretize the differential equations and IPOPT was used to solve this set of problems using the PF(ρ) formulation. The model was initially solved with 10 finite elements per time interval, 10 time intervals, and 3 tanks, as the base case scenario. The state trajectories for the solution of the base case problem are plotted in

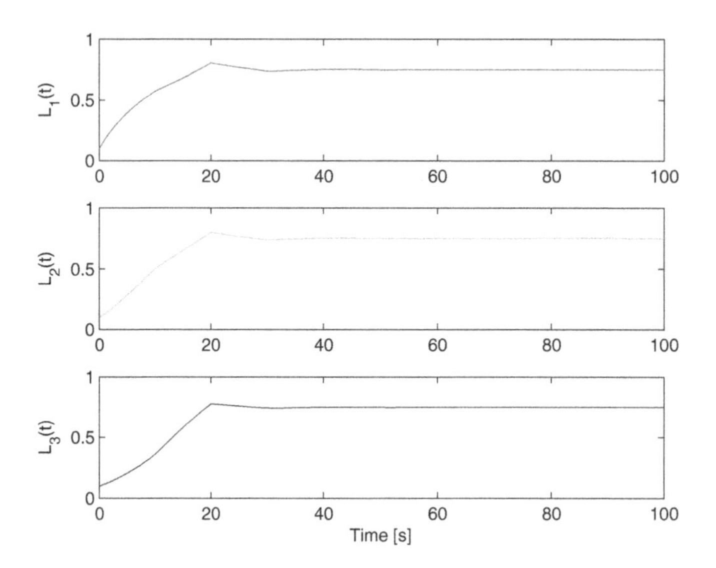

Figure 11.9. *Plot of state trajectories of tank levels L_i for base case.*

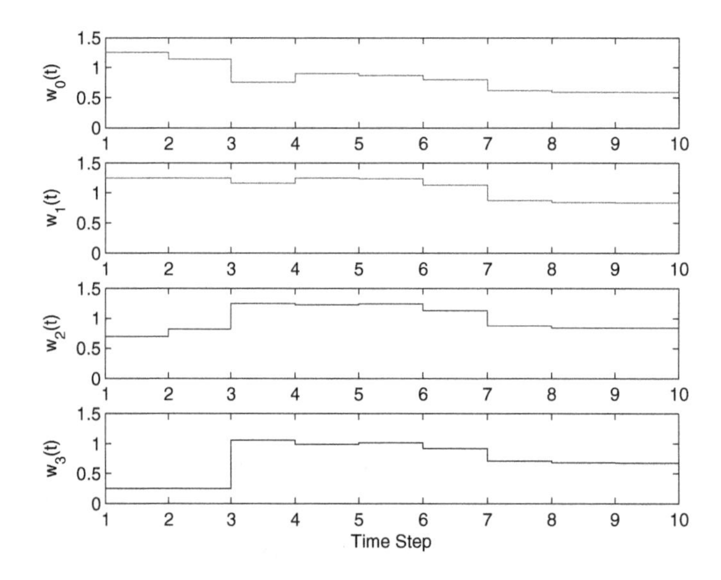

Figure 11.10. *Plot of optimal control profiles of valve openings $w(t)$.*

Figure 11.9, while the corresponding control profiles are given in Figure 11.10. The tank levels reach the target during the first two time intervals and remain at the target setpoint for all subsequent intervals. Note that the valve openings remain *at intermediate values* for these target levels.

Table 11.6. *Scaling of solution time with the number of finite elements* (N) *with* 3 *tanks and* 10 *time intervals.*

N	Time (s)	Iterations
10	1.703	28
20	7.297	42
30	12.125	41
40	23.641	48
50	28.078	43
60	40.219	47
70	46.063	43
80	63.094	45
90	80.734	48
100	90.234	44

Table 11.7. *Scaling of solution time with the number of tanks with* 10 *time intervals and* 10 *elements per interval.*

Tanks	Time (s)	Iterations
3	1.703	28
4	4.157	45
5	9.984	69
6	7.484	40
7	14.968	61
8	11.156	40
9	13.906	41
10	78.703	122
11	20.266	43
12	28.203	52

From this base case the number of finite elements per time interval and the number of tanks are varied independently, with the solution time on a Pentium 4 CPU with 1.8 GHz, 992 MB RAM) and iteration results given in Tables 11.6 and 11.7 for complementarities representing up to 3000 discrete switches. The solution time grows linearly with increasing numbers of tanks, and grows between linearly and quadratically with the number of finite elements. The increase in solution times is largely the result of the increased cost of the sparse linear factorization of the KKT matrix, which increases linearly with the size of the problem. As both the number of finite elements and the number of tanks are increased, the solution times never increase more than quadratically.

11.7 Summary and Conclusions

Mathematical programs with complementarity constraints (MPCCs) represent an interesting extension of nonlinear programming (NLP) to include a class of discrete decisions. Such formulations are meant to complement mixed integer nonlinear programming strategies and have the advantage of solving only a single NLP problem. On the other hand,

MPCCs do not satisfy constraint qualifications and have nonunique and unbounded multipliers. Consequently, the MPCC must be reformulated to a well-posed nonlinear program. This chapter describes a number of regularization and penalty formulations for MPCCs. In particular, the penalty formulation is advantageous, as it can be solved with standard NLP algorithms and does not require a sequence of nonlinear programs to be solved. In addition to summarizing convergence properties of these formulations, a small numerical comparison is presented on a collection of MPCC test problems.

The modeling of complementarity constraints has a strong influence on the successful solution of the MPCC. Here we motivate the development of complementarity models through formulation of bilevel optimization problems, where the lower level problem is convex, and present guidelines for the formulation of complementarity constraints for applications that include flow reversals, check valves, controller saturation, and phase equilibrium. These are demonstrated on the optimization of two distillation case studies.

The MPCC-based formulation is also extended to a class of hybrid dynamic optimization problems, where the differential states are continuous over time. These include differential inclusions of the Filippov type. Using direct transcription formulations, the location of switching points is modeled by allowing variable finite elements and by enforcing only one branch of the complementarity condition across the finite element. Any nonsmooth behavior from a change in these conditions is thus forced to occur only at the finite element boundary. As in Chapter 10, high-order IRK discretization is used to ensure accurate solutions in each finite element.

Two case studies illustrate this approach. The first demonstrates the MPCC formulation to deal with hybrid systems as well as the complexity of the MPCC solution strategy with increases in problem size. In particular, with fixed finite elements the problem can exhibit nonsmoothness in the parametric sensitivities, while moving finite elements lead to smooth solutions. This study also demonstrates that solution times grow polynomially with the number of complementarity constraints and problem size. In the second case study an optimal control profile was determined for a set of cascading tanks with check valves. Here the solution time grows no worse than quadratically with respect to number of tanks, and to number of finite elements per time interval.

11.8 Notes for Further Reading

This chapter provides only a brief introduction to complementarity systems and MPCCs for process optimization. Complementarity constraints play a significant role in game theory and economic, transportation, and mechanical engineering applications. The use of complementarity constraints in process engineering models is a relatively new and exciting field of research. As complementarities represent a set of nonsmooth equations, their treatment has motivated the development of a number of specialized algorithms based on smoothed reformulation of complementarity constraints, as well as interior point and pivoting techniques. A survey of these algorithms can be found in [187]. Studies specialized to chemical problems can be found in [167], [335].

Moreover, the incorporation of complementarity constraints within NLP formulations is a key part of engineering and economic systems. The pioneering monograph of Luo, Pang, and Ralph [267] considered the analysis, solution, and application of MPEC systems. Compared to NLP algorithms, the development of numerical methods for the solution of MPECs is still in its infancy. Several specialized algorithms have been proposed for solving MPECs [130, 267]. In a parallel development, researchers have also investigated the applicability of general NLP algorithms for the solution of MPECs. Anitescu [9] analyzed convergence behavior of a certain class of NLP algorithms when applied to MPECs, and Ralph and Wright [326] examined convergence properties for a wide range of NLP formulations for MPCCs. Moreover, the modeling of complementarities and MPCCs has become a very useful feature in optimization modeling environments. Ferris et al. [129] surveyed software advances for MPCCs and discussed extended mathematical programming frameworks for this class of problems.

MPCC formulations that arise in process engineering are described in Baumrucker, Renfro, and Biegler [33]. Applications of MPCCs in process engineering frequently stem from multilevel optimization problems. Clark and Westerberg [96] were among the first to reformulate such problems as an MPCC by replacing the inner minimization problem with its stationary conditions. Moreover, flexibility and resilience problems involved in design and operation under uncertainty have been formulated and solved as bilevel and trilevel optimization problems [376, 176]. Complementarity formulations of vapor-liquid phase equilibrium have been derived from the minimization of Gibbs free energy [167], and these complementarity conditions have been extended to the optimization of trayed distillation columns [320]. Complementarity models have also been applied to parameter estimation of core flood and reservoir models [221]. Other MPCC applications include optimization of metabolic flux networks [323], real-time optimization (RTO) of chemical processes [363], hybrid dynamic systems [191], dynamic startup of fuel cell systems [218], and the design of robust control systems [416]. In particular, this approach extends to optimization problems with DAE and PDE models that include equilibrium phase transitions [320, 221]. Moreover, Raghunathan et al. [322, 321] and Hjersted and Henson [198] considered optimization problems for metabolic network flows in cellular systems (modeled as linear programs in the inner problem).

MPCC formulations that arise in hybrid dynamic systems are described in [32, 320, 40]. For the application of complementarity constraints to hybrid dynamic systems, a survey of differential inclusions (DIs) along with an analysis of several classes of DIs is given by Dontchev and Lempio [117]. More recently, Pang and Stewart [303] introduced and analyzed differential variational inequalities (DVIs). Their study includes development of time-stepping methods along with conditions for existence of solutions, convergence of particular methods, and rates of convergence with respect to time steps. Consistency of time-stepping methods for DVIs and complementarity systems was also investigated in [191, 192, 193]. DVIs are a broad problem class that unify several other fields including ODEs with smooth or discontinuous right-hand sides, DAEs, and dynamic complementarity systems. Leine and Nijmeijer [249] considered dynamics and bifurcation of nonsmooth Filippov systems with applications in mechanics. Control and optimization of hybrid systems have been active application areas for DVIs, and smoothing methods, nonsmooth solution techniques, and mixed integer approaches have been considered in a number of studies including [371, 29, 30, 374, 18, 297]. In particular, the last two approaches deal with optimization formulations with discrete decisions.

11.9 Exercises

1. Show the equivalence between the Fischer–Burmeister function and the RegEq(ϵ) formulation. Additional smoothing functions can be derived by (i) noting the

equivalence of $0 \leq x \perp y \geq 0$ and $x - \max(0, x - y) = 0$, and (ii) replacing the max operator with a smooth approximation.

2. From the analysis of [326], show that the properties of the cones S, S^*, T, and T^* lead to the following relations among the second order conditions: $RNLP - SSOSC \Longrightarrow RNLP - SOSC \Longrightarrow MPEC - SOSC$, and $MPEC - SSOSC \Longrightarrow MPEC - SOSC$.

3. Apply a barrier method to RegComp(ϵ) and PF(ρ) and compare the KKT conditions of the corresponding subproblems.

4. Simplify the complementarity conditions for the following nonsmooth elements in process models:

   (a) the signum function, $z = \text{sgn}(x)$,

   (b) conditions for the relief valve,

   (c) conditions for compressor kick-back.

5. Extend the Gibbs free energy minimization (11.41) to $p > 2$ phases and derive a complementarity system that is analogous to (11.42).

6. Reformulate the distillation model in Section 11.4.1 to deal with the tray optimization extensions in Section 11.4.2.

7. Consider the differential inclusion in Section 11.6.1.

   (a) Find the analytic solution to this problem.

   (b) Resolve this problem using the formulation (11.51), $z(0) = -1$, and Radau collocation with $K = 2$ and $K = 3$

8. Consider the multitank regulation problem in Section 11.6.2. Resolve the base case problem using the formulation (11.51) and Radau collocation with $K = 2$ and $K = 3$.

Bibliography

[1] *NEOS Wiki.* Mathematicals and Computer Science Division, Argonne National Laboratory, http://wiki.mcs.anl.gov/NEOS/index.php/NEOS_wiki, (1996).

[2] Aspen Custom Modeler User's Guide. Technical report, Aspen Technology, http://www.aspentech.com, 2002.

[3] gPROMS User's Guide. Technical report, Process Systems Enterprises Limited, http://www.psenterprise.com, 2002.

[4] JACOBIAN dynamic modeling and optimization software. Technical report, Numerica Technology LLC, http://www.numericatech.com, 2005.

[5] J. Abadie and J. Carpentier. Generalization of the Wolfe reduced gradient method to the case of nonlinear constraints. In R. Fletcher, editor, *Optimization*, pages 37–49. Academic Press, New York, 1969.

[6] N. Adhya and N. Sahinidis. A Lagrangian Approach to the Pooling Problem. *Ind. Eng. Chem. Res.*, 38:1956–1972, 1999.

[7] J. Albuquerque and L.T. Biegler. Decomposition Algorithms for On-Line Estimation with Nonlinear DAE Models. *Comput. Chem. Eng.*, 21:283–294, 1997.

[8] G. M. Aly and W. C. Chan. Application of a Modified Quasilinearization Technique to Totally Singular Optimal Problems. *Int. J. Control*, 17:809–815, 1973.

[9] M. Anitescu. On Using the Elastic Mode in Nonlinear Programming Approaches to Mathematical Programs with Complementarity Constraints. *SIAM J. Optim.*, 15:1203–1236, 2005.

[10] M. Anitescu, P. Tseng, and S.J. Wright. Elastic-Mode Algorithms for Mathematical Programs with Equilibrium Constraints: Global Convergence and Stationarity Properties. *Math. Program.*, 110:337–371, 2007.

[11] N. Arora and L. T. Biegler. A Trust Region SQP Algorithm for Equality Constrained Parameter Estimation with Simple Parametric Bounds. *Comput. Optim. Appl.*, 28:51–86, 2004.

[12] J. J. Arrieta-Camacho and L. T. Biegler. Real Time Optimal Guidance of Low-Thrust Spacecraft: An Application of Nonlinear Model Predictive Control. *Ann. N.Y. Acad. Sci.*, 1065:174, 2006.

[13] U. Ascher, J. Christiansen, and R. Russell. Collocation Software for Boundary Value ODEs. *ACM Trans. Math. Software*, 7:209–222, 1981.

[14] U. M. Ascher, R. M. M. Mattheij, and R. D. Russell. *Numerical Solution of Boundary Value Problems for Ordinary Differential Equations*. Classics in Appl. Math. 13 SIAM, Philadelphia, 1995.

[15] U. M. Ascher and R. J. Spiteri. Collocation Software for Boundary Value Differential-Algebraic Equations. *SIAM J. Sci. Comput.*, 15:938–952, 1994.

[16] U. M. Ascher and L. R. Petzold. *Computer Methods for Ordinary Differential Equations and Differential-Algebraic Equations*. SIAM, Philadelphia, 1998.

[17] C. Audet, P. Hansen, and J. Brimberg. Pooling Problem: Alternate Formulations and Solutions Methods. *Les Cahiers du GERAD G*, 23–31, 2000.

[18] M. Avraam, N. Shah, and C. C. Pantelides. Modeling and Optimization of General Hybrid Systems in Continuous Time Domain. *Comput. Chem. Eng.*, 225:221–224, 1998.

[19] R. Bachmann, L. Bruell, T. Mrziglod, and U. Pallaske. On Methods of Reducing the Index of Differential-Algebraic Equations. *Comput. Chem. Eng.*, 14:1271–1276, 1990.

[20] J. K. Bailey, A. N. Hrymak, S. S. Treiber, and R. B. Hawkins. Nonlinear Optimization of a Hydrocracker Fractionation Plant. *Comput. Chem. Eng.*, 17:123–130, 1993.

[21] S. Balakrishna and L.T. Biegler. A Unified Approach for the Simultaneous Synthesis of Reaction, Energy, and Separation Systems. *Ind. Eng. Chem. Res.*, 32:1372–1382, 1993.

[22] J. R. Banga and W. D. Seider. Global optimization of chemical processes using stochastic algorithms. In C. Floudas and P. Pardalos, editors, *State of the Art in Global Optimization, Kluwer, Dordrecht*, page 563, 1996.

[23] Y. Bard. *Nonlinear Parameter Estimation*. Academic Press, New York, 1974.

[24] R. A. Bartlett. MOOCHO: Multi-functional Object-Oriented arCHitecture for Optimization. *http://trilinos.sandia.gov/packages/moocho/*, 2005.

[25] R. A. Bartlett and L. T. Biegler. rSQP++ : An Object-Oriented Framework for Successive Quadratic Programming. In O. Ghattas, M. Heinkenschloss, D. Keyes, L. Biegler, and B. van Bloemen Waanders, editors, *Large-Scale PDE-Constrained Optimization*, page 316. Lecture Notes in Computational Science and Engineering, Springer, Berlin, 2003.

[26] R. A. Bartlett and L. T. Biegler. QPSchur: A Dual, Active Set, Schur Complement Method for Large-Scale and Structured Convex Quadratic Programming Algorithm. *Optim. Eng.*, 7:5–32, 2006.

[27] P. I. Barton and C. C. Pantelides. The Modeling of Combined Discrete/Continuous Processes. *AIChE J.*, 40:966–979, 1994.

[28] P. I. Barton, R. J. Allgor, W. F. Feehery, and S. Galan. Dynamic Optimization in a Discontinuous World. *Ind. Eng. Chem. Res.*, 37:966–981, 1998.

[29] P. I. Barton, J. R. Banga, and S. Galan. Optimization of Hybrid Discrete:continuous Dynamic Systems. *Comput. Chem. Eng.*, 24:2171–2182, 2000.

[30] P. I. Barton, C. K. Lee, and M. Yunt. Optimization of Hybrid Systems. *Comput. Chem. Eng.*, 30:1576–1589, 2006.

[31] R. D. Bartusiak. NLMPC: A platform for optimal control of feed- or product-flexible manufacturing. In *Assessment and Future Directions of NMPC*, pp. 338–347, Springer, Berlin, 2007.

[32] B. T. Baumrucker and L. T. Biegler. MPEC Strategies for Optimization of Hybrid Dynamic Systems. *J. Process Control*, 19:1248–1256, 2009.

[33] B. T. Baumrucker, J. G. Renfro, and L. T. Biegler. MPEC Problem Formulations and Solution Strategies with Chemical Engineering Applications. *Comput. Chem. Eng.*, 32:2903–2913, 2008.

[34] J. Bausa and G. Tsatsaronis. Dynamic Optimization of Startup and Load-Increasing Processes in Power Plants—Part I: Method. *ASME J. Eng. Gas Turbines Power*, 123:246–250, 2001.

[35] E. M. L. Beale. Numerical Methods. In J. Abadie, editor, *Nonlinear Programming*, pages 135–205. North–Holland, Amsterdam, 1967.

[36] H. Y. Benson, D. F. Shanno, and R. J. Vanderbei. Interior-Point Methods for Nonconvex Nonlinear Programming: Filter Methods and Merit Functions. Technical report, Operations Research and Financial Engineering, Princeton University, Princeton, NJ, 2000.

[37] H. Y. Benson and R. J. Vanderbei. Using LOQO to Solve Second-Order Cone Programming Problems. Technical Report SOR-98-09, Department of Operations Research and Financial Engineering, Princeton University, NJ, 1998.

[38] T. J. Berna, M. H. Locke, and A. W. Westerberg. A New Approach to Optimization of Chemical Processes. *AIChE J.*, 26:37–43, 1980.

[39] D. P. Bertsekas. *Constrained Optimization and Lagrange Multiplier Methods*. Academic Press, New York, 1982.

[40] J. T. Betts. *Practical Methods for Optimal Control and Estimation Using Nonlinear Programming*. 2nd ed., Advances in Design and Control 19, SIAM, Philadelphia, 2010.

[41] J. T. Betts and W. P. Huffman, Sparse Optimal Control Software (SOCS). Mathematics and Engineering Analysis Technical Document, MEA-LR-085, Boeing Information and Support Services, The Boeing Company, PO Box 3707, Seattle, WA 98124, 1997.

[42] J. T. Betts and E. J. Cramer. Application of Direct Transcription to Commercial Aircraft Trajectory Optimization. *J. Guid. Contr. and Dynam.*, 18/1:151–159, 1995.

[43] J. T. Betts. Optimal Interplanetary Orbit Transfers by Direct Transcription. *J. Astronaut. Sci.*, 42, 3:247–268, 1994.

[44] J. T. Betts, N. Biehn, and S. L. Campbell. Convergence of nonconvergent IRK discretizations of optimal control problems with state inequality constraints. *Technical report, Department of Mathematics, North Carolina State University*, NC, 2000.

[45] J. T. Betts and S. L. Campbell. Discretize then Optimize. *M&CT-TECH-03-01 Technical report, The Boeing Company*, Chicago, IL, 2003.

[46] T. Bhatia and L. T. Biegler. Dynamic Optimization in the Design and Scheduling of Multiproduct Batch Plants. *Ind. Eng. Chem. Res.*, 35, 7:2234, 1996.

[47] L. T. Biegler. Efficient Solution of Dynamic Optimization and NMPC Problems. In F. Allgoewer and A. Zheng, editors, *Nonlinear Model Predictive Control, Birkhaeuser, Basel*, pages 219–245, 2000.

[48] L. T. Biegler, A. M. Cervantes, and A. Wächter. Advances in Simultaneous Strategies for Dynamic Process Optimization. *Chem. Eng. Sci.*, 57(4):575–593, 2002.

[49] L. T. Biegler, J. J. Damiano, and G. E. Blau. Nonlinear Parameter Estimation: A Case Study Comparison. *AIChE J.*, 32:29–40, 1986.

[50] L. T. Biegler, O. Ghattas, M. Heinkenschloss, D. Keyes, and B. van Bloemen Waanders (eds.). *Real-Time PDE-Constrained Optimization*. Computational Science and Engineering 3, SIAM, Philadelphia, 2007.

[51] L. T. Biegler, O. Ghattas, M. Heinkenschloss, and B. van Bloemen Waanders (eds.). *Large-Scale PDE Constrained Optimization*. Lecture Notes in Computational Science and Engineering 30, Springer, Berlin, 2003.

[52] L. T. Biegler and I. E. Grossmann. Part I: Retrospective on Optimization. *Comput. Chem. Eng.*, 28, 8:1169–1192, 2004.

[53] L. T. Biegler, I. E. Grossmann, and A. W. Westerberg. *Systematic Methods of Chemical Process Design*. Prentice–Hall, Upper Saddle River, NJ, 1997.

[54] L. T. Biegler, J. Nocedal, and C. Schmid. A Reduced Hessian Method for Large-Scale Constrained Optimization. *SIAM J. Optim.*, 5:314–347, 1995.

[55] L. T. Biegler. Solution of Dynamic Optimization Problems by Successive Quadratic Programming and Orthogonal Collocation. *Comput. Chem. Eng.*, 8, 3/4:243–248, 1984.

[56] L. T. Biegler, A. M. Cervantes, and A. Wächter. Advances in Simultaneous Strategies for Dynamic Process Optimization. *Chem. Eng. Sci.*, 57:575–593, 2002.

[57] J. R. Birge and F. Louveaux. *Introduction to Stochastic Programming*. Springer Series in Operations Research and Financial Engineering, Springer, Heidelberg, 1997.

[58] C. Bischof, A. Carle, G. Corliss, and A. Griewank. ADIFOR—Generating Derivative Codes from Fortran Programs. *Sci. Programm.*, 1:1–29, 1992.

[59] G. A. Bliss. *Lectures of the Calculus of Variations*. University of Chicago Press, Chicago, IL, 1946.

[60] H. G. Bock. Recent advances in parameter identification techniques for ODE *Numerical Treatment of Inverse Problem in Differential and Integral Equation*, 1983.

[61] H. G. Bock. Recent advances in parameter identification techniques for ODE. In *Numerical Treatment of Inverse Problems in Differential and Integral Equations*, pages 95–121, Birkhäuser Boston, Boston, MA, 1983.

[62] H. G. Bock and K. J. Plitt. A Multiple Shooting Algorithm for Direct Solution of Optimal Control Problems. *Ninth IFAC World Congress, Budapest*, 1984.

[63] P. T. Boggs, J. W. Tolle, and P. Wang. On the Local Convergence of Quasi-Newton Methods for Constrained Optimization. *SIAM J. Control Optim.*, 20:161–171, 1982.

[64] I. Bongartz, A. R. Conn, N. I. M. Gould, and Ph. L. Toint. CUTE: Constrained and Unconstrained Testing Environment. *ACM Trans. Math. Software*, 21:123–142, 1995.

[65] J. F. Bonnans and J. Laurent-Varin. Computation of Order Conditions for Symplectic Partitioned Runge-Kutta Schemes with Application to Optimal Control. *Numer. Math.*, 103:1–10, 2006.

[66] A. J. Booker, Jr., J. E. Dennis, P. D. Frank, D. B. Serafini, V. Torczon, and M. W. Trosset. A Rigorous Framework for Optimization of Expensive Functions by Surrogates. *CRPC Technical Report 98739, Rice University*, Houston, TX, 1998.

[67] E. A. Boss, R. Maciel Filho, V. De Toledo, and E. Coselli. Freeze Drying Process: Real Time Model and Optimization. *Chem. Eng. Process.*, 12:1475–1485, 2004.

[68] C. L. Bottasso and A. Croce. Optimal Control of Multibody Systems Using an Energy Preserving Direct Transcription Method. *Multibody Sys. Dyn.*, 12/1:17–45, 2004.

[69] G. E. P. Box. Evolutionary Operation: A Method for Increasing Industrial Productivity. *Applied Statistics*, 6:81–101, 1957.

[70] K. E. Brenan, S. L. Campbell, and L. R. Petzold. *Numerical Solution of Initial-Value Problems in Differential-Algebraic Equations*. North–Holland, New York, 1989.

[71] A. Brooke, D. Kendrick, A. Meeraus, and R. Raman. GAMS Home Page. *http://www.gams.com*, 1998.

[72] A. E. Bryson and Y. C. Ho. *Applied Optimal Control*. Hemisphere, Washington, DC, 1975.

[73] D. S. Bunch, D. M. Gay, and R. E. Welsch. Algorithm 717: Subroutines for Maximum Likelihood and Quasi-Likelihood Estimation of Parameters in Nonlinear Regression Models. *ACM Trans. Math. Software*, 19:109–120, 1993.

[74] J. R. Bunch and L. Kaufman. Some Stable Methods for Calculating Inertia and Solving Symmetric Linear Systems. *Math. Comput.*, 31:163–179, 1977.

[75] J. V. Burke and S. P. Han. A Robust Sequential Quadratic Programming Method. *Math. Programm.*, 43:277–303, 1989.

[76] C. Büskens and H. Maurer. Real-Time Control of an Industrial Robot. In M. Grötschel, S. Krumke, and J. Rambau, editors, *Online Optimization of Large Systems*, pages 57–68. Springer, Berlin, 2001.

[77] J. C. Butcher. *Numerical Methods for Ordinary Differential Equations*. Wiley, Chichester, 2008.

[78] R. H. Byrd, F. E. Curtis, and J. Nocedal. An inexact SQP Method for Equality Constrained Optimization. *SIAM J. Optim.*, 19:351–369, 2008.

[79] R. Byrd, J. Nocedal, and R. Waltz. KNITRO: An integrated package for nonlinear optimization. In G. Di Pillo and M. Roma, editors, *Large-Scale Nonlinear Optimization*, pages 35–60. Springer, Berlin, 2006.

[80] R. H. Byrd, J. C. Gilbert, and J. Nocedal. A Trust Region Method Based on Interior Point Techniques for Nonlinear Programming. *Math. Program.*, 89:149–185, 2000.

[81] R. H. Byrd, M. E. Hribar, and J. Nocedal. An Interior Point Algorithm for Large-Scale Nonlinear Programming. *SIAM J. Optim.*, 9:877–900, 1999.

[82] R. H. Byrd and J. Nocedal. An Analysis of Reduced Hessian Methods for Constrained Optimization. *Math. Programm.*, 49:285–323, 1991.

[83] R. H. Byrd and J. Nocedal. Representations of Quasi- Newton Matrices and Their Use in Limited Memory Methods. *Math. Program.*, 63:129–156, 1994.

[84] M. Canon, C. Cullum, and E. Polak. *Theory of Optimal Control and Mathematical Programming*. McGraw–Hill, New York, 1970.

[85] M. Caracotsios and W. E. Stewart. Sensitivity Analysis of Initial Value Problems with Mixed ODEs and Algebraic Equations. *Comput. Chem. Eng.*, 9:350–365, 1985.

[86] R. G. Carter. A worst-case example using linesearch methods for numerical optimization using inexact gradients. Technical Report MCS-P283-1291, Argonne National Laboratory, 1991.

[87] C. Cartis, N. I. M. Gould, and Ph. L. Toint. Adaptive cubic overestimation methods for unconstrained optimization. Technical Report 07/05, FUNDP, University of Namur, 2008.

[88] A. Cervantes and L. T. Biegler. Large-Scale DAE Optimization Using Simultaneous Nonlinear Programming Formulations. *AIChE J.*, 44:1038, 1998.

[89] A. M. Cervantes and L. T. Biegler. A Stable Elemental Decomposition for Dynamic Process Optimization. *Comput. Optim. Appl.*, 120:41–57, 2000.

[90] A. M. Cervantes and L. T. Biegler. Optimization Strategies for Dynamic Systems. In C. Floudas and P. Pardalos, editors, *Encyclopedia of Optimization*. Kluwer Academic Publishers, Norwell, MA, 2000.

[91] A. M. Cervantes, S. Tonelli, A. Brandolin, J. A. Bandoni, and L. T. Biegler. Large-Scale Dynamic Optimization for Grade Transitions in a Low Density Polyethylene Plant. *Comput. Chem. Eng.*, 26:227–238, 2002.

[92] A. M. Cervantes. Stable Large-Scale DAE Optimization Using Simultaneous Approaches. *Ph.D. Thesis*. Chemical Engineering Department, Carnegie Mellon University, Pittsburgh, PA, 2000.

[93] R. Chamberlain, C. Lemarechal, H. C. Pedersen, and M. J. D. Powell. The Watchdog Technique for Forcing Convergence in Algorithms for Constrained Optimization. *Math. Programm.*, 16:1–17, 1982.

[94] X. Chen and M. Fukushima. A Smoothing Method for a Mathematical Program with P-Matrix Linear Complementarity Constraints. *Comput. Optim. Appl.*, 27:223–246, 2004.

[95] M. Cizniar, D. Salhi, M. Fikar, and M. A. Latifi. A MATLAB Package for Orthogonal Collocations on Finite Elements in Dynamic Optimisation. *Proceedings of the 15th Int. Conference Process Control '05*, 2005.

[96] P. A. Clark and A. W. Westerberg. Bilevel Programming for Steady State Chemical Process Design I. *Comput. Chem. Eng.*, 14:87–97, 1990.

[97] F. H. Clarke. *Optimization and Nonsmooth Analysis*. Wiley, New York, 1983.

[98] T. F. Coleman and A. R. Conn. On the Local Convergence of a Quasi-Newton Method for the Nonlinear Programming Problem. *SIAM J. Numer. Anal.*, 21:755–769, 1984.

[99] A. R. Conn, N. I. Gould, and P. L. Toint. *LANCELOT: A Fortran Package for Large-Scale Nonlinear Optimization (Release A)*, volume 17. Springer Series in Computational Mathematics, Springer, Berlin, 1992.

[100] A. R. Conn, N. I. M. Gould, and Ph. L. Toint. *Trust-Region Methods*. MPS-SIAM Series on Optimization 1, SIAM, Philadelphia, 2000.

[101] A. R. Conn, K. Scheinberg, and P. Toint. Recent Progress in Unconstrained Nonlinear Optimization without Derivatives. *Math. Programm., Series B*, 79:397, 1997.

[102] A. R. Conn, K. Scheinberg, and L. N. Vicente. *Introduction to Derivative-Free Optimization*. SIAM, Philadelphia, 2009.

[103] B. A. Conway. Optimal Low-Thrust Interception of Earth-Crossing Asteroids. *J. Guid. Control Dyn.*, 5:995–1002, 1997.

[104] R. Courant and D. Hilbert. *Methods of Mathematical Physics*. Interscience Press, New York, 1953.

[105] F. E. Curtis and J. Nocedal. Flexible Penalty Functions for Nonlinear Constrained Optimization. *IMA J. Numer. Anal.*, 28:749–769, 2008.

[106] J. E. Cuthrell and L. T. Biegler. On the Optimization of Differential-Algebraic Process Systems. *AIChE J.*, 33:1257–1270, 1987.

[107] J. E. Cuthrell and L. T. Biegler. Simultaneous Optimization and Solution Methods for Batch Reactor Control Profiles. *Comput. Chem. Eng.*, 13:49–62, 1989.

[108] G. B. Dantzig. *Linear Programming and Extensions*. Princeton University Press, Princeton, NJ, 1963.

[109] F. R. de Hoog and R. M. M. Mattheij. On Dichotomy and Well Conditioning in BVP. *SIAM J. Numer. Anal.*, 24:89–105, 1987.

[110] J. E. Dennis, Jr., and R. B. Schnabel. *Numerical Methods for Unconstrained Optimization and Nonlinear Equations*. Prentice–Hall, Englewood Cliffs, NJ, 1983.

[111] J. E. Dennis and V. Torczon. Direct Search Methods on Parallel Machines. *SIAM J. Optim.*, 1:448–474, 1991.

[112] J. E. Dennis, M. Heinkenschloss, and L. N. Vicente. Trust-Region Interior-Point SQP Algorithms for a Class of Nonlinear Programming Problems. *SIAM J. Control Optim.*, 36:1750–1794, 1998.

[113] P. Deuflhard. *Newton Methods for Nonlinear Problems: Affine Invariance and Adaptive Algorithms*. Springer, Berlin, 2004.

[114] W. C. DeWitt, L. S. Lasdon, D. A. Waren, A. Brenner, and D. Omega. An Improved Gasoline Blending System for Texaco. *Interfaces*, 19:85–101, 1989.

[115] M. Diehl, H. G. Bock, and J. P. Schlöder. A Real-Time Iteration Scheme for Nonlinear Optimization in Optimal Feedback Control. *SIAM J. Control Optim.*, 43:1714–1736, 2005.

[116] E. D. Dolan and J. Moré. Benchmarking Optimization Software with COPS. Technical report, ANL/MCS-246, Argonne National Laboratory, 2001.

[117] A. Dontchev and F. Lempio. Difference Methods for Differential Inclusions: A Survey. *SIAM Review*, 34:263–294, 1992.

[118] J. J. Downs and E. F. Vogel. A Plant-Wide Industrial Process Control Problem. *Comput. Chem. Eng.*, 17–28, 1993.

[119] A. Drud. CONOPT: A System for Large Scale Nonlinear Optimization, Reference Manual for CONOPT. *ARKI Consulting and Development A/S.*

[120] A. Drud. CONOPT—A Large Scale GRG Code. *ORSA J. Comput.*, 6:207–216, 1994.

[121] D. W. Green (ed.). *Perry's Chemical Engineers' Handbook*. 8 edition, McGraw–Hill, New York, 2008.

[122] T. F. Edgar, D. M. Himmelblau, and L. S. Lasdon. *Optimization of Chemical Processes*. McGraw–Hill, New York, 2001.

[123] A. S. El-Bakry, R. A. Tapia, T. Tsuchiya, and Y. Zhang. On the Formulation and Theory of the Newton Interior-Point Method for Nonlinear Programming. *J. Optim. Theory Appl.*, 89:507–541, 1996.

[124] M. Eldred. DAKOTA: A Multilevel Parallel Object-Oriented Framework for Design Optimization, Parameter Estimation, Uncertainty Quantification, and Sensitivity Analysis. http://www.cs.sandia.gov/dakota/licensing/release/Users4.2.pdf (2008).

[125] A. M. Eliceche, S. M. Corvalan, M. F. S. Roman, and I. Ortiz. Minimum Membrane Area of an Emulsion Pertraction Process for Cr and Removal and Recovery. *Comput. Chem. Eng.*, 6:1483–90, 2005.

[126] W. R. Esposito and C. A. Floudas. Global Optimization for the Parameter Estimation of Differential-Algebraic Systems. *Ind. Eng. Chem. Res.*, 5:1291–1310, 2000.

[127] L. C. Evans. An Introduction to Mathematical Optimal Control Theory. *Department of Mathematics, University of California, Berkeley, math.berkeley.edu/~evans/ control.course.pdf.*

[128] W. Feehery, J. Tolsma, and P. Barton. Efficient Sensitivity Analysis of Large-Scale Differential-Algebraic Systems. *Appl. Numer. Math.*, 25:41–54, 1997.

[129] M. C. Ferris, S. P. Dirkse, J. H. Jagla, and A. Meeraus. An Extended Mathematical Programming Framework. *Comput. Chem. Eng.*, 33:1973–1982, 2009.

[130] M. C. Ferris and J. S. Pang (eds.). *Complementarity and Variational Problems: State of the Art.* Proceedings of the International Conference on Complementarity Problems SIAM, Philadelphia, (Baltinore, MD, 1995), 1997.

[131] A. Fiacco and G. McCormick. *Nonlinear Programming: Sequential Unconstrained Minimization Techniques.* John Wiley and Sons, New York, 1968.

[132] A. F. Filippov. Differential Equations with Discontinuous Right-Hand Sides. *Matematicheskii Sbornik*, 51(93):99–128, 1960.

[133] W. Fleming and R. Rishel. *Deterministic and Stochastic Optimal Control.* Springer, Berlin, 1975.

[134] R. Fletcher. *Practical Methods of Optimization.* Wiley, Chichester, 1987.

[135] R. Fletcher, N. I. M. Gould, S. Leyffer, Ph. L. Toint, and A. Wächter. Global Convergence of a Trust-Region SQP-Filter Algorithm for General Nonlinear Programming. *SIAM J. Optim.*, 13:635–659, 2002.

[136] R. Fletcher and S. Leyffer. User Manual for FilterSQP. Technical Report, Numerical Analysis Report, NA/181, University of Dundee, 1999.

[137] R. Fletcher and S. Leyffer. Nonlinear Programming without a Penalty Function. *Math. Programm.*, 91:239–269, 2002.

[138] R. Fletcher and S. Leyffer. Solving Mathematical Programs with Complementarity Constraints as Nonlinear Program. *Optim. Methods Softw.*, 19:15–40, 2004.

[139] R. Fletcher, S. Leyffer, and Ph. L. Toint. A Brief History of Filter Methods. *SIAG/Optimization Views-and-News*, 18:2–12, 2007.

[140] R. Fletcher and W. Morton. Initialising Distillation Column Models. *Comput. Chem. Eng.*, 23:1811–1824, 2000.

[141] A. Flores-Tlacuahuac, L. T. Biegler, and E. Saldivar-Guerra. Dynamic Optimization of HIPS Open-Loop Unstable Polymerization Reactors. *Ind. Eng. Chem. Res.*, 8:2659–2674, 2005.

[142] A. Flores-Tlacuahuac and I. E. Grossmann. Simultaneous Cyclic Scheduling and Control of a Multiproduct CSTR. *Ind. Eng. Chem. Res.*, 20:6698–6712, 2006.

[143] C. A. Floudas. *Nonlinear and Mixed Integer Optimization: Fundamentals and Applications*. Oxford University Press, New York, 1995.

[144] C. A. Floudas. *Deterministic Global Optimization: Theory, Algorithms and Applications*. Kluwer Academic Publishers, Norwell, MA, 2000.

[145] fmincon: An SQP Solver. MATLAB User's Guide. *The MathWorks*, 2009.

[146] J. F. Forbes and T. E. Marlin. Design Cost: A Systematic Approach to Technology Selection for Model-Based Real-Time Optimization Systems. *Comput. Chem. Eng.*, 20:717–734, 1996.

[147] A. Forsgren, P. E. Gill, and M. H. Wright. Interior Methods for Nonlinear Optimization. *SIAM Review*, 44:525–597, 2002.

[148] R. Fourer, D. M. Gay, and B. W. Kernighan. *AMPL: A Modeling Language for Mathematical Programming*. Duxbury Press/Brooks/Cole Publishing Company, 2002.

[149] R. Franke and J. Doppelhamer. Real-time implementation of nonlinear model predictive control of batch processes in an industrial framework. In *Assessment and Future Directions of NMPC*, pp. 465–472. Springer, Berlin, 2007.

[150] R. Franke and J. Engell. Integration of advanced model based control with industrial IT. In R. Findeisen, F. Allgöwer, and L.T. Biegler, editors, *Assessment and Future Directions of Nonlinear Model Predictive Control*, pages 399–406. Springer, 2007.

[151] S. E. Gallun, R. H. Luecke, D. E. Scott, and A. M. Morshedi. Use Open Equations for Better Models. *Hydrocarbon Processing*, pages 78–90, 1992.

[152] GAMS. Model Library Index. *GAMS Development Corporation* (2010).

[153] M. A. Gaubert and X. Joulia. Tools for Computer Aided Analysis and Interpretation of Process Simulation Results. *Comput. Chem. Eng.*, 21:S205–S210, 1997.

[154] D. M. Gay. Computing Optimal Locally Constrained Steps. *SIAM J. Sci. Statist. Comput.*, 2:186–197, 1981.

[155] D. M. Gay. A Trust-Region Approach to Linearly Constrained Optimization. *Numerical Analysis Proceedings (Dundee, 1983), D. F. Griffiths, editor*, Springer, 1983.

[156] D. M. Gay. Algorithm 611: Subroutine for Unconstrained Minimization Using a Model/Trust-Region Approach. *ACM Trans. Math. Software*, 9:503–524, 1983.

[157] I. M. Gelfand and S. V. Fomin. *Calculus of Variations*. Dover Publications, New York, 2000.

[158] GESOP. Graphical Environment for Simulation and Optimization. *http://www.astos.de/products/gesop*.

[159] P. E. Gill, W. Murray, and M. A. Saunders. User's guide for SNOPT: A FORTRAN package for large-scale nonlinear programming. Technical Report SOL 96-0, Department of Mathematics, University of California, San Diego, 1996.

[160] P. E. Gill, W. Murray, and M. A. Saunders. SNOPT: An SQP Algorithm for Large-Scale Constrained Optimization. *SIAM Review*, 47:99–131, 2005.

[161] P. E. Gill, W. Murray, M. A. Saunders, and M. H.Wright. User's guide for NPSOL (Version 4.0): A Fortran package for nonlinear programming. Report SOL 86-2, *Department of Operations Research, Stanford University*, 1986.

[162] P. E. Gill, W. Murray, and M. H. Wright. *Practical Optimization*. Academic Press, New York, 1981.

[163] P. E. Gill, M. A. Saunders, and M. H. Wright. A Schur-Complement Method for Sparse Quadratic Programming. In *Reliable Numerical Computation*, pages 113–138, Oxford University Press, New York, 1990.

[164] C. J. Goh and K. L. Teo. Control Parametrization: A Unified Approach to Optimal Control Problems with General Constraints. *Automatica*, 24:3–18, 1988.

[165] D. Goldfarb and A. Idnani. A Numerically Stable Dual Method for Solving Strictly Convex Quadratic Programs. *Math. Programm.*, 27:1–33, 1983.

[166] G. H. Golub and C. F. Van Loan. *Matrix Computations* 3 edition. The Johns Hopkins University Press, Baltimore, MD, 1996.

[167] V. Gopal and L. T. Biegler. Smoothing Methods for Complementarity Problems in Process Engineering. *AIChE J.*, 45:1535, 1999.

[168] R. Goulcher and J. Cesares Long. The Solution of Steady State Chemical Engineering Optimization Problems Using a Random Search Algorithm. *Comput. Chem. Engr.*, 2:23–30, 1978.

[169] N. I. M. Gould. On Practical Conditions for the Existence and Uniqueness of Solutions to the General Equality Quadratic Programming Problem. *Math. Programm.*, 32:90–99, 1985.

[170] N. I. M. Gould, S. Lucidi, M. Roma, and Ph. L. Toint. Solving the Trust-Region Subproblem Using the Lanczos Method. *SIAM J. Optim.*, 9:504–525, 1999.

[171] H. J. Greenberg. Analyzing the Pooling Problem. *ORSA J. Comput.*, 7:206–217, 1995.

[172] R. Griesse and A. Walther. Evaluating Gradients in Optimal Control—Continuous Adjoints versus Automatic Differentiation. *J. Optim. Theory Appl.*, 122:63–86, 2004.

[173] A. Griewank, D. Juedes, and J. Utke. ADOL-C: A Package for the Automatic Differentiation of Algorithms written in C/C++. *ACM Trans. Math. Software*, 22:131–167, 1996.

[174] L. Grippo, F. Lampariello, and S. Lucidi. A Nonmonotone Line Search Technique for Newton's Method. *SIAM J. Numer. Anal*, 23:707–716, 1986.

[175] I. E. Grossmann and L. T. Biegler. Part II: Future Perspective on Optimization. *Comput. Chem. Eng.*, 8:1193–1218, 2004.

[176] I. E. Grossmann and C. A. Floudas. Active Constraint Strategy for Flexibility Analysis in Chemical Process. *Comput. Chem. Eng.*, 11:675–693, 1987.

[177] M. Grötschel, S. Krumke, and J. Rambau (eds.). *Online Optimization of Large Systems*. Springer, Berlin, 2001.

[178] W. W. Hager. Rates of Convergence for Discrete Approximations to Unconstrained Control Problems. *SIAM J. Numer. Anal.*, 13:449–472, 1976.

[179] W. W. Hager. Runge-Kutta Methods in Optimal Control and the Transformed Adjoint System. *Numer. Math.*, 87:247–282, 2000.

[180] W. W. Hager. Minimizing a Quadratic over a Sphere. *SIAM J. Optim.*, 12:188–208, 2001.

[181] W. W. Hager and S. Park. Global Convergence of SSM for Minimizing a Quadratic over a Sphere. *Math Comp.*, 74:1413–1423, 2005.

[182] E. Hairer, S. P. Norsett, and G. Wanner. *Solving Ordinary Differential Equations I: Nonstiff Problems*. Springer Series in Computational Mathematics, Springer, Berlin, 2008.

[183] E. Hairer and G. Wanner. *Solving Ordinary Differential Equations II: Stiff and Differential-Algebraic Problems*. Springer Series in Computational Mathematics, Springer, Berlin, 2002.

[184] S. P. Han. Superlinearly Convergent Variable Metric Algorithms for General Nonlinear Programming Problems. *Math. Programm.*, 11:263–282, 1976.

[185] S. P. Han and O. L. Mangasarian. Exact Penalty Functions in Nonlinear Programming. *Math. Programm.*, 17:251–269, 1979.

[186] R. Hannemann and W. Marquardt. Continuous and Discrete Composite Adjoints for the Hessian of the Lagrangian in Shooting Algorithms for Dynamic Optimization. *SIAM J. Sci. Comput.*, 31:4675–4695, 2010.

[187] P. T. Harker and J. S. Pang. Finite-Dimensional Variational Inequalities and Complementarity Problems: A Survey of Theory, Algorithms and Applications. *Math. Programm.*, 60:161–220, 1990.

[188] R. F. Hartl, S. P. Sethi, and R. G. Vickson. A Survey of the Maximum Principles for Optimal Control Problems with State Constraints. *SIAM Review*, 37:181–218, 1995.

[189] L. Hasdorff. *Gradient Optimization and Nonlinear Control.* Wiley, New York, 1976.

[190] C. A. Haverly. Studies of the Behavior of Recursion for the Pooling Problem. *SIGMAP Bull.*, page 25, 1978.

[191] M. Heemels. Linear Complementarity Systems: A Study in Hybrid Dynamics. *Ph.D. Thesis*. Technische Universiteit Eindhoven, 1999.

[192] W. P. Heemels, B. DeSchutter, and A. Bemporad. On the equivalence of hybrid dynamical models. *40th IEEE Conference on Decision and Control*, pages 364–369, 2001.

[193] W. P. M. H. Heemels and B. Brogliato. The Complementarity Class of Hybrid Dynamical Systems. *European J. Control*, 9:322–360, 2003.

[194] G. A. Hicks and W. H. Ray. Approximation Methods for Optimal Control Synthesis. *Can. J. Chem. Eng.*, 40:522–529, 1971.

[195] F. Hillier and G. J. Lieberman. *Introduction to Operations Research.* Holden-Day, San Francisco, CA, 1974.

[196] A. C. Hindmarsh, P. N. Brown, K. E. Grant, S. L. Lee, R. Serban, D. E. Shumaker, and C. S. Woodward. SUNDIALS: Suite of Nonlinear and Differential/Algebraic Equation Solvers. *ACM Trans. Math. Software*, 31:363–396, 2005.

[197] A. C. Hindmarsh and R. Serban. User Documentation for CVODES, An ODE Solver with Sensitivity Analysis Capabilities. *LLNL Report UCRL-MA-148813*, 2002.

[198] J. L. Hjersted and M. A. Henson. Optimization of Fed-Batch Saccharomyces cerevisiae Fermentation Using Dynamic Flux Balance Models. *Biotechnol. Prog.*, 22:1239–1248, 2006.

[199] W. Hock and K. Schittkowski. Test Examples for Nonlinear Programming Codes. *J. Optim. Theory Appl.*, 30:127–129 (1980). http://www.math.uni-bayreuth.de/~kschittkowski/tp_coll1.htm.

[200] J. H. Holland. *Adaptations in Natural and Artificial Systems.* University of Michigan Press, Ann Arbor, MI, 1975.

[201] W. Hong, S. Wang, P. Li, G. Wozny, and L. T. Biegler. A Quasi-Sequential Approach to Large-Scale Dynamic Optimization Problems. *AIChE J.*, 52, 1:255–268, 2006.

[202] R. Hooke and T. A. Jeeves. Direct Search Solution of Numerical and Statistical Problems. *J. ACM*, 8:212–220, 1961.

[203] R. Horst and P. M. Tuy. *Global Optimization: Deterministic Approaches.* Springer, Berlin, 1996.

[204] X. M. Hu and D. Ralph. Convergence of a Penalty Method of Mathematical Programming with Complementarity Constraints. *J. Optim. Theory Appl.*, 123:365–390, 2004.

[205] R. R. Hughes. *Mathematical Modeling and Optimization.* AIChE Continuing Education Series, 1973.

[206] IPOPT (Interior Point OPTimizer). https://projects.coin-or.org/Ipopt.

[207] R. Jackson. Optimal Use of Mixed Catalysts for Two Successive Chemical Reactions. *J. Optim. Theory Appl.*, 2/1:27–39, 1968.

[208] D. H. Jacobson, S. B. Gershwin, and M. M. Lele. Computation of Optimal Singular Control. *IEEE Trans. Autom. Control*, AC-15:67–73, 1970.

[209] Shi-Shang Jang and Pin-Ho Lin. Discontinuous Minimum End-Time Temperature/Initiator Policies for Batch Emulsion Polymerization of Vinyl Acetate. *Chem. Eng. Sci.*, 46:12–19, 1991.

[210] L. Jiang, L. T. Biegler, and V. G. Fox. Simulation and Optimization of Pressure Swing Adsorption Systems for Air Separation. *AIChE J.*, 49, 5:1140, 2003.

[211] T. Jockenhövel, L. T. Biegler, and A. Wächter. Dynamic Optimization of the Tennessee Eastman Process Using the OptControlCentre. *Comput. Chem. Eng.*, 27:1513–1531, 2003.

[212] D. I. Jones and J. W. Finch. Comparison of Optimization Algorithms. *Int. J. Control*, 40:747–761, 1984.

[213] B. S. Jung, W. Mirosh, and W. H. Ray. Large Scale Process Optimization Techniques Applied to. Chemical and Petroleum Processes. *Can. J. Chem. Eng.*, 49:844–851, 1971.

[214] J. Kadam and W. Marquardt. Sensitivity-Based Solution Updates in Closed-Loop Dynamic Optimization. In *Proceedings of the DYCOPS 7 Conference*. Elsevier, 2004.

[215] J. Kadam and W. Marquardt. Integration of economical optimization and control for intentionally transient process operation. In R. Findeisen, F. Allgöwer, and L. Biegler, editors, *Assessment and future directions of nonlinear model predictive control*, pages 419–434. Springer, Berlin, 2007.

[216] P. Kall and S. W. Wallace. *Stochastic Programming*. John Wiley and Sons, Chichester, 1994.

[217] S. Kameswaram and L. T. Biegler. Convergence Rates for Direct Transcription of Optimal Control Problems Using Collocation at Radau Points. *Comput. Optim. Appl.*, 41:81–126, 2008.

[218] S. Kameswaran, L. T. Biegler, S. Tobias Junker, and H. Ghezel-Ayagh. Optimal Off-Line Trajectory Planning of Hybrid Fuel Cell/Gas Turbine Power Plants. *AIChE J.*, 53:450–474, 2007.

[219] S. Kameswaran and L.T. Biegler. Convergence Rates for Direct Transcription of Optimal Control Problems with Final-Time Equality Constraints Using Collocation at Radau Points. In *Proc. 2006 American Control Conference*, pages 165–171, 2006.

[220] S. Kameswaran and L.T. Biegler. Advantages of Nonlinear-Programming-Based Methodologies for Inequality Path-Constrained Optimal Control Problems—A Numerical Study. *SIAM J. Sci. Comput.*, 30:957–981, 2008.

[221] S. Kameswaran, G. Staus, and L. T. Biegler. Parameter Estimation of Core Flood and Reservoir Models. *Comput. Chem. Eng.*, 29:1787–1800, 2005.

[222] S. K. Kameswaran. Analysis and Formulation of a Class of Complex Dynamic Optimization Problems. *Ph.D. Thesis*. Chemical Engineering Department, Carnegie Mellon University, Pittsburgh, PA, 2007.

[223] N. Karmarkar. A New Polynomial-Time Algorithm for Linear Programming. *Combinatorics*, 4:373–395, 1984.

[224] W. Karush. *Minima of Functions of Several Variables with Inequalities as Side Constraints*. M.Sc. Dissertation, Department of Mathematics, University of Chicago, Chicago, IL, 1939.

[225] Y. Kawajiri and L. T. Biegler. Optimization Strategies for Simulated Moving Bed and PowerFeed Processes. *AIChE J.*, 52:1343–1350, 2006.

[226] C. T. Kelley. *Iterative Methods for Linear and Nonlinear Equations*. Frontiers in Appl. Math. 16, SIAM, Philadelphia, 1995.

[227] C. T. Kelley. *Iterative Methods for Optimization*. Frontiers in Appl. Math. 18, SIAM, Philadelphia, 1999.

[228] J. D. Kelly and J. L. Mann. Crude Oil Blend Scheduling Optimization: An Application with Multimillion Dollar Benefits. *Hydrocarbon Process*, pages 47–53, 2002.

[229] C. Kiparissides, P. Seferlis, G. Mourikas, and A. J. Morris. Online Optimizing Control of Molecular Weight Properties in Batch Free-Radical Polymerization Reactors. *Ind. Eng. Chem. Res.*, 41:6120–6131, 2002.

[230] E. Kloppenburg and E. D. Gilles. A New Concept for Operating Simulated Moving Bed Processes. *Chem. Eng. Technol.*, 10, 22:813–781, 1999.

[231] M. Kocvara and M. Stingl. PENNON, A Code for Convex Nonlinear and Semidefinite Programming. *Optim. Methods Softw.*, 8:317–333, 2003.

[232] L. G. Koster, E. Gazi, and W. D. Seider. Finite Elements for Near-Singular Systems: Application to Bacterial Chemotaxis. *Comput. Chem. Eng.*, 176:485–503, 1993.

[233] M. Kristic and H. H. Wang. Stability of Extremum Seeking Feedback for General Nonlinear Dynamic Systems. *Automatica*, 36:595–601, 2000.

[234] A. Kröner, W. Marquardt, and E. D. Gilles. Computing Consistent Initial Conditions for Differential—Algebraic Equations. *Comput. Chem. Eng.*, 16:S131, 1990.

[235] T. Kronseder, O.V. Stryk, R. Bulirsch, and A. Kröner. Towards nonlinear model-based predictive optimal control of large-scale process models with application to air separation unit. In M. Grötschel and S. O. Krumke, editors, *Online Optimization of Large Scale Systems: State of the Art*, pages 385–412, Springer, Berlin, 2001.

[236] H. W. Kuhn and A. W. Tucker. Nonlinear programming. In J. Neyman, editor, *Proceedings of the Second Berkeley Symposium on Mathematical Statistics and Probability*, pages 481–492, Berkeley, CA, 1951. University of California Press.

[237] P. Kunkel and V. Mehrmann. *Differential-Algebraic Equations, Analysis and Numerical Solution*. EMS Publishing House, Zürich, Switzerlad, 2006.

[238] P. Kunkel, V. Mehrmann, and I. Seufer. GENDA: A software package for the numerical solution of general nonlinear differential-algebraic equations. Technical Report 730–2002, Institut für Mathematik, Zürich, 2002.

[239] P. J. M. van Laarhoven and E. H. L. Aarts. *Simulated Annealing: Theory and Applications*. Reidel Publishing, Dordrecht, 1987.

[240] Y. D. Lang and L. T. Biegler. A Software Environment for Simultaneous Dynamic Optimization. *Comput. Chem. Eng.*, 31:931–942, 2007.

[241] Y. D. Lang and L.T. Biegler. Distributed Stream Method for Tray Optimization. *AIChE J.*, 48:582–595, 2002.

[242] Y. D. Lang, A. Cervantes, and L. T. Biegler. Dynamic Optimization of a Batch Crystallization Process. *Ind. Eng. Chem. Res.*, 38, 4:1469, 1998.

[243] Y. D. Lang, A. Malacina, L. T. Biegler, S. Munteanu, J. I. Madsen, and S. E. Zitney. Reduced Order Model Based on Principal Component Analysis For Process Simulation and Optimization. *Energy Fuels*, 23:1695–1706, 2009.

[244] L. S. Lasdon, S. K. Mitter, and A. D. Waren. The Conjugate Gradient Method for Optimal Control Problems. *IEEE Trans. Autom. Control*, AC-12:132, 1967.

[245] L. S. Lasdon, A. Waren, A. Jain, and M. Ratner. Design and Testing of a Generalized Reduced Gradient Code for Nonlinear Programming. *ACM Trans. Math. Software*, 4:34–50, 1978.

[246] C. T. Lawrence, J. L. Zhou, and A. L. Tits. User's Guide for CFSQP Version 2.5: A C Code for Solving (Large Scale) Constrained Nonlinear (Minimax) Optimization Problems, Generating Iterates Satisfying All Inequality Constraints. Technical Report TR-94-16r1, Institute for Systems Research, University of Maryland, College Park, MD, 1997.

[247] C. T. Lawrence and A. L. Tits. A Computationally Efficient Feasible Sequential Quadratic Programming Algorithm. *SIAM J. Optim.*, 11:1092–1118, 2001.

[248] E. B. Lee and L. Markus. *Foundations of Optimal Control Theory*. John Wiley and Sons, New York, 1967.

[249] R. I. Leine and H. Nijmeijer. *Dynamics and Bifurcations of Non-Smooth Mechanical Systems*. Lecture Notes in Applied and Computational Mechanics, 18, Springer, Berlin, 2004.

[250] D. B. Leineweber. Efficient Reduced SQP Methods for the Optimization of Chemical Processes Described by Large Sparse DAE Models. *University of Heidelberg, Heidelberg, Germany*, 1999.

[251] D. B. Leineweber, H. G. Bock, J. P. Schlöder, J. V. Gallitzendörfer, A. Schäfer, and P. Jansohn. A Boundary Value Problem Approach to the Optimization of Chemical Processes Described by DAE Models. *IWR-Preprint 97-14, Universität Heidelberg*, 1997.

[252] J. Leis and M. Kramer. The Simultaneous Solution and Sensitivity Analysis of Systems Described by Ordinary Differential Equations. *ACM Trans. Math. Software*, 14:45–60, 1988.

[253] M. Lentini and V. Pereyra. PASVA4: An Ordinary Boundary Solver for Problems with Discontinuous Interfaces and Algebraic Parameters. *Mat. Apl. Comput.*, 2:103–118, 1983.

[254] D. R. Lewin, W. D. Seider, J. D. Seader, E. Dassau, J. Golbert, D. N. Goldberg, M. J. Fucci, and R. B. Nathanson. *Using Process Simulators in Chemical Engineering: A Multimedia Guide for the Core Curriculum—Version 2.0.* John Wiley, New York, 2003.

[255] S. Leyffer, G. López-Calva, and J. Nocedal. Interior Methods for Mathematical Programs with Complementarity Constraints. *SIAM J. Optim.*, 17:52–77, 2006.

[256] S. Leyffer and T. S. Munson. A globally convergent filter method for MPECs. Technical Report ANL/MCS-P1457-0907, Argonne National Laboratory, Mathematics and Computer Science Division, 2007.

[257] P. Li, H. Arellano-Garcia, and G. Wozny. Optimization of a Semibatch Distillation Process with Model Validation on the Industrial Site. *Ind. Eng. Chem. Res.*, 37:1341–1350, 1998.

[258] S. Li and L. Petzold. Design of new DASPK for sensitivity analysis. Technical Report, Dept. of Computer Science, UCSB, 1999.

[259] S. Li and L. Petzold. Description of DASPKADJOINT: An adjoint sensitivity solver for differential-algebraic equations. Technical report, University of California, Department of Computer Science, Santa Barbara, CA, 2001.

[260] J. Liebman, L. Lasdon, L. Schrage, and A. Waren. *Modeling and Optimization with GINO.* The Scientific Press, Palo Alto, CA, 1986.

[261] C. J. Lin and J. J. Moré. Newton's Method for Large Bound-Constrained Optimization problems. *SIAM J. Optim.*, 9:1100–1127, 1999.

[262] M. H. Locke, R. Edahl, and A. W. Westerberg. An Improved Successive Quadratic Programming Algorithm for Engineering Design Problems. *AIChE J.*, 29(5):871, 1983.

[263] C. Loeblein, J. D. Perkins, B. Srinivasan, and D. Bonvin. Economic Performance Analysis in the Design of On-Line Batch Optimization Systems. *J. Process Control*, 9:61–78, 1999.

[264] J. S. Logsdon and L. T. Biegler. Accurate Determination of Optimal Reflux Policies for the Maximum Distillate Problem in Batch Distillation. *Ind. Eng. Chem. Res.*, 32, 4:692, 1993.

[265] A. Lucia and J. Xu. Methods of Successive Quadratic Programming. *Comput. Chem. Eng.*, 18:S211–S215, 1994.

[266] A. Lucia, J. Xu, and M. Layn. Nonconvex Process Optimization. *Comput. Chem. Eng.*, 20:1375–13998, 1996.

[267] Z.-Q. Luo, J.-S. Pang, and D. Ralph. *Mathematical Programs with Equilibrium Constraints*. Cambridge University Press, Cambridge, 1996.

[268] R. Luus and T. H. I. Jaakola. Direct Search for Complex Systems. *AIChE J.* 19:645–646, 1973.

[269] J. Macki and A. Strauss. *Introduction to Optimal Control Theory*. Springer, Berlin, 1982.

[270] L. Magni and R. Scattolini. Robustness and robust design of MPC for nonlinear systems. In *Assessment and Future Directions of NMPC*, pp. 239–254. Springer, Berlin, 2007.

[271] R. Mahadevan, J. S. Edwards, and F. J. Doyle, III F. J. Doyle, Dynamic Flux Balance Analysis of Diauxic Growth in Escherichia coli. *Biophys. J.* 83:1331–1340, 2002.

[272] T. Maly and L. R. Petzold. Numerical Methods and Software for Sensitivity Analysis of Differential-Algebraic Systems. *Appl. Numer. Math.*, to appear.

[273] O. L. Mangasarian. *Nonlinear Programming*. McGraw–Hill, New York, 1969.

[274] M. Mangold, O. Angeles-Palacios, M. Ginkel, R. Waschler, A. Kienle, and E. D. Gilles. Computer Aided Modeling of Chemical and Biological Systems—Methods, Tools, and Applications. *Ind. Eng. Chem. Res.*, 44:2579–2591, 2005.

[275] H. M. Markowitz. *Portfolio Selection, Efficient Diversification of Investments*. John Wiley & Sons, New York, 1959.

[276] T. E. Marlin and A. N. Hrymak. Real-time operations optimization of continuous processes. In B. Carnahan J. C. Kantor, C. E. Garcia, editor, *Chemical Process Control V*, pages 156–164. CACHE, AIChE, 1997.

[277] S. E. Mattsson and G. Söderlind. Index Reduction in Differential-Algebraic Equations Using Dummy Derivatives. *SIAM J. Sci. Comput.*, 14:677–692, 1993.

[278] D. Q. Mayne, J. R. Rawlings, C. V. Rao, and P. O. M. Scokaert. Constrained Model Predictive Control: Stability and Optimality. *Automatica*, 36:789–814, 2000.

[279] A. Miele. Gradient algorithms for the optimization of dynamic systems. In Leondes C.T., editor, *Control and Dynamic Systems: Advances in Theory and Applications*, volume 16, pages 1–52. Academic Press, New York, 1980.

[280] H. D. Mittelmann. *Benchmarks for Optimization Software*. http://plato.asu.edu/bench.html; http://plato.asu.edu/ftp/ampl-nlp.html, 2009.

[281] J. J. Moré. Recent Developments in Algorithms and Software for Trust Region Methods. *Mathematical Programming: State of the Art, Springer, Berlin*, pages 258–287, 1983.

[282] J. J. Moré and D. C. Sorensen. Computing a Trust Region Step. *SIAM J. Sci. Comput.*, 4:553–572, 1983.

[283] J. J. Moré and S. J. Wright. *Optimization Software Guide*. Frontiers in Appl. Math. 14, SIAM, Philadelphia, 1993.

[284] MPEC Library. *http://www.gamsworld.eu/mpec/mpeclib.htm*. GAMS Development Corporation, Washington, DC, 2006.

[285] B. A. Murtagh and M. A. Saunders. MINOS 5.1 User's Guide. *Technical Report SOL 83-20R, Stanford University*, 1987.

[286] Z. K. Nagy, R. Franke, B. Mahn, and F. Allgöwer. Real-time implementation of nonlinear model predictive control of batch processes in an industrial framework. In *Assessment and Future Directions of NMPC*, pp. 465–472. Springer, Berlin, 2006.

[287] S. Nash and A. Sofer. *Linear and Nonlinear Programming*. McGraw–Hill, New York, 1996.

[288] J. A. Nelder and R. Mead. A Simplex Method for Function Minimization. *Computer Journal*, 7:308–320, 1965.

[289] Y. Nesterov and A. Nemirovskii. *Interior-Point Polynomial Methods in Convex Programming*. SIAM, Philadelphia, 1994.

[290] C. P. Neuman and A. Sen. A Suboptimal Control Algorithm for Constrained Problems Using Cubic Splines. *Automatica*, 9:601–613, 1973.

[291] S. Nilchan and C. C. Pantelides. On the Optimisation of Periodic Adsorption Processes. *Adsorption*, 4:113–14, 1998.

[292] J. Nocedal and M. L. Overton. Projected Hessian Updating Algorithms for Nonlinearly Constrained Optimization. *SIAM J. Numer. Anal.*, 22:821–850, 1985.

[293] J. Nocedal, A. Wächter, and R. A. Waltz. Adaptive Barrier Strategies for Nonlinear Interior Methods. Technical Report RC 23563, IBM T. J. Watson Research Center, Yorktown, 2006.

[294] J. Nocedal and S. Wright. *Numerical Optimization*. second edition, Springer, New York, 2006.

[295] I. Nowak. *Relaxation and Decomposition Methods for Mixed Integer Nonlinear Programming*. Birkhäuser, Boston, MA, 2005.

[296] S. H. Oh and R. Luus. Use of Orthogonal Collocation Method in Optimal Control Problems. *Int. J. Control*, 26:657–673, 1977.

[297] J. Oldenburg and W. Marquardt. Disjunctive Modeling for Optimal Control of Hybrid Systems. *Comput. Chem. Eng.*, 32:2346–2364, 2008.

[298] J. Oldenburg, W. Marquardt, D. Heinz, and B. Leineweber. Mixed-Logic Dynamic Optimization Applied to Batch Distillation Process Design. *AIChE J.* 49,11:2900–2917, 2003.

[299] N. M. C. Oliveira and L. T. Biegler. Newton-Type Algorithms for Nonlinear Process Control: Algorithm and Stability Results. *Automatica*, 31:281–286, 1995.

[300] G. M. Ostrovskii, T. A. Berejinski, K. A. Aliev, and E. M. Michailova. Steady State Simulation of Chemical Plants. *Chem. Eng. Commun.*, 23:181–190, 1983.

[301] D. B. Özyurt and P. I. Barton. Cheap Second Order Directional Derivatives of Stiff ODE Embedded Functionals. *SIAM J. Sci. Comput.*, 26:1725–1743, 2005.

[302] B. Pahor and Z. Kravanja. Simultaneous Solution and MINLP Synthesis of DAE Process Problems: PFR Networks in Overall Processes. *Comput. Chem. Eng.*, 19:S181–S188, 1995.

[303] J. S. Pang and D. Stewart. Differential Variational Inequalities. *Math. Programm.*, 113:345–424, 2008.

[304] C. C. Pantelides. The Consistent Initialization of Differential-Algebraic Systems. *SIAM J. Sci. Comput.*, 9:213–231, 1988.

[305] P. Papalambros and D. Wilde. *Principles of Optimal Design*. Cambridge University Press, Cambridge, 1988.

[306] V. Pascual and L. Hascoët. Extension of TAPENADE toward Fortran 95. In H. M. Bücker, G. Corliss, P. Hovland, U. Naumann, and B. Norris, editors, *Automatic Differentiation: Applications, Theory, and Implementations*, Lecture Notes in Computational Science and Engineering, pages 171–179. Springer, 2005.

[307] H. J. Pesch. A practical guide to the solution of real-life optimal control problems. *Control Cybernetics*, 23:7–60, 1994.

[308] L. R. Petzold. A Description of DASSL: A Differential/Algebraic System Solver. Technical Report SAND82-8637, Sandia National Laboratory, 1982.

[309] L. Pibouleau, P. Floquet, and S. Domenech. Optimisation de produit chimiques par une methode de gradient reduit: Partie 1: Presentation de l'algorithme. *RAIRO Rech.*, 19:247–274, 1985.

[310] M. B. Poku, J. D. Kelly, and L. T. Biegler. Nonlinear Programming Algorithms for Process Optimization with Many Degrees of Freedom. *Ind. Eng. Chem. Res.*, 43:6803–6812, 2004.

[311] V. V. Pontryagin, Y. Boltyanskii, R. Gamkrelidze, and E. Mishchenko. *The Mathematical Theory of Optimal Processes*. Interscience Publishers, New York, 1962.

[312] M. J. D. Powell. An Efficient Method for Finding the Minimum of a Function of Several Variables without Calculating Derivatives. *Computer Journal*, 7:155–162, 1964.

[313] M. J. D. Powell. A fast algorithm for nonlinearly constrained optimization calculations. In G. A.Watson, editor, *Numerical Analysis Dundee 1977*, pages 144–157. Springer, Berlin, 1977.

[314] M. J. D. Powell. VF13: Minimize a General Function, General Constraints, SQP Method. *HSL Archive*, 1977.

[315] M. J. D. Powell. UOBYQA: Unconstrained Optimization by Quadratic Approximation. *Math. Programm.*, 92:555–582, 2002.

[316] M. J. D. Powell. Developments of NEWUOA for minimization without derivatives. Technical Report DAMTP 2007/NA05, Dept. of Applied Mathematics and Theoretical Physics, Cambridge University, 2007.

[317] A. Prata, J. Oldenburg, A. Kroll, and W. Marquardt. Integrated Scheduling and Dynamic Optimization of Grade Transitions for a Continuous Polymerization Reactor. *Comput. Chem. Eng.*, 32:463–476, 2008.

[318] R. Pytlak. *Numerical Methods for Optimal Control Problems with State Constraints.* Lecture Notes in Mathematics, Vol. 1707, Springer, Berlin, 1999.

[319] A. U. Raghunathan and L. T. Biegler. An Interior Point Method for Mathematical Programs with Complementarity Constraints (MPCCs). *SIAM J. Optim.*, 15:720–750, 2005.

[320] A. U. Raghunathan, M. S. Diaz, and L. T. Biegler. An MPEC Formulation for Dynamic Optimization of Distillation Operation. *Comput. Chem. Eng.*, 28:2037–2052, 2004.

[321] A. U. Raghunathan, J. R. Perez-Correa, E. Agosin, and L. T. Biegler. Parameter Estimation in Metabolic Flux Balance Models for Batch Fermentation—Formulations and Solution Using Differential Variational Inequalities. *Ann. Oper. Res.*, 148:251–270, 2006.

[322] A. U. Raghunathan, J. R. Perez-Correa, and L. T. Biegler. Data Reconciliation and Parameter Estimation in Flux-Balance Analysis. *Biotechnol. Bioeng.*, 84:700–709, 2003.

[323] A. U. Raghunathan. Mathematical Programs with Equilibrium Constraints (MPECs) in Process Engineering. *Ph.D. Thesis*. Department of Chemical Engineering, Carnegie Mellon University, 2004.

[324] A. U. Raghunathan and L. T. Biegler. MPEC Formulations and Algorithms in Process Engineering. *Comput. Chem. Eng.*, 27:1381–1392, 2003.

[325] A. U. Raghunathan, V. Gopal, D. Subramanian, L. T. Biegler, and T. Samad. Dynamic Optimization Strategies for Three-Dimensional Conflict Resolution of Multiple Aircraft. *J. Guid. Control Dynam.*, 27:586–594, 2004.

[326] D. Ralph and S. J. Wright. Some Properties of Regularization and Penalization Schemes for MPECs. *Optim. Methods Softw.*, 19:527–556, 2004.

[327] A. V. Rao, C. L. Darby, M. Patterson, C. Francolin, I. Sanders, and G. Huntington. GPOPS: A MATLAB Software for Solving Non-Sequential Multiple-Phase Optimal Control Problems Using the Gauss Pseudospectral Method. *ACM Trans. Math. Software*, 37:Article 22, 2010.

[328] W. H. Ray. On the Mathematical Modeling of Polymerization Reactors. *J. Macromol. Sci.-Revs. Macromol. Chem.*, C8:1–56, 1972.

[329] W. H. Ray. *Advanced Process Control.* McGraw–Hill, New York, 1981.

[330] W. H. Ray and J. Szekely. *Process Optimization.* John Wiley and Sons, New York, 1973.

[331] G. W. Reddien. Collocation at Gauss Points as a Discretization in Optimal Control. *SIAM J. Control Optim*, 17:298–306, 1979.

[332] G. Reklaitis, A. Ravindran, and K. Ragsdell. *Engineering Optimization.* Wiley, New York, 1983.

[333] J. G. Renfro, A. M. Morshedi, and O. A. Asbjornsen. Simultaneous Optimization and Solution of Systems Described by DAEs. *Comput. Chem. Eng.*, 11:503–517, 1987.

[334] C. A. M. Riascos and J. M. Pinto. Optimal Control of Bioreactors: A Simultaneous Approach for Complex Systems. *Chem. Eng. J.*, 99:23–34, 2004.

[335] V. Rico-Ramírez and A. W. Westerberg. Interior Point Methods for the Solution of Conditional Models. *Comput. Chem. Eng.*, 26:375–383, 2002.

[336] S. M. Robinson. A Quadratically Convergent Algorithm for General Nonlinear Programming Problems. *Math. Program.*, 3:145–156, 1972.

[337] A. Romanenko and L. O. Santos. A nonlinear model predictive control framework as free software: outlook and progress report. In R. Findeisen, F. Allgöwer, and L. Biegler, editors, *Assessment and Future Directions of Nonlinear Model Predictive Control*, pages 507–514. Springer, Berlin, 2007.

[338] W. C. Rooney and L. T. Biegler. Optimal Process Design with Model Parameter Uncertainty and Process Variability. *AIChE J.*, 49, 2:438, 2003.

[339] J. B. Rosen. The Gradient Projection Method for Nonlinear Programming. Part I. Linear Constraints *SIAM J. Appl. Math.*, 8:181–217, 1960.

[340] J. B. Rosen. The Gradient Projection Method for Nonlinear Programming. Part II. Nonlinear Constraints *SIAM J. Appl. Math.*, 9:514–532, 1961.

[341] R. D. Russell and J. Christiansen. Adaptive Mesh Selection Strategies for Solving Boundary Value Problems. *SIAM J. Numer. Anal.*, 15:59–80, 1978.

[342] P. E. Rutquist and M. M. Edvall. *PROPT—Matlab Optimal Control Software*. TOMLAB Optimization, 2009.

[343] H. Sagan. *Introduction to the Calculus of Variations*. Dover Publications, New York, 1992.

[344] S. I. Sandler. *Chemical, Biochemical, and Engineering Thermodynamics*. 4th edition, Wiley, New York, 2006.

[345] L. O. Santos, N. de Oliveira, and L.T. Biegler. Reliable and efficient optimization strategies for nonlinear model predictive control. In J. B. Rawlings, W. Marquardt, D. Bonvin, and S. Skogestad, editors, *Proc. Fourth IFAC Symposium on Dynamics and Control of Chemical Reactors, Distillation Columns and Batch Processes (DYCORD '95)*, page 33. Pergamon, 1995.

[346] L. O. Santos, P. Afonso, J. Castro, N. Oliveira, and L. T. Biegler. On-Line Implementation of Nonlinear MPC: An Experimental Case Study. *Control Eng. Pract.*, 9:847–857, 2001.

[347] R. W. H. Sargent. Reduced gradient and projection methods for nonlinear programming. In P. E. Gill and W. Murray, editors, *Numerical Methods in Constrained Optimization*, pages 140–174. Academic Press, New York, 1974.

[348] R. W. H. Sargent and B. A. Murtagh. Projection Methods for Nonlinear Programming. *Math. Programm.*, 4:245–268, 1973.

[349] R. W. H. Sargent and G. R. Sullivan. Development of Feed Changeover Policies for Refinery Distillation Units. *Ind. Eng. Chem. Res.*, 18:113, 1979.

[350] H. Scheel and S. Scholtes. Mathematical Programs with Complementarity Constraints: Stationarity, Optimality, and Sensitivity. *Math. Oper. Res.*, 25:1–22, 2000.

[351] W. A. Scheel and B. A. Conway. Optimization of Very-Low-Thrust, Many-Revolution Spacecraft Trajectories. *J. Guid. Control Dyn.*, 17:1185–92, 1994.

[352] A. Schiela and M. Weiser. Superlinear Convergence of the Control Reduced Interior Point Method for PDE Constrained Optimization. *Comput. Optim. Appl.*, 39:369–393, 2008.

[353] K. Schittkowski. On the Convergence of a Sequential Quadratic Programming Method with an Augmented Lagrangian Line Search Function. *Optimization*, 14:197–216, 1983.

[354] K. Schittkowski. NLPQL: A FORTRAN Subroutine Solving Constrained Nonlinear Programming Problems. *Ann. Oper. Res.*, 5:485–500, 1985/86.

[355] M. Schlegel, K. Stockmann, T. Binder, and W. Marquardt. Dynamic Optimization Using Adaptive Control Vector Parameterization. *Comput. Chem. Eng.*, 29:1731–1751, 2005.

[356] C. Schmid and L. T. Biegler. Quadratic Programming Methods for Tailored Reduced Hessian SQP. *Comput. Chem. Eng.*, 18:817, 1994.

[357] A. Schwartz. Theory and Implementation of Numerical Methods based on Runge–Kutta Integration for Solving Optimal Control Problems. *Ph.D. Thesis*. Department of Electrical Engineering and Computer Science, University of California, Berkeley, 1996.

[358] W. D. Seider, J. D. Seader, and D. R. Lewin. *Product and Process Design Principles: Synthesis, Analysis, and Evaluation*. John Wiley, New York, 2004.

[359] S. P. Sethi and G. L. Thompson. *Optimal Control Theory: Applications to Management Science and Economics*. Kluwer Academic Publishers, Dordrecht, Netherlands, 2000.

[360] L. F. Shampine. *Numerical Solution of Ordinary Differential Equations*. Chapman and Hall, London, 1994.

[361] L. Shrage. *Optimization Modeling with LINGO*. sixth edition, LINDO Systems, Inc., Chicago, 2006.

[362] J. D. Simon and H. M. Azma. Exxon Experience with Large Scale Linear and Nonlinear Programming Applications. *Comput. Chem. Eng.*, 7:605–614, 1983.

[363] SimSci-Esscor. Getting Started with ROMeo. Technical report, Invensys, 2004.

[364] P. B. Sistu, R. S. Gopinath, and B. W. Bequette. Computational Issues in Nonlinear Predictive Control. *Comput. Chem. Eng.*, 17:361–366, 1993.

[365] I. Spangelo and I. Egeland. Trajectory Planning and Collision Avoidance for Underwater Vehicles Using Optimal Control. *IEEE J. Oceanic Eng.*, 19:502–511, 1994.

[366] P. Spellucci. An SQP Method for General Nonlinear Programs Using only Equality Constrained Subproblems. *Math. Programm.*, 82:413–448, 1998.

[367] P. Spellucci. DONLP2 User's Guide. Technical report, Technical University at Darmstadt, Department of Mathematics, 1999.

[368] B. Srinivasan, S. Palanki, and D. Bonvin. Dynamic Optimization of Batch Processes: I. Characterization of the Nominal Solution. *Comput. Chem. Eng.*, 44:1–26, 2003.

[369] O. Stein, J. Oldenburg, and W. Marquardt. Continuous Reformulations of Discrete-Continuous Optimization Problems. *Comput. Chem. Eng.*, 28:1951–1966, 2004.

[370] M. C. Steinbach, H. G. Bock, G. V. Kostin, and R. W. Longman. Mathematical Optimization in Robotics: Towards Automated High Speed Motion Planning. *Math. Ind.*, 7:303–340, 1997.

[371] D. Stewart and M. Anitescu. Optimal Control of Systems with Discontinuous Differential Equations. *Numer. Math.*, 114:653–695, 2010.

[372] W. E. Stewart and M. C. Caracotsios. Athena Visual Studio. *http://www.athenavisual.com*.

[373] M. Stingl. *On the Solution of Nonlinear Semidefinite Programs by Augmented Lagrangian Methods. Ph.D. Thesis*, Department of Mathematics, University of Erlangen, 2005.

[374] O. Stursberg, S. Panek, J. Till, and S. Engell. Generation of optimal control policies for systems with switched hybrid dynamics. *Modelling, Analysis, and Design of Hybrid Systems*, 279/LNCIS:337–352, 2002.

[375] M. C. Svensson. Process Identification in On-Line Optimizing Control, an Application to a Heat Pump. *Modeling, Identification and Control*, 17/ 4:261–78, 1996.

[376] R. E. Swaney and I. E. Grossmann. An Index for Operational Flexibility in Chemical Process Design. *AIChE J.*, 31:621–630, 1985.

[377] P. Tanartkit. and L. T. Biegler. A Nested, Simultaneous Approach for Dynamic Optimization Problems. *Comput. Chem. Eng.*, 20:735, 1996.

[378] P. Tanartkit and L. T. Biegler. A Nested Simultaneous Approach for Dynamic Optimization Problems II: The Outer Problem. *Comput. Chem. Eng.*, 21:1365, 1997.

[379] M. Tawarmalani and N. Sahinidis. *Convexification and Global Optimization in Continuous and Mixed-Integer Nonlinear Programming: Theory, Algorithms, Software, and Applications*. Kluwer Academic Publishers, Dordrecht, The Netherlands, 2002.

[380] D. Ternet and L. T. Biegler. Recent Improvements to a Multiplier Free Reduced Hessian Successive Quadratic Programming Algorithm. *Comput. Chem. Eng.*, 22:963–978, 1998.

[381] J. Till, S. Engell, S. Panek, and O. Stursberg. Applied Hybrid System Optimization: An Empirical Investigation of Complexity. *Control Eng. Prac.*, 12:1291–1303, 2004.

[382] A. L. Tits, A. Wächter, S. Bakhtiari, T. J. Urban, and C. T. Lawrence. A Primal-Dual Interior-Point Method for Nonlinear Programming with Strong Global and Local Convergence Properties. *SIAM J. Optim.*, 14:173–199, 2003.

[383] I-B. Tjoa. *Simultaneous Solution and Optimization Strategies for Data Analysis. Ph.D. Thesis*, Carnegie Mellon University, 1991.

[384] I.-B. Tjoa and L. T. Biegler. Simultaneous Solution and Optimization Strategies for Parameter Estimation of Differential-Algebraic Equation Systems. *Ind. Eng. Chem. Res.*, 30:376–385, 1991.

[385] M. J. Todd and Y. Ye. A Centered Projective Algorithm for Linear Programming. *Math. Oper. Res.*, 15:508–529, 1990.

[386] J. Tolsma and P. Barton. DAEPACK: An Open Modeling Environment for Legacy Models. *Ind. Eng. Chem. Res.*, 39:1826–1839, 2000.

[387] A. Toumi, M. Diehl, S. Engell, H. G. Bock, and J. P. Schlöder. Finite Horizon Optimizing Control of Advanced SMB Chromatographic Processes. In *16th IFAC World Congress, Prague*, 2005.

[388] T. H. Tsang, D. M. Himmelblau, and T. F. Edgar. Optimal Control via Collocation and Non-Linear Programming. *Int. J. Control*, 24:763–768, 1975.

[389] M. Ulbrich, S. Ulbrich, and L. Vicente. A Globally Convergent Primal-Dual Interior-Point Filter Method for Nonlinear Programming. *Math. Programm.*, 100:379–410, 2004.

[390] S. Ulbrich. On the Superlinear Local Convergence of a Filter-SQP Method. *Math. Programm.*, 100(1):217–245, 2004.

[391] R. J. Vanderbei and D. F. Shanno. An interior point algorithm for nonconvex nonlinear programming. *Technical Report SOR-97-21, CEOR, Princeton University, Princeton, NJ*, 1997.

[392] R. J. Vanderbei and D. F. Shanno. An Interior-Point Algorithm for Nonconvex Nonlinear Programming. *Comput. Optim. Appl.*, 23:257–272, 1999.

[393] J. C. Varazaluce-Garcśa, A. A. Flores-Tlacuahuac, and E. Saldśvar-Guerra. Steady-State Nonlinear Bifurcation Analysis of a High- Impact Polystyrene Continuous Stirred Tank Reactor. *Ind. Eng. Chem. Res.*, 39:1972, 2000.

[394] S. Vasantharajan. Large-Scale Optimization of Differential-Algebraic Systems. *Ph.D. Thesis*. Chemical Engineering Department, Carnegie Mellon University, Pittsburgh, PA, 1989.

[395] S. Vasantharajan and L. T. Biegler. Large-Scale Decomposition for Successive Quadratic Programming. *Comput. Chem. Eng.*, 12:1087, 1988.

[396] S. Vasantharajan and L. T. Biegler. Simultaneous Strategies for Parameter Optimization and Accurate Solution of Differential-Algebraic Systems. *Comput. Chem. Eng.*, 14:1083, 1990.

[397] V. S. Vassiliadis, E. B. Canto, and J. R. Banga. Second-Order Sensitivities of General Dynamic Systems with Application to Optimal Control Problems. *Chem. Eng. Sci.*, 54:3851–3860, 1999.

[398] V. S. Vassiliadis, R. W. H. Sargent, and C. C. Pantelides. Solution of a Class of Multistage Dynamic Optimization Problems. Part I—Algorithmic Framework. *Ind. Eng. Chem. Res.*, 33:2115–2123, 1994.

[399] V. S. Vassiliadis, R. W. H. Sargent, and C. C. Pantelides. Solution of a Class of Multistage Dynamic Optimization Problems. Part II—Problems with Path Constraints. *Ind. Eng. Chem. Res.*, 33:2123–2133, 1994.

[400] O. von Stryk and R. Bulirsch. Direct and Indirect Methods for Trajectory Optimization. *Ann. Oper. Res.*, 37:357–373, 1992.

[401] A. Wächter. *An Interior Point Algorithm for Large-Scale Nonlinear Optimization with Applications in Process engineering. Ph.D. Thesis*, Carnegie Mellon University, 2001.

[402] A. Wächter and L. T. Biegler. Line Search Filter Methods for Nonlinear Programming: Local Convergence. *SIAM J. Optim.*, 16:32–48, 2005.

[403] A. Wächter and L. T. Biegler. Line Search Filter Methods for Nonlinear Programming: Motivation and Global Convergence. *SIAM J. Optim.*, 16:1–31, 2005.

[404] A. Wächter and L. T. Biegler. On the Implementation of a Primal-Dual Interior Point Filter Line Search Algorithm for Large-Scale Nonlinear Programming. *Math. Programm.*, 106:25–57, 2006.

[405] A. Wächter, C. Visweswariah, and A. R. Conn. Large-Scale Nonlinear Optimization in Circuit Tuning. *Future Generat. Comput. Syst.*, 21:1251–1262, 2005.

[406] M. Weiser. *Function Space Complementarity Methods for Optimal Control Problems. Ph.D. Thesis*, Freie Universität Berlin, 2001.

[407] M. Weiser and P. Deuflhard. Inexact Central Path Following Algorithms for Optimal Control Problems. *SIAM J. Control Optim.*, 46:792–815, 2007.

[408] T. Williams and R. Otto. A generalized Chemical Processing. Model for the Investigation of Computer Control. *AIEE Trans.*, 79:458, 1960.

[409] R. B. Wilson. *A Simplicial Algorithm for Concave Programming. Ph.D. Thesis*, Graduate School of Business Administration, Harvard University, 1963.

[410] D. Wolbert, X. Joulia, B. Koehret, and L. T. Biegler. Flowsheet Optimization and Optimal Sensitivity Analysis Using Exact Derivatives. *Comput. Chem. Eng.*, 18:1083–1095, 1994.

[411] S. J. Wright. *Primal-Dual Interior-Point Methods*. SIAM, Philadelphia, 1997.

[412] C. Xu, P. M. Follmann, L. T. Biegler, and M. S. Jhon. Numerical Simulation and Optimization of a Direct Methanol Fuel Cell. *Comput. Chem. Eng.*, 29, 8:1849–1860, 2005.

[413] H. Yamashita. A Globally Convergent Primal-Dual Interior-Point Method for Constrained Optimization. *Optim. Methods Softw.*, 10:443–469, 1998.

[414] B. P. Yeo. A Modified Quasilinearization Algorithm for the Computation of Optimal Singular Control. *Int. J. Control*, 32:723–730, 1980.

[415] W. S. Yip and T. Marlin. The Effect of Model Fidelity on Real-Time Optimization Performance. *Comput. Chem. Eng.*, 28:267, 2004.

[416] J. C. C. Young, R. Baker, and C. L. E. Swartz. Input Saturation Effects in Optimizing Control: Inclusion within a Simultaneous Optimization Framework. *Comput. Chem. Eng.*, 28:1347–1360, 2004.

[417] R. E. Young, R. D. Bartusiak, and R. W. Fontaine. Evolution of an industrial nonlinear model predictive controller. In J. W. Eaton, J. B. Rawlings, B. A. Ogunnaike, editor, *Chemical Process Control VI: Sixth International Conference on Chemical Process Control*, pages 342–351. AIChE Symposium Series, Volume 98, Number 326, 2001.

[418] V. M. Zavala and L. T. Biegler. Large-Scale Parameter Estimation in Low-Density Polyethylene Tubular Reactors. *Ind. Eng. Chem. Res.*, 45:7867–7881, 2006.

[419] V. M. Zavala, C. D. Laird, and L. T. Biegler. Interior-Point Decomposition Approaches for Parallel Solution of Large-Scale Nonlinear Parameter Estimation Problems. *Chem. Eng. Sci.*, 63:4834–4845, 2008.

[420] C. Zhu, R. H. Byrd, P. Lu, and J. Nocedal. Algorithm 778: L-BFGS-B, FORTRAN Subroutines for Large Scale Bound Constrained Optimization. *ACM Trans. Math. Software*, 23:550–560, 1997.

[421] S. E. Zitney and M. Syamlal. Integrated Process Simulation and CFD for Improved Process Engineering. In *Proc. of the European Symposium on Computer Aided Process Engineering -12*, pages 397–402, 2002.

[422] G. Zoutendijk. Nonlinear Programming, Computational Methods. In J. Abadie, editor, *Integer and Nonlinear Programming*, pages 37–86. North–Holland, Amsterdam, 1970.

Index